CONTROLLING
TECHNOLOGY

CONTROLLING
TECHNOLOGY
Contemporary Issues

SECOND EDITION

EDITED BY

ERIC KATZ, ANDREW LIGHT,
AND
WILLIAM THOMPSON

Prometheus Books
59 John Glenn Drive
Amherst, New York 14228-2197

Published 2003 by Prometheus Books

Inquiries should be addressed to
Prometheus Books
59 John Glenn Drive
Amherst, New York 14228–2197
VOICE: 716–691–0133, ext. 207
FAX: 716–564–2711
WWW.PROMETHEUSBOOKS.COM

07 06 05 04 03   5 4 3 2 1

Library of Congress Cataloging-in-Publication Data

Controlling technology : contemporary issues / edited by Eric Katz, Andrew Light, William Thompson.—2nd ed.
        p. cm.
    Includes bibliographical references.
    ISBN 1–57392–983–2 (alk. paper)
    1. Technology. I Katz, Eric, 1952- II. Light, Andrew, 1966- III. Thompson, William B., 1929-
T185 .C68 2002
303.48'3—dc21

                                                                                    2002069815

Printed in the United States of America on acid-free paper

Contents

PART 3. THE BLESSINGS OF TECHNOLOGY

PART 4. THE AUTONOMY OF TECHNOLOGY AND ITS PHILOSOPHICAL CRITICS

PART 5. DEMYSTIFYING AUTONOMOUS TECHNOLOGY THROUGH THE HISTORY OF TECHNOLOGY

Contents   9

PART 6. TECHNOLOGY, ETHICS, AND POLITICS

PART 7. APPROPRIATE TECHNOLOGY

PART 8. COMPUTERS, INFORMATION, AND VIRTUAL REALITY

Preface to the
Second Edition

In this second edition we have made several changes to update the selections of readings. First, we have added new readings in almost all parts of the book to reflect more recent scholarship of the central issues. The addition of new readings has, by necessity, meant that some of the older readings had to be eliminated. Second, we have added readings that reflect nontraditional viewpoints such as feminist and non-Western analyses of the role of technology in society. Third, we have added a part focused specifically on computers and information technology.

Although this book is edited by three philosophers, it should not be considered a "philosophy reader." It fits more neatly into that relatively new academic field called "STS"—Science and Technology Studies—a field that has developed quite rapidly in the decade since the first edition appeared. STS combines the philosophical study of the meaning of technology in human life with the study of the ethics, politics, history, and sociology of science and technology. It is a truly interdisciplinary and multidisciplinary field of study, and we believe that the readings that we have selected for this second edition reflect the wide range of approaches and methods that can be used to study the role of technology in society.

The two new editors for this second edition wish to thank Steven L. Mitchell of Prometheus Books for inviting us to join this project, and Chris Hubbard and Lucy Frazier at NYU for support and extremely helpful editorial work in putting this volume together. But our deepest appreciation goes to William Thompson, the editor of the first edition, for creating a comprehensive survey of the issues, a book which we found difficult to improve.

E. K. and A. L.
Bay Shore, New York, and New York City
September 2001

11

Preface to the
First Edition

The title of this book is quite deliberately ambiguous. This is to reflect a central issue in the philosophy of technology. Does technology control us, or do we control technology? What is, and what ought to be, the role of technology in human affairs?

The study of technology has not traditionally been thought of as a part of a liberal arts curriculum and certainly not a legitimate topic for philosophical reflection. However, it is clear that modern technology, with its grounding in scientific knowledge, is a major force shaping the society in which we live and the way we experience the world. If we understand philosophy as the attempt to analyze and make sense of our experience, we can no longer neglect a critical examination of technology. In the Atomic Age this is quite literally a matter of life and death.

One of the reasons for this neglect can be attributed to philosophers' traditional disdain for *praxis*, for practical applications and actions, which has its roots in the ancient Greek evaluation of the manual arts and fabrication as fitting only for slaves. Philosophers, as free men, should deal with purely abstract and theoretical matters.

Another factor lies in the modern fragmentation of human knowledge, which is itself an expression of a technological mode of thinking. All problems, we have believed, ought to be divided, and subdivided, into ever smaller parts to be investigated only by experts with the relevant specializing training. By this mode of thinking, such practical considerations as the safety of nuclear power are purely technical and therefore ought to be left to nuclear engineers. The philosopher should limit himself to his own area of expertise—conceptual analysis—and not attempt to assess highly technical issues where he lacks the requisite specialized knowledge—as though there were no important conceptual issues about the nature and direction of technology.

This prejudice is reinforced by the prevalent view that technology is value-

13

neutral, that whether it is good or bad simply depends upon how it is used. If this view were correct, there would, indeed, be no philosophical or ethical issues about technology per se. However, this commonsense notion will not stand critical analysis. All technologies, from hammers and MX missiles to organizational techniques such as the division of labor, are designed for a purpose. Thus, engineering and management design cannot be divorced from the larger ethical issues confronting our society. While we may argue about how technology ought to be defined, on no account is it to be regarded as neutral.

This book is addressed to both those in the arts and sciences and those in technical fields. Its aim, to paraphrase Plato, is to see technology writ large. In the sense that all the readings concern fundamental issues, they are all of philosophical interest. However, the readings are not all written by professional philosophers. Some of the more seminal, if less careful, thinking has been done by those outside the discipline of philosophy. I have excluded treatments that are overly technical, so this material should be accessible to undergraduates. The selections are intended to be read in order, for the later readings typically discuss authors and arguments presented earlier in the text.

The issues discussed in this volume are organized in such a way as to bring into sharp focus the basic and opposing positions concerning the relationships between technology and fundamental human values—especially those ideals that center around the concept of democracy: human autonomy and freedom, human equality and the right of all persons to exercise some control over the factors that determine their destiny, and respect for the universal aspiration for a rich and meaningful existence.

I would like to thank the National Endowment for the Humanities for a grant awarded to me over ten years ago to develop a course dealing with technology and human values. It was this grant that stimulated my interest in the issues discussed in this volume. Additional thanks are due to Potsdam College of the State University of New York for a sabbatical leave to work on this book.

William Thompson
1991

General Introduction

*"And what rough beast, its hour come round at last,
Slouches toward Bethlehem to be born?"*
                                        —W. B. Yeats, "The Second Coming"

In the summer 1998 blockbuster *Godzilla*, technology runs amuck. Literally. And it isn't a pretty sight. Technology comes in the guise of a giant lizard, the product (in this resurrected version) of radioactive experiments by the French in the South Pacific, dating back to the 1960s. Moving quickly from its home in French Polynesia, this unintentional by-product of technology makes its way to New York City and begins the process of reproducing itself more efficiently than a photocopier (as well as trashing most of midtown and lower Manhattan). Our hero, this time, comes in the form of Matthew Broderick, playing a biologist/geneticist working for the Nuclear Regulatory Commission. Fresh from assignment studying the effects of the Chernobyl radiation on the local worms (we learn that they've increased in size by 17 percent), Broderick spends the next ninety minutes figuring out the puzzle of the giant reptile, finally succeeding in directing a high-tech U.S. military force in the destruction of the mutant Gila monster on the Brooklyn Bridge (just in time to save the outer boroughs). In essence, a terror produced by nuclear technology is ultimately destroyed by more conventional military technology, all under the watchful eye of a twentieth-century scientist-hero.

But the ending to Godzilla this time around is rather odd. Lost are the implicit and explicit cautionary tales about the horror of nuclear technology that occasionally surfaced in the earlier films. At the end of this movie, when Broderick and Godzilla have a soulful human-to-giant-lizard moment of eye contact, only a brief pause is taken before the crowds watching this struggle at home live on television erupt in applause. The good guys having won, the film wraps up the sappy love story thrown in for character development and goes to the credits.

15

There is no moral to this story. One form of technology is destroyed by another form of technology under the watchful eye of technology. In some ways, *Godzilla* takes us back to more than the nostalgic horror movies of our youth; it reminds us that technology is still largely conceived as a neutral tool created by scientists and engineers for our continual enjoyment. At most, we can be dazzled by what sort of technique produces more power for destruction or creation. But otherwise there is nothing to discuss here but a purely technical issue.

At about the same time that Takeo Murata and Ishiro Honda were planning the first Godzilla movie—which, after all, did in some ways try to open the question of the moral responsibility of the United States for the A-bomb—sociologists, historians, and philosophers were beginning to ask whether modern technology, and the world it had created, contained issues of theoretical importance beyond their mere technical challenges. For example, sociologist Jacques Ellul and philosopher Martin Heidegger both urged that we think of technology, in Ellul's words, as "the stake of the century." Both argued that modern technology, although it may seem to be just a more efficient means of doing what humans have always done, confronts humanity with issues that go to the very core of who we are and how we live. If Ellul and Heidegger were even partly right, then there was an issue here worthy of everyone's attention, and not just a handful of technical experts.

This book considers the unbridled optimism of technological development evinced in popular culture, but more important in the halls of policymakers and the laboratories of scientists. Although people have always displayed a certain ambivalence toward technology, the most prevalent attitude in the twentieth century, especially in North America, has been a quasi-religious faith in the capacity of technology to provide the good life for all. Surely, the rising standard of living in the industrialized countries gives some grounds for this confidence. Many people believe that it is technology, in conjunction with the free-enterprise system, that has made the free world great. The dominant view in our society equates technological innovation and growth in the gross national product (GNP) with human progress. This belief is so common that *Webster's New Collegiate Dictionary* defines "technology" as "the totality of the means employed to provide objects necessary for human sustenance and comfort." The positive value of technology is thus built into the very meaning of the term.

It is no wonder then, that technology is seen as the panacea for all social ills: genetic- and bioengineering will create new agricultural products and new forms of livestock (through cloning) that will solve the problem of food distribution and availability; the universal use of computers and the Internet in schools will solve the educational crisis as well as increase worker productivity; the harvesting of rare tropical plants will lead to the development of pharmaceutical drugs to cure human diseases; the human genome project will even prevent diseases before they begin; new forms of energy—nuclear and solar power, cold fusion—will ease our dependence on a slowly vanishing supply of oil and natural gas; the development of high-speed trains will make intercity and cross-country travel a pleasure; improvements in telecommunications will lead to new social relations and stronger family ties; new reproductive technologies will increase human fertility and happiness. The list goes on and on as all these new technologies increase material wealth and human freedom, while they decrease poverty, disease, and all other obstacles to human well-being.

This technological optimism, however, cannot be accepted uncritically. It has to be analyzed in the light of the many negative and harmful consequences of technology that have increasingly appeared throughout the twentieth century. In Nazi Germany in the 1940s science and technology were used to create a chillingly efficient form of mass-produced death, the extermination camps where six million human beings were murdered. Beginning with the explosion of the atomic bomb on Hiroshima on August 6, 1945, we created a world in which foreign policy was dominated by the accelerating arms race between the United States and the Soviet Union, and where new military strategies between the major superpowers were based on theories of massive retaliation and mutual assured destruction (MAD). Although the Cold War has been declared over since the fall of the Iron Curtain in 1991, there are still on this planet enough nuclear weapons to destroy all life on earth many times over.

Technology has also been a major cause of the ecological crisis that surrounds us—the profusion of toxic chemicals in the air and water, acid rain, global warming, the depletion of the ozone layer, the loss of biodiversity, new carcinogens, soil erosion, deforestation, and overpopulation. The technological promise of cheap, safe nuclear power has been dimmed by escalating costs and events like the near disaster at Three Mile Island in 1979 and the actual nuclear catastrophe at Chernobyl in the Soviet Union in 1986. In the field of medicine, technology seems to have taken over the fundamental decisions of medical personnel; we submit to more and more tests, drugs, and machines without clear benefits in quality of life. Perhaps the most dramatic failure of modern technology, watched by millions on television, was the explosion of the space shuttle *Challenger*. In recognizing these problems, can we continue to believe in the essential goodness of technology?

These contemporary crises have led many to conclude that our technology has escaped rational human control. Rather than being essentially good, technological advancement appears to be more akin to a cancerous growth endemic to all technological societies. This conclusion gives credence to a philosophical view formulated between the world wars that technology has become autonomous, and that it evolves according to its own internal logic, which operates independently of human purposes. Those who hold that technology has become autonomous argue that, barring a nuclear holocaust, we are being ineluctably propelled toward a "brave new world," an all-pervasive totalitarian state in which persons and their values are subservient to the requirements of technological development. On this view, the "controlling" in this volume's title should be understood as an adjective: it is technology that is controlling us. Questions remain however on the veracity of this thesis: Is technology controlling us in a way akin to how one human can control another, or have we simply been too sloppy in our application of the tools that we have produced? Do bad consequences really mean that there is something wrong with the technology itself or is there something wrong with the users?

Our optimism in technology is, however, very much alive. The naïve response is to blindly assert that tools are neutral artifacts, good or bad according to the intentions of their users. Actor and National Rifle Association spokesperson Charleton Heston often rehearsed the old wag, "Guns don't kill people. People kill people." Restraint should not be placed on the production or availability of guns, but only on the "bad people" who use guns for bad reasons.

A slightly more subtle response to our technological crises is to keep faith with the notion of technological progress, but in addition to admitting that along with the intended good consequences there may develop some unintended bad ones, and that further technological growth is necessary to eliminate these undesirable "side effects." Such thinking underlies the nuclear strategy that proposed a Strategic Defense Initiative during the Reagan administration in the United States: while offensive missiles now threaten the human race, our country's salvation was to lie in the deployment of a new technology, the so-called Star Wars defense. If we do not lose the old American "can-do spirit," the promise of technology will continue to be fulfilled, and we will be saved from our enemies. No radical rethinking of the very nature of technology is necessary. For every technological problem—indeed for every human problem—there is an appropriate technological solution. Accordingly, it does not make sense to blame technology for the problems we face; technology provides the very means of solving those problems. We humans can and do "control" technology—here the word "controlling" in the title is a verb.

Of course, if technology is autonomous and *necessarily* brings about the good life for us all, we need not worry—or think. But if technology is autonomous and *necessarily* leads to a degradation of the quality of our lives, stoic resignation seems the only appropriate attitude. However, there is reason to believe that there are points in the system where values can, and do, exert leverage. The esoteric complexity of large-scale technical systems discourages many ordinary citizens from attempting to evaluate them, leaving these most important issues in the hands of experts. But, while expert knowledge about probable consequences of our technology is important, when available, the fundamental questions still involve basic human values. Indeed, as we shall see, moral, or more generally "normative," choices (choices about how we ought to live rather than mere descriptions of how we do live) based on personal values frequently masquerade as purely technical decisions.

Although it may be proved that some technological developments are, in an important sense, out of human control, it does not follow that this is an inevitable or universal condition. Once we untangle the multiple elements that propel technical change, we shall see that human purposes are at work here and that technology typically does serve human interests. However, "human" interests are not necessarily "humane" or just, nor can we move automatically from the interests of some to those of all humanity. Thus, one crucial moral issue is concerned with distributive justice: Who profits and who loses as a consequence of new technologies? Are the burdens and benefits of technological change distributed fairly? Even more fundamental, perhaps, is the question of the good itself. Precisely which goods and harms are produced by technological development? Are these technological goods constitutive of the good life for human beings? Do those who control technological development have the appropriate vision of the good for human life, and do they know how to create and use technology to achieve the good? Who in fact controls, and ought to control, technology? These are the most basic questions addressed by the readings in this volume.

We have divided the readings in this book into eight parts. Each part is preceded by its own introduction that will summarize the essays and place them in their appropriate context. Here we present merely a brief outline of the text as a whole.

Part 1, "The Modern Predicament," presents two essays that together serve as bookends to the range of issues confronting humanity as it seeks control of technology—the deadly serious issue of nuclear disarmament and the more light-hearted issue of the technological flavor industry. Part 2, "Defining Technology," contains sociological and philosophical analyses of the meaning of "technology" with the intent to broaden the commonsense notion that technology simply means machines or hardware. Part 3, "The Blessings of Technology," offers several optimistic views of technological development, including the importance of science and technology for the nonindustrialized world. Part 4, "The Autonomy of Technology and Its Philosophical Critics," presents six essays that explain and debate one of the crucial philosophical issues in this volume, the question of whether the development of technology is autonomous. In part 5, "Demystifying Autonomous Technology through the History of Technology," we continue the discussion of autonomous technology with essays that focus primarily on historical examples of technological development. Part 6, "Technology, Ethics, and Politics," raises a second crucial philosophical issue, the relationship between technological development and the development of specific moral problems and political structures and institutions. Part 7, "Appropriate Technology," contains three essays that examine the prospects for developing an alternative technology that overcomes some of the problems raised in the earlier essays. Part 8, "Computers, Information, and Virtual Reality," presents five essays concerning the development of computers and how this affects our view of labor, information, and reality itself.

Following the first nuclear explosion, Albert Einstein commented that "the unleashed power of the atom has changed everything save our modes of thinking; thus we drift toward catastrophe." This volume challenges us to reevaluate our habitual modes of thinking in order to gain some insight into how this drift might be prevented. Although many of the essays in this volume are concerned with theory, the book's purpose overall is crucially practical: to understand and to control technological development. The readings in this volume contain no definitive answers to the problems facing us, but it is our hope that the reader will come away with a clearer understanding of the issues and of the questions that we all ought to be asking. In the ideal of a democratic society, the management of technology must be everybody's business.

PART 1

THE MODERN PREDICAMENT

Power makes the modern predicament unique. Throughout recorded history there have been wars and rumors of war, but the effects of human aggression have been circumscribed by the limited means available for the infliction of death and destruction. Environmental pollution is not a new phenomenon, but until the development of modern industry it remained a localized problem and did not affect the biosphere in what are now seen to be potentially irreversible ways. Nature once ruled supreme: humans could cause harm and death to local populations of other humans and local ecosystems and natural areas, but the overall functioning of the natural system remained inviolate and impervious to human activity. Modern technology has radically changed this situation. In the contemporary world, the balance of power has decisively shifted from nature to humanity. As the two essays in this first part demonstrate, nature no longer threatens humanity—humanity threatens nature. The starting point for an inquiry into the control of technology is an understanding that science and technology confer awesome power on humanity.

The first essay approaches the issue from a deadly serious perspective. Jonathan Schell wrote *The Fate of the Earth* at the height of the arms race between the United States and the Soviet Union. The collapse of communism and the end of the Cold War have now reduced, if not eliminated, our anxiety about an all-out exchange of nuclear weapons. However, we need to realize that most of these weapons are still in place and that Americans and the former Soviets do not have a monopoly on their manufacture or use. Indeed, one can plausibly argue that the likelihood of nuclear weapons being used by a Third World country or terrorist organization has increased. Even if the dream of eliminating all nuclear weapons is realized, the threat of nuclear annihilation cannot be eliminated. As Schell makes clear, the only *final* solution to the nuclear peril is the destruction

21

of the civilization that has the knowledge to produce such weapons of mass destruction. Scientific and technological "progress" is, in this sense, irreversible.

Although nuclear weapons pose the most obvious and dramatic threat to life on earth, Schell argues that they are to be seen as only part of a more general ecological crisis that may be even more intractable. Men do not actively conspire to destroy the planet that sustains them; toxic wastes, acid rain, ozone depletion, and changes wrought by the greenhouse effect are unintended "side effects" of our intended goals. Nonetheless, we may, by incremental steps, so alter the ecology of the biosphere that life on earth is no longer viable. Our world may end in a whimper, not a bang.

Most important for our purposes is Schell's understanding of the causes of the modern predicament. On his account, the origins of our peril lie in basic scientific understanding, not in social circumstances. Scientific revolutions are to be sharply distinguished from social revolutions, the former being sprung on the world by relatively few individuals working in the privacy of their laboratories. Science is also a collective enterprise, in which each generation of scientists builds on the work of previous generations in progressively increasing our understanding of nature. This linear view of scientific development contrasts sharply with the circular nature of social revolutions that require the tacit consent of large numbers of ordinary citizens. These revolutions are reversible and occur only in societies where there has been a long period of preparation.

The critical problem, which many writers have described as cultural lag, is the failure of our understanding of social and political relations—our wisdom— to keep pace with the explosive increase in our scientific knowledge of physical reality. Science and technology continue to develop without waiting for philosophers, political leaders, and other social experts to form a consensus about what is morally correct or incorrect about the new scientific and technological developments. The recent controversies over cloning humans and other animals and the medical research involving human stem cells are examples of this cultural lag: the science is possible but the ethics are unclear.

After considering the "death of death" in the total extinction of all life on earth, it may seem somewhat frivolous to read Eric Schlosser's "Why McDonald's French Fries Taste So Good." But Schlosser's essay, excerpted from his book, *Fast Food Nation*, revolves around the same general issue raised by Schell: For what purpose do we use the immense power of contemporary science and technology? Schlosser examines the "flavor" industry in the United States and discovers that we now have the ability to chemically reproduce any natural flavor at all: the proper mixture of chemicals on a white strip of paper will smell and taste exactly like fried onions, black olives, or fresh cherries. The ability to reproduce chemically any and all flavors means that we can move closer to the goal of "designer" food. Technology can also determine the chemicals and ingredients to create the appropriate texture of foods—how they feel in one's mouth as they are eaten. The result is that food becomes a "highly processed industrial commodity" far removed from its natural or agricultural origins. The industrialization of food removes food from its natural roots, and this process is only possible through the discoveries of science and technology.

Whether we consider nuclear war or the taste of french fries, we see that science and technology are the principal causes of the modern predicament. But scientists, engineers, and other technologists are not normally thought to be primarily responsible in a moral sense, and therefore we should not look to them for a solution to the problem of technological control. (There is a countervailing view that was evidenced in the movement for "Scientists for Social Responsibility," which is beyond the scope of this section of the book. Nonetheless, such movements are still dependent on those outside science to help inform the moral views of scientists.) According to both Schell and Schlosser, scientists aim solely at understanding, at discovering the laws of nature. In this sense science is autonomous. But science and technology are embedded in a particular social context, and to some extent this social environment determines the direction of scientific research and technological development. In a world without processed food, there would be no need to develop chemical substitutes for natural flavors, and hence no scientific curiosity to discover the chemical constituents of the flavor of grilled hamburgers. The moral responsibility for the control of science and technology rests within the political and social system—in a democracy, it rests with the citizens.

1

The Fate of the Earth

Jonathan Schell

If a council were to be empowered by the people of the earth to do whatever was necessary to save humanity from extinction by nuclear arms, it might well decide that a good first step would be to order the destruction of all the nuclear weapons in the world. When the order had been carried out, however, warlike or warring nations might still rebuild their nuclear arsenals—perhaps in a matter of months. A logical second step, accordingly, would be to order the destruction of the factories that make the weapons. But, just as the weapons might be rebuilt, so might the factories, and the world's margin of safety would not have been increased by very much. A third step, then, would be to order the destruction of the factories that make the factories that make the weapons—a measure that might require the destruction of a considerable part of the world's economy. But even then lasting safety would not have been reached because in some number of years—at most, a few decades—everything could be rebuilt, including the nuclear arsenals, and mankind would again be ready to extinguish itself. A determined council might next decide to try to arrest the world economy in a pre-nuclear state by throwing the blueprints and technical manuals for reconstruction on the bonfires that had by then consumed everything else, but that recourse, too, would ultimately fail, because the blueprints and manuals could easily be redrawn and rewritten. As long as the world remained acquainted with the basic physical laws that underlie the construction of nuclear weapons—and these laws include the better part of physics as physics is understood in our century—mankind would have failed to put many years between itself and its doom. For the fundamental origin of the peril of human extinction by nuclear arms lies not in any particular social or political circumstances of our time but in the attainment by mankind as a whole, after millennia of scientific progress, of a certain level of knowledge of the physical universe. As long as that knowledge is in our possession, the atoms themselves, each one stocked with its prodigious supply of energy, are, in a manner of speaking, in a perilously advanced state of mobilization for nuclear hostilities, and any conflict anywhere in the world can become a nuclear one. To return to safety through technical measures alone, we would have to disarm matter itself, converting it back into its relatively safe, inert, nonexplosive nineteenth-century Newtonian state—something that not even the physics of our time can teach us how to do. (I mention these farfetched, wholly imaginary programs of demolition and suppression in part because the final destruction of all mankind is so much more farfetched, and therefore seems to give us license to at least consider extreme alternatives, but mainly because their obvious inadequacy serves to demonstrate how deeply the nuclear peril is ingrained in our world.)

It is fundamental to the shape and character of the nuclear predicament that its origins lie in scientific knowledge rather than in social circumstances. Revolutions born in the laboratory are to be sharply distinguished from revolutions born in society. Social revolutions are usually born in the minds of millions, and are led up to by what the Declaration of Independence calls "a long train of abuses," visible to all; indeed, they usually cannot occur unless they are widely understood by and supported by the public. By contrast, scientific revolutions usually take shape quietly in the minds of a few men, under cover of the impenetrability to most laymen of scientific theory, and thus catch the world by surprise. In the case of nuclear weapons, of course, the surprise was greatly increased by the governmental secrecy that surrounded the construction of the first bombs. When the world learned of their existence, Mr. Fukai [a person described by John Hersey in his book *Hiroshima*] had already run back into the flames of Hiroshima, and tens of thousands of people in that city had already been killed. Even long after scientific discoveries have been made and their applications have transformed our world, most people are likely to remain ignorant of the underlying principles at work, and this has been particularly true of nuclear weapons, which, decades after their invention, are still surrounded by an aura of mystery, as though they had descended from another planet. (To most people, Einstein's famous formula $E = mc^2$, which defines the energy released in nuclear explosions, stands as a kind of symbol of everything that is esoteric and incomprehensible.)

But more important by far than the world's unpreparedness for scientific revolutions are their universality and their permanence once they have occurred. Social revolutions are restricted to a particular time and place; they arise out of particular circumstances, last for a while, and then pass into history. Scientific revolutions, on the other hand, belong to all places and all times. In the words of Alfred North Whitehead, "Modern science was born in Europe, but its home is the whole world." In fact, of all the products of human hands and minds, scientific knowledge has proved to be the most durable. The physical structures of human life—furniture, buildings, paintings, cities, and so on—are subject to inevitable natural decay, and human institutions have likewise proved to be transient. Hegel, whose philosophy of history was framed in large measure in an attempt to redeem the apparent futility of the efforts of men to found something enduring in their midst, once wrote, "When we see the evil, the vice, the ruin that has befallen the most flourishing kingdoms which the mind of man ever created, we can scarce avoid being filled with sorrow at this universal taint of corruption; and, since this decay is not the work of mere Nature, but of Human Will—a moral embitterment—a revolt of the Good Spirit (if it have a place within us) may well be the result of our reflections." Works of thought and many works of art have a better chance of surviving, since new copies of a book or a symphony can be transcribed from old ones, and so can be preserved indefinitely; yet these works, too, can and do go out of existence, for if every copy is lost, then the work is also lost. The subject matter of these works is man, and they seem to be touched with his mortality. The results of scientific work, on the other hand, are largely immune to decay and disappearance. Even when they are lost, they are likely to be rediscovered, as is shown by the fact that several scientists often make the same discovery independently. (There is no record of several poets' having independently written the same poem, or of several composers' having independently written the same symphony.) For both the subject matter and the method of science are available to all capable minds in a way that the subject matter and the method of the arts are not. The human experiences that art deals with are, once over, lost forever, like the people who undergo them, whereas matter, energy, space, and time, alike everywhere and in all ages, are always available for fresh inspection. The subject matter of science is the physical world, and its findings seem to share in the immortality of the physical world. And artistic vision grows out of the unrepeatable individuality of each artist, whereas the reasoning power of the mind—its ability to add two and two and get four—is the same in all competent persons. The rigorous exactitude of scientific methods does not mean that creativity is any less individual, intuitive, or mysterious in great scientists than in great artists, but it does mean that scientific findings, once arrived at, can be tested and confirmed by shared canons of logic and experimentation. The agreement among scientists thus achieved permits science to be a collective enterprise, in which each generation, building on the accepted findings of the generations before, makes amendments and additions, which in their turn become the starting point for the next generation. (Philosophers, by contrast, are constantly tearing down the work of their predecessors, and circling back to re-ask questions that have been asked and answered countless times before. Kant once wrote in despair, "It seems ridiculous that while every science moves forward ceaselessly, this [metaphysics], claiming to be wisdom itself, whose oracular pronouncements everyone consults, is continually revolving in one spot, without advancing one step.") Scientists, as they erect the steadily growing structure of scientific knowledge, resemble nothing so much as a swarm of bees working harmoniously together to construct a single, many-chambered hive, which grows more elaborate and splendid with every year that passes. Looking at what they have made over the centuries, scientists need feel no "sorrow" or "moral embitterment" at any "taint of corruption" that sup-posedly undoes all human achievements. When God, alarmed that the builders of the Tower of Babel would reach heaven with their construction, and so become as God, put an end to their undertaking by making them all speak different lan-guages, He apparently overlooked the scientists, for they, speaking what is often called the "universal language" of their disciplines from country to country and generation to generation, went on to build a new tower—the edifice of scientific knowledge. The phenomenal success, beginning not with Einstein but with Euclid and Archimedes, has provided the unshakable structure that supports the world's nuclear peril. So durable is the scientific edifice that if we did not know that human beings had constructed it we might suppose that the findings on which our whole technological civilization rests were the pillars and crossbeams of an invulnerable, inhuman order obtruding into our changeable and perishable human realm. It is the crowning irony of this lopsided development of human abilities that the only means in sight for getting rid of the knowledge of how to destroy ourselves would be to do just that—in effect, to remove the knowledge by removing the knower.

Although it is unquestionably the scientists who have led us to the edge of the nuclear abyss, we would be mistaken if we either held them chiefly respon-sible for our plight or looked to them, particularly, for a solution. Here, again, the difference between scientific evolutions and social revolutions shows itself, for the notion that scientists bear primary responsibility springs from a tendency to confuse scientists with political actors. Political actors, who, of course, include ordinary citizens as well as government officials, act with definite social ends in view, such as the preservation of peace, the establishment of a just society, or, if they are corrupt, their own aggrandizement; and they are accordingly held responsible for the consequences of their actions, even when these are unin-tended ones, as they so often are. Scientists, on the other hand (and here I refer to the so-called pure scientists, who search for the laws of nature for the sake of knowledge itself, and not to the applied scientists, who make use of already dis-covered natural laws to solve practical problems), do not aim at social ends, and, in fact, usually do not know what the social results of their findings will be; for that matter, they cannot know what the findings themselves will be, because sci-ence is a process of discovery, and it is in the nature of discovery that one cannot know beforehand what one will find. This element of the unexpected is present when a researcher sets out to unravel some small, carefully defined mystery—say, the chemistry of a certain enzyme—but it is most conspicuous in the syn-thesis of the great laws of science and in the development of science as a whole, which, over decades and centuries, moves toward destinations that no one can predict. Thus, only a few decades ago it might have seemed that physics, which had just placed nuclear energy at man's disposal, was the dangerous branch of science while biology, which underlay improvements in medicine and also helped us to understand our dependence on the natural environment, was the beneficial branch; but now that biologists have begun to fathom the secrets of genetics, and to tamper with the genetic substance of life directly, we cannot be so sure. The most striking illustration of the utter disparity that may occur between the wishes of the scientist as a social being and the social results of his scientific findings is certainly the career of Einstein. By nature, he was, according to all accounts, the gentlest of men, and by conviction he was a pacifist, yet he made intellectual discoveries that led the way to the invention of weapons with which the species could exterminate itself. Inspired wholly by a love of knowledge for its own sake, and by an awe at the creation which bordered on the religious, he made possible an instrument of destruction with which the terrestrial creation could be disfigured.

A disturbing corollary of the scientists' inability even to foresee the path of science, to say nothing of determining it, is that while science is without doubt the most powerful revolutionary force in our world, no one directs that force. For science is a process of submission, in which the mind does not dictate to nature but seeks out and then bows to nature's laws, letting its conclusions be guided by that which *is*, independent of our will. From the political point of view, therefore, scientific findings, some lending themselves to evil, some to good, and some to both, simply pour forth from the laboratory in senseless profusion, offering the world now a neutron bomb, now bacteria that devour oil, now a vaccine to prevent polio, now a cloned frog. It is not until the pure scientists, seekers of knowledge for its own sake, turn their findings over to the applied scientists that social intentions begin to guide the results. The applied scientists do indeed set out to make a better vaccine or a bigger bomb, but even they, perhaps, deserve less credit or blame than we are sometimes inclined to give them. For as soon as our intentions enter the picture we are in the realm of politics in the broadest sense, and in politics it is ultimately not technicians but governments and citizens who are in charge. The scientists in the Manhattan Project could not decide to make the first atomic bomb; only President Roosevelt, elected to office by the American people, could do that.

If scientists are unable to predict their discoveries, neither can they cancel them once they have been made. In this respect, they are like the rest of us, who are asked not whether we would like to live in a world in which we can convert matter into energy but only what we want to do about it once we have been told that we do live in such a world. Science is a tide that can only rise. The individual human mind is capable of forgetting things, and mankind has collectively forgotten many things, but we do not know how, as a species, to *deliberately* set out to forget something. A basic scientific finding, therefore, has the character of destiny for the world. Scientific discovery is in this regard like any other form of discovery; once Columbus had discovered America, and had told the world about it, America could not be hidden again.

Scientific progress (which can and certainly will occur) offers little more hope than scientific regression (which probably cannot occur) of giving us relief from the nuclear peril. It does not seem likely that science will bring forth some new invention—some antiballistic missile or laser beam—that will render nuclear weapons harmless (although the unpredictability of science prevents any categorical judgment on this point). In the centuries of the modern scientific revolution, scientific knowledge has steadily increased the destructiveness of warfare, for it is in the very nature of knowledge, apparently, to increase our might rather than to diminish it. One of the most common forms of the hope for deliverance from the nuclear peril by technical advances is the notion that the species will be spared extinction by fleeing in spaceships. The thought seems to be that while the people on earth are destroying themselves communities in space will be able to survive and carry on. This thought does an injustice to our birthplace and habitat, the earth. It assumes that if only we could escape the earth we would find safety—as though it were the earth and its plants and animals that threatened us, rather than the other way around. But the fact is that wherever human beings went there also would go the knowledge of how to build nuclear weapons, and, with it, the peril of extinction. Scientific progress may yet deliver us from many evils, but there are at least two evils that it cannot deliver us from: its own findings and our own destructive and self-destructive bent. This is a combination that we will have to learn to deal with by some other means.

We live, then, in a universe whose fundamental substance contains a supply of energy with which we can extinguish ourselves. We shall never live in any other. We now know that we live in such a universe, and we shall never stop knowing it. Over the millennia, this truth lay in waiting for us, and now we have found it out, irrevocably. If we suppose that it is an integral part of human existence to be curious about the physical world we are born into, then, to speak in the broadest terms, the origin of the nuclear peril lies, on the one hand, in our nature as rational and inquisitive beings and, on the other, in the nature of matter. Because the energy that nuclear weapons release is so great, the whole species is threatened by them, and because the spread of scientific knowledge is unstoppable, the whole species poses the threat: in the last analysis, it is all of mankind that threatens all of mankind. (I do not mean to overlook the fact that at present it is only two nations—the United States and the Soviet Union—that possess nuclear weapons in numbers great enough to possibly destroy the species, and that they thus now bear the chief responsibility for the peril. I only wish to point out that, regarded in its full dimensions, the nuclear peril transcends the rivalry between the present superpowers.)

The fact that the roots of the nuclear peril lie in basic scientific knowledge has broad political implications that cannot be ignored if the world's solution to the predicament is to be built on a solid foundation, and if futile efforts are to be avoided. One such effort would be to rely on secrecy to contain the peril—that is, to "classify" the "secret" of the bomb. The first person to try to suppress knowledge of how nuclear weapons can be made was the physicist Leo Szilard, who in 1939, when he first heard that a nuclear chain reaction was possible, and realized that a nuclear bomb might be possible, called on a number of his colleagues to keep the discovery secret from the Germans. Many of the key scientists refused. His failure foreshadowed a succession of failures, by whole governments, to restrict the knowledge of how the weapons are made. The first, and most notable, such failure was the United States' inability to monopolize nuclear weapons, and prevent the Soviet Union from building them. And we have subsequently witnessed the failure of the entire world to prevent nuclear weapons from spreading. Given the nature of scientific thought and the very poor record of past attempts to suppress it, these failures should not have surprised anyone. (The Catholic Church succeeded in making Galileo recant his view that the earth revolves around the sun, but we do not now believe that the sun revolves around the earth.) Another, closely related futile effort—the one made by our hypothetical council—would be to try to resolve the nuclear predicament through disarmament alone, without accompanying political measures. Like the hope that the knowledge can be classified, this hope loses sight of the fact that the nuclear predicament consists not in the possession of nuclear weapons at a particular moment by certain nations but in the circumstance that mankind as a whole has now gained possession once and for all of the knowledge of how to make them, and that all nations—and even some groups of people which are not nations, including terrorist groups—can potentially build them. Because the nuclear peril, like the scientific knowledge that gave rise to it, is probably global and everlasting, our solution must at least aim at being global and everlasting. And the only kind of solution that holds out this promise is a global political one. In defining the task so broadly, however, I do not mean to argue against short-term palliatives, such as the Strategic Arms Limitation Talks between the United States and the Soviet Union, or nuclear-nonproliferation agreements, on the ground that they are short-term. If a patient's life is in danger, as mankind's now is, no good cause is served by an argument between the nurse who wants to give him an aspirin to bring down his fever and the doctor who wants to perform the surgery that can save his life; there is need for an argument only if the nurse is claiming that the aspirin is all that is necessary. If, given the world's discouraging record of political achievement, a lasting political solution seems almost beyond human powers, it may give us confidence to remember that what challenges us is simply our extraordinary success in another field of activity—the scientific. We have only to learn to live politically in the world in which we already live scientifically.

Since 1947, the *Bulletin of the Atomic Scientists* has included a "doomsday clock" in each issue. The editors place the hands farther away from or closer to midnight as they judge the world to be farther from or closer to a nuclear holocaust. A companion clock can be imagined whose hands, instead of metaphorically representing a judgment about the likelihood of a holocaust, could represent an estimate of the amount of time that, given the world's technical and political arrangements, the people of the earth can be sure they have left before they are destroyed in a holocaust. At present, the hands would stand at, or a fraction of a second before, midnight, because none of us can be sure that at any second we will not be killed in a nuclear attack. If, by treaty, all nuclear warheads were removed from their launchers and stored somewhere else, and therefore could no longer descend on us at any moment without warning, the clock would show the amount of time that it would take to put them back on. If all the nuclear weapons in the world were destroyed, the clock would show the time that it would take to manufacture them again. If in addition confidence-inspiring political arrangements to prevent rearmament were put in place, the clock would show some estimate of the time that it might take for the arrangements to break down. And if these arrangements were to last for hundreds or thousands of years (as they must if mankind is to survive this long), then some generation far in the future might feel justified in setting the clock at decades, or even centuries, before midnight. But no generation would ever be justified in retiring the clock from use altogether, because, as far as we can tell, there will never again be a time when self-extinction is beyond the reach of our species. An observation that Plutarch made about politics holds true also for the task of survival, which has now become the principal obligation of politics: "They are wrong who think that politics is like an ocean voyage or a military campaign, something to be done with some end in view, something which levels off as soon as that end is reached. It is not a public chore, to be got over with; it is a way of life."

The scientific principles and techniques that make possible the construction of nuclear weapons are, of course, only one small portion of mankind's huge reservoir of scientific knowledge, and, as I have mentioned, it has always been known that scientific findings can be made use of for evil as well as for good, according to the intentions of the user. What is new to our time is the realization that, acting quite independently of any good or evil intentions of ours, the human enterprise as a whole has begun to strain and erode the natural terrestrial world on which human and other life depends. Taken in its entirety, the increase in mankind's strength has brought about a decisive, many-sided shift in the balance of strength between man and the earth. Nature, once a harsh and feared master, now lies in subjection, and needs protection against man's powers. Yet because man, no matter what intellectual and technical heights he may scale, remains embedded in nature, the balance has shifted against him, too, and the threat that he poses to the earth is a threat to him as well. The peril to nature was difficult to see at first, in part because its symptoms made their appearance as unintended "side effects" of our intended goals, on which we had fixed most of our attention. In economic production, the side effects are the peril of gradual pollution of the natural environment—by, for example, global heating through an increased "greenhouse effect." In the military field, the side effects, or prospective side effects—sometimes referred to by the strategists as the "collateral effects"—include the possible extinction of the species through sudden severe harm to the ecosphere, caused by global radioactive contamination, ozone depletion, climatic change, and the other known and unknown possible consequences of a nuclear holocaust.

Though from the point of view of the human actor there might be a clear difference between the "constructive" economic applications of technology and the
"destructive" military ones, nature makes no such distinction: both are beach-
heads of human mastery in a defenseless natural world. (For example, the ozone
doesn't care whether oxides of nitrogen are injected into it by the use of supersonic transports or by nuclear weapons; it simply reacts according to the appropriate chemical laws.) It was not until recently that it became clear that often the
side effects of both the destructive and the constructive applications were really
the main effects. And now the task ahead of us can be defined as one of giving
the "side effects," including, above all, the peril of self-extinction, the weight
they deserve in our judgments and decisions. To use a homely metaphor, if a man
discovers that improvements he is making to his house threaten to destroy its
foundation he is well advised to rethink them.

A nuclear holocaust, because of its unique combination of immensity and
suddenness, is a threat without parallel; yet at the same time it is only one of
countless threats that the human enterprise, grown mighty through knowledge,
poses to the natural world. Our species is caught in the same tightening net of
technical success that has already strangled so many other species. (At present, it
has been estimated, the earth loses species at the rate of about three per day.) The
peril of human extinction, which exists not because every single person in the
world would be killed by the immediate explosive and radioactive effects of a
holocaust—something that is exceedingly unlikely, even at present levels of
armament—but because a holocaust might render the biosphere unfit for human
survival, is, in a word, an *ecological* peril. The nuclear peril is usually seen in isolation from the threats to other forms of life and their ecosystems, but in fact it
should be seen as the very center of the ecological crisis—as the cloud-covered
Everest of which the more immediate, visible kinds of harm to the environment
are the mere foothills. Both the effort to preserve the environment and the effort
to save the species from extinction by nuclear arms would be enriched and
strengthened by this recognition. The nuclear question, which now stands in eerie
seclusion from the rest of life, would gain a context, and the ecological movement, which, in its concern for plants and animals, at times assumes an almost
misanthropic posture, as though man were an unwanted intruder in an otherwise
unblemished natural world, would gain the humanistic intent that should stand at
the heart of its concern.

Seen as a planetary event, the rising tide of human mastery over nature has
brought about a categorical increase in the power of death on earth. An organism's
ability to renew itself during its lifetime and to reproduce itself depends on the
integrity of what biologists call "information" stored in its genes. What endures—
what lives—in an organism is not any particular group of cells but a configuration of cells which is dictated by the genetic information. What survives in a
species, correspondingly, is a larger configuration, which takes in all the individuals in the species. An ecosystem is a still larger configuration, in which a whole

constellation of species forms a balanced, self-reproducing, slowly changing whole. The ecosphere of the earth—Dr. Lewis Thomas's "cell"—is finally, the largest of the living configurations, and is a carefully regulated and balanced, self-perpetuating system in its own right. At each of these levels, life is coherence, and the loss of coherence—the sudden slide toward disorder—is death. Seen in this light, life is information, and death is the loss of information, returning the substance of the creature to randomness. However, the death of a species or an ecosystem has a role in the natural order that is very different from that of the death of an organism. Whereas an individual organism, once born, begins to proceed inevitably toward death, a species is a source of new life that has no fixed term. An organism is a configuration whose demise is built into its plan, and within the life of a species the death of individual members normally has a fixed, limited, and necessary place, so that as death moves through the ranks of the living its pace is roughly matched by the pace of birth, and populations are kept in a rough balance that enables them to coexist and endure in their particular ecosystem. A species, on the other hand, can survive as long as environmental circumstances happen to permit. An ecosystem, likewise, is indefinitely self-renewing. But when the pace of death is too much increased, either by human intervention in the environment or by some other event, death becomes an extinguishing power, and species and ecosystems are lost. Then not only are individual creatures destroyed but the sources of all future creatures of those kinds are closed down, and a portion of the diversity and strength of terrestrial life in its entirety vanishes forever. And when man gained the ability to intervene directly in the workings of the global "cell" as a whole, and thus to extinguish species wholesale, his power to encroach on life increased by still another order of magnitude, and came to threaten the balance of the entire planetary system of life.

Hence, there are two competing forces at work in the terrestrial environment—one natural, which acts over periods of millions of years to strengthen and multiply the forms of life, and the other man-made and man-operated, which, if it is left unregulated and unguided, tends in general to deplete life's array of forms. Indeed, it is a striking fact that both of these great engines of change on earth depend on stores of information that are passed down from generation to generation. There is, in truth, no closer analogy to scientific progress, in which a steadily growing pool of information makes possible the creation of an ever more impressive array of artifacts, than evolution, in which another steadily growing pool of information makes possible the development of ever more complex and astonishing creatures—culminating in human beings, who now threaten to raze both the human and natural structures to their inanimate foundations. One is tempted to say that only the organic site of the evolutionary information has changed—from genes to brains. However, because of the extreme rapidity of technological change relative to natural evolution, evolution is unable to refill the vacated niches of the environment with new species, and, as a result, the genetic pool of life as a whole is imperilled. Death, having been augmented by human strength, has lost its appointed place in the natural order and become a counter-evolutionary force, capable of destroying in a few years, or even in a few hours, what evolution has built up over billions of years. In doing so, death threatens even itself, since death, after all, is a part of life: stones may be lifeless but they do not die. The question now before the human species, therefore, is whether life or death will prevail on the earth. This is not metaphorical language but a literal description of the present state of affairs.

One might say that after billions of years nature, by creating a species equipped with reason and will, turned its fate, which had previously been decided by the slow, unconscious movements of natural evolution, over to the conscious decisions of just one of its species. When this occurred, human activity, which until then had been confined to the historical realm—which, in turn, had been supported by the broader biological current—spilled out of its old boundaries and came to menace both history and biology. Thought and will became mightier than the earth that had given birth to them. Now human beings became actors in the geological time span, and the laws that had governed the development and the survival of life began to be superseded by processes in the mind of man. Here, however, there were no laws; there was only choice, and the thinking and feeling that guide choice. The reassuring, stable, self-sustaining prehistoric world of nature dropped away, and in its place mankind's own judgments, moods, and decisions loomed up with an unlooked-for, terrifying importance.

Regarded objectively, as an episode in the development of life on earth, a nuclear holocaust that brought about the extinction of mankind and other species by mutilating the ecosphere would constitute an evolutionary setback of possibly limited extent—the first to result from a deliberate action taken by the creature extinguished but perhaps no greater than any of several evolutionary setbacks, such as the extinction of the dinosaurs, of which the geological record offers evidence. (It is, of course, impossible to judge what course evolution would take after human extinction, but the past record strongly suggests that the reappearance of man is not one of the possibilities. Evolution has brought forth an amazing variety of creatures, but there is no evidence that any species, once extinguished, has ever evolved again. Whether or not nature, obeying some law of evolutionary progress, would bring forth another creature equipped with reason and will, and capable of building, and perhaps then destroying, a world, is one more unanswerable question, but it is barely conceivable that some gifted new animal will pore over the traces of our self-destruction, trying to figure out what went wrong and to learn from our mistakes. If this should be possible, then it might justify the remark once made by Kafka: "There is infinite hope, but not for us." If, on the other hand, as the record of life so far suggests, terrestrial evolution is able to produce only once the miracle of the qualities that we now associate with human beings, then all hope rides with human beings.) However, regarded subjectively, from within human life, where we are all actually situated, and as something that would happen to us, human extinction assumes awesome, inapprehensible proportions. It is of the essence of the human condition that we are born, live for a while, and then die. Through mishaps of all kinds, we may also suffer untimely death, and in extinction by nuclear arms the number of untimely deaths would reach the limit for any one catastrophe: everyone in the world would die. But although the untimely death of everyone in the world would in itself constitute an unimaginably huge loss, it would bring with it a separate, distinct loss that would be in a sense even huger—the cancellation of all future generations of human beings. According to the Bible, when Adam and Eve ate the fruit of the tree of knowledge God punished them by withdrawing from them the privilege of immortality and dooming them and their kind to die. Now our species has eaten more deeply of the fruit of the tree of knowledge, and has brought itself face to face with a second death—the death of mankind. In doing so, we have caused a basic change in the circumstances in which life was given to us, which is to say that we have altered the human condition. The distinctness of this second death from the deaths of all the people on earth can be illustrated by picturing two different global catastrophes. In the first, let us suppose that most of the people on earth were killed in a nuclear holocaust but that a few million survived and the earth happened to remain habitable by human beings. In this catastrophe, billions of people would perish, but the species would survive, and perhaps one day would even repopulate the earth in its former numbers. But now let us suppose that a substance was released into the environment which had the effect of sterilizing all the people in the world but otherwise leaving them unharmed. Then, as the existing population died off, the world would empty of people, until no one was left. Not one life would have been shortened by a single day, but the species would die. In extinction by nuclear arms, the death of the species and the death of all the people in the world would happen together, but it is important to make a clear distinction between the two losses; otherwise, the mind, overwhelmed by the thought of the deaths of the billions of living people, might stagger back without realizing that behind this already ungraspable loss there lies the separate loss of the future generations.

The possibility that the living can stop the future generations from entering into life compels us to ask basic new questions about our existence, the most sweeping of which is what these unborn ones, most of whom we will never meet even if they are born, mean to us. No one has ever thought to ask this question before our time, because no generation before ours has ever held the life and death of the species in its hands. But if we hardly know how to comprehend the possible deaths in a holocaust of the billions of people who are already in life how are we to comprehend the life or death of the infinite number of possible people who do not yet exist at all? How are we, who are a part of human life, to step back from life and see it whole, in order to assess the meaning of its disappearance? To kill a human being is murder, and there are those who believe that to abort a fetus is also murder, but what crime is it to cancel the numberless multitude of unconceived people? In what court is such a crime to be judged? Against whom is it committed? And what law does it violate? If we find the nuclear peril to be somehow abstract, and tend to consign this whole elemental issue to "defense experts" and other dubiously qualified people, part of the reason, certainly, is that the future generations really are abstract—that is to say, without the tangible existence and the unique particularities that help to make the living real to us. And if we find the subject strangely "impersonal" it may be in part because the unborn, who are the ones directly imperilled by extinction, are not yet persons. What are they, then? They lack the individuality that we often associate with the sacredness of life, and may at first thought seem to have only a shadowy, mass existence. *Where* are they? Are they to be pictured lined up in a sort of fore-life, waiting to get into life? Or should we regard them as nothing more than a pinch of chemicals in our reproductive organs, toward which we need feel no special obligations? What standing should they have among us? How much should their needs count in competition with ours? How far should the living go in trying to secure their advantage, their happiness, their existence?

The individual person, faced with the metaphysical-seeming perplexities involved in pondering the possible cancellation of people who do not yet exist—an apparently extreme effort of the imagination, which seems to require one first to summon before the mind's eye the countless possible people of the future generations and then to consign these incorporeal multitudes to a more profound nothingness—might well wonder why, when he already has his own death to worry about, he should occupy himself with this other death. Since our own individual death promises to inflict a loss that is total and final, we may find the idea of a second death merely redundant. After all, can everything be taken away from us twice? Moreover, a person might reason that even if mankind did perish he wouldn't have to know anything about it, since in that event he himself would perish. There might actually be something consoling in the idea of having so much company in death. In the midst of universal death, it somehow seems out of order to want to go on living oneself. As Randall Jarrell wrote in his poem "Losses," thinking back to his experience in the Second World War, "it was not dying: everybody died."

However, the individual would misconceive the nuclear peril if he tried to understand it primarily in terms of personal danger, or even in terms of danger to the people immediately known to him, for the nuclear peril threatens life, above all, not at the level of individuals, who already live under the sway of death, but at the level of everything that individuals hold in common. Death cuts off life; extinction cuts off birth. Death dispatches into the nothingness after life each person who has been born; extinction in one stroke locks up in the nothingness before life all the people who have not yet been born. For we are finite beings at both ends of our existence—natal as well as mortal—and it is the natality of our kind that extinction threatens. We have always been able to send people to their death, but only now has it become possible to prevent all birth and so doom all future human beings to uncreation. The threat of the loss of birth—a beginning that is over and done with for every living person—cannot be a source of immediate, selfish concern; rather, this threat assails everything that people hold in common, for it is the ability of our species to produce new generations which assures the continuation of the world in which all our common enterprises occur and have their meaning. Each death belongs unalienably to the individual who must suffer it, but birth is our common possession. And the meaning of extinction is therefore to be sought first not in what each person's own life means to him but in what the world and the people in it mean to him.

In its nature, the human world is, in Hannah Arendt's words, a "common world," which she distinguishes from the "private realm" that belongs to each person individually. (Somewhat surprisingly, Arendt, who devoted so much of her attention to the unprecedeted evils that have appeared in our century, never addressed the issue of nuclear arms; yet I have discovered her thinking to be an indispensable foundation for reflection on this question.) The private realm, she writes in *The Human Condition*, a book published in 1958, is made up of "the passions of the heart, the thoughts of the mind, the delights of the senses," and terminates with each person's death, which is the most solitary of all human experiences. The common world, on the other hand, is made up of all institutions, all cities, nations, and other communities, and all works of fabrication, art, thought, and science, and it survives the death of every individual. It is basic to the common world that it encompasses not only the present but all past and future generations. "The common world is what we enter when we are born and what we leave behind when we die," Arendt writes. "It transcends our life-span into past and future alike; it was there before we came and will outlast our brief sojourn in it. It is what we have in common not only with those who live with us, but also with those who were here before and with those who will come after us." And she adds, "Without this trancendence into a potential earthly immortality, no politics, strictly speaking, no common world, and no public realm is possible." The creation of a common world is the use that we human beings, and we alone among the earth's creatures, have made of the biological circumstance that while each of us is mortal, our species is biologically immortal. If mankind had not established a common world, the species would still outlast its individual members and be immortal, but this immortality would be unknown to us and would go for nothing, as it does in the animal kingdom, and the generations, unaware of one another's existence, would come and go like waves on the beach, leaving everything just as it was before. In fact, it is only because humanity has built up a common world that we can fear our destruction as a species. It may even be that man, who has been described as the sole creature that knows that it must die, can know this only because he lives in a common world, which permits him to imagine a future beyond his own life. This common world, which is unharmed by individual death but depends on the survival of the species, has now been placed in jeopardy by nuclear arms. Death and extinction are thus complementary, dividing between them the work of undoing, or threatening to undo, everything that human beings are or can ever become, with death terminating the life of each individual and extinction imperilling the common world shared by all. In one sense, extinction is less terrible than death, since extinction can be avoided, while death is inevitable; but in another sense extinction is more terrible—is the more radical nothingness—because extinction ends death just as surely as it ends birth and life. Death is only death; extinction is the death of death.

2

Why McDonald's French Fries Taste So Good

Eric Schlosser

The taste of McDonald's french fries has long been praised by customers, competitors, and even food critics. James Beard loved McDonald's fries. Their distinctive taste does not stem from the type of potatoes that McDonald's buys, the technology that processes them, or the restaurant equipment that fries them. Other chains buy their french fries from the same large processing companies, use Russet Burbanks, and have similar fryers in their restaurant kitchens. The taste of a fast food fry is largely determined by the cooking oil. For decades, McDonald's cooked its french fries in a mixture of about 7 percent cottonseed oil and 93 percent beef tallow. The mix gave the fries their unique flavor—and more saturated beef fat per ounce than a McDonald's hamburger.

Amid a barrage of criticism over the amount of cholesterol in their fries, McDonald's switched to pure vegetable oil in 1990. The switch presented the company with an enormous challenge: how to make fries that subtly taste like beef without cooking them in tallow. A look at the ingredients now used in the preparation of McDonald's french fries suggests how the problem was solved. Toward the end of the list is a seemingly innocuous, yet oddly mysterious phrase: "natural flavor." That ingredient helps to explain not only why the fries taste so good, but also why most fast food—indeed, most of the food Americans eat today—tastes the way it does.

Open your refrigerator, your freezer, your kitchen cupboards, and look at the labels on your food. You'll find "natural flavor" or "artificial flavor" in just about every list of ingredients. The similarities between these two broad categories of flavor are far more significant than their differences. Both are man-made additives that give most processed food most of its taste. The initial purchase of a food item may be driven by its packaging or appearance, but subsequent purchases are

From *Fast Food Nation: The Dark Side of the All-American Meal*, by Eric Schlosser. Excerpted and reprinted by permission of Houghton Mifflin Company. All rights reserved. First published in the *Atlantic Monthly*, January 2001.

determined mainly by its taste. About 90 percent of the money that Americans spend on food is used to buy processed food. But the canning, freezing, and dehydrating techniques used to process food destroy most of its flavor. Since the end of World War II, a vast industry has arisen in the United States to make processed food palatable. Without this flavor industry, today's fast food industry could not exist. The names of the leading American fast food chains and their best-selling menu items have become famous worldwide, embedded in our popular culture. Few people, however, can name the companies that manufacture fast food's taste.

The flavor industry is highly secretive. Its leading companies will not divulge the precise formulas of flavor compounds or the identities of clients. The secrecy is deemed essential for protecting the reputation of beloved brands. The fast food chains, understandably, would like the public to believe that the flavors of their food somehow originate in their restaurant kitchens, not in distant factories run by other firms.

The New Jersey Turnpike runs through the heart of the flavor industry, an industrial corridor dotted with refineries and chemical plants. International Flavors & Fragrances (IFF), the world's largest flavor company, has a manufacturing facility off Exit 8A in Dayton, New Jersey; Givaudan, the world's second-largest flavor company, has a plant in East Hanover. Haarmann & Reimer, the largest German flavor company, has a plant in Teterboro, as does Takasago, the largest Japanese flavor company. Flavor Dynamics has a plant in South Plainfield; Frutarom is in North Bergen; Elan Chemical is in Newark. Dozens of companies manufacture flavors in the corridor between Teaneck and South Brunswick. Indeed, the area produces about two-thirds of the flavor additives sold in the United States.

The IFF plant in Dayton is a huge pale blue building with a modern office complex attached to the front. It sits in an industrial park, not far from a BASF plastics factory, a Jolly French Toast factory, and a plant that manufactures Liz Claiborne cosmetics. Dozens of tractor-trailers were parked at the IFF loading dock the afternoon I visited, and a thin cloud of steam floated from the chimney. Before entering the plant, I signed a nondisclosure form, promising not to reveal the brand names of products that contain IFF flavors. The place reminded me of Willy Wonka's chocolate factory. Wonderful smells drifted through the hallways, men and women in neat white lab coats cheerfully went about their work, and hundreds of little glass bottles sat on laboratory tables and shelves. The bottles contained powerful but fragile flavor chemicals, shielded from light by the brown glass and the round plastic caps shut tight. The long chemical names on the little white labels were as mystifying to me as medieval Latin. They were the odd-sounding names of things that would be mixed and poured and turned into new substances, like magic potions.

I was not invited to see the manufacturing areas of the IFF plant, where it was thought I might discover trade secrets. Instead, I toured various laboratories and pilot kitchens, where the flavors of well-established brands are tested or adjusted, and where whole new flavors are created. IFF's snack and savory lab is responsible for the flavor of potato chips, corn chips, breads, crackers, breakfast cereals, and pet food. The confectionery lab devises the flavor for ice cream, cookies, candies, toothpastes, mouthwashes, and antacids. Everywhere I looked, I saw famous, widely advertised products sitting on laboratory desks and tables. The beverage lab is full of brightly colored liquids in clear bottles. It comes up with the flavor for popular soft drinks, sport drinks, bottled teas, and wine coolers, for all-natural juice drinks, organic soy drinks, beers, and malt liquors. In one pilot kitchen I saw a dapper food technologist, a middle-aged man with an elegant tie beneath his lab coat, carefully preparing a batch of cookies with white frosting and pink-and-white sprinkles. In another pilot kitchen I saw a pizza oven, a grill, a milk-shake machine, and a french fryer identical to those I'd seen behind the counter at countless fast food restaurants.

In addition to being the world's largest flavor company, IFF manufactures the smell of six of the ten best-selling fine perfumes in the United States, including Estée Lauder's Beautiful, Clinique's Happy, Lancôme's Trésor, and Calvin Klein's Eternity. It also makes the smell of household products such as deodorant, dishwashing detergent, bath soap, shampoo, furniture polish, and floor wax. All of these aromas are made through the same basic process: the manipulation of volatile chemicals to create a particular smell. The basic science behind the scent of your shaving cream is the same as that governing the flavor of your TV dinner.

The aroma of a food can be responsible for as much as 90 percent of its flavor. Scientists now believe that human beings acquired the sense of taste as a way to avoid being poisoned. Edible plants generally taste sweet; deadly ones, bitter. Taste is supposed to help us differentiate food that's good for us from food that's not. The taste buds on our tongues can detect the presence of half a dozen or so basic tastes, including: sweet, sour, bitter, salty, astringent, and umami (a taste discovered by Japanese researchers, a rich and full sense of deliciousness triggered by amino acids in foods such as shellfish, mushrooms, potatoes, and seaweed). Taste buds offer a relatively limited means of detection, however, compared to the human olfactory system, which can perceive thousands of different chemical aromas. Indeed "flavor" is primarily the smell of gases being released by the chemicals you've just put in your mouth.

The act of drinking, sucking, or chewing a substance releases its volatile gases. They flow out of the mouth and up the nostrils, or up the passageway in the back of the mouth, to a thin layer of nerve cells called the olfactory epithelium, located at the base of the nose, right between the eyes. The brain combines the complex smell signals from the epithelium with the simple taste signals from the tongue, assigns a flavor to what's in your mouth, and decides if it's something you want to eat.

Babies like sweet tastes and reject bitter ones; we know this because scientists have rubbed various flavors inside the mouths of infants and then recorded their facial reactions. A person's food preferences, like his or her personality, are formed during the first few years of life, through a process of socialization. Toddlers can learn to enjoy hot and spicy food, bland health food, or fast food, depending upon what the people around them eat. The human sense of smell is still not fully understood and can be greatly affected by psychological factors and expectations. The color of a food can determine the perception of its taste. The mind filters out the overwhelming majority of chemical aromas that surround us, focusing intently on some, ignoring others. People can grow accustomed to bad smells or good smells; they stop noticing what once seemed overpowering. Aroma and memory are somehow inextricably linked. A smell can suddenly evoke a long-forgotten moment. The flavors of childhood foods seem to leave an indelible mark, and adults often return to them, without always knowing why. These "comfort foods" become a source of pleasure and reassurance, a fact that fast food chains work hard to promote. Childhood memories of Happy Meals can translate into frequent adult visits to McDonald's, like those of the chain's "heavy users," the customers who eat there four or five times a week.

The human craving for flavor has been a largely unacknowledged and unexamined force in history. Royal empires have been built, unexplored lands have been traversed, great religions and philosophies have been forever changed by the spice trade. In 1492 Christopher Columbus set sail to find seasoning. Today the influence of flavor in the world marketplace is no less decisive. The rise and fall of corporate empires—of soft drink companies, snack food companies, and fast food chains—is frequently determined by how their products taste.

The flavor industry emerged in the mid-nineteenth century, as processed foods began to be manufactured on a large scale. Recognizing the need for flavor additives, the early food processors turned to perfume companies that had years of experience working with essential oils and volatile aromas. The great perfume houses of England, France, and the Netherlands produced many of the first flavor compounds. In the early part of the twentieth century, Germany's powerful chemical industry assumed the technological lead in flavor production. Legend has it that a German scientist discovered methyl anthranilate, one of the first artificial flavors, by accident while mixing chemicals in his laboratory. Suddenly the lab was filled with the sweet smell of grapes. Methyl anthranilate later became the chief flavoring compound of grape Kool-Aid. After World War II, much of the perfume industry shifted from Europe to the United States, settling in New York City near the garment district and the fashion houses. The flavor industry came with it, subsequently moving to New Jersey to gain more plant capacity. Man-made flavor additives were used mainly in baked goods, candies, and sodas until the 1950s, when sales of processed food began to soar. The invention of gas chromatographs and mass spectrometers—machines capable of detecting volatile gases at low levels—vastly increased the number of flavors that could be synthesized. By the mid-1960s the American flavor industry was churning out compounds to supply the taste of Pop Tarts, Bac-Os, Tab, Tang, Filet-O-Fish sandwiches, and literally thousands of other new foods.

The American flavor industry now has annual revenues of about $1.4 billion. Approximately ten thousand new processed food products are introduced every year in the United States. Almost all of them require flavor additives. And about nine out of every ten of these new food products fail. The latest flavor innovations and corporate realignments are heralded in publications such as *Food Chemical News*, *Food Engineering*, *Chemical Market Reporter*, and *Food Product Design*. The growth of IFF has mirrored that of the flavor industry as a whole. IFF was formed in 1958, through the merger of two small companies. Its annual revenues have grown almost fifteenfold since the early 1970s, and it now has manufacturing facilities in twenty countries.

The quality that people seek most of all in a food, its flavor, is usually present in a quantity too infinitesimal to be measured by any traditional culinary terms such as ounces or teaspoons. Today's sophisticated spectrometers, gas chromatographs, and headspace vapor analyzers provide a detailed map of a food's flavor components, detecting chemical aromas in amounts as low as one part per billion. The human nose, however, is still more sensitive than any machine yet invented. A nose can detect aromas present in quantities of a few parts per trillion—an amount equivalent to 0.000000000003 percent. Complex aromas, like those of coffee or roasted meat, may be composed of volatile gases from nearly a thousand different chemicals. The smell of a strawberry arises from the interaction of at least 350 different chemicals that are present in minute amounts. The chemical that provides the dominant flavor of bell pepper can be tasted in amounts as low as .02 parts per billion; one drop is sufficient to add flavor to five average size swimming pools. The flavor additive usually comes last, or second to last, in a processed food's list of ingredients (chemicals that add color are frequently used in even smaller amounts). As a result, the flavor of a processed food often costs less than its packaging. Soft drinks contain a larger proportion of flavor additives than most products. The flavor in a twelve-ounce can of Coke costs about half a cent.

The Food and Drug Administration does not require flavor companies to disclose the ingredients of their additives, so long as all the chemicals are considered by the agency to be GRAS (Generally Regarded As Safe). This lack of public disclosure enables the companies to maintain the secrecy of their formulas. It also hides the fact that flavor compounds sometimes contain more ingredients than the foods being given their taste. The ubiquitous phrase "artificial strawberry flavor" gives little hint of the chemical wizardry and manufacturing skill that can make a highly processed food taste like a strawberry.

A typical artificial strawberry flavor, like the kind found in a Burger King strawberry milk shake, contains the following ingredients: amyl acetate, amyl butyrate, amyl valerate, anethol, anisyl formate, benzyl acetate, benzyl isobutyrate, butyric acid, cinnamyl isobutyrate, cinnamyl valerate, cognac essential oil, diacetyl, dipropyl ketone, ethyl acetate, ethyl amylketone, ethyl butyrate, ethyl cinnamate, ethyl heptanoate, ethyl heptylate, ethyl lactate, ethyl methylphenylglycidate, ethyl nitrate, ethyl propionate, ethyl valerate, heliotropin, hydrox-yphenyl-2-butanone (10 percent solution in alcohol), α–ionone, isobutyl anthranilate, isobutyl butyrate, lemon essential oil, maltol, 4-methylacetophe-none, methyl anthranilate, methyl benzoate, methyl cinnamate, methyl heptine carbonate, methyl naphthyl ketone, methyl salicylate, mint essential oil, neroli essential oil, nerolin, neryl isobutyrate, orris butter, phenethyl alcohol, rose, rum ether, γ–undecalactone, vanillin, and solvent.

Although flavors usually arise from a mixture of many different volatile chemicals, a single compound often supplies the dominant aroma. Smelled alone, that chemical provides an unmistakable sense of the food. Ethyl-2-methyl butyrate, for example, smells just like an apple. Today's highly processed foods offer a blank palette: whatever chemicals you add to them will give them specific tastes. Adding methyl-2-peridylketone makes something taste like popcorn. Adding ethyl-3-hydroxybutanoate makes it taste like marshmallow. The possibil-ities are now almost limitless. Without affecting the appearance or nutritional value, processed foods could even be made with aroma chemicals such as hexanal (the smell of freshly cut grass) or 3-methyl butanoic acid (the smell of body odor).

The 1960s were the heyday of artificial flavors. The synthetic versions of flavor compounds were not subtle, but they did not need to be, given the nature of most processed food. For the past twenty years food processors have tried hard to use only "natural flavors" in their products. According to the FDA, these must be derived entirely from natural sources—from herbs, spices, fruits, vegetables, beef, chicken, yeast, bark, roots, etc. Consumers prefer to see natural flavors on a label, out of a belief that they are healthier. The distinction between artificial and natural flavors can be somewhat arbitrary and absurd, based more on how the flavor has been made than on what it actually contains. "A natural flavor," says Terry Acree, a professor of food science at Cornell University, "is a flavor that's been derived with an out-of-date technology." Natural flavors and artificial fla-vors sometimes contain exactly the same chemicals, produced through different methods. Amyl acetate, for example, provides the dominant note of banana flavor. When you distill it from bananas with a solvent, amyl acetate is a natural flavor. When you produce it by mixing vinegar with amyl alcohol, adding sul-furic acid as a catalyst, amyl acetate is an artificial flavor. Either way it smells and tastes the same. The phrase "natural flavor" is now listed among the ingredi-ents of everything from Stonyfield Farm Organic Strawberry Yogurt to Taco Bell Hot Taco Sauce.

A natural flavor is not necessarily healthier or purer than an artificial one. When almond flavor (benzaldehyde) is derived from natural sources, such as peach and apricot pits, it contains traces of hydrogen cyanide, a deadly poison. Benzalde-hyde derived through a different process—by mixing oil of clove and the banana flavor, amyl acetate—does not contain any cyanide. Nevertheless, it is legally con-sidered an artificial flavor and sells at a much lower price. Natural and artificial fla-vors are now manufactured at the same chemical plants, places that few people would associate with Mother Nature. Calling any of these flavors "natural" requires a flexible attitude toward the English language and a fair amount of irony.

The small and elite group of scientists who create most of the flavor in most of the food now consumed in the United States are called "flavorists." They draw upon a number of disciplines in their work: biology, psychology, physiology, and organic chemistry. A flavorist is a chemist with a trained nose and a poetic sensibility. Flavors are created by blending scores of different chemicals in tiny amounts, a process governed by scientific principles but demanding a fair amount of art. In an age when delicate aromas, subtle flavors, and microwave ovens do not easily coexist, the job of the flavorist is to conjure illusions about processed food and, in the words of one flavor company's literature, to ensure "consumer likeability." The flavorists with whom I spoke were charming, cosmopolitan, and ironic. They were also discreet, in keeping with the dictates of their trade. They were the sort of scientist who not only enjoyed fine wine, but could also tell you the chemicals that gave each vintage its unique aroma. One flavorist compared his work to composing music. A well-made flavor compound will have a "top note," followed by a "dry-down," and a "leveling-off," with different chemicals responsible for each stage. The taste of a food can be radically altered by minute changes in the flavoring mix. "A little odor goes a long way," one flavorist said.

In order to give a processed food the proper taste, a flavorist must always consider the food's "mouthfeel"—the unique combination of textures and chemical interactions that affects how the flavor is perceived. The mouthfeel can be adjusted through the use of various fats, gums, starches, emulsifiers, and stabilizers. The aroma chemicals of a food can be precisely analyzed, but mouthfeel is much harder to measure. How does one quantify a french fry's crispness? Food technologists are now conducting basic research in rheology, a branch of physics that examines the flow and deformation of materials. A number of companies sell sophisticated devices that attempt to measure mouthfeel. The TA.XT2i Texture Analyzer, produced by the Texture Technologies Corporation, performs calculations based on data derived from as many as 250 separate probes. It is essentially a mechanical mouth. It gauges the most important rheological properties of a food—the bounce, creep, breaking point, density, crunchiness, chewiness, gumminess, lumpiness, rubberiness, springiness, slipperiness, smoothness, softness, wetness, juiciness, spreadability, springback, and tackiness.

Some of the most important advances in flavor manufacturing are now occurring in the field of biotechnology. Complex flavors are being made through fermentation, enzyme reactions, fungal cultures, and tissue cultures. All of the flavors being created through these methods—including the ones being synthesized by funguses—are considered natural flavors by the FDA. The new enzyme-based processes are responsible for extremely lifelike dairy flavors. One company now offers not just butter flavor, but also fresh creamy butter, cheesy butter, milky butter, savory melted butter, and super-concentrated butter flavor, in liquid or powder form. The development of new fermentation techniques, as well as new techniques for heating mixtures of sugar and amino acids, have led to the creation of much more realistic meat flavors. The McDonald's Corporation will not reveal the exact origin of the natural flavor added to its french fries. In response to inquiries from *Vegetarian Journal*, however, McDonald's did acknowledge that its fries derive some of their characteristic flavor from "animal products."

Other popular fast foods derive their flavor from unexpected sources. Wendy's Grilled Chicken Sandwich, for example, contains beef extracts. Burger King's BK Broiler Chicken Breast Patty contains "natural smoke flavor." A firm called Red Arrow Products Company specializes in smoke flavor, which is added to barbecue sauces and processed meats. Red Arrow manufactures natural smoke flavor by charring sawdust and capturing the aroma chemicals released into the air. The smoke is captured in water and then bottled, so that other companies can sell food which seems to have been cooked over a fire.

In a meeting room at IFF, Brian Grainger let me sample some of the company's flavors. It was an unusual taste test; there wasn't any food to taste. Grainger is a senior flavorist at IFF, a soft-spoken chemist with graying hair, an English accent, and a fondness for understatement. He could easily be mistaken for a British diplomat or the owner of a West End brasserie with two Michelin stars. Like many in the flavor industry, he has an Old World, old-fashioned sensibility which seems out of step with our brand-conscious, egocentric age. When I suggested that IFF should put its own logo on the products that contain its flavors—instead of allowing other brands to enjoy the consumer loyalty and affection inspired by those flavors—Grainger politely disagreed, assuring me such a thing would never be done. In the absence of public credit or acclaim, the small and secretive fraternity of flavor chemists praises one another's work. Grainger can often tell, by analyzing the flavor formula of a product, which of his counterparts at a rival firm devised it. And he enjoys walking down supermarket aisles, looking at the many products that contain his flavors, even if no one else knows it.

Grainger had brought a dozen small glass bottles from the lab. After he opened each bottle, I dipped a fragrance testing filter into it. The filters were long white strips of paper designed to absorb aroma chemicals without producing off-notes. Before placing the strips of paper before my nose, I closed my eyes. Then I inhaled deeply, and one food after another was conjured from the glass bottles. I smelled fresh cherries, black olives, sautéed onions, and shrimp. Grainger's most remarkable creation took me by surprise. After closing my eyes, I suddenly smelled a grilled hamburger. The aroma was uncanny, almost miraculous. It smelled like someone in the room was flipping burgers on a hot grill. But when I opened my eyes, there was just a narrow strip of white paper and a smiling flavorist.

PART 2
DEFINING TECHNOLOGY

INTRODUCTION

There are considerable confusions about the meaning of the word "technology." It might seem that these could be cleared up by simply referring to a good dictionary. However, since dictionaries report on the ways we do in fact use language, one finds the same multiple, vague, and ambiguous meanings that we find in ordinary discourse. Thus, these definitions do little to clarify our thinking.

It is also the case that there is no consensus among the authors in this part, indeed throughout this volume, as to how "technology" ought to be defined. This is not, however, because these writers do not know what they are talking about. The concepts of technology they propose are used to introduce theories of technology. Disputes that may appear to be a matter of semantics are, at a deeper level, disputes about the meaning of technology in our lives. Particular definitions of technology often incorporate specific value judgments about technology. Some definitions may hide, while others may uncover, important points of view regarding technology. It will thus be important to not ask for *the* meaning of technology, but as philosopher Ludwig Wittgenstein might say, to look for the way a thinker is using the term.

It might be argued that if we define "technology" in a narrow sense as hardware, machines, tools, or material artifacts, then technology is value-neutral in two senses. (1) Since technology simply provides us with means, it does not affect our lives in any significant way. New technologies simply provide more efficient means for accomplishing preexisting ends. A calculator is just a faster way to balance the checkbook. (2) Technology is ethically neutral, for its value depends on how it is used. A hammer can be used to build a house or to bash your neighbor's head in.

If, however, we focus on the design of even simple tools, a very different pic-

49

ture emerges. Tools are created for specific purposes. A carpenter's hammer is designed to drive nails, not to kill your neighbor—handguns are designed to kill. On this view the ends are built into the very construction of the implement. Thus, a technology *is* a use, and in this sense is thoroughly intentional. Therefore, the definition of technology must include its purpose, its role in a particular society or human community or institution.

A further difficulty is that some authors talk of technology in the abstract, arguing that there is an essential nature to all technology, while others will focus on the operation of this or that particular technology and may deny that there is any such thing as technology in general.

One last cautionary note. The reader should be aware of persuasive definitions! Some ways of defining "technology" attempt to smuggle into the very meaning of the word important theoretical and evaluative conclusions. The definition cited in the general introduction, "the totality of the means employed to provide objects necessary for human sustenance and comfort," is a prime example. The Star Wars missile defense shield or the cloning of human beings may be necessary for human sustenance and comfort, but surely some argument is necessary to prove that this is true. We should not be seduced into accepting this conclusion without good reasons.

In "Technology: Practice and Culture," Arnold Pacey's thinking about technology is rooted in case studies of how specific technologies have functioned in various cultural contexts. Modeled on the idea of medical practice, his lucid concept of technology-practice allows us a comprehensive view of technology without obscuring important distinctions. Technology in the traditional narrow sense, as machines or hardware, is to be understood as it interacts with cultural and organizational factors. Thinking of technology only in the traditional, restricted sense, which Pacey refers to as the technical aspect, leads not only to intellectual confusions but also to misapplications of technology—especially in the transfer of technology to "underdeveloped" societies. The wider perspective reveals that technology is neither culturally nor ethically neutral, but constitutes a way of life. Pacey thus gives us a dynamic model, which he thinks more accurately reflects the multiplicity of causes of technology. The technology-practice perspective shows that there is no "technological imperative," no inevitable "one best way." There is a place, on his view, for human choices; important ethical and political decisions have to be made. The question is, Who will make them, and for whom will they be made?

The second selection, Jacques Ellul's "Defining Technique," is an excerpt from Ellul's classic study of technology in human life, *The Technological Society*. Ellul's fundamental point is that the contemporary world is completely determined—one might even say, dominated—by our technologies. All that matters for modern humanity is the efficiency of our technological means. Thus, Ellul argues that no distinctions can be made in the contemporary world between science and technology. On his view, "this traditional distinction is radically false. . . . It is true only for the physical sciences and for the nineteenth century." While a few intellectuals may continue to value knowledge as an end, their views are irrelevant, for our technological society values science only as a means. To raise questions about ends is considered irrational in a society dominated by what Ellul calls "technique," which "is nothing more than means and the ensemble of means."

If we are to understand the current situation, Ellul thinks we need a very broad notion of technology. Although the machine remains the model for technique, the machine is not the problem. The basic problem is the all-embracing "ensemble of means" that integrates and coordinates all human activities, eliminating all that is spontaneous, free, and distinctively human. This "technical phenomenon" is the quest for the one best means in every field: economics, social and political organization—even for humanity itself. We cannot, Ellul claims, bring ourselves to question the universal applicability of the machine model that divides, quantifies, and rationalizes all activities in seeking the most efficient means, for our very mode of thinking is determined by the model of technique. As an analogy, Ellul might point to the popularity of sex manuals that prescribe the steps to follow in our most intimate, human relationships. The human person "becomes the object of technique." The machine is the model for dealing with both mice and men.

Ellul's conception of technology as a broadly conceived "technique" that penetrates all aspects of human life has had an enormous influence on the social analysis of technology in society, particularly in the theory that technological development is, in some sense, autonomous, beyond human control. The essays in part 4 will focus on the subject of autonomous technology. The remaining selections in this part look at more specific problems in the definition of technology and its relationship to science and human life in general. In the third reading, "The Knowing World of Things," Davis Baird examines the way in which scientific and technological instruments, tools, and material products actually constitute a form of knowledge—what Baird calls "thing knowledge." In looking at the history of the development of the electromagnetic motor, Baird considers it important that Michael Faraday not only reported what he had invented but also sent ready-made versions of the motor to his colleagues. Later, one of the inventors of the electric motor, Henry Davenport, built a motor without any knowledge of the science of electromagnetism. For Baird, these examples show that there is more to knowledge than scientific theories—words and equations; epistemology, the theory of knowledge, must find a place for material instruments, tools, and technology. If this argument is correct, then we must redraw the boundary lines between science and technology—both things and theories are equally important for the progress of knowledge.

The final essay in part 2 is by Albert Borgmann, one of the leading philosophers of technology in the world today. "The Device Paradigm" is a short excerpt from Borgmann's full-length treatment of the philosophy of technology, *Technology and the Character of Contemporary Life*, now considered a classic in the field. In this selection, Borgmann presents a clear distinction between two kinds of technological artifacts: things and devices. A thing, for Borgmann, is inseparable from its context in the world. When we make use of a thing, we engage the world and the thing in a network of bodily and social relationships. A device, in contrast, is just a tool, a machine that performs a function; the machinery of the device is unimportant (and is hidden from us) as long as the function is maintained; the device becomes the function. Borgmann uses the illustrative example of a comparison between a hearth and a central heating plant. The hearth in a household is a central focus of family life, involving different family members in various tasks, different bodily functions, and different kinds of engagement with the natural world and the rhythms of the seasons. A central heating system does none of these things. Its machinery is completely hidden from the user; all it provides is heat. Technological devices become mere commodities to be used and discarded. The intentions of their creators are forever hidden from the users of the technology. The implication is that we must reengage with technological things, objects that serve as focal points of human social interaction. If we live in a world of devices, we will not be able to understand the purpose and value of technology, for we will be cut off from everything except the mere functionality of the object. To judge technology we must engage it in the world and in human society.

The readings in this part offer a wide-ranging selection of definitions of technology. It is important to remember that how we define technology may have serious consequences for how we evaluate specific technological objects and how we conceive of our ability to control the process of technological development.

3

Technology
Practice and Culture

Arnold Pacey

QUESTIONS OF NEUTRALITY

Winter sports in North America gained a new dimension during the 1960s with the introduction of the snowmobile. Ridden like a motorcycle, and having handlebars for steering, this little machine on skis gave people in Canada and the northern United States extra mobility during their long winters. Snowmobile sales doubled annually for a while, and in the boom year of 1970–71 almost half a million were sold. Subsequently the market dropped back, but snowmobiling had established itself, and organized trails branched out from many newly prosperous winter holiday resorts. By 1978, there were several thousand miles of public trails, marked and maintained for snowmobiling, about half in the province of Quebec.

Although other firms had produced small motorized toboggans, the type of snowmobile which achieved this enormous popularity was only really born in 1959, chiefly on the initiative of Joseph-Armand Bombardier of Valcourt, Quebec.[1] He had experimented with vehicles for travel over snow since the 1920s, and had patented a rubber-and-steel crawler track to drive them. His first commercial success, which enabled his motor repair business to grow into a substantial manufacturing firm, was a machine capable of carrying seven passengers which was on the market from 1936. He had other successes later, but nothing that caught the popular imagination like the little snowmobile of 1959, which other manufacturers were quick to follow up.

However, the use of snowmobiles was not confined to the North American tourist centers. In Sweden, Greenland, and the Canadian Arctic, snowmobiles have now become part of the equipment on which many communities depend for their livelihood. In Swedish Lapland they are used for reindeer herding. On

Canada's Banks Island they have enabled Eskimo trappers to continue providing their families' cash income from the traditional winter harvest of fox furs.

Such use of the snowmobile by people with markedly different cultures may seem to illustrate an argument very widely advanced in discussions of problems associated with technology. This is the argument which states that technology is culturally, morally, and politically neutral—that it provides tools independent of local value-systems which can be used impartially to support quite different kinds of lifestyle.

Thus in the world at large, it is argued that technology is "essentially amoral, a thing apart from values, an instrument which can be used for good or ill."[2] So if people in distant countries starve; if infant mortality within the inner cities is persistently high; if we feel threatened by nuclear destruction or more insidiously by the effects of chemical pollution, then all that, it is said, should not be blamed on technology, but on its misuse by politicians, the military, big business, and others.

The snowmobile seems the perfect illustration of this argument. Whether used for reindeer herding or for recreation, for ecologically destructive sport, or to earn a basic living, it is the same machine. The engineering principles involved in its operation are universally valid, whether its users are Lapps or Eskimos, Dene (Indian) hunters, Wisconsin sportsmen, Quebecois vacationists, or prospectors from multinational oil companies. And whereas the snowmobile has certainly had a social impact, altering the organization of work in Lapp communities, for example, it has not necessarily influenced basic cultural values. The technology of the snowmobile may thus appear to be something quite independent of the lifestyles of Lapps or Eskimos or Americans.

One look at a modern snowmobile with its fake streamlining and flashy colors suggests another point of view. So does the advertising which portrays virile young men riding the machines with sexy companions, usually blonde and usually riding pillion. The Eskimo who takes a snowmobile on a long expedition in the Arctic quickly discovers more significant discrepancies. With his traditional means of transport, the dog-team and sledge, he could refuel as he went along by hunting for his dogs' food. With the snowmobile he must take an ample supply of fuel and spare parts; he must be skilled at doing his own repairs and even then he may take a few dogs with him for emergency use if the machine breaks down. A vehicle designed for leisure trips between well-equipped tourist centers presents a completely different set of servicing problems when used for heavier work in more remote areas. One Eskimo "kept his machine in his tent so it could be warmed up before starting in the morning, and even then was plagued by mechanical failures."[3] There are stories of other Eskimos, whose mechanical aptitude is well known, modifying their machines to adapt them better to local use.

So is technology culturally neutral? If we look at the construction of a basic machine and its working principles, the answer seems to be yes. But if we look at the web of human activities surrounding the machine, which include its practical uses, its role as a status symbol, the supply of fuel and spare parts, the organized tourist trails, and the skills of its owners, the answer is clearly no. Looked at in this second way, technology is seen as a part of life, not something that can be kept in a separate compartment. If it is to be of any use, the snowmobile must fit into a pattern of activity which belongs to a particular lifestyle and set of values.

The problem here, as in much public discussion, is that "technology" has become a catchword with a confusion of different meanings. Correct usage of the word in its original sense seems almost beyond recovery, but consistent distinction between different levels of meaning is both possible and necessary. In medicine, a distinction of the kind required is often made by talking about "medical practice" when a general term is required, and employing the phrase "medical science" for the more strictly technical aspects of the subject. Sometimes, references to "medical practice" only denote the organization necessary to use medical knowledge and skill for treating patients. Sometimes, however, and more usefully, the term refers to the whole activity of medicine, including its basis in technical knowledge, its organization, and its cultural aspects. The latter comprise the doctor's sense of vocation, his personal values and satisfactions, and the ethical code of his profession. Thus "practice" may be a broad and inclusive concept.

Once this distinction is established, it is clear that although medical practice differs quite markedly from one country to another, medical science consists of knowledge and techniques which are likely to be useful in many countries. It is true that medical science in many Western countries is biased by the way that most research is centered on large hospitals. Even so, most of the basic knowledge is widely applicable and relatively independent of local cultures. Similarly, the design of snowmobiles reflects the way technology is practiced in an industrialized country—standardized machines are produced which neglect some of the special needs of Eskimos and Lapps. But one can still point to a substratum of knowledge, technique, and underlying principle in engineering which has universal validity, and which may be applied anywhere in the world.

We would understand much of this more clearly, I suggest, if the concept of practice were to be used in all branches of technology as it has traditionally been used in medicine. We might then be better able to see which aspects of technology are tied up with cultural values, and which aspects are, in some respects, value-free. We would be better able to appreciate technology as a human activity and as part of life. We might then see it not only as comprising machines, techniques, and crisply precise knowledge, but also as involving characteristic patterns or organization and imprecise values.

Medical practice may seem a strange exemplar for the other technologies, distorted as it so often seems to be by the lofty status of the doctor as an expert. But what is striking to anybody more used to engineering is that medicine has at least got concepts and vocabulary which allows vigorous discussion to take place about different ways of serving the community. For example, there are phrases such as "primary health care" and "community medicine" which are sometimes emphasized as the kind of medical practice to be encouraged wherever the emphasis on hospital medicine has been pushed too far. There are also some interesting adaptations of the language of medical practice. In parts of Asia, paramedical workers, or paramedics, are now paralleled by "para-agros" in agriculture, and the Chinese barefoot doctors have inspired the suggestion that barefoot technicians could be recruited to deal with urgent problems in village water supply. But despite these occasional borrowings, discussion about practice in most branches of technology has not progressed very far.

PROBLEMS OF DEFINITION

In defining the concept of technology-practice more precisely, it is necessary to think with some care about its human and social aspect. Those who write about the social relations and social control of technology tend to focus particularly on organization. In particular, their emphasis is on planning and administration, the management of research, systems for regulation of pollution and other abuses, and professional organization among scientists and technologists. These are important topics, but there is a wide range of other human content in technology-practice which such studies often neglect, including personal values and individual experience of technical work.

To bring all these things into a study of technology-practice may seem likely to make it bewilderingly comprehensive. However, by remembering the way in which medical practice has a technical and ethical as well as an organizational element, we can obtain a more orderly view of what technology-practice entails. To many politically-minded people, the *organizational aspect* may seem most crucial. It represents many facets of administration, and public policy; it relates to the activities of designers, engineers, technicians, and production workers, and also concerns the users and consumers of whatever is produced. Many other people, however, identify technology with its *technical aspect*, because that has to do with machines, techniques, knowledge, and the essential activity of making things work.

Beyond that, though, there are values which influence the creativity of designers and inventors. These, together with the various beliefs and habits of thinking which are characteristic of technical and scientific activity, can be indicated by talking about an ideological or *cultural aspect* of technology-practice. There is some risk of ambiguity here, because strictly speaking, ideology, organization, technique, and tools are all aspects of the culture of a society. But in common speech, culture refers to values, ideas, and creative activity, and it is convenient to use the term with this meaning. It is in this sense that the title of this [chapter] refers to the cultural aspect of technology-practice.

All these ideas are summarized by figure 1, in which the whole triangle stands for the concept of technology-practice and the corners represent its organizational, technical, and cultural aspects. This diagram is intended to illustrate how the word technology is sometimes used by people in a restricted sense, and sometimes with a more general meaning. When technology is discussed in the more restricted way, cultural values and organizational factors are regarded as external to it. Technology is then identified entirely with its technical aspects, and the words "technics" or simply "technique" might often be more appropriately used. The more general meaning of the word, however, can be equated with technology-practice, which clearly is not value-free and politically neutral, as some people say it should be.

Some formal definitions of technology hover uncertainly between the very general and the more restricted usage. Thus J. K. Galbraith defines technology as "the systematic application of scientific or other organized knowledge to practical tasks."[4] This sounds a fairly narrow definition, but on reading further one finds that Galbraith thinks of technology as an activity involving complex organizations and value-systems. In view of this, other authors have extended Galbraith's wording.

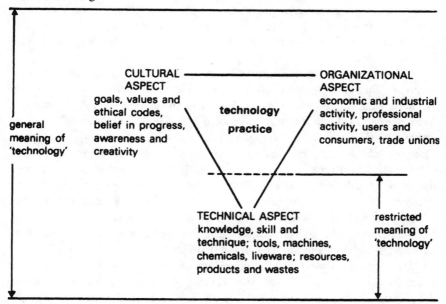

Figure 1. *Diagrammatic definition of "techology" and "technology practice"*

For them a definition which makes explicit the role of people and organizations as well as hardware is one which describes the technology as "the application of scientific and other organized knowledge to practical tasks by . . . ordered systems that involve people and machines."[5] In most respects, this sums up technology-practice very well. But some branches of technology deal with processes dependent on living organisms. Brewing, sewage treatment, and the new biotechnologies are examples. Many people also include aspects of agriculture, nutrition, and medicine in their concept of technology. Thus our definition needs to be enlarged further to include "liveware" as well as hardware; technology-practice is thus *the application of scientific and other knowledge to practical tasks by ordered systems that involve people and organizations, living things and machines.*

This is a definition which to some extent includes science within technology. That is not, of course, the same as saying that science is merely one facet of technology with no purpose if its own. The physicist working on magnetic materials or semiconductors may have an entirely abstract interest in the structure of matter, or in the behavior of electrons in solids. In that sense, he may think of himself as a pure scientist, with no concern at all for industry and technology. But it is no coincidence that the magnetic materials he works on are precisely those that are used in transformer cores and computer memory devices, and that the semiconductors investigated may be used in microprocessors. The scientist's choice of research subjects is inevitably influenced by technological requirements, both through material pressures and also via a climate of opinion about what subjects are worth pursuing. And a great deal of science is like this, with goals that are definitely outside the technology-practice, but with a practical function within it.

Given the confusion that surrounds usage of the word "technology," it is not surprising that there is also confusion about the two adjectives "technical" and "technological." Economists make their own distinction, defining change of technique as a development based on choice from a range of known methods, and technological change as involving fundamentally new discovery or invention. This can lead to a distinctive use of the word "technical." However, I shall employ this adjective when I am referring solely to the technical aspects of practice as defined by figure 1. For example, the application of a chemical water treatment to counteract river pollution is described here as a "technical fix" (not a "technological fix"). It represents an attempt to solve a problem by means of technique alone, and ignores possible changes in practice that might prevent the dumping of pollutants in the river in the first place.

By contrast, when I discuss developments in the practice of technology which include its organizational aspects, I shall describe these as "technological developments," indicating that they are not restricted to technical form. The terminology that results from this is usually consistent with everyday usage, though not always with the language of economics.

EXPOSING BACKGROUND VALUES

One problem arising from habitual use of the word technology in its more restricted sense is that some of the wider aspects of technology-practice have come to be entirely forgotten. Thus behind the public debates about resources and the environment, or about world food supplies, there is a tangle of unexamined beliefs and values, and a basic confusion about what technology is for. Even on a practical level, some projects fail to get more than halfway to solving the problems they address, and end up as unsatisfactory technical fixes, because important organizational factors have been ignored. Very often the users of equipment (figure 2) and their patterns of organization are largely forgotten.

Part of [our] aim is to strip away some of the attitudes that restrict our view of technology in order to expose these neglected cultural aspects. With the snowmobile, a first step was to look at different ways in which the use and maintenance of the machine is organized in different communities. This made it clear that a machine designed in response to the values of one culture needed a good deal of effort to make it suit the purposes of another.

A further example concerns the apparently simple handpumps used at village wells in India. During a period of drought in the 1960s, large power-driven drilling rigs were brought in to reach water at considerable depths in the ground by means of bore-holes. It was at these new wells that most of the handpumps were installed. By 1975 there were some 150,000 of them, but surveys showed that at any one time as many as two-thirds had broken down. New pumps sometimes failed within three or four weeks of installation. Engineers identified several faults, both in the design of the pumps and in standards of manufacture. But although these defects were corrected, pumps continued to go wrong. Eventually it was realized that the breakdowns were not solely an engineering problem. They were also partly an administrative or management issue, in that arrangements for servicing the pumps were not very effective. There was another difficulty, too, because in many villages, nobody felt any personal responsibility for looking after the pumps. It was only when these things were tackled together that pump performance began to improve.

This episode and the way it was handled illustrates very well the importance of an integrated appreciation of technology-practice. A breakthrough only came when all aspects of the administration, maintenance, and technical design of the pump were thought out in relation to one another. What at first held up solution of the problem was a view of technology which began and ended with the machine—a view which, in another similar context, had been referred to as tunnel vision in engineering.

Any professional in such a situation is likely to experience his own form of tunnel vision. If a management consultant had been asked about the handpumps, he would have seen the administrative failings of the maintenance system very quickly, but might not have recognized that mechanical improvements to the pumps were required. Specialist training inevitably restricts people's approach to problems. But tunnel vision in attitudes to technology extends far beyond those who have had specialized training; it also affects policymaking, and influences popular expectations. People in many walks of life tend to focus on the tangible, technical aspect of any practical problem, and then to think that the extraordinary capabilities of modern technology ought to lead to an appropriate "fix." This atti-

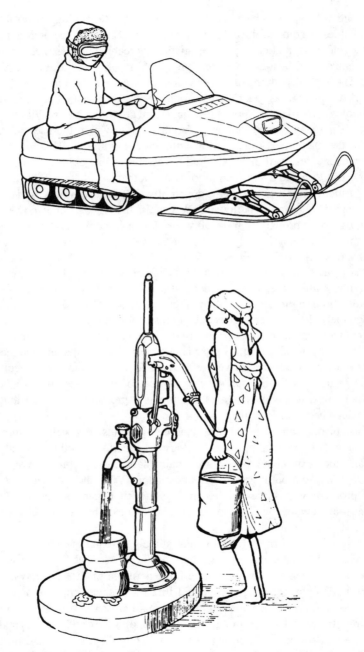

Figure 2. *Technology is about "systems that involve people and machines," and many of the people concerned are users of machines, such as handpumps or snowmobiles.*

tude seems to apply to almost everything from inner city decay to military security, and from pollution to a cure for cancer. But all these issues have a social component. To hope for a technical fix for any of them that does not also involve social and cultural measures is to pursue an illusion.

So it was with the handpumps. The technical aspect of the problem was exemplified by poor design and manufacture. There was the organizational difficulty about maintenance. Also important, though, was the cultural aspect of technology as it was practiced by the engineers involved. This refers, firstly, to the engineers' way of thinking, and the tunnel vision it led to; secondly, it indicates conflicts of values between highly trained engineers and the relatively uneducated people of the Indian countryside whom the pumps were meant to benefit. The local people probably had exaggerated expectations of the pumps as the products of an all-powerful, alien technology, and did not see them as vulnerable bits of equipment needing care in use and protection from damage; in addition, the local people would have their own views about hygiene and water use.

Many professionals in technology are well aware that the problems they deal with have social implications, but feel uncertainty about how these should be handled. To deal only with the technical detail and leave other aspects on one side is the easier option, and after all, is what they are trained for. With the handpump problem, an important step forward came when one of the staff of a local water development unit started looking at the case-histories of individual pump breakdowns. It was then relatively easy for him to pass from a technical review of components which were worn or broken to looking at the social context of each pump. He was struck by the way some pumps had deteriorated but others had not. One well-cared-for pump was locked up during certain hours; another was used by the family of a local official; others in good condition were in places where villagers had mechanical skills and were persistent with improvised repairs. It was these specific details that enabled suggestions to be made about the reorganization of pump maintenance.[6]

A first thought prompted by this is that a training in science and technology tends to focus on general principles, and does not prepare one to look for specifics in quite this way. But the human aspect of technology—its organization and culture—is not easily reduced to general principles, and the investigator with an eye for significant detail may sometimes learn more than the professional with a highly systematic approach.

A second point concerns the way in which the cultural aspect of technology-practice tends to be hidden beneath more obvious and more practical issues. Behind the tangible aspect of the broken handpumps lies an administrative problem concerned with maintenance. Behind that lies a problem of political will—the official whose family depended on one of the pumps was somehow well served. Behind that again were a variety of questions concerning cultural values regarding hygiene, attitudes to technology, and the outlook of the professionals involved.

This need to strip away the more obvious features of technology-practice to expose the background values is just as evident with new technology in Western countries. Very often concern will be expressed about the health risk of a new device when people are worried about more intangible issues, because health risk is partly a technical question that is easy to discuss openly. A relatively minor technical problem affecting health may thus become a proxy for deeper worries about the way technology is practiced which are more difficult to discuss.

An instance of this is the alleged health risks associated with visual display units (VDUs) in computer installations. Careful research has failed to find any real hazard except that operators may suffer eyestrain and fatigue. Yet complaints about more serious problems continue, apparently because they can be discussed seriously with employers while misgivings about the overall systems are more difficult to raise. Thus a negative reaction to new equipment may be expressed in terms of a fear of "blindness, sterility, etc.," because in our society, this is regarded as a legitimate reason for rejecting it. But to take such fears at face value will often be to ignore deeper, unspoken anxieties about "deskilling, inability to handle new procedures, loss of control over work."[7]

Here, then, is another instance where, beneath the overt technical difficulty there are questions about the organizational aspect of technology—especially the organization of specific tasks. These have political connotations, in that an issue about control over work raises questions about where power lies in the workplace, and perhaps ultimately, where it lies within industrial society. But beyond arguments of that sort, there are even more basic values about creativity in work and the relationship of technology and human need.

In much the same way as concern about health sometimes disguises workplace issues, so the more widely publicized environmental problems may also hide underlying organizational and political questions. C. S. Lewis once remarked that "Man's power over Nature often turns out to be a power exerted by some men over other men with Nature as its instrument," and a commentator notes that this, "and not the environmental dilemma as it is usually conceived," is the central issue for technology.[8] As such, it is an issue whose political and social ramifications have been ably analyzed by a wide range of authors.[9]

Even this essentially political level of argument can be stripped away to reveal another cultural aspect of technology. If we look at the case made out in favor of almost any major project—a nuclear energy plant, for example—there are nearly always issues concerning political power behind the explicit arguments about tangible benefits and costs. In a nuclear project, these may relate to the power of management over trade unions in electricity utilities; or to prestige of governments and the power of their technical advisers. Yet those who operate these levers of power are able to do so partly because they can exploit deeper values relating to the so-called technological imperative, and to the basic creativity that makes innovation possible. This, I argue, is a central part of the culture of technology. If these values underlying the technological imperative are understood, we may be able to see that here is a stream of feeling which politicians can certainly manipulate at times, but which is stronger than their short-term purposes, and often runs away beyond their control.

NOTES

1. M. B. Doyle, *An Assessment of the Snowmobile Industry and Sport* (Washington, D.C.: International Snowmobile Industry Association, 1978), pp. 14, 47; on Joseph-Armand Bombardier, see Alexander Ross, *The Risk Takers* (Toronto: Macmillan and the Financial Post, 1978), p. 155.

2. R. A. Buchanan, *Technology and Social Progress* (Oxford: Permagon Press, 1965), p. 163.

3. P. J. Usher, "The Use of Snowmobiles for Trapping on Banks Island," *Arctic* (Arctic Institute of North America) 25 (1972): 173.

4. J. K. Galbraith, *The New Industrial State*, 2d British edition (London: André Deutsch, 1972), chap. 2.

5. John Naughton, "Introduction: Technology and Human Values," in *Living with Technology: A Foundation Course* (Milton Keynes: The Open University Press, 1979).

6. Charles Heineman, "Survey of Handpumps in Vellakovil," unpublished report, January 1975, quoted by Arnold Pacey, *Handpump Maintenance* (London: Intermediate Technology Publications, 1977).

7. Leela Damodaran, "Health Hazards of VDUs?—Chairman's Introduction," conference at Loughborough University of Technology, December 11, 1980.

8. Quoted by Peter Hartley, "Educating Engineers," *The Ecologist* 10 (December 1980): 353.

9. For example, David Elliott and Ruth Elliott, *The Control of Technology* (London and Winchester: Wykeham, 1976).

4
Defining Technique

Jacques Ellul

DEFINITIONS

Once we stop identifying technique and machine, the definitions of technique we find are inadequate to the established facts. Marcel Mauss, the sociologist, understands the problem admirably, and has given various definitions of technique, some of which are excellent. Let us take one that is open to criticism and, by criticizing it, state our ideas more precisely: "Technique is a group of movements, of actions generally and mostly manual, organized, and traditional, all of which unite to reach a known end, for example, physical, chemical, or organic."

This definition is perfectly valid for the sociologist who deals with the primitive. It offers, as Mauss shows, numerous advantages. For example, it eliminates from the realm of techniques questions of religion or art. But these advantages apply only in a historical perspective. In the modern perspective, this definition is insufficient.

Can it be said that the technique of elaboration of an economic plan (purely a technical operation) is the result of such movements as Mauss describes? No particular motion or physical act is involved. An economic plan is purely an intellectual operation, which nevertheless is a technique.

When we consider Mauss's statement that technique is restricted to manual activity, the inadequacy of his definition is even more apparent. Today most technical operations are not manual. Whether machines are substituted for men, or technique becomes intellectual, the most important sphere in the world today (because in it lie the seeds of future development) is scarcely that of manual labor. True, manual labor is still the basis of mechanical operation, and we would

From *The Technological Society*, by Jacques Ellul, translated by John Wilkinson (1964), published by Jonathan Cape. Reprinted by permission of the Random House Group Ltd.

do well to recall [Friedrich Georg] Jünger's principal argument against the illusion of technical progress. He holds that the more technique is perfected, the more it requires secondary manual labor; and, furthermore, that the volume of manual operations increases faster than the volume of mechanical operations. This may be so, but the most important feature of techniques today is that they do not depend on manual labor but on organization and on the arrangement of machines.

I am willing to accept the term *organized*, as Mauss uses it in his definition, but I must part company with him in respect to his use of the term *traditional*. And this differentiates the technique of today from that of previous civilizations. It is true that in all civilizations technique has existed as tradition, that is, by the transmission of inherited processes that slowly ripen and are even more slowly modified; that evolve under the pressure of circumstances along with the body social; that create automatisms which become hereditary and are integrated into each new form of technique.

But how can anyone fail to see that none of this holds true today? Technique has become autonomous; it has fashioned an omnivorous world which obeys its own laws and which has renounced all tradition. Technique no longer rests on tradition, but rather on previous technical procedures; and its evolution is too rapid, too upsetting, to integrate the older traditions. This fact, which we shall study at some length later on,* also explains why it is not quite true that a technique assures a result known in advance. It is true if one considers only the user: the driver of an automobile knows that he can expect to go faster when he steps on the accelerator. But even in the field of the mechanical, with the advent of the technique of servo-mechanisms,† this axiom does not hold true. In these cases the machine itself adapts as it operates; this very fact makes it difficult to predict the final result of its activity. This becomes clear when one considers not use but technical progress—although, at the present time, the two are closely associated. It is less and less exact to maintain that the user remains for very long in possession of a technique the results of which he can predict; constant invention ceaselessly upsets his habits.

Finally, Mauss appears to think that the goal attained is of a physical order. But today we recognize that techniques go further. Psychoanalysis and sociology have passed into the sphere of technical application; one example of this is propaganda. Here the operation is of a moral, psychic, and spiritual character. However, that does not prevent it from being a technique. But what we are talking about is a world once given over to the pragmatic approach and now being taken over by method. We can say, therefore, that Mauss's definition, which was valid for technique until the eighteenth century, is not applicable to our times. In this

---

*See part 4, chapter 10.

†Mechanisms which involve so-called feedback, in which information measuring the degree to which an effector (e.g., an oil furnace) is in error with respect to producing a desired value (e.g., a fixed room temperature) is "fed back" to the effector by a monitor (e.g., a thermostat). (Trans.)

respect Mauss has been the victim of his own sociological studies of primitive people, as his classification of techniques (food gathering, the making of garments, transport, etc.) clearly shows.

Further examples of inadequate definition are those supplied by Jean Fourastié and others who pursue the same line of research as he. For Fourastié, technical progress is "the growth of the volume of production obtained through a fixed quantity of raw material or human labor"—that is, technique is uniquely that which promotes this increase in yield. He then goes on to say that it is possible to analyze this theorem under three aspects. In *yield in kind*, technique is that which enables raw materials to be managed in order to obtain some predetermined product; in *financial yield*, technique is that which enables the increase in production to take place through the increase of capital investment; in *yield of human labor*, technique is that which increases the quantity of work produced by a fixed unit of human labor. In this connection we must thank Fourastié for correcting Jünger's error—Jünger opposes technical progress to economic progress because they would be, in his opinion, contradictory; Fourastié shows that, on the contrary, the two coincide. However, we must nevertheless challenge his definition of technique on the ground that it is completely arbitrary.

It is arbitrary, first of all, because it is purely economic and contemplates only economic yield. There are innumerable traditional techniques which are not based on a quest for economic yield and which have no economic character. It is precisely these which Mauss alludes to in his definition; and they still exist. Among the myriad modern techniques, there are many that have nothing to do with economic life. Take, for example, a technique of mastication based on the science of nutrition, or techniques of sport, as in the Boy Scout movement—in these cases we can see a kind of yield, but this yield has little to do with economics.

In other cases, there are economic results, but these results are secondary and cannot be said to be characteristic. Take, for example, the modern calculating machine. The solving of equations in seventy variables, required in certain econometric research, is impossible except with an electronic calculating machine. However, it is not the economic productivity that results from the utilization of this machine by which its importance is measured.

A second criticism of Fourastié's definition is that he assigns an exclusively productive character to technique. The growth of the volume of production is an even narrower concept than yield. The techniques which have shown the greatest development are not techniques of production at all. For example, techniques in the care of human beings (surgery, psychology, and so on) have nothing to do with productivity. The most modern techniques of destruction have even less to do with productivity; the atomic and hydrogen bombs and the Germans' V1 and V2 weapons are all examples of the most powerful technical creations of man's mind. Human ingenuity and mechanical skill are today being exploited along lines which have little reference to productivity.

Nothing equals the perfection of our war machines. Warships and warplanes are vastly more perfect than their counterparts in civilian life. The organization of the army—its transport, supplies, administration—is much more precise than any civilian organization. The smallest error in the realm of war would cost countless lives and would be measured in terms of victory or defeat. What is the yield there? Very poor, on the whole. Where is the productivity? There is none.

[André] Vincent, in his definition, likewise refers to productivity: "Technical progress is the relative variation in world production in a given sphere between two given periods." This definition, useful of course from the economic point of view, leads him at once into a dilemma. He is obliged to distinguish technical progress from progress of technique (which corresponds to the progression of techniques in all fields) and to distinguish these two from "technical progress, properly speaking," which concerns variations in productivity. This is an inference made from natural phenomena for, in his definition, Vincent is obliged to recognize that technical progress includes *natural* phenomena (the greater or lesser richness of an ore, of the soil, etc.) by definition the very contrary of *technique*!

These linguistic acrobatics and hairsplittings suffice to prove the inanity of such a definition, which aims at a single aspect of technical progress and includes elements which do not belong to technique. From this definition, Vincent infers that technical progress is slow. But what is true of economic productivity is not true of technical progress in general. If one considers technique shorn of one whole part, and that its most progressive, one can indeed assert that it is slow in its progress. This abstraction is even more illusory when one claims to measure technical progress. The definition proposed by Fourastié is inexact because it excludes everything that does not refer to production, and all effects that are not economic.

This tendency to reduce the technical problem to the dimensions of the technique of production is also present in the works of so enlightened a scholar as Georges Friedmann. In his introduction to the UNESCO* Colloquium on technique, he appears to start out with a very broad definition. But in the second paragraph, without warning, he begins to reduce everything to the level of economic production.

What gives rise to this limitation of the problem? One factor might be a tacit optimism, a need to hold that technical progress is unconditionally valid—which leads to the selection of the most positive aspect of technical progress, as though it were its only one.

This may have guided Fourastié, but it does not seem to hold true in Friedmann's case. I believe that the reasoning behind Friedmann's way of thinking is to be found in the turn of the scientific mind. All aspects—mechanical, economic, psychological, sociological—of the techniques of production have been

---

*United Nations Educational, Scientific, and Cultural Organization (Ed.)

subjected to innumerable specialized studies; as a result, we are beginning to learn in a more precise and scientific way about the relationships between man and the industrial machine. Since the scientist must use the materials he has at hand; and since almost nothing is known about the relationship of man to the automobile, the telephone, or the radio, and absolutely nothing about the relationship of man to the *Apparat* or about the sociological effects of other aspects of technique, the scientist moves unconsciously toward the sphere of what is known scientifically, and tries to limit the whole question to that.

There is another element in this scientific attitude: only that is knowable which is expressed (or, at least, can be expressed) in numbers. To get away from the so-called "arbitrary and subjective," to escape ethical or literary judgments (which, as everyone knows, are trivial and unfounded), the scientist must get back to numbers. What, after all, can one hope to deduce from the purely qualitative statement that the worker is fatigued? But when biochemistry makes it possible to measure fatigability numerically, it is at last possible to take account of the worker's fatigue. Then there is hope of finding a solution. However, an entire realm of effects of technique—indeed, the largest—is not reducible to numbers; and it is precisely that realm which we are investigating in this [chapter]. Yet, since what can be said about it is apparently not to be taken seriously, it is better for the scientist to shut his eyes and regard it as a realm of pseudo-problems, or simply as nonexistent. The "scientific" position frequently consists of denying the existence of whatever does not belong to current scientific method. The problem of the industrial machine, however, is a numerical one in nearly all its aspects. Hence, all of technique is unintentionally reduced to a numerical question. In the case of Vincent, this is intentional, as his definition shows: "We embrace in technical progress all kinds of progress . . . *provided* that they are treatable numerically in a reliable way."

H. D. Lasswell's definition of technique as "the ensemble of practices by which one uses available resources in order to achieve certain valued ends" also seems to follow the conventions cited above, and to embrace only industrial technique. Here it might be contested whether technique does indeed permit the realization of values. However, to judge from Lasswell's examples, he conceives the terms of his definition in an extremely broad manner. He gives a list of values and the corresponding techniques. As values, for example, he lists riches, power, well-being, affection; and as techniques, the techniques of government, production, medicine, the family, and so on. Lasswell's conception of *value* may seem somewhat strange; the term is obviously not apt. But what he has to say indicates that he gives techniques their full scope. Moreover, he makes it quite clear that it is necessary to show the effects of technique not only on inanimate objects but also on people. I am, therefore, in substantial agreement with this conception.

TECHNICAL OPERATION AND
TECHNICAL PHENOMENON

With the use of these few guideposts, we can now try to formulate, if not a full definition, at least an approximate definition of technique. But we must keep this in mind: we are not concerned with the different individual techniques. Everyone practices a particular technique, and it is difficult to come to know them all. Yet in this great diversity we can find certain points in common, certain tendencies and principles shared by them all. It is clumsy to call these common features Technique with a capital T; no one would recognize his particular technique behind this terminology. Nevertheless, it takes account of a reality—the technical phenomenon—which is worldwide today.

If we recognize that the method each person employs to attain a result is, in fact, his particular technique, the problem of means is raised. In fact, technique is nothing more than *means* and the *ensemble of means*. This, of course, does not lessen the importance of the problem. Our civilization is first and foremost a civilization of means; in the reality of modern life, the means, it would seem, are more important than the ends. Any other assessment of the situation is mere idealism.

Techniques considered as methods of operation present certain common characteristics and certain general tendencies, but we cannot devote ourselves exclusively to them. To do this would lead to a more specialized study than I have in mind. The technical phenomenon is much more complex than any synthesis of characteristics common to individual techniques. If we desire to come closer to a definition of technique, we must in fact differentiate between the technical operation and the technical phenomenon.

The technical operation includes every operation carried out in accordance with a certain method in order to attain a particular end. It can be as rudimentary as splintering a flint or as complicated as programming an electronic brain. In every case, it is the method which characterizes the operation. It may be more or less effective or more or less complex, but its nature is always the same. It is this which leads us to think that there is a continuity in technical operations and that only the great refinement resulting from scientific progress differentiates the modern technical operation from the primitive one.

Every operation obviously entails a certain technique, even the gathering of fruit among primitive peoples—climbing the tree, picking the fruit as quickly and with as little effort as possible, distinguishing between the ripe and the unripe fruit, and so on. However, what characterizes technical action within a particular activity is the search for greater efficiency. Completely natural and spontaneous effort is replaced by a complex of acts designed to improve, say, the yield. It is this which prompts the creation of technical forms, starting from simple forms of activity. These technical forms are not necessarily more complicated than the spontaneous ones, but they are more efficient and better adapted.

Thus, technique creates means, but the technical operation still occurs on the same level as that of the worker who does the work. The skilled worker, like the primitive huntsman, remains a technical operator; their attitudes differ only to a small degree.

But two factors enter into the extensive field of technical operation: consciousness and judgment. This double intervention produces what I call the technical phenomenon. What characterizes this double intervention? Essentially, it takes what was previously tentative, unconscious, and spontaneous and brings it into the realm of clear, voluntary, and reasoned concepts.

When André Leroi-Gourhan tabulates the efficiency of Zulu swords and arrows in terms of the most up-to-date knowledge of weaponry, he is doing work that is obviously different from that of the swordsmith of Bechuanaland who created the form of the sword. The swordsmith's choice of form was unconscious and spontaneous; although it can now be justified by numerical calculations, such calculations had no place whatever in the technical operation he performed. But reason did, inevitably, enter into the process because man spontaneously imitates nature in his activities. Accomplishments that merely copy nature, however, have no future (for instance, the imitation of birds' wings from Icarus to Ader). Reason makes it possible to produce objects in terms of certain features, certain abstract requirements; and this in turn leads, not to the imitation of nature, but to the ways of technique.

The intervention of rational judgment in the technical operation has important consequences. Man becomes aware that it is possible to find new and different means. Reason upsets pragmatic traditions and creates new operational methods and new tools; it [generates] experimentation. Reason in these ways multiplies technical operations to a high degree of diversity. But it also operates in the opposite direction: it considers results and takes account of the fixed end of technique—efficiency. It notes what every means devised is capable of accomplishing and selects from the various means at its disposal with a view to securing the ones that are the most efficient, the best adapted to the desired end. Thus the multiplicity of means is reduced to one: the most efficient. And here reason appears clearly in the guise of technique.

In addition, there is the intervention of consciousness. Consciousness shows clearly, and to everybody, the advantages of technique and what it can accomplish. The technician takes stock of alternative possibilities. The immediate result is that he seeks to apply the new methods in fields which traditionally had been left to chance, pragmatism, and instinct. The intervention of consciousness causes a rapid and far-flung extension of technique.

The twofold intervention of reason and consciousness in the technical world, which produces the technical phenomenon, can be described as the quest of the one best means in every field. And this "one best means" is, in fact, the technical means. It is the aggregate of these means that produces technical civilization.

The technical phenomenon is the main preoccupation of our time; in every field men seek to find the most efficient method. But our investigations have reached a limit. It is no longer the best relative means that counts, as compared to other means also in use. The choice is less and less a subjective one among several means that are potentially applicable. It is really a question of finding the best means in the absolute sense, on the basis of numerical calculation.

It is, then, the specialist who chooses the means; he is able to carry out the calculations that demonstrate the superiority of the means chosen over all others. Thus a science of means comes into being—a science of techniques, progressively elaborated.

This science extends to greatly diverse areas; it ranges from the act of shaving to the act of organizing the landing in Normandy, or to cremating thousands of deportees. Today no human activity escapes this technical imperative. There is a technique of organization (the great fact of organization described by Toynbee fits very well into this conception of the technical phenomenon); just as there is a technique of friendship and a technique of swimming. Under the circumstances, it is easy to see how far we are from confusing technique and machine. And, if we examine the broader areas where this search for means is taking place, we find three principal subdivisions of modern technique, in addition to the mechanical (which is the most conspicuous but which I shall not discuss because it is so well known) and to the forms of intellectual technique (card indices, libraries, and so on).

(1) *Economic technique* is almost entirely subordinated to production, and ranges from the organization of labor to economic planning. This technique differs from the others in its object and goal. But its problems are the same as those of all other technical activities.

(2) *The technique of organization* concerns the great masses and applies not only to commercial or industrial affairs of magnitude (coming, consequently, under the jurisdiction of the economic) but also to states and to administration and police power. This organizational technique is also applied to warfare and insures the power of an army at least as much as its weapons. Everything in the legal field also depends on organizational technique.

(3) *Human technique* takes various forms, ranging all the way from medicine and genetics to propaganda (pedagogical techniques, vocational guidance, publicity, etc.). Here man himself becomes the object of technique.

We observe, in the case of each of these subdivisions, that the subordinate techniques may be very different in kind and not necessarily similar one to another as techniques. They have the same goal and preoccupation, however, and are thus related. The three subdivisions show the wide extent of the technical phenomenon. In fact, nothing at all escapes technique today. There is no field where technique is not dominant—this is easy to say and is scarcely surprising. We are so habituated to machines that there seems to be nothing left to discover.

REFERENCES

Fourastié, Jean. *Révolution à l'Ouest* [*Revolution in the West*]. Paris: Presses Universitaires de France, 1957.

Friedmann, Georges. *La crise du progrès* [*The Crisis of Progress*]. Paris: Gallimard, 1936.

Jünger, Friedrich Georg *Die Perfection der Technik* [*The Perfection of the Technology*]. Frankfurt: Klostermann, 1949.

Lasswell, Harold Dwight. *Power and Personality*. New York: Academy of Medicine, Compass Books edition, 1962.

Leroi-Gourhan, André. *Milieu et techniques. Évolution et techniques* [*Environment and Techniques. Evolution and Techniques*]. A. Michel, 1945.

Mauss, Marcel. *Sociologie et anthropologie* [*Sociology and Anthropology*]. Paris: Presses Universitaires de France, 1949–1950.

Vincent, André L. A., and René Froment. *Le progrès technique en France dupuis cent aus* [*A Century of Technical Progress in France*]. Imprimerie Nationale, 1944.

5

The Knowing World of Things

Davis Baird

> If your knowledge of fire has been turned
> to certainty by words alone,
> then seek to be cooked by the fire itself.
> Don't abide in borrowed certainty.
> There is no real certainty until you burn;
> if you wish for this, sit down in the fire.
> —Mevlana Jalaluddin Rumi, *Masnavi*, BK II

EPISTEMOLOGY AND DEFINITION

Knowledge has been understood to be an affair of the mind. To know is to think, and in particular, to think thoughts expressible in words. Nonverbal creations—from diagrams to densitometers—are excluded as merely "instrumental"; they are pragmatic crutches that help thinking—in the form of theory construction and interpretation. In *Thing Knowledge: A Philosophy of Scientific Instruments*, I urge a different view. I argue for a materialist conception of knowledge. Along with theories, the material products of science and technology constitute knowledge. I focus on scientific instruments such as cyclotrons and spectrometers, but I would also include recombinant DNA enzymes, "wonder" drugs, and robots, among other things, as other material products of science and technology that constitute our knowledge. These material products are constitutive of scientific knowledge in a manner different from theory, and not simply "instrumental to" theory.

The materialist epistemology I offer contributes to and is motivated by the need for an epistemology of technology. Any definition of technology will be contentious, and one source of controversy is the lack of an epistemology of technology. Few will argue with the claim that technology is involved with making things. Perhaps there are technologies that have little to do with making things. Computer software comes to mind. When running on a computer, however, we are in the realm of things again, and the epistemology I offer has something to say. My materialist epistemology is concerned with material things, of course, but my object is epistemology: knowing things—and I don't mean how people subjectively or mentally know about things. I mean things themselves bearing knowledge of the world. One important part of technology in general and engineering in particular is that it is a domain of knowledge. But engineers frequently express themselves in their making, not their saying. My materialist epistemology aims to explain the epistemic commerce of making things and the knowledge borne by the things made. It should promote a less contentious definition of technology as a domain of a particular kind of knowledge, what I call thing knowledge.

MICHAEL FARADAY'S FIRST ELECTRIC MOTOR

To begin to put some flesh on these abstract generalities, consider the discovery (or invention?) of Michael Faraday's first electric motor. On September 3 and 4, 1821, Faraday, age thirty, performed a series of experiments that ultimately produced what were called "electromagnetic rotations." Faraday showed how an appropriately organized combination of electric and magnetic elements would produce rotary motion. He invented the first electromagnetic motor.

Faraday's work resulted in several "products." He published several papers describing his discovery.[1] He wrote letters to many scientific colleagues.[2] He built, or had built, several copies of an apparatus that, requiring no experimental knowledge or dexterity on the part of its user, would display the notable rotations, and he shipped these to his scientific colleagues.[3]

A permanent magnet is cemented vertically in the center of a mercury bath. A wire, with one end immersed a little into the mercury, is suspended over the magnet in such a way as to allow for free motion around the magnet. The suspension of the wire is such that contact can be made with it and one pole of a battery. The other pole of the battery is connected to the magnet that carries the current to the mercury bath, and thence to the other end of the wire, completing the circuit. See figure 1.

The apparatus produces a striking phenomenon: When an electric current is run through the wire, via the magnet and the mercury bath, the wire spins around the magnet. The observed behavior of Faraday's apparatus requires no interpretation. While there was considerable disagreement over the explanation for this phenomenon, no one contested what the apparatus did: It exhibited (still does) rotary motion as a consequence of a suitable combination of electric and magnetic elements.

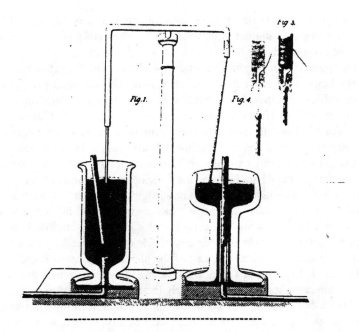

Figure 1: From *Faraday, Experimental Researches in Electricity* (London: Richard and John Edward Taylor, 1844), pl. iv.

DEVICE EPISTEMOLOGY

How should we understand Faraday's device? One could say that it justifies assertions such as "A current-carrying wire will rotate around a magnet in a mercury bath as shown in figure 1." One could say, and Faraday did say, that the phenomenon exhibited by the device articulates Oersted's 1820 discovery of the magnetic effects of an electric current.[4] One could speculate—and several did— that the device showed that all forces were convertible.[5] Are such theoretical moves all that is important about the device? Why did Faraday think it necessary to ship ready-made versions of this motor to his colleagues?

Moving immediately from the device to its importance for these various theoretical issues misses its immediate importance. When Faraday made the device, there was considerable disagreement over how it worked. Today many people do not know the physics that explains how it works. Both then and now, however, no one denies *that* it works. When Faraday built it, this phenomenon was striking and proved to be very important for the future development of science and technology. Whatever explanations would be offered for the device and more generally for the nature of "electromagnetical motions" would have to recognize the motions Faraday produced. We don't need a load of theory (or indeed any "real" theory) to learn something from the construction and demonstration of Faraday's device. Or to put it another way, we learn by interacting with bits of the world even when our words for how these bits work are inadequate.

This point is more persuasive when confronted with the actual device. Unfortunately, I cannot build a Faraday motor into this chapter; reader imagination will have to suffice. But it is significant that Faraday did not depend on the imaginations of his readers. He made and shipped "pocket editions" of his newly created phenomenon to his colleagues. He knew from his own experience how difficult it is to interpret descriptions of experimental discoveries. He also knew how difficult it is to fashion even a simple device like his motor and have it work reliably. The material product Faraday sent his colleagues encapsulated his considerable manipulative skill—his "fingertip knowledge"—in such a way that someone without the requisite skill could still experience firsthand the new phenomenon. He did not have to depend on the skills of his colleagues, nor on their ability to interpret a verbal description of his device. He could depend on the ability of the device itself to communicate the fact of the phenomenon it exhibited.

I conclude from this that there is something in the device itself that is epistemologically important, something that a purely literary description misses. The epistemological products of science and technology must include such stuff, not simply words and equations. In particular, they must include instruments such as Faraday's motor.

Understanding instruments as bearers of knowledge conflicts with any of the more-or-less standard views that take knowledge as a subspecies of belief.[6] Instruments, whatever they may be, are not beliefs. A different approach to epistemology, characterized under the heading "growth of scientific knowledge," also does not accommodate instruments; such work inevitably concentrates on *theory* change.[7] While some instruments—models such as Watson and Crick's "ball and stick" model of DNA—might be understood in terms similar to theories, generally speaking instruments cannot be understood in such terms. Even recent work on the philosophy of experiment that has focused on the literally material aspects of science, either has adopted a standard proposition-based epistemology, or has not addressed epistemology.[8] My materialist epistemology aims to correct this failure, to present instruments and technological devices generally epistemologically.

This project raises a variety of problems at the outset. There are conceptual problems that, for many, seem immediately to refute the very possibility that instruments are a kind of scientific knowledge. We are strongly wedded to connections between the concepts of knowledge, truth, and justification. It is hard to fit concepts such as truth and justification around instruments. Even work that drops these connections finds substitutes. Work on the growth of scientific knowledge does not require truth—"every theory is born refuted." Instead, we have the "growth of scientific knowledge" expressed in terms of verisimilitude,[9] progressive research programs,[10] increasing problem-solving effectiveness of research traditions.[11] Substitutes for truth and justification that work with instruments are necessary for my materialist epistemology.

Prior to these philosophical problems are problems with the very concept of a scientific instrument. At the most basic level, this is not a unitary concept. There are many different kinds of scientific instrument. What is worse, the different kinds work differently epistemologically. Thus, models—Watson and Crick's model of DNA—clearly have a representative function. Yet, devices such as Faraday's motor do not; they perform. Measuring instruments, such as thermometers, in many ways are hybrids; they perform to produce representations. Consequently, in addition to philosophical issues of truth and justification, my materialist epistemology has to examine the epistemological differences between these three types of instrument: models, devices that create a phenomenon, and measuring instruments. I do not claim that this is a philosophically exhaustive or fully articulated typology of instruments or instrumental functions. I do claim significant epistemological differences for each type, differences requiring special treatment.

These categories have histories. Indeed, the very category of scientific instrument has its own history.[12] The self-conscious adoption of instruments, along with theories, as scientific knowledge has a history. A major epistemological event of the mid–twentieth century has been the recognition by the scientific community of the centrality of instruments to the epistemological project of technology and science.[13] My arguments, then, for understanding instruments as scientific knowledge, have to be understood historically. While I use examples scattered through history, my goal is neither to provide a history of scientific instruments, nor to argue for the timeless significance of this category. To understand technology and science *now*, however, we need to construct an epistemology capable of including instruments.

SEMANTIC ASCENT

Instrument epistemology confronts a long history of what I call text bias, dating back at least to Plato, with what is commonly taken as his definition of knowledge in terms of justified true belief. To do proper epistemology, we have to "ascend" from the material world to the "Platonic world" of thought. This may reflect Plato's concern with the impermanence of the material world and what he saw as the unchanging eternal perfection of the realm of forms. If knowledge is timeless, it cannot exist in the corruptible material realm.

This strikes me simply as prejudice. Derek de Solla Price captured my reaction exactly when he noted that "it is unfortunate that so many historians of science and virtually all of the philosophers of science are born-again theoreticians instead of bench scientists."[14] Philosophers and historians express themselves in words, not things, and so it is not surprising that those who hold a virtual monopoly over saying (words!) what scientific knowledge is, characterize it in terms of the kind of knowledge with which they are familiar—words.

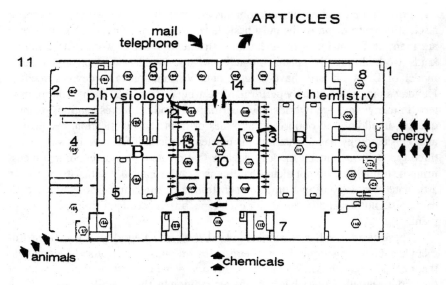

Figure 2: Latour and Woolar, *Laboratory Life*, fig. 2.1, p. 46.

Prejudice it may be, but powerfully entrenched it is, too. The logical posi-tivists were obsessed with "the languages of science."[15] But text bias did not die with them. Consider the following figure from Bruno Latour and Steve Woolgar's seminal postpositivist book, *Laboratory Life*.[16] See figure 2.

Here is the function of the laboratory. Animals, chemicals, mail, telephone, and energy go in; articles go out. The picture Latour and Woolgar present of sci-ence is thoroughly literary. "Nature," with the help of inscription devices (i.e., instruments), produces literary outputs for scientists; scientists use these outputs, plus other literary resources (mail, telephone, preprints, etc.), to produce their own literary outputs. The material product the scientists happened to be investi-gating in Latour and Woolgar's study—a substance called "TRF"—becomes, on their reading, merely an instrumental good, "just one more of the many tools uti-lized as part of long research programs."[17]

This picture of the function of a laboratory is a travesty. There is a long his-tory of scientists sharing material other than words. William Thompson sent elec-tric coils to colleagues as part of his measurement of the ohm. Henry Rowland's fame rests on the gratings he ruled and sent to colleagues. Chemists share chemi-cals. Biologists share biologically active chemicals—enzymes, etc.—as well as prepared animals for experiments. When it is hard to share devices, scientists with the relevant expertise are shared; such is the manner in which E. O. Lawrences's cyclotron moved beyond Berkeley. Laboratories do not simply produce words.

There is much to learn from Latour and Woolgar's *Laboratory Life*, and from the subsequent work of these authors. Indeed, Latour and Woolgar are important because they do attend to the material context of laboratory life. But, continuing a long tradition of text bias, they misdescribe the telos of science and technology exclusively in literary terms. While the rhetoric with which they introduce their "literary" framework for analysis seems new, even "postmodern," it is very old. Once again scholars—wordsmiths—have reduced science to the mode in which they are most familiar, words.

A considerable portion of David Gooding's *Experiment and the Making of Meaning* focuses on Michael Faraday's experimental production of electromagnetic rotations—the motor I started with. Given this focus, one might suspect that Gooding would see the making of phenomena—such as Faraday's motor—as one of the key epistemological *ends* of science. On the contrary, the first sentences of his book are instructive:

> It is inevitable that language has, as Ian Hacking put it, mattered to philosophy. It is not inevitable that practices—especially extralinguistic practices—have mattered so little. Philosophy has not yet addressed an issue that is central to any theory of the *language* of observation and therefore, to any theory of science: how do observers *ascend* from the world to talk, thought, and argument about the world.[18]

Scientists "ascend" from the world to talk about the world, from instruments to words, from the material realm to the literary realm. Semantic ascent is the key move in experimental science, according to Gooding. Words are above things.

As with Latour and Woolgar, I mention Gooding's use of "semantic ascent" not so much to criticize Gooding, for the problem of how words get tied to new bits of the world is important and Gooding has much of interest and value to say about it. But thinking in terms of the metaphor of ascent implies a hierarchy of ultimate values. It turns our attention from other aspects of science and technology that are equally important.

It is instructive to see how Gooding discusses Faraday's literary and material products. Faraday accomplished two feats. He built a reliable device, and he described this device's operation. Gooding writes:

> [T]he literary account places phenomena in an objective relationship to theories just as the material embodiment of the skills places phenomena in an objective relation to human experience.[19]

Faraday's descriptions—his literary "ascent"—"places phenomena in an objective relationship to theories." Analogously, his material work—his device— "places phenomena in an objective relation to human experience."

But "human experience" is the wrong concept. Faraday's descriptions could speak to theory. In doing so they could call on the power of logic and contribute

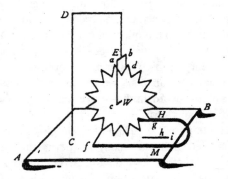

Figure 3: Letter from P. Barlow to M. Faraday, March 14, 1822,
in *The Selected Correspondence of Michael Faraday, Volume 1, 1812–48*
(Cambridge: Cambridge University Press, 1971), p. 133.

to knowledge. We need an analogously detailed articulation of how Faraday's material work could contribute to knowledge. "Human experience" ducks this responsibility. We can and should say more and in more detail about what the material work had "objective relations" with. Avoiding doing so is a symptom of the disease of semantic ascent.

Faraday's device had a good bit "to say." The apparatus "spoke" objectively about the possibilities for producing rotary motion from electromagnetism; these possibilities could be developed through material manipulations, starting with the apparatus as a material given. Six months after Faraday made his device, Peter Barlow produced a variant. See figure 3.

Current runs from one "voltaic pole" to the star's suspension [*abcd*], through the star to the mercury bath [*fg*], and thence to the other voltaic pole. A strong horseshoe magnet [*HM*] surrounds the mercury bath and, as Barlow put it in a letter to Faraday, "the wheel begins to rotate, with an astonishing velocity, and thus exhibits a very pretty appearance."[20]

It is another step to figure out how to create such rotary motion without the use of mercury. Then we might have something useful. This is a story not primarily about the evolution of our words and equations, but a story about material manipulations. The story involves many players and extends over many years. A full telling would not serve here.[21] It involves the invention of the electromagnet—developed by William Sturgeon among others, and considerably improved by early American physicist Joseph Henry. From the electromagnet to the electric motor is another step, one taken by several people independently.[22]

The story of one of the claimants to inventing the electric motor, Thomas Davenport, a Vermont blacksmith, is instructive.[23] In 1834, Davenport was intrigued by news of a powerful electromagnet built by Professor Henry, capable of lifting a common blacksmith's anvil. Davenport traveled some distance from his home in

DAVENPORT MOTOR MODEL NOW IN THE
SMITHSONIAN INSTITUTION, WASHINGTON, D. C.

Figure 4: From W. R. Davenport, *Biography of Thomas Davenport: The
"Brandon Blacksmith," Inventor of the Electric Motor* (Montpelier: Vermont
Historical Society, 1929), p. 144.

Vermont to Rennselaer in Troy, New York, to see a demonstration of the electromagnet. He was amazed and entranced with the possibilities it possessed. A year
later, Davenport succeeded in building a motor capable of driving a seven-inch
diameter wheel at a rate of thirty revolutions per minute. See figure 4.

The motor works by switching the polarity of four electromagnets in syn-
chronicity with the motion of the wheel so that the wheel is always drawn forward. All of this was accomplished despite the fact that Davenport did not know
electromagnetic theory. When he first saw Henry's electromagnet, he had never

heard of any of the main contributors to the science of electromagnetism. But he did have an appreciation for the phenomenon exhibited by the electromagnet, and he could use this knowledge—*presented by the device itself*—to make other devices. Davenport was interested in developing devices that would have practical utility, and he did succeed in using his motor to drive a printing press.[24] But Davenport's motor also expresses a further articulation of knowledge of electromagnetic phenomena.

Semantic "ascent" prevents us from attending to those pieces of the history of science and technology that do not immediately speak to theory. Yet, as this example—and many others I've not discussed—makes clear, maneuvers in the material realm are central to the progress of science and technology. The more basic point here is that the material realm provides a space within which work can be done. Exactly what is done in this space frequently—although not always—depends on available theory. But, also frequently, this theory turns out to be erroneous. This does not bring work to a halt. On the contrary, work goes forward independent of theory or with controversial and/or erroneous theory.

MULTIPLE EPISTEMOLOGIES

A primary consequence of the epistemological picture I advance is that no single unified account of knowledge will serve science and technology. In advancing a materialist account of epistemology—thing knowledge—I do not also argue for the negative doctrines that propositional and/or mentalistic accounts of knowledge are wrong. On their own, however, they do not provide a sufficient framework for an adequate epistemology of technology and science. More is needed and a critical part of this is an articulation of how the material dimensions of science and technology do epistemological work. Things and theory can both constitute our knowledge of the world. But I deny that there is a unified epistemological treatment for both. Even within my materialist epistemology, there are fundamentally different ways that different kinds of instruments constitute knowledge.

Material models work epistemologically in ways that are very similar to theory. They provide representations, and in providing representations, they can be assessed in terms of the virtues and vices that are used to assess theoretical representations: explanatory and predictive power, simplicity, accuracy, etc.

Instruments that create phenomena, such as Faraday's motor, are different and constitute knowledge in a different nonrepresentational way. Such instruments work epistemologically in a manner that draws on pragmatist conceptions of knowledge as effective action. A fundamental difference, however, is that with instruments the action has been separated from human agency and built into the reliable behavior of an artifact. I call this kind of knowledge "working knowledge."[25] When we have made an instrument to do something in a particular way and it does it successfully and reliably, we say the instrument works. It is *working knowledge*, and this knowledge is different from the knowledge constituted by models—model knowledge.

Measuring instruments present a third kind of material knowledge that is a hybrid of the representational and effective action senses of knowledge. Measurement presupposes representation, for measuring something locates it in an ordered space of possible measurement outcomes. A representation—or model—of this ordered space has to be built into a measuring instrument. This can be as simple as a scale on a thermometer. At the same time, a measuring instrument has to do something, and do it reliably. It has to work. Presented with the same object for measurement, the instrument must yield outcomes that are the same or can be understood to be the same given an analysis of error. That is, the instrument has to present a phenomenon[26] in the sense of constituting "working knowledge." Measuring instruments integrate both model knowledge and working knowledge. I call this integration "encapsulated knowledge" where effective action and accurate representation work together in a material instrument to provide measurement.

SUBJECTIVE AND OBJECTIVE

Larry Bucciarelli begins *Designing Engineers*[27] with a question raised at a conference he attended on technological literacy: Do you know how your telephone works? A speaker at the conference noted with alarm that less than 20 percent of Americans knew how their telephones worked. But, Bucciarelli notes, the question is ambiguous. Some people (although perhaps fewer than 20 percent) may have an inkling of how sound waves can move a diaphragm and drive a coil back and forth in a magnetic field to create an electric current. But there is more to telephony than such simple physics. Bucciarelli wonders whether the conference speaker knows how his phone works:

> Does he know about the heuristics used to achieve optimum routing for long-distance calls? Does he know about the intricacies of the algorithms used for echo and noise suppression? Does he know how a signal is transmitted to and retrieved from a satellite in orbit? Does he know how AT&T, MCI, and the local phone companies are able to use the same network simultaneously? Does he know how many operators are needed to keep the system working, or what these repair people actually do when they climb a telephone pole? Does he know about corporate financing, capital investment strategies, or the role of regulation in the functioning of this expansive and sophisticated communication system?[28]

Indeed, Bucciarelli concluded, "Does anyone know how their telephone works?"[29]

Here, following the conference speaker, Bucciarelli uses "know" in a subjective sense. He makes a persuasive case that, in this sense, no one knows how his or her phone works. In the first place, the phone system is too big to be comprehended by a single "subjective knower." In the second place, the people who developed pieces of the hardware and software that constitute the phone system may have moved onto other concerns and forgotten the hows and whys of the pieces they developed. Their "subjective knowledge" is lost. In the third place, complicated systems with many interacting parts do not always behave in ways we can predict in detail. Despite having created them, programmers cannot always predict, and in this sense do not "subjectively know," how their complicated computer programs will behave.

It is, of course, well and proper to engage in what might be called subjective epistemology. This is the attempt to understand that aspect of knowledge that is a species of subjective belief. But if we want to understand technological and scientific knowledge, this is the wrong place to look. This is true for several reasons, the first of which is made clear by Bucciarelli's telephones. If no one—subjectively—knows how the phone system works, the situation with all scientific and technological knowledge is radically worse. The epistemological world of technology and science is too big for a single person to comprehend. People change the focus of their research and forget. Expert knowledge systems transcend their makers.

There is a second important reason that the epistemology of technology and science should not be sought at the level of individual belief. One of the important defining characteristics of scientific and technological knowledge is that it cannot be private. A scientist may do some research that provides strong evidence—in the scientist's view—for some claim. But the claim is not scientific knowledge until it has been subjected to scrutiny by the relevant scientific community and accepted by that community. Scientific and technological knowledge is public in the sense that the knowledge has passed review by peers. With respect to theoretical knowledge, publication in a book or journal article (or preprint, etc.) is the significant point when knowledge claims pass into the public realm of scientific and technological knowledge. In addition to these literary domains of scientific and technological knowledge, there are material domains. When Faraday sent copies of his motor to his colleagues, he was making it available for peer review.

We may be interested in what, for example, Faraday knew—subjectively—when he sent around copies of his motor. This can be important for understanding the history of electromagnetism. We can uncover evidence concerning the papers Faraday read. We can read Faraday's own notes. We thereby can develop an appreciation for his subjective theoretical knowledge. But we also can uncover evidence about Faraday's tactile and visual skills in eliciting the phenomenon that he ultimately built into his motor.[30] We thereby develop an appreciation for Faraday's embodied skills, his know-how and tacit knowledge. Taken together we come to understand the full spectrum of Faraday's subjective knowledge that went into both the writing of his articles and the making of his motor.

Once out of his hands and subject to review by his peers, the articles *and* the motor both pass into the public domain, the domain of possible objective knowledge. An adequate epistemology of science and technology has to include such public objective knowledge. Here are the epistemological products of the subjective engagements of scientists, engineers, and others. These products include theories and the like, written products that occupy the pages of professional journals. But they also include the material artifacts that I consider under the headings of model knowledge, working knowledge, and encapsulated knowledge, in short, thing knowledge.

ARGUMENTS

The multiple material epistemologies that I articulate as thing knowledge rest on several interconnected and mutually supporting arguments. In my extended treatment of the subject, *Thing Knowledge*, these arguments are developed in detail. Here I summarize. There are four types of argument that support my materialist epistemology: arguments from analogy, arguments from cognitive autonomy, arguments from history, and finally what I call arguments by articulation. The specific instances of each type of argument are different from each other in detail since they serve different epistemological conceptions—model knowledge, working knowledge, and so on. While all the arguments stand as integral parts to the overall picture I present of thing knowledge, it is useful to disentangle the strands and explain my basic argumentative strategies.

I use arguments by analogy to demonstrate that the material products of science bear knowledge. I show, for instance, how, in several epistemologically important respects, material models function analogously to theoretical contributions to science and technology. Material models can provide explanations and predictions. They can be confirmed or refuted by empirical evidence. These points can be developed by appeal to a version of the semantic account of theories where a theory is identified with a class of abstract structures called models. Material models satisfy all the requirements for abstract models in the sense of the semantic view of theories.

There is a distinct argument from analogy that deals with "working knowledge." Take Faraday's motor. We say someone knows how to ride a bicycle when he or she can consistently and successfully accomplish the task. A phenomenon, such as that exhibited by Faraday's motor, shares these features of consistency and success with what is usually called know-how or skill knowledge. To over-anthropomorphize the situation, one might say Faraday's motor "knows how to make rotations." I prefer to say the motor bears knowledge of a kind of material agency, and I call such knowledge "working knowledge." The analogy runs deeper. Frequently enough we are unable to articulate in words our knowledge of, for example, how to ride a bicycle. The knowledge is tacit. We find a similar situation with instruments such as Faraday's motor, and from two points of view. From an anthropomorphic point of view, the motor articulates nothing in words.

But also from the point of view of its maker—Faraday in this example—it was difficult to articulate how the phenomenon comes about. Yet, as in the case of bicycle riding, it is clear that the instrument presents a phenomenon, that it works. The action is effective, in a general sense, even lacking a verbal articulation for it. The knowledge resides in the regular controlled action of the instrument. The instrument bears this tacit "working knowledge."

A different collection of arguments turns on what can be called the cognitive autonomy of instruments. Davenport learned something from his examination of Henry's electromagnet. He then took what he learned and turned it into another, potentially commercially useful, device. He did this while ignorant of theory and unable to express in words both what Henry's electromagnet taught him and what he was doing with this knowledge. A variant of this argument applies to Watson and Crick's discovery of the structure of DNA. Here we see how Watson's ability to physically manipulate cardboard models of DNA base pairs led to his discovery of base-pair bonding. Watson employed a distinct "cognitive channel" from the consideration and manipulation of theoretical or propositional material. In a nutshell, the point is that "making" is different from "saying," and yet we learn from made things and from our actions of making. Cognitive content is not exhausted by theory. Epistemic content should not be exhausted by theory for the same reason. This is, perhaps, the core meaning to a note found on Richard Feynman's blackboard at the time of his death, "What I cannot create I do not understand."[31] Feynman subjectively knew something through his efforts to create it. His creation then carries the objective content of this knowledge in a way that might be subjectively recovered by someone else, just as Henry's electromagnet had meaning for Davenport.

A third kind of argument draws on history to support the epistemological standing of instruments. There is, in the first place, the argument that we miss a tremendous amount of what is epistemologically significant in the history of science and technology if we limit our examination to the history of theory. Carnot cycles in thermodynamics are the cycles that were being traced out by steam engine indicators in the twenty years preceding Sadi Carnot's and Émile Clapeyron's work on thermodynamics.[32] During the middle years of the twentieth century there was a fundamental transformation in the history of science and technology. Scientists and engineers came to understand that the development of instruments was a central component to the progress in our knowledge of the world. This is the time when instrument maker and analytical chemist Ralph Müller wrote these lines:

> That the history of physical science is largely the history of instruments and their intelligent use is well known. The broad generalizations and theories which have arisen from time to time have stood or fallen on the basis of accurate measurement, and in several instances new instruments have had to be devised for the purpose. There is little evidence to show that the mind of modern man is supe-

rior to that of the ancients, his tools are incomparably better. . . . Although the modern scientist accepts and welcomes new instruments, he is less tolerant of instrumentation. He is likely to regard preoccupation with instruments and their design as "gadgeteering" and distinctly inferior to the mere use of instruments in pure research: Thus, Lord Rutherford once said of Callender, the father of recording potentiometers, "He seems to be more interested in devising a new instrument than in discovering a fundamental truth.". . .

Fortunately, there is a great body of earnest workers, oblivious to these jibes, devoted to these pursuits, whose handiwork we may examine. They are providing means with which the "Olympians" may continue to study nature.[33]

At the end of the day, the fundamental argument for the epistemological place of instruments is my articulation of how instruments do epistemological work. I cannot make good on this argument in the scope of this chapter. Further detail, as developed in *Thing Knowledge*, is necessary. There I articulate three different ways in which instruments bear knowledge: first as a material mode of representation, then as a material mode of effective action, and finally as a material mode of encapsulated knowledge synthesizing representation and action. There I present the historical evidence of the coming to scientific self-awareness that instruments bear scientific knowledge. There I develop a neo-Popperian account of objective knowledge to accommodate instruments and the material products of science and technology generally.

Collectively, the point of the various chapters of *Thing Knowledge* is to articulate a picture of why and how instruments should be understood epistemologically on a par with theory. While the various arguments aim to persuade readers of this conclusion, it is the overall picture that must seal the deal. Beyond why instruments should be understood as knowledge bearers, I show how they do this and what the consequences are of their doing this.

BEYOND SCIENCE TO TECHNOLOGY

The kind of epistemology that I advocate here brings out relationships, which, while of recognized importance, have not found a comfortable place in the philosophy of science and technology. The idea that engineers and industrialists simply take and materially instantiate the knowledge provided by science cannot stand up to even the most cursory historical study. James Watt's work on steam engine instrumentation—specifically the indicator diagram—made a seminal contribution to the development of thermodynamics. Yet without a broader understanding of epistemology, where instruments themselves express knowledge of the world, alternatives to this notion of "applied science," to the idea of engineering and industry as epistemological hangers-on, are difficult to develop.

"Craft knowledge," "fingertip knowledge," "tacit knowledge," and "know-how" are useful concepts in that they remind us that there is more to knowing than saying. But they tend to render this kind of knowledge ineffable. Instruments have a kind of public existence that allows for more explicit study. My intention is not to downplay the significance of "craft knowledge" and the rest. On the contrary, I believe that an analysis of instruments as knowledge provides insight into this difficult and important epistemological territory.

The most immediate consequence of recognizing instruments as knowledge is that the boundary between science and technology changes. Recent science studies scholarship has recognized a more fluid relationship between science and technology than earlier positivist and postpositivist philosophies of science. Still, theoretical science is where one turns to examine *knowledge*. Previously ignored contributions of craftsmen and engineers now are understood to have provided important, and in many cases essential, contributions to the growth of scientific knowledge. But it is theory that is seen to be growing. Davenport's story is a sidebar.

The picture I offer here is different. I see developments of things and of theory on a par. In many cases, they interact, sometimes with beneficial results all around. But in many cases they develop independently, again sometimes with beneficial results. Work done in industry, putting together bits of the material world, is as constitutive of knowledge as work done by "theoretical scientists." Some of it is fundamental (John Harrison's seaworthy chronometer, perhaps);[34] some of it is less so (the translucent case for Apple's iMac, perhaps). In this sense, material contributions are not different from theoretical contributions— which run the gamut from Einstein's general theory of relativity to psychotherapeutic notions that subliminal exposure to the words "Mommy and I are one" will improve behavior.[35]

There are, however, important differences between work with theory and work with things. Things are not as tidy as ideas. Plato was exactly right on this point. Things are impermanent, impure, and imperfect. Yet, I argue, many instruments hide their very materiality. The ideal measuring instrument provides information about the world that can be trusted and acted upon. The instrument performs semantic ascent for us, providing output that is useful in the commerce of ideas. The instrument renders the materiality of the world transparent, and, indeed, it renders the materiality of thing knowledge transparent. In the information age we like to pretend we can live entirely in our heads, or, rather, in the data.

But this is an illusion. We live, work, and know in a material world. Technology, as I view it, is a particular and special epistemological domain, a domain where making things and made things are epistemological. Technology, as developed in science, engineering, and other life pursuits, is the domain of a special kind of material knowledge. I call it thing knowledge.

NOTES

1. Michael Faraday, "On Some New Electromagnetical Motions, and on the Theory of Magnetism," *Quarterly Journal of Science* 12 (1821): 74–96; Faraday, "Historical Sketch of Electro-magnetism," *Annals of Philosophy* 18 (1821): 195–200, 274–90; Faraday, "Historical Sketch of Electro-magnetism," *Annals of Philosophy* 19 (1822): 107–21; Faraday, "Note on New Electro-magnetical Motions," *Quarterly Journal of Science* 12 (1822): 416–21.

2. Michael Faraday, *The Selected Correspondence of Michael Faraday, Volume 1, 1812–1848,* eds., L. Pearce Williams with Rosemary FitzGerald and Oliver Stallybrass (Cambridge: Cambridge University Press, 1971), pp. 122–39.

3. Michael Faraday, "Electro-magnetic Rotations Apparatus," *Quarterly Journal of Science* 12 (1822): 186; Faraday, "Description of an Electro-magnetical Apparatus for the Exhibition of Rotary Motion," *Quarterly Journal of Science* 12 (1822): 283–85; Faraday, *Selected Correspondence of Michael Faraday,* pp. 128–29.

4. Michael Faraday, *Experimental Researches in Electricity* (London: Richard and John Edward Taylor, 1844), p. 129.

5. L. P. Williams, *Michael Faraday: A Biography* (New York: Basic Books, 1964), p. 157.

6. L. Bonjour, *The Structure of Empirical Knowledge* (Cambridge: Harvard University Press, 1985); A. Goldman, *Epistemology and Cognition* (Cambridge: Harvard University Press, 1986); R. Audi, *Epistemology: A Contemporary Introduction to the Theory of Knowledge* (London: Routledge, 1998).

7. I. Lakatos, "Falsification and the Methodology of Scientific Research Programmes," in *Criticism and the Growth of Knowledge*, eds. I. Lakatos and A Musgrave (Cambridge: Cambridge University Press, 1970) I. Lakatos and A. Musgrave, eds. *Criticism and the Growth of Knowledge*, eds. I. Lakatos and A Musgrave (Cambridge: Cambridge University Press, 1970); K. Popper, *Objective Knowledge: An Evolutionary Approach* (Oxford: Oxford University Press, 1972); L. Laudan, *Progress and Its Problems* (Berkeley: University of California Press, 1977).

8. Among recent work on the philosophy of experiment, the following have tended to focus on the literally material aspects of science: D. d. S. Price, "Philosophical Mechanism and Mechanical Philosophy: Some Notes toward a Philosophy of Scientific Instruments," *Annali Dell 'Instituto é Muséo di Storia Della Scienza di Firenze* 5 (1980): 75–85; Price, "Scientists and Their Tools," in *Frontiers of Science: On the Brink of Tomorrow*, ed. D. d. S. Price (Washington, D.C.: National Geographic Special Publications, 1982), pp. 1–23; Price, "Notes toward a Philosophy of the Science/Technology Interaction," in *The Nature of Technological Knowledge: Are Models of Scientific Change Relevant?* ed. R. Laudan (Dordrecht: Reidel, 1984), pp. 105–14; Ian Hacking, *Representing and Intervening* (Cambridge: Cambridge University Press, 1983); S. Shapin and S. Schaffer, *Leviathan and the Air-Pump* (Princeton: Princeton University Press, 1985); A. Franklin, *Experiment: Right or Wrong* (Cambridge: Cambridge University Press, 1990); H. Radder, *The Material Realization of Science* (Assen, Netherlands: Van Gorcum, 1988); D. Baird and T. Faust, "Scientific Instruments, Scientific Progress and the Cyclotron," *British Journal for the Philosophy of Science* 41 (1990): 147–75; A. Franklin, *The Neglect of Experiment* (Cambridge: Cambridge University Press, 1986); D. Gooding, *Experiment and the Making of Meaning* (Dordrecht: Kluwer Academic Publishers, 1990); D. Ihde,

Instrumental Realism (Evanston: Northwestern University Press, 1991); R. G. W. Anderson, J. A. Bennett, and W. F. Ryan, eds., *Making Instruments Count* (Aldershot, Hampshire, England: Variorum Ashgate Publishing, 1993); D. Baird and A. Nordmann, "Facts-Well-Put," *British Journal for the Philosophy of Science* 45 (1994): 37–77; J. Z. Buchwald, *The Creation of Scientific Effects* (Chicago: University of Chicago Press, 1994); A. van Helden and T. Hankins, eds., *Instruments,* Osiris, Chicago: University of Chicago Press 9 (1994): 67–84; T. Hankins and R. Silverman, eds., *Instruments and the Imagination* (Princeton: Princeton University Press, 1995); A. Pickering, *The Mangle of Practice: Time, Agency, and Science* (Chicago: University of Chicago Press, 1995); N. Wise, ed., *The Values of Precision* (Princeton: Princeton University Press, 1995); P. Galison, *Image and Logic: A Material Culture of Microphysics* (Chicago: University of Chicago Press, 1997).

9. Popper, *Objective Knowledge.*

10. Lakatos, "Falsification and the Methodology of Scientific Research Programmes."

11. L. Laudan, *Progress and Its Problems.*

12. D. Warner, "Terrestrial Magnetism: For the Glory of God and the Benefit of Mankind," in *Instruments,* eds. A. van Helden and T. Hankins, Osiris, Chicago: University of Chicago Press 9 (1994): 67–84.

13. D. Baird, *Thing Knowledge: A Philosophy of Scientific Instruments* (Berkeley: University of California Press, 2003).

14. Price, "Philosophical Mechanism and Mechanical Philosophy."

15. F. Suppe, ed., *The Structure of Scientific Theories* (Urbana: University of Illinois Press, 1994).

16. B. Latour and S. Woolgar, *Laboratory Life* (Beverly Hills: Sage Press, 1979).

17. Ibid., p. 148.

18. Gooding, *Experiment and the Making of Meaning,* p. 3 [emphasis added].

19. Ibid., p. 177.

20. Faraday, *Selected Correspondence of Michael Faraday,* p. 133, letter dated March, 14, 1822.

21. W. J. King, "The Development of Electrical Technology in the 19th Century," *Contributions from the Museum of History and Technology*, Washington, D.C., Smithsonian Institution 19–30 (1963): 231–407; B. Gee, "Electromagnetic Engines: Pre-technology and Development Immediately Following Faraday's Discovery of Electromagnetic Rotations," *History of Technology* 13 (1991): 41–72.

22. King, "Development of Electrical Technology in the 19th Century," pp. 260–71.

23. W. R. Davenport, *Biography of Thomas Davenport: The "Brandon Blacksmith," Inventor of the Electric Motor* (Montpelier: Vermont Historical Society, 1929); M. Schiffer, "The Blacksmith's Motor," *Invention and Technology* 9, no. 3 (1994): 64.

24. Schiffer, "The Blacksmith's Motor," p. 64.

25. In coining this neologism, I call on our use of "working" to describe an instrument or machine that performs regularly and reliably. I also draw on the phrase "to have a working knowledge." Someone with a working knowledge of something has knowledge that is sufficient to do something. My neologism, "working knowledge," draws attention to the connection between knowledge and effective action.

26. On this point, see Hacking, *Representing and Intervening,* chap. 14.

27. L. L. Bucciarelli, *Designing Engineers* (Cambridge: MIT Press, 1994), p. 3.

28. Ibid., p. 3.

29. Ibid. [emphasis in the original]

30. See Gooding, *Experiment and the Making of Meaning.*

31. J. Gleick, *Genius: The Life and Science of Richard Feynman* (New York: Pantheon Books, 1993), p. 437.

32. D. Baird, "Instruments on the Cusp of Science and Technology: The Indicator Diagram," *Knowledge and Society: Studies in the Sociology of Science Past and Present* 8 (1989): 107–22.

33. Ralph Müller, "American Apparatus, Instruments, and Instrumentation," *Industrial and Engineering Chemistry, Analytical Edition* 12, no. 10 (1940): 571–72.

34. D. Sobel, *Longitude* (New York: Walker & Company, 1995).

35. "Mommy and I Are One," *Science News* 156 (1986).

6

The Device Paradigm

Albert Borgmann

We must now provide an explicit account of the pattern or paradigm of technology. I begin with two clear cases and analyze them in an intuitive way to bring out the major features of the paradigm. And I attempt to raise those features into sharper relief against the sketch of a pretechnological setting and through the consideration of objections that may be advanced against the distinctiveness of the pattern.

Technology, as we have seen, promises to bring the forces of nature and culture under control, to liberate us from misery and toil, and to enrich our lives. To speak of technology making promises suggests a substantive view of technology and is misleading. But the parlance is convenient and can always be reconstructed to mean that implied in the technological mode of taking up with the world there is a promise that this approach to reality will, by way of the domination of nature, yield liberation and enrichment. Who issues the promise to whom is a question of political responsibility; and who the beneficiaries of the promise are is a question of social justice. What we must answer first is the question of how the promise of liberty and prosperity was specified and given a definite pattern of implementation.

As a first step let us note that the notions of liberation and enrichment are joined in that of availability. Goods that are available to us enrich our lives and, if they are technologically available, they do so without imposing burdens on us. Something is available in this sense if it has been rendered instantaneous, ubiquitous, safe, and easy.[1] Warmth, e.g., is now available. We get a first glimpse of the distinctiveness of availability when we remind ourselves that warmth was not available, e.g., in Montana a hundred years ago. It was not instantaneous because in the

Technology and the Character of Contemporary Life, by Albert Borgmann, "The Device Paradigm," pp. 40–48 (Copyright © 1984 University of Chicago Press). Published by the University of Chicago Press. Reprinted by permission.

morning a fire first had to be built in the stove or the fireplace. And before it could be built, trees had to be felled, logs had to be sawed and split, the wood had to be hauled and stacked. Warmth was not ubiquitous because some rooms remained unheated, and none was heated evenly. The coaches and sleighs were not heated, nor were the boardwalks or all of the shops and stores. It was not entirely safe because one could get burned or set the house on fire. It was not easy because work, some skills, and attention were constantly required to build and sustain a fire.

Such observations, however, are not sufficient to establish the distinctiveness of availability. In the common view, technological progress is seen as a more or less gradual and straightforward succession of lesser by better implements.[2] The wood-burning stove yields to the coal-fired central plant with heat distribution by convection, which in turn gives way to a plant fueled by natural gas and heating through forced air, and so on.[3] To bring the distinctiveness of availability into relief we must turn to the distinction between things and devices. A thing, in the sense in which I want to use the word here, is inseparable from its context, namely, its world, and from our commerce with the thing and its world, namely, engagement. The experience of a thing is always and also a bodily and social engagement with the thing's world. In calling forth a manifold engagement, a thing necessarily provides more than one commodity. Thus a stove used to furnish more than mere warmth. It was a *focus*, a hearth, a place that gathered the work and leisure of a family and gave the house a center. Its coldness marked the morning, and the spreading of its warmth the beginning of the day. It assigned to the different family members tasks that defined their place in the household. The mother built the fire, the children kept the firebox filled, and the father cut the firewood. It provided for the entire family a regular and bodily engagement with the rhythm of the seasons that was woven together of the threat of cold and the solace of warmth, the smell of wood smoke, the exertion of sawing and of carrying, the teaching of skills, and the fidelity to daily tasks. These features of physical engagement and of family relations are only first indications of the full dimensions of a thing's world. Physical engagement is not simply physical contact but the experience of the world through the manifold sensibility of the body. That sensibility is sharpened and strengthened in skill. Skill is intensive and refined world engagement. Skill, in turn, is bound up with social engagement. It molds the person and gives the person character.[4] Limitations of skill confine any one person's primary engagement with the world to a small area. With the other areas one is mediately engaged through one's acquaintance with the characteristic demeanor and habits of the practitioners of the other skills. That acquaintance is importantly enriched through one's use of their products and the observation of their working. Work again is only one example of the social context that sustains and comes to be focused in a thing. If we broaden our focus to include other practices, we can see similar social contexts in entertainment, in meals, in the celebration of the great events of birth, marriage, and death. And in these wider horizons of social engagement we can see how the cultural and natural dimensions of the world open up.

We have now sketched a background against which we can outline a specific notion of the device. We have seen that a thing such as a fireplace provides warmth, but it inevitably provides those many other elements that compose the world of the fireplace. We are inclined to think of these additional elements as burdensome, and they were undoubtedly often so experienced. A device such as a central heating plant procures mere warmth and disburdens us of all other elements. These are taken over by the machinery of the device. The machinery makes no demands on our skill, strength, or attention, and it is less demanding the less it makes its presence felt. In the progress of technology, the machinery of a device has therefore a tendency to become concealed or to shrink. Of all the physical properties of a device, those alone are crucial and prominent which constitute the commodity that the device procures. Informally speaking, the commodity of a device is "what a device is there for." In the case of a central heating plant it is warmth, with a telephone it is communication, a car provides transportation, frozen food makes up a meal, a stereo set furnishes music. "Commodity" for the time being is to be taken flexibly. The emphasis lies on the commodious way in which devices make goods and services available. There are at first unavoidable ambiguities in the notion of the device and the commodity; they can gradually be resolved through substantive analyses and methodological reflections.[5] Tentatively, then, those aspects or properties of a device that provide the answer to "What is the device for?" constitute its commodity, and they remain relatively fixed. The other properties are changeable and are changed, normally on the basis of scientific insight and engineering ingenuity, to make the commodity still more available. Hence every device has functional equivalents, and equivalent devices may be physically and structurally very dissimilar from one another.

The development of television provides an illustration of these points. The bulky machinery of the first sets was obtrusive in relation to the commodity it procured, namely, the moving two-dimensional picture which appeared in fuzzy black and white on a screen with the size and shape of a bull's-eye. Gradually the screens became larger, more rectangular; the picture became sharper and eventually colored. The sets became relatively smaller and less conspicuous in their machinery. And this development continues and has its limit in match-box-sized sets which provide arbitrarily large and most finely grained moving and colored pictures. The example also shows how radical changes in the machinery amounted to continuous improvements of the function as tubes gave way to transistors and these yielded to silicon chips. Cables and satellites were introduced as communication links. Pictures could be had in recorded rather than transmitted form, and recordings can be had on tapes or discs. These considerations in turn show how the technical development of a device increases availability. Increasingly, video programs can be seen nearly everywhere—in bars, cars, in every room of a home. Every conceivable film can be had. A program broadcast at an inconvenient time can be recorded and played later. The constraints of time and place are more and more dissolved. It is an instructive exercise to see how in the implements that surround us daily the machinery becomes less conspicuous, the function more prominent, how radical technical changes in the machinery are but degrees of advancement in the commodity, and how the availability of the commodities increases all the while.

The distinction in the device between its machinery and its function is a specific instance of the means-ends distinction. In agreement with the general distinction, the machinery or the means is subservient to and validated by the function or the end. The technological distinction of means and ends differs from the general notion in two respects. In the general case, it is very questionable how clearly and radically means and ends can be distinguished without doing violence to the phenomena.[6] In the case of the technological device, however, the machinery can be changed radically without threat to the identity and familiarity of the function of the device. No one is confused when one is invited to replace one's watch, powered by a spring, regulated by a balance wheel, displaying time with a dial and pointers, with a watch that is powered electrically, is regulated by a quartz crystal, and displays time digitally. This concomitance of radical variability of means and relative stability of ends is the first distinguishing feature. The second, closely tied to the first, is the concealment and unfamiliarity of the means and the simultaneous prominence and availability of the ends.[7]

The concealment of the machinery and the disburdening character of the device go hand in hand. If the machinery were forcefully present, it would *eo ipso* make claims on our faculties. If claims are felt to be onerous and are therefore removed, then so is the machinery. A commodity is truly available when it can be enjoyed as a mere end, unencumbered by means. It must be noted that the disburdenment resting on a feudal household is ever incomplete. The lord and the lady must always reckon with the moods, the insubordination, and the frailty of the servants.[8] The device provides social disburdenment, i.e., anonymity. The absence of the master-servant relation is of course only one instance of social anonymity. The starkness of social anonymity in the technological universe can be gauged only against a picture of the social relations in a world of things. Such a picture will also show that social anonymity necessarily shades off into one of nature, culture, and history.

Since the transformative power of technology is very uneven chronologically, settings that approached the character of a world of things still prevailed at the beginning of this century. Here it pays to look closely, to see in one case and in detail how nature and culture were interwoven and how this texture was rent by the advance of technology and overtaken by anonymity. The case I want to consider is that of a wheelwright's shop just prior to its dissolution. A moving account has been given by George Sturt, the last in a succession of wheelwrights.

Since the web of relations is so tight and manifold, it is difficult to present it in an abstract and summary way. But let us begin with those aspects in which the relation of humans to nature is singled out. The experience of cultivating the land is still alive at this time in England, and Sturt speaks repeatedly of "the age-old effort of colonizing England."[9] But he does not understand colonizing as the domination of nature, i.e., as conquering and subduing, but as an adaption of people to the land, and he paraphrases it as the "age-long effort of Englishmen to get themselves close and ever closer into England."[10] As people adjust to the land, the land discloses itself to the people. There is "a close relationship between the tree-clad country-side and the English who dwelt there." Sturt speaks of "the affection and the reverence bred of this."[11] But it is impossible to abstract a relationship in this pretechnological setting that obtains merely between human beings and nature. What takes the wheelwright into "sunny woodland solitudes," "into winter woods or along leafless hedgerows," and "across wet water-meadows in February" is the search for timber.[12] But "timber was far from being a prey, a helpless victim, to a machine," Sturt says, and continues: "Rather it would lend its subtle virtues to the man who knew how to humor it: with him, as with an understanding friend, it would co-operate."[13] This is a relationship not of domination but of mastery. If the wheelwright, Sturt says elsewhere, "was really master of his timber, if he knew what he had already got in stock and also what was likely to be wanted in years to come, he kept a watch always for timber with special curve, suitable for hames, or shaft-braces, or waggon-heads, or hounds, or tailboard rails, or whatever else the tree-shape might suggest."[14] Such respectful working with nature is not just as close to nature as conservation; it opens up dimensions that remain otherwise closed. "Under the plane (it is little used now)," Sturt says, "or under the axe (if it is all but obsolete) timber disclosed qualities hardly to be found otherwise."[15] And elsewhere he says:

> With the wedges cleaving down between the clinging fibres—as he let out the wood-scent, listened to the tearing splitting sounds—the workman found his way into a part of our environment—felt the laws of woodland vitality—not otherwise visited or suspected.[16]

But again the intimacy of the wheelwright with nature did not stop with the materials but embraced his entire world by way of the needs of his customers. Sturt puts it this way:

> And so we got curiously intimate with the peculiar needs of the neighbourhood. In farm-waggon or dung-cart, barley-roller, plough, water-barrel or what not, the dimensions we chose, the curves we followed (and almost every piece of timber was curved) were imposed upon us by the nature of the soil in this or that farm, the gradient of this or that hill, the temper of this or that customer or his choice perhaps in horseflesh.[17]

And similarly he says in another place:

> The field, the farm-yard, the roads and hills, the stress of weather, the strength and shape of horses, the lifting power of men, all were factors which had deter-

mined in the old villages how the farm tackle must be made, of what timber and shape and of what dimensions, often to the sixteenth of an inch.[18]

This web of relations had, finally, its social aspects. It contained different guilds or groups, but no classes, i.e., divisions of people whose political and especially economic interests were opposed to one another.[19] The different groups had their character from their work and their relation to nature. In his search for timber, the wheelwright found not only trees but also "country men of a shy type, good to meet."[20] And back at his shop he was met by the carters, "a whole country-side of strong and good-tempered Englishmen. With the timber and the horses they seemed to bring the lonely woodlands, the far-off roads into the little town."[21] The social network was sustained by fidelity, by wagons that were built to last a lifetime and that were carefully repaired when they had broken down.[22] Prices were charged by tradition and not by calculation of costs and profits.[23] The tie between employer and employed was one of "kindly feeling" as Sturt puts it, a relation of resourcefulness and trust.[24]

Sturt's account is remarkable not only for its portrayal of the strength and character of a pretechnological world of things. It is also painfully aware of the rise of technology and the destruction of the pretechnological setting. This process, too, becomes visible at the reference points of nature, materials, and social relations. Accelerated by the demands of the First World War, a "sort of greedy prostitution desecrated the ancient woods. . . I resented it," Sturt says, "resented seeing the fair timber callously felled at the wrong time of year, cut too soon, not 'seasoned' at all."[25] The conquest of nature is not confined to the treatment of the forests but moves into the wheelwright's shop, too, replacing skill with mechanical power which can "drive, with relentless unintelligence, through every resistance."[26] As said before, domination is not an end in itself but serves to secure more radically the products of labor. Thus, as Sturt points out, "work was growing less interesting to the workman, although far more sure in its results."[27] And domination provides more income for the purchase of commodities, but at the same time it disengages the worker from the world. This is Sturt's experience in the following passage:

> Of course wages are higher—many a workman today receives a larger income than I was able to get as "profit" when I was an employer. But no higher wage, no income, will buy for men that satisfaction which of old—until machinery made drudges of them—streamed into their muscles all day long from close contact with iron, timber, clay, wind and wave, horse-strength.[28]

These transformations finally touched the social relations as well. "The Men," Sturt says of his employees, "though still my friends, as I fancied, became machine 'hands.'"[29] The loss of skill went hand in hand with the loss of rustic village life, and the change in the living situation upset the old social relations. Sturt, speaking of the changes in the life of one wheelwright in particular, says:

I was not in touch, through him, with the quiet dignified country life of England and I was more of a capitalist. Each of us had slipped a little nearer to the igno- minious class division of these present times—I to the employer's side, he to the disregarded workman's.[30]

Sturt had an uncanny sense for the transformative power that changed the face of his world. He recognized its concealment, the semblance, i.e., as though tech- nology were only a more efficient way of doing what had been done throughout the ages.[31] And he recognized its radical novelty, the fact that technology upsets the tradition from the ground up. The technological changes forced him to intro- duce modern machines and take in a partner who could supervise the new ways of working. "Neither my partner nor myself," he says in retrospect, "realised at all that a new world (newer than ever America was to the Pilgrim Fathers) had begun even then to form all around us."[32] This returns us to the difficulty of bringing the distinctive features of this "new world" into relief. As was argued above, these features become visible when we learn to see how the presence of things is replaced with the availability of commodities and how availability is procured through devices. Devices, that was the claim, dissolve the coherent and engaging character of the pretechnological world of things. In a device, the relat- edness of the world is replaced by a machinery, but the machinery is concealed, and the commodities, which are made available by a device, are enjoyed without the encumbrance of or the engagement with a context.

But this analysis of the distinctiveness of the device is still deficient, and the deficiency can be brought into relief through two objections. Is not, one may ask, the concealment of the machinery and the lack of engagement with our world due to widespread scientific, economic, and technical illiteracy?[33] And quite apart from one's level of education, is not everyone in his or her work directly and explicitly engaged with the machinery of devices?

We can approach the first point through one of its companion phenomena, peoples alleged unwillingness and inability to maintain and repair technological devices.[34] How well-founded is this allegation? One way in which commodities are made available is that of making them discardable. It is not just unnecessary but impossible to maintain and repair paper napkins, cans, Bic ball points, or any of the other one-way or one-time devices. Another way to availability is that of making products carefree. Stainless steel tableware requires no polishing, plastic dishes need not be handled carefully. In other cases maintenance and repair become impossible because of the sophistication of the product. Microcomputers are becoming increasingly common and influential as devices that free us of the tasks of allocation, record keeping, and control. The theories and technical processes that underlie the production of microcircuits are too complicated and too much in flux to be known in detail by more than a handful of people. And the microcircuits themselves are realized at a functional level so minute and dense that it does not permit the intrusions necessary for repairs even if structure and functions are fully understood.[35] Finally, microcomputers are being used more and more widely because they are becoming "friendly," i.e., easy to operate and understand.[36] But such "friendliness" is just the mark of how wide the gap has become between the function accessible to everyone and the machinery known by nearly no one. And not only lay people are confined to the side of ignorance of this gap, but so are many, perhaps most, of the professional programmers.[37]

Still, education in engineering and in the natural and social sciences would make much of the machinery, i.e., of the context, of technological devices, perspicuous. But even if such education were to become more common, the context of functions and commodities would remain different from the world of things for two reasons. First, the presence of that context would remain entirely cerebral since it increasingly resists, as we have seen, appropriation through care, repair, the exercise of skill, and bodily engagement. Second, the context would remain anonymous in the senses indicated above. The machinery of a device does not of itself disclose the skill and character of the inventor and producer, it does not reveal a region and its particular orientation within nature and culture. In sum, the machinery of devices, unlike the context of things, is either entirely occluded or only cerebrally and anonymously present. It is in this sense necessarily unfamiliar.

The function of the device, on the other hand, and the commodity it provides are available and enjoyed in consumption. The peculiar presence of the end of the device is made possible by means of the device and its concealment. Everyone understands that the former rests on the latter, and everyone understands as well that the enjoyment of ends requires some kind of attention to the means. Only in magic are ends literally independent of means. The inevitable explicit concern with the machinery takes place in labor. But labor does not in general lift the veil of unfamiliarity from the machinery of devices. The labor process is itself transformed according to the paradigm of the device.

NOTES

1. Earlier versions of this notion of technology can be found in "Technology and Reality," *Man and World* 4 (1971): 59–69; "Orientation in Technology," *Philosophy Today* 16 (1972): 135–47; "The Explanation of Technology," *Research in Philosophy and Technology* 1 (1978): 99–118. Daniel J. Boorstin similarly describes the character of everyday America in terms of availability and its constituents. See his *Democracy and Its Discontents: Reflections on Everyday America* (New York: Vintage Books, 1975).

2. See Emmanuel G. Mesthene, *Technological Change: Its Impact on Man and Society* (New York: New American Library, 1970), p. 28.

3. See Melvin M. Rotsch, "The Home Environment," in *Technology in Western Civilization*, eds. Melvin Kranzberg and Carroll W. Pursell Jr., 2 vols. (New York: Oxford University Press, 1967), 2: 226–28. For the development of the kitchen stove (the other branch into which the original fireplace or stove developed), see Siegfried Giedion, *Mechanization Takes Command* (New York: Oxford University Press, 1969 [first published in 1948]), pp. 527–47.

4. See George Sturt's description of the sawyers in *The Wheelwright's Shop* (Cambridge: Cambridge University Press, 1974 [first published in 1923]), pp. 32–40.

5. In economics, "commodity" is a technical term for a tradable (and usually movable) economic good. In social science, it has become a technical term as a translation of Marx's *Ware* (merchandise). Marx's use and the use here suggested and to be developed agree inasmuch as both are intended to capture a novel and ultimately detrimental transformation of a traditional (pretechnological) phenomenon. For Marx, a commodity of the negative sort is the result of the reification of social relations, in particular of the reification of the workers' labor power, into something tradable and exchangeable which is then wrongfully appropriated by the capitalists and used against the workers. This constitutes the exploitation of the workers and their alienation from their work. It finally leads to their pauperization. I disagree that this transformation is at the center of gravity of the modern social order. The crucial change is rather the splitting of the pretechnological fabric of life into machinery and commodity according to the device paradigm. Though I concede and stress the tradable and exchangeable character of commodities, as I use the term, their primary character, here intended, is their commodious and consumable availability with the technological machinery as their basis and with disengagement and distraction as their recent consequences. On Marx's notion of commodity and commodity fetishism, see Paul M. Sweezy, *The Theory of Capitalist Development* (New York: Monthly Review Press, 1968), pp. 34–40.

6. See Morton Kaplan, "Means/Ends Rationality," *Ethics* 87 (1976): 61–65.

7. Martin Heidegger gives a careful account of the interpenetration of means and ends in the pretechnological disclosure of reality. But when he turns to the technological disclosure of being (*das Gestell*) and to the device in particular (*das Gerät*), he never points out the peculiar technological diremption of means and ends though he does mention the instability of the machine within technology. Heidegger's emphasis is perhaps due to his concern to show that technology as a whole is not a means or an instrument. See his "The Question Concerning Technology," in *The Question Concerning Technology and Other Essays*, trans. William Lovitt (New York: Harper and Row, 1977), pp. 3–35, pp. 6–12 and 17 in particular.

8. It also turns out that a generally rising standard of living makes personal services disproportionately expensive. See Staffan B. Linder, *The Harried Leisure Class* (New York: Columbia University Press, 1970), pp. 34–37.

9. See Sturt, *The Wheelwright's Shop*, p. 132; see also pp. 31, 38.

10. Ibid., p. 66.

11. Ibid., p. 23.

12. Ibid., p. 25.

13. Ibid., p. 45.

14. Ibid., p. 31.

15. Ibid., p. 24.

16. Ibid., p. 192.

17. Ibid., pp. 17–18.

18. Ibid., p. 41.

19. See Peter Laslett, *The World We Have Lost* (New York: Scribner, 1965), pp. 22–52.

20. See Sturt, *The Wheelwright's Shop*, p. 25.

21. Ibid., p. 28. See also the portrait of the sawyers, pp. 32–40.

22. Ibid., pp. 30, 43, 175–81.

23. Ibid., pp. 53, 200.

24. Ibid., pp. 53–55.

25. Ibid., p. 23.

26. Ibid., p. 45.

27. Ibid., p. 153; see also pp. 201–202.

28. Ibid., pp. 201–202.

29. Ibid., p. 201.

30. Ibid., p. 113.

31. Ibid., pp. 154, 201.

32. Ibid., p. 201.

33. A sketch and an analysis of technological illiteracy can be found in Langdon Winner, *Autonomous Technology: Technics-out-of-Control as a Theme in Political Thought* (Cambridge: MIT Press, 1977), pp. 282–95. While ignorance (of the machinery) is to be admitted and stressed, one must add that this ignorance goes hand in hand with an understanding (discussed below) of the overall pattern of technology.

34. Robert M. Pirsig describes this aversion to technology and contends that we can find wholeness at the center of technology if we begin to understand, maintain, and care for our devices. See his *Zen and the Art of Motorcycle Maintenance: An Inquiry into Values* (New York: William Morrorw and Company, 1974), pp. 11–35, 49–50, 97–106, 276, 290–92, 300–326.

35. Joseph Weizenbaum argues that certain computer programs have altogether escaped comprehensibility. See his *Computer Power and Human Reason: From Judgment to Calculation* (San Francisco: W. H. Freeman, 1976), pp. 228–57.

36. See "Wonders of '89," *Newsweek*, 19 November 1979, p. 151; and "And Man Created the Chip," *Newsweek*, 30 June 1980, p. 50.

37. See Weizenbaum, *Computer Power*, p. 103.

PART 3

THE BLESSINGS OF TECHNOLOGY

INTRODUCTION

In the previous part we saw that debates over the precise meaning of technology are important for an evaluation of the benefits and harms of technological development. This part contains essays by partisans of modern technology. The authors here can be regarded as being protechnology. All believe that the application of technology has been and will continue to be beneficial to humanity. In general, these advocates favor the current direction of technological development. We cannot turn the clock back; we must go forward. To borrow General Electric's phrase, "Progress is our most important product." While modern technology does produce some undesirable side effects, these can be rectified only through continual technological development.

As we have seen, Arnold Pacey, Jacques Ellul, and Albert Borgmann take a broad view of technology that includes cultural and organizational aspects. Without denying the social impact of technical inventions (the "material works" of technology), the authors in this part insist that cultural and organizational factors be kept separate from the material parts. Technology, pure and simple, is the application of scientific knowledge to produce good for human society. If we keep our focus on creating and improving instruments and tools for the use of human beings, we will greatly increase our chances of creating a good society.

One strong advocate for the beneficial uses of technology is Alvin Weinberg. In his essay, "Can Technology Replace Social Engineering?" Weinberg argues that technology is the foundation for providing a better life for humanity. Technological solutions are available for social problems and should be the primary focus of our efforts. In contrast to social engineering, which aims at the tremendously complex problem of changing human behavior, "technical fixes eliminate the original social problem" or "so alter the problem as to make its resolution more feasible." Technology, by increasing our productive capacity, can obviate

105

questions of social justice by providing more than enough goods to go around. The H-bomb has brought us peace without making people good, or even more rational. If we don't lose faith in the idea of progress and put our minds and money into technological development, we can develop efficient nuclear desalinization plants to produce all the water people want. We would thus not need to urge conservation or deal with questions of distributive justice.

Weinberg admits that "technological solutions to social problems tend to be incomplete and 'metastable,' to replace one social problem with another." The precarious peace brought about by nuclear weapons is clearly "metastable." However, even here technical fixes are relevant. Weinberg would, we think, agree that President George W. Bush's attempt to complete President Ronald Reagan's Strategic Defense Initiative was a proper policy initiative that will lead to more international security and peace. Further, cheap nuclear energy would go a long way in reducing tensions between nations of haves and have-nots. "Technology will never *replace* social engineering," but through a cooperative effort between technologists and social engineers we can hope to bring about a better society and a better life for us all.

In the second reading in this part, "The Role of Technology in Society," Emmanuel Mesthene presents what he regards as a balanced view of the role of technology: it is neither an unalloyed blessing nor an unmitigated evil. To comprehend the importance of technology in society, we need a concept that goes beyond the idea of simple hardware, for example, Weinberg's notion of a "technical fix." Mesthene defines "technology" as "the organization of knowledge for practical purposes." "Practical," however, does not mean beneficial, for "It has both positive and negative effects, and it usually has the two *at the same time and in virtue of each other*."

Mesthene argues that the fundamental characteristic of technology in human society is that it increases the choices open to us. The problem is to reorganize society to take maximum advantage of technological opportunities. Traditional structures and values tend, however, to impede technological and human progress. Thus, "technical fixes" alone are insufficient. We also require "human engineering."

The trick, Mesthene argues, is to do cost-benefit analyses in order to maximize the social benefits of new technologies while at the same time minimizing the costs. For example, the perceived benefits of the chemical Alar in producing beautiful apples must be weighed against the possibility that it causes cancer. Control of technology may clash with the ideals of individual freedom and the free market. But despite the tension between these forces, Mesthene clearly remains, despite his reservations, an advocate of laissez-faire capitalism and what may be called laissez-faire technology. Advancing technology makes possible great advances in the direction of equality among men and more democratic societies.

The final essay in this part, Kwame Gyekye's "Technology and Culture in a

Developing Country," presents a view of science and technology from a non-Western country and culture. Gyekye is concerned with the problems that may hinder the technological advancement of African countries, an advancement he see as a crucial good for the people of Africa. The fundamental problem for the development of technology is the historical lack of science in African cultures. Traditional technologies developed by empirical practical means closely tied to religion and a subjective agent-based view of causation. There was no interest in developing universal scientific principles—for example, to understand herbal medicines; instead, the power of a particular medicine man to use herbs to call forth the healing spirits was all that mattered. The idea that science should be pursued to discover knowledge for its own sake was not an idea present in many if not most African cultures—knowledge was practically oriented and the idea of knowledge for its own sake was unknown.

Although an adequate technological system developed in African culture, the lack of scientific understanding to explain the technology led, Gyekye argues, to a lack of interest in innovation and refinement of technologies. Technology in traditional cultures was stuck at a simple level because there was no science and no ideal of experimentation to be used to improve the methods and procedures. For modern Africa to join the world in developing sophisticated technologies will thus require an intensive effort to introduce the scientific worldview of the Western developed countries into the culture of Africa. But Gyekye claims that this transfer of Western technology and science must take place within the context of African cultural values—thus he argues that technology *appropriation* is a more precise term for what is needed than technology transfer. Western technology can and should be adopted in Africa to improve human life, but it must be adopted without neglecting, discarding, or harming the values of African culture.

1

Can Technology Replace Social Engineering?

Alvin M. Weinberg

During World War II, and immediately afterward, our federal government mobilized its scientific and technical resources, such as the Oak Ridge National Laboratory [ORNL], around great technological problems. Nuclear reactors, nuclear weapons, radar, and space are some of the miraculous new technologies that have been created by this mobilization of federal effort. In the past few years there has been a major change in focus of much of our federal research. Instead of being preoccupied with technology, our government is now mobilizng around problems that are largely social. We are beginning to ask what we can do about world population, about the deterioration of our environment, about our educational system, our decaying cities, race relations, poverty. Recent administrations have dedicated the power of a scientifically oriented federal apparatus to finding solutions for these complex social problems.

Social problems are much more complex than are technological problems. It is much harder to identify a social problem than a technological problem: how do we know when our cities need renewing, or when our population is too big, or when our modes of transportation have broken down? The problems are, in a way, harder to identify just because their solutions are never clear-cut: how do we know when our cities are renewed, or our air clean enough, or our transportation convenient enough? By contrast, the availability of a crisp and beautiful technological *solution* often helps focus on the problem to which the new technology is the solution. I doubt that we would have been nearly as concerned with an eventual shortage of energy as we now are if we had not had a neat solution—nuclear energy—available to eliminate the shortage.

There is a more basic sense in which social problems are much more difficult than are technological problems. A social problem exists because many people

Reprinted by permission of *The University of Chicago Magazine*, Alvin M. Weinberg, "Can Technology Replace Social Engineering?" No. 54 (October 1966): 6–10, 31–39.

behave, individually, in a socially unacceptable way. To solve a social problem one must induce social change—one must persuade many people to behave differently than they have behaved in the past. One must persuade many people to have fewer babies, or to drive more carefully, or to refrain from disliking blacks. By contrast, resolution of a technological problem involves many fewer individual decisions. Once President Roosevelt decided to go after atomic energy, it was by comparison a relatively simple task to mobilize the Manhattan Project.

The resolution of social problems by the traditional methods—by motivating or forcing people to behave more rationally—is a frustrating business. People don't behave rationally; it is a long, hard business to persuade individuals to forgo immediate personal gain or pleasure (as seen by the individual) in favor of longer term social gain. And indeed, the aim of social engineering is to invent the social devices—usually legal, but also moral and educational and organizational—that will change each person's motivation and redirect his activities along ways that are more acceptable to the society.

The technologist is appalled by the difficulties faced by the social engineer; to engineer even a small social change by inducing individuals to behave differently is always hard even when the change is rather neutral or even beneficial. For example, some rice eaters in India are reported to prefer starvation to eating wheat which we send to them. How much harder it is to change motivations where the individual is insecure and feels threatened if he acts differently, as illustrated by the poor white's reluctance to accept the black as an equal. By contrast, technological engineering is simple: the rocket, the reactor, and the desalination plants are devices that are expensive to develop, to be sure, but their feasibility is relatively easy to assess, and their success relatively easy to achieve once one understands the scientific principles that underlie them. It is, therefore, tempting to raise the following question: In view of the simplicity of technological engineering, and the complexity of social engineering, to what extent can social problems be circumvented by reducing them to technological problems? Can we identify Quick Technological Fixes for profound and almost infinitely complicated social problems, "fixes" that are within the grasp of modern technology, and which would either eliminate the original social problem without requiring a change in the individual's social attitudes, or would so alter the problem as to make its resolution more feasible? To paraphrase Ralph Nader, to what extent can technological *remedies* be found for social problems without first having to remove the *causes* of the problem? It is in this sense that I ask, "Can technology replace social engineering?"

THE MAJOR TECHNOLOGICAL FIXES OF THE PAST

To better explain what I have in mind, I shall describe how two of our profoundest social problems—poverty and war—have in some limited degree been solved by the Technological Fix, rather than by the methods of social engineering. Let me begin with poverty.

The traditional Marxian view of poverty regarded our economic ills as being primarily a question of maldistribution of goods. The Marxist recipe for elimination of poverty, therefore, was to eliminate profit, in the erroneous belief that it was the loss of this relatively small increment from the worker's paycheck that kept him poverty-stricken. The Marxist dogma is typical of the approach of the social engineer: one tries to convince or coerce many people to forgo their short-term profits in what is presumed to be the long-term interest of the society as a whole.

The Marxian view seems archaic in this age of mass production and automation, not only to us but apparently to many Eastern bloc economists. For the brilliant advances in the technology of energy, of mass production, and of automation have created the affluent society. Technology has expanded our productive capacity so greatly that even though our distribution is still inefficient, and unfair by Marxian precept, there is more than enough to go around. Technology has provided a "fix"—greatly expanded production of goods—which enables our capitalistic society to achieve many of the aims of the Marxist social engineer without going through the social revolution Marx viewed as inevitable. Technology has converted the seemingly intractable social problem of *widespread* poverty into a relatively tractable one.

My second example is war. The traditional Chinese position views war as primarily a moral issue: if men become good, and model themselves after the Prince of Peace, they will live in peace. This doctrine is so deeply ingrained in the spirit of all civilized men that I suppose it is a blasphemy to point out that it has never worked very well—that men have not been good, and that they are not paragons of virtue or even of reasonableness.

Though I realize it is terribly presumptuous to claim, I believe that Edward Teller may have supplied the nearest thing to a quick Technological Fix to the problem of war. The hydrogen bomb greatly increases the provocation that would precipitate large-scale war—and not because men's motivations have been changed, not because men have become more tolerant and understanding, but rather because the appeal to the primitive instinct of self-preservation has been intensified far beyond anything we could have imagined before the H-bomb was invented. To point out these things today [1966], with the United States involved in a shooting war [in Vietnam], may sound hollow and unconvincing; yet the desperate and partial peace we have now is much better than a full-fledged exchange of thermonuclear weapons. One cannot deny that the Soviet leaders now recognize the force of H-bombs, and that this has surely contributed to the less militant attitude of the USSR. One can only hope that the Chinese leadership, as it acquires familiarity with H-bombs, will also become less militant. If I were to be asked who has given the world a more effective means of achieving peace, our great religious leaders who urge men to love their neighbors and, thus, avoid fights, or our weapons technologists who simply present men with no rational alternative to peace, I would vote for the weapons technologists. That the peace we get is at best terribly fragile, I cannot deny; yet, as I shall explain, I think technology can help stabilize our imperfect and precarious peace.

THE TECHNOLOGICAL FIXES OF THE FUTURE

Are there other Technological Fixes on the horizon, other technologies that can reduce immensely complicated social questions to a matter of "engineering"? Are there new technologies that offer society ways of circumventing social problems and at the same time do *not* require individuals to renounce short-term advantage for long-term gain?

Probably the most important new Technological Fix is the Intra-Uterine Device for birth control. Before the IUD was invented, birth control demanded very strong motivation of countless individuals. Even with the pill, the individual's motivation had to be sustained day in and day out; should it flag even temporarily, the strong motivation of the previous month might go for naught. But the IUD, being a one-shot method, greatly reduces the individual motivation required to induce a social change. To be sure, the mother must be sufficiently motivated to accept the IUD in the first place, but, as experience in India already seems to show, it is much easier to persuade the Indian mother to accept the IUD once, than it is to persuade her to take a pill every day. The IUD does not completely replace social engineering by technology; and indeed, in some Spanish American cultures where the husband's manliness is measured by the number of children he has, the IUD attacks only part of the problem. Yet, in many other situations, as in India, the IUD so reduces the social component of the problem as to make an impossibly difficult social problem much less hopeless.

Let me turn now to problems which from the beginning have had both technical and social components—broadly, those concerned with conservation of our resources: our environment, our water, and our raw materials for production of the means of subsistence. The social issue here arises because many people by their individual acts cause shortages and, thus, create economic, and ultimately social, imbalance. For example, people use water wastefully, or they insist on moving to California because of its climate, and so we have water shortages; or too many people drive cars in Los Angeles with its curious meteorology, and so Los Angeles suffocates from smog.

The water resources problem is a particularly good example of a complicated problem with strong social and technological connotations. Our management of water resources in the past has been based largely on the ancient Roman device, the aqueduct: every water shortage was to be relieved by stealing water from someone else who at the moment didn't need the water or was too poor or too weak to prevent the steal. Southern California would steal from northern California, New York City from upstate New York, the farmer who could afford a cloud-seeder from the farmer who could not afford a cloud-seeder. The social engineer insists that such shortsighted expedients have got us into serious trouble; we have no water resources policy, we waste water disgracefully, and, perhaps, in denying the ethic of thriftiness in using water, we have generally undermined our moral fiber. The social engineer, therefore, views such technological shenanigans as being shortsighted, if not downright immoral. Instead, he says, we should persuade or force people to use less water, or to stay in the cold Middle West where water is plentiful instead of migrating to California where water is scarce.

The water technologist, on the other hand, views the social engineer's approach as rather impractical. To persuade people to use less water, to get along with expensive water, is difficult, time-consuming, and uncertain in the extreme. Moreover, say the technologists, what right does the water resources expert have to insist that people use water less wastefully? Green lawns and clean cars and swimming pools are part of the good life, American style, . . . and what right do we have to deny this luxury if there is some alternative to cutting down the water we use?

Here we have a sharp confrontation of the two ways of dealing with a complex social issue: the social engineering way which asks people to behave more "reasonably," the technologists' way which tries to avoid changing people's habits or motivation. Even though I am a technologist, I have sympathy for the social engineer. I think we must use our water as efficiently as possible, that we ought to improve people's attitudes toward the use of water, and that everything that can be done to rationalize our water policy will be welcome. Yet as a technologist, I believe I see ways of providing more water more cheaply than the social engineers may concede is possible.

I refer to the possibility of nuclear desalination. The social engineer dismisses the technologist's simpleminded idea of solving a water shortage by transporting more water primarily because, in so doing, the water user steals water from someone else—possibly foreclosing the possibility of ultimately uitilizing land now only sparsely settled. But surely water drawn from the sea deprives no one of his share of water. The whole issue is then a technological one; can fresh water be drawn from the sea cheaply enough to have a major impact on our chronically water-short areas like southern California, Arizona, and the Eastern seaboard?

I believe the answer is yes, though much hard technical work remains to be done. A large program to develop cheap methods of nuclear desalting has been undertaken by the United States, and I have little doubt that within the next ten to twenty years we shall see huge dual-purpose desalting plants springing up on many parched seacoasts of the world. At first these plants will produce water at municipal prices. But I believe, on the basis of research now in progress at ORNL and elsewhere, water from the sea at a cost acceptable for agriculture—less than ten cents per 1,000 gallons—is eventually in the cards. In short, for areas close to the seacoasts, technology can provide water without requiring a great and diffi-cult-to-accomplish change in people's attitudes toward the utilization of water.*

The Technological Fix for water is based on the availability of extremely cheap energy from very large nuclear reactors. What other social consequences can one foresee flowing from really cheap energy eventually available to every country regardless of its endowment of conventional resources? Though we now see only vaguely the outlines of the possibilities, it does seem likely that from very cheap nuclear energy we shall get hydrogen by electrolysis of water, and, thence, the all-important ammonia fertilizer necessary to help feed the hungry of the world; we shall reduce metals without requiring coking coal; we shall even power automobiles with electricity, via fuel cells or storage batteries, thus reducing our world's dependence on crude oil, as well as eliminating our air pol-lution insofar as it is caused by automobile exhaust or by the burning of fossil fuels. In short, the widespread availability of very cheap energy everywhere in the world ought to lead to an energy autarky in every country of the world; and eventually to an autarky in the many staples of life that should flow from really cheap energy.

WILL TECHNOLOGY REPLACE
SOCIAL ENGINEERING?

I hope these examples suggest how social problems can be circumvented or at least reduced to less formidable proportions by the application of the Technolog-ical Fix. The examples I have given do not strike me as being fanciful, nor are they at all exhaustive. I have not touched, for example, upon the extent to which really cheap computers and improved technology of communication can help improve elementary teaching without having first to improve our elementary teachers. Nor have I mentioned Ralph Nader's brilliant observation that a safer car, and even its development and adoption by the auto company, is a quicker and probably surer way to reduce traffic deaths than is a campaign to teach people to drive more carefully. Nor have I invoked some really fanciful Technological Fixes: like providing air conditioners and free electricity to operate them for every black family in Watts on the assumption (suggested by Huntington) that race rioting is correlated with hot, humid weather, or the ultimate Technological Fix, Aldous Huxley's soma pills that eliminate human unhappiness without improving human relations in the usual sense.

My examples illustrate both the strength and the weakness of the Technolog-ical Fix for social problems. The Technological Fix accepts man's intrinsic short-comings and circumvents them or capitalizes on them for socially useful ends.

---

*That this has not been realized should, perhaps, temper our enthusiasm for technological fixes. (Ed.).

The Fix is, therefore, eminently practical and, in the short term, relatively effective. One does not wait around trying to change people's minds: if people want more water, one gets them more water rather than requiring them to reduce their use of water; if people insist on driving autos while they are drunk, one provides safer autos that prevent injuries even after a severe accident.

But the technological solutions to social problems tend to be incomplete and metastable, to replace one social problem with another. Perhaps the best example of this instability is the peace imposed upon us by the H-bomb. Evidently the *pax hydrogenica* is metastable in two senses: in the short term, because the aggressor still enjoys such an advantage; in the long term, because the discrepancy between have and have-not nations must eventually be resolved if we are to have permanent peace. Yet, for these particular shortcomings, technology has something to offer. To the imbalance between offense and defense, technology says let us devise passive defense which redresses the balance. A world with H-bombs and adequate civil defense is less likely to lapse into thermonuclear war than a world with H-bombs alone, at least if one concedes that the danger of the thermonuclear war mainly lies in the acts of irresponsible leaders. Anything that deters the irresponsible leader is a force for peace: a technologically sound civil defense therefore would help stabilize the balance of terror.

To the discrepancy between haves and have-nots, technology offers the nuclear energy revolution, with its possibility of autarky for haves and have-nots alike. How this might work to stabilize our metastable thermonuclear peace is suggested by the possible political effect of the recently proposed Israeli desalting plant. The Arab states I should think would be much less set upon destroying the Jordan River Project if the Israelis had a desalination plant in reserve that would nullify the effect of such action. In this connection, I think countries like ours can contribute very much. Our country will soon have to decide whether to continue to spend 5.5×10^9* per year for space exploration after our lunar landing. Is it too outrageous to suggest that some of this money be devoted to building huge nuclear desalting complexes in the arid ocean rims of the troubled world? If the plants are empowered with breeder reactors, the out-of-pocket costs, once the plants are built, should be low enough to make large-scale agriculture feasible in these areas. I estimate that for 4×10^9† we could build enough desalting capacity to feed more than ten million new mouths per year (provided we use agricultural methods that husband water), and we would, thereby, help stabilize the metastable, bomb-imposed balance of terror.

Yet, I am afraid we technologists shall not satisfy our social engineers, who tell us that our Technological Fixes do not get to the heart of the problem; they are at best temporary expedients; they create new problems as they solve old ones; to put a Technological Fix into effect requires a positive social action.

---

*$5.5 billion, equivalent to approximately $17 billion in today's dollars.
†$4 billion, equivalent to approximately $12 billion in today's dollars.

Eventually, social engineering, like the Supreme Court decision on desegregation, must be invoked to solve social problems. And, of course, our social engineers are right. Technology will never *replace* social engineering. But technology has provided and will continue to provide to the social engineer broader options, to make intractable social problems less intractable; perhaps, most of all, technology will buy time—that precious commodity that converts violent social revolution into acceptable social evolution.

Our country now recognizes and is mobilizing around the great social problems that corrupt and disfigure our human existence. It is natural that in this mobilization we should look first to the social engineer. But, unfortunately, the apparatus most readily available to the government, like the great federal laboratories, is technologically oriented, not socially oriented. I believe we have a great opportunity here; for, as I hope I have persuaded [my readers], many of our seemingly social problems do admit of partial technological solutions. Our already deployed technological apparatus can contribute to the resolution of social questions. I plead, therefore, first for our government to deploy its laboratories, its hardware contractors, and its engineering universities around social problems. And I plead, secondly, for understanding and cooperation between technologist and social engineer. Even with all the help he can get from the technologist, the social engineer's problems are never really solved. It is only by cooperation between technologist and social engineer that we can hope to achieve what is the aim of all technologists and social engineers—a better society, and thereby, a better life, for all of us who are part of society.

8

The Role of Technology in Society

Emmanuel G. Mesthene

SOCIAL CHANGE

Three Unhelpful Views about Technology

While a good deal of research is aimed at discerning the particular effects of technological change on industry, government, or education, systematic inquiry devoted to seeing these effects together and to assessing their implications for contemporary society as a whole is relatively recent and does not enjoy the strong methodology and richness of theory and data that mark more established fields of scholarship. It therefore often has to contend with facile or one-dimensional views about what technology means for society. Three such views, which are prevalent at the present time, may be mildly caricatured somewhat as follows.

The first holds that technology is an unalloyed blessing for man and society. Technology is seen as the motor of all progress, as holding the solution to most of our social problems, as helping to liberate the individual from the clutches of a complex and highly organized society, and as the source of permanent prosperity; in short, as the premise of utopia in our time. This view has its modern origins in the social philosophies of such nineteenth-century thinkers as Saint-Simon, Karl Marx, and Auguste Comte. It tends to be held by many scientists and engineers, by many military leaders and aerospace industrialists, by people who believe that man is fully in command of his tools and his destiny, and by many of the devotees of modern techniques of "scientific management."

A second view holds that technology is an unmitigated curse. Technology is said to rob people of their jobs, their privacy, their participation in democratic

Emmanuel G. Mesthene, "The Role of Technology in Society," *Technology and Culture* 10, no. 4 (1969): 489–513. © Society for the History of Technology. Reprinted with permission of the Johns Hopkins University Press.

government, and even, in the end, their dignity as human beings. It is seen as autonomous and uncontrollable, as fostering materialistic values and as destructive of religion, as bringing about a technocratic society and bureaucratic state in which the individual is increasingly submerged, and as threatening, ultimately, to poison nature and blow up the world. This view is akin to historical "back-to-nature" attitudes toward the world and is propounded mainly by artists, literary commentators, popular social critics, and existentialist philosophers. It is becoming increasingly attractive to many of our youth, and it tends to be held, understandably enough, by segments of the population that have suffered dislocation as a result of technological change.

The third view is of a different sort. It argues that technology as such is not worthy of special notice, because it has been well recognized as a factor in social change at least since the Industrial Revolution, because it is unlikely that the social effects of computers will be nearly so traumatic as the introduction of the factory system in eighteenth-century England, because research has shown that technology has done little to accelerate the rate of economic productivity since the 1800s, because there has been no significant change in recent decades in the time period between invention and widespread adoption of new technology, and because improved communications and higher levels of education make people much more adaptable than heretofore to new ideas and to new social reforms required by technology.

While this view is supported by a good deal of empirical evidence, however, it tends to ignore a number of social, cultural, psychological, and political effects of technological change that are less easy to identify with precision. It thus reflects the difficulty of coming to grips with a new or broadened subject matter by means of concepts and intellectual categories designed to deal with older and different subject matters. This view tends to be held by historians, for whom continuity is an indispensable methodological assumption, and by many economists, who find that their instruments measure some things quite well while those of the other social sciences do not yet measure much of anything.

Stripped of caricature, each of these views contains a measure of truth and reflects a real aspect of the relationship of technology and society. Yet they are oversimplifications that do not contribute much to understanding. One can find empirical evidence to support each of them without gaining much knowledge about the actual mechanism by which technology leads to social change or significant insight into its implications for the future. All three remain too uncritical or too partial to guide inquiry. Research and analysis lead to more differentiated conclusions and reveal more subtle relationships.

Some Countervailing Considerations

Two of the projects of the Harvard University Program on Technology and Society serve, respectively, to temper some exaggerated claims made for tech-

nology and to replace gloom with balanced judgment. Professor Anthony G. Oettinger's study of information technology in education* has shown that, in the
schools at least, technology is not likely to bring salvation with it quite so soon
as the U.S. Office of Education, leaders of the education industry, and enthusiastic computermen and systems analysts might wish. Neither educational technology nor the school establishment seems ready to consummate the revolution
in learning that will bring individualized instruction to every child, systematic
planning and uniform standards across 25,000 separate school districts, an
answer to bad teachers and unmovable bureaucracies, and implementation of a
national policy to educate every American to his full potential for a useful and
satisfying life. Human fallibility and political reality are still here to keep utopia
at bay, and neither promises soon to yield to a quick technological fix. Major
institutional change that can encourage experimentation, flexibility, variety, and
competition among educational institutions seems called for before the new technology can contribute significantly to education. Application of the technology
itself, moreover, poses problems of scale-up, reliability, and economics that have
scarcely been faced as yet.

By contrast, Professor Manfred Stanley's study of the value presuppositions
that underlie the pessimistic arguments about technology suggests that predictions of inevitable doom are premature and that a number of different social outcomes are potential in the process of technological change. In other words, the
range of possibility and of human choice implicit in technology is much greater
than most critics assume. The problem—here, as well as in the application of educational technology—is how to organize society to free the possibility of choice.

Finally, whether modern technology and its effects constitute a subject
matter deserving of special attention is largely a matter of how technology is
defined. The research studies of the Harvard Program on Technology and Society
reflect an operating assumption that the meaning of technology includes more
than machines. As most serious investigators have found, understanding is not
advanced by concentrating single-mindedly on such narrowly drawn yet imprecise questions as "What are the social implications of computers, or lasers, or
space technology?" Society and the influences of technology upon it are much
too complex for such artificially limited approaches to be meaningful. The opposite error, made by some, is to define technology too broadly by identifying it
with rationality in the broadest sense. The term is then operationally meaningless
and unable to support fruitful inquiry.

We have found it more useful to define technology as tools in a general
sense, including machines, but also including linguistic and intellectual tools and
contemporary analytic and mathematical techniques. That is, we define technology as the organization of knowledge for practical purposes. It is in this

---

*Unless otherwise noted, studies such as Oettinger's ,which are referred to in this [chapter], are
described in the Fourth Annual Report (1967–68) of the Harvard University Program on Technology
and Society.

broader meaning that we can best see the extent and variety of the effects of technology on our institutions and values. Its pervasive influence on our very culture
would be unintelligible if technology were understood as no more than hardware.

It is in the pervasive influence of technology that our contemporary situation
seems qualitatively different from that of past societies, for three reasons. (1) Our
tools are more powerful than any before. The rifle wiped out the buffalo, but
nuclear weapons can wipe out man. Dust storms lay whole regions waste, but too
much radioactivity in the atmosphere could make the planet uninhabitable. The
domestication of animals and the invention of the wheel literally lifted the burden
from man's back, but computers could free him from all need to labor. (2) This
quality of finality of modern technology has brought our society, more than any
before, to explicit awareness of technology as an important determinant of our
lives and institutions. (3) As a result, our society is coming to a deliberate decision to understand and control technology to good social purpose and is therefore
devoting significant effort to the search for ways to measure the full range of its
effects rather than only those bearing principally on the economy. It is this prominence of technology in many dimensions of modern life that seems novel in our
time and deserving of explicit attention.

How Technological Change Impinges on Society

It is clearly possible to sketch a more adequate hypothesis about the interaction
of technology and society than the partial views outlined above. Technological
change would appear to induce or "motor" social change in two principal ways.
New technology creates new opportunities for men and societies, and it also generates new problems for them. It has both positive and negative effects, and it
usually has the two *at the same time and in virtue of each other*. Thus, industrial
technology strengthens the economy, as our measures of growth and productivity
show. As Dr. Anne P. Carter's study on structural changes in the American
economy has helped to demonstrate, however, it also induces changes in the relative importance of individual supplying sectors in the economy as new techniques of production alter the amounts and kinds of materials, parts and components, energy, and service inputs used by each industry to produce its output. It
thus tends to bring about dislocations of businesses and people as a result of
changes in industrial patterns and in the structure of occupations.

The close relationship between technological and social change itself helps
to explain why any given technological development is likely to have both positive and negative effects. The usual sequence is that (1) technological advance
creates a new opportunity to achieve some desired goal; (2) this requires (except
in trivial cases) alterations in social organization if advantage is to be taken of the
new opportunity, (3) which means that the functions of existing social structures
will be interfered with, (4) with the result that other goals which were served by
the older structures are now only inadequately achieved.

As the Meyer-Kain study has shown, for example, improved transportation technology and increased ownership of private automobiles have increased the mobility of businesses and individuals. This has led to altered patterns of industrial and residential location, so that older unified cities are being increasingly transformed into larger metropolitan complexes. The new opportunities for mobility are largely denied to the poor and black populations of the core cities, however, partly for economic reasons, and partly as a result of restrictions on choice of residence by blacks, thus leading to persistent black unemployment despite a generally high level of economic activity. Cities are thus increasingly unable to perform their traditional functions of providing employment opportunities for all segments of their populations and an integrated social environment that can temper ethnic and racial differences. The new urban complexes are neither fully viable economic units nor effective political organizations able to upgrade and integrate their core populations into new economic and social structures. The resulting instability is further aggravated by modern mass communications technology, which heightens the expectations of the poor and the fears of the well-to-do and adds frustration and bitterness to the urban crisis.

An almost classic example of the sequence in which technology impinges on society is provided by Professor Mark Field's study of changes in the system and practice of medical care. Recent advances in biomedical science and technology have created two new opportunities: (1) they have made possible treatment and cures that were never possible before, and (2) they provide a necessary condition for the delivery of adequate medical care to the population at large as a matter of right rather than privilege. In realization of the first possibility, the medical profession has become increasingly differentiated and specialized and is tending to concentrate its best efforts in a few major, urban centers of medical excellence. This alters the older social organization of medicine that was built around the general practitioner. The second possibility has led to big increases in demand for medical services, partly because a healthy population has important economic advantages in a highly industrialized society. This increased demand accelerates the process of differentiation and multiplies the levels of paramedical personnel between the physician at the top and the patient at the bottom of the hospital pyramid.

Both of these changes in the medical system are responsive to the new opportunities for technical excellence that have been created by biomedical technology. Both also involve a number of well-known costs in terms of some older desiderata of medical care. The increasing scarcity of the general practitioner in many sections of the country means that people in need often have neither easy access to professional care nor the advantage of a "medical general manager" who can direct them to the right care at the right place at the right time, which can result both in poor treatment and a waste of medical resources. Also, too exclusive a concentration on technical excellence can lead to neglect of the patient's psychological well-being, and even the possibility of technical error increases as the "medical assembly line" gets longer.

The pattern illustrated by the preceding examples tends to be the general one. Our most spectacular technological successes in America in the last quarter of a century have been in national defense and in space exploration. They have brought with them, however, enthusiastic advocates and vested interests who claim that the development of sophisticated technology is an intrinsic good that should be pursued for its own sake. They thus contribute to the self-reinforcing quality of technological advance and raise fears of an autonomous technology uncontrollable by man. Mass communications technology has also made rapid strides since World War II, with great benefit to education, journalism, commerce, and sheer convenience. It has also been accompanied by an aggravation of social unrest, however, and may help to explain the singular rebelliousness of a youth who can find out what the world is like from television before home and school have had the time to instill some ethical sense of what it could or should be like.

In all such cases, technology creates a new opportunity and a new problem at the same time. That is why isolating the opportunity or the problem and construing it as the whole answer is ultimately obstructive of rather than helpful to understanding.

How Society Reacts to Technological Change

The heightened prominence of technology in our society makes the interrelated tasks of profiting from its opportunities and containing its dangers a major intellectual and political challenge of our time.

Failure of society to respond to the opportunities created by new technology means that much actual or potential technology lies fallow, that is, is not used at all or is not used to its full capacity. This can mean that potentially solvable problems are left unsolved and potentially achievable goals unachieved, because we waste our technological resources or use them inefficiently. A society has at least as much stake in the efficient utilization of technology as in that of its natural or human resources.

There are often good reasons, of course, for not developing or utilizing a particular technology. The mere fact that it can be developed is not sufficient reason for doing so. The costs of development may be too high in the light of the expected benefits, as in the case of the project to develop a nuclear-powered aircraft. Or, a new technological device may be so dangerous in itself or so inimical to other purposes that it is never developed, as in the cases of Herman Kahn's "Doomsday Machine" and the recent proposal to "nightlight" Vietnam by reflected sunlight.

But there are also cases where technology lies fallow because existing social structures are inadequate to exploit the opportunities it offers. This is revealed clearly in the examination of institutional failure in the ghetto by Professor Richard S. Rosenbloom and his colleagues. At point after point, their analyses confirm what has been long suspected, that is, that existing institutions and traditional approaches are by and large incapable of coming to grips with the new problems of our cities—many of them caused by technological change, as the Meyer-Kain study has reminded us—and unable to realize the possibilities for resolving them that are also inherent in technology. Vested economic and political interests serve to obstruct adequate provision of low-cost housing. Community institutions wither for want of interest and participation by residents. City agencies are unable to marshal the skills and take the systematic approach needed to deal with new and intensified problems of education, crime control, and public welfare. Business corporations, finally, which are organized around the expectation of private profit, are insufficiently motivated to bring new technology and management know-how to bear on urban projects where the benefits will be largely social. All these factors combine to dilute what may otherwise be a genuine desire to apply our best knowledge and adequate resources to the resolution of urban tensions and the eradication of poverty in the nation.

There is also institutional failure of another sort. Government in general and agencies of public information in particular are not yet equipped for the massive task of public education that is needed if our society is to make full use of its technological potential, although the federal government has been making significant strides in this direction in recent years. Thus, much potentially valuable technology goes unused because the public at large is insufficiently informed about the possibilities and their costs to provide support for appropriate political action. As noted, we have done very well with our technology in the face of what were or were believed to be crisis situations, as with our military technology in World War II and with our space efforts when beating the Russians to the moon was deemed a national goal of first priority. We have also done very well when the potential benefits of technology were close to home or easy to see, as in improved health care and better and more varied consumer goods and services. We have done much less well in developing and applying technology where the need or opportunity has seemed neither so clearly critical nor so clearly personal as to motivate political action, as in the instance of urban policy already cited. Technological possibility continues to lie fallow in those areas where institutional and political innovation is a precondition of realizing it.

Containing the Negative Effects of Technology

The kinds and magnitude of the negative effects of technology are no more independent of the institutional structures and cultural attitudes of society than is realization of the new opportunities that technology offers. In our society, there are individuals or individual firms always on the lookout for new technological opportunities, and large corporations hire scientists and engineers to invent such opportunities. In deciding whether to develop a new technology, individual entrepreneurs engage in calculations of expected benefits and expected costs to themselves, and proceed if the former are likely to exceed the latter. Their calculations do not take adequate account of the probable benefits and costs of the new developments to others than themselves or to society generally. These latter are what economists call external benefits and costs.

The external-benefits potential in new technology will thus not be realized by the individual developer and will rather accrue to society as a result of deliberate social action, as has been argued above. Similarly with the external costs. In minimizing only expected costs to himself, the individual decision maker helps to contain only some of the potentially negative effects of the new technology. The external costs and therefore the negative effects on society at large are not of principal concern to him and, in our society, are not expected to be.

Most of the consequences of technology that are causing concern at the present time—pollution of the environment, potential damage to the ecology of the planet, occupational and social dislocations, threats to the privacy and political significance of the individual, social and psychological malaise—are negative externalities of this kind. They are with us in large measure because it has not been anybody's explicit business to foresee and anticipate them. They have fallen between the stools of innumerable individual decisions to develop individual technologies for individual purposes without explicit attention to what all these decisions add up to for society as a whole and for people as human beings. This freedom of individual decision making is a value that we have cherished and that is built into the institutional fabric of our society. The negative effects of technology that we deplore are a measure of what this traditional freedom is beginning to cost us. They are traceable, less to some mystical autonomy presumed to lie in technology, and much more to the autonomy that our economic and political institutions grant to individual decision making.

When the social costs of individual decision making in the economic realm achieved crisis proportions in the great depression of the 1930s, the federal government introduced economic policies and measures many of which had the effect of abridging the freedom of individual decision. Now that some of the negative impacts of technology are threatening to become critical, the government is considering measures of control that will have the analogous effect of constraining the freedom of individual decision makers to develop and apply new technologies irrespective of social consequence. Congress is actively seeking to establish technology-assessment boards of one sort or another which it hopes may be able to foresee potentially damaging effects of technology on nature and man. In the executive branch, attention is being directed (1) to development of a system of social indicators to help gauge the social effects of technology, (2) to establishment of some body of social advisers to the president to help develop policies in anticipation of such effects, and generally (3) to strengthening the role of the social sciences in policy making.

Measures to control and mitigate the negative effects of technology, however, often appear to threaten freedoms that our traditions still take for granted as inalienable rights of men and good societies, however much they may have been tempered in practice by the social pressures of modern times: the freedom of the market, the freedom of private enterprise, the freedom of the scientist to follow truth wherever it may lead, and the freedom of the individual to pursue his fortune and decide his fate. There is thus set up a tension between the need to control technology and our wish to preserve our values, which leads some people to conclude that technology is inherently inimical to human values. The political effect of this tension takes the form of inability to adjust our decision-making structures to the realities of technology so as to take maximum advantage of the opportunities it offers and so that we can act to contain its potential ill effects before they become so pervasive and urgent as to seem uncontrollable.

To understand why such tensions are so prominent a social consequence of technological change, it becomes necessary to look explicitly at the effects of technology on social and individual values.

VALUES

Technology's Challenge to Values

Despite the practical importance of the techniques, institutions, and processes of knowledge in contemporary society, political decision making and the resolution of social problems are clearly not dependent on knowledge alone. Numerous commentators have noted that ours is a "knowledge" society, devoted to rational decision making and an "end of ideology," but none would deny the role that values play in shaping the course of society and the decisions of individuals. On the contrary, questions of values become more pointed and insistent in a society that organizes itself to control technology and that engages in deliberate social planning. Planning demands explicit recognition of value hierarchies and often brings into the open value conflicts which remain hidden in the more impersonal working of the market.

In economic planning, for example, we have to make choices between the values of leisure and increased productivity, without a common measure to help us choose. In planning education, we come face to face with the traditional American value dilemma of equality versus achievement: do we opt for equality and nondiscrimination and give all students the same basic education, or do we foster achievement by tailoring education to the capacity for learning, which is itself often conditioned by socioeconomic background?

The new science-based decision-making techniques also call for clarity: in the specification of goals, thus serving to make value preferences explicit. The effectiveness of systems analysis, for example, depends on having explicitly stated objectives and criteria of evaluation to begin with, and the criteria and objectives of specific actions invariably relate to the society's system of values.

That, incidentally, is why the application of systems analysis meets with less relative success in educational or urban planning than in military planning: the value conflicts are fewer in the latter and the objectives and criteria easier to specify and agree on. This increased awareness of conflicts among our values contributes to a general questioning attitude toward traditional values that appears to be endemic to a hightechnology, knowledge-based society: [according to Robin Williams] "A society in which the store of knowledge concerning the consequences of action is large and is rapidly increasing is a society in which received norms and their 'justifying' values will be increasingly subjected to questioning and reformulation."[1]

This is another way of pointing to the tension alluded to earlier, between the need for social action based on knowledge on the one hand, and the pull of our traditional values on the other. The increased questioning and reformulation of values that Williams speaks of, coupled with a growing awareness that our values are in fact changing under the impact of technological change, leads many people to believe that technology is by nature destructive of values. But this belief presupposes a conception of values as eternal and unchanging and therefore tends to confuse the valuable with the stable. The fact that values come into question as our knowledge increases and that some traditional values cease to function adequately when technology leads to changes in social conditions does not mean that values per se are being destroyed by knowledge and technology.

What does happen is that values change through a process of accommodation between the system of existing values and the technological and social changes that impinge on it. The projects of the Harvard Program in the area of technology and values are devoted to discovering the specific ways in which this process of accommodation occurs and to tracing its consequences for value changes in contemporary American society. The balance of this section is devoted to a more extended discussion of the first results of these projects.

Technology as a Cause of Value Change

Technology has a direct impact on values by virtue of its capacity for creating new opportunities. By making possible what was not possible before, it offers individuals and society new options to choose from. For example, space technology makes it possible for the first time to go to the moon or to communicate by satellite and thereby adds those two new options to the spectrum of choices available to society. By adding new options in this way, technology can lead to changes in values in the same way that the appearance of new dishes on the heretofore standard menu of one's favorite restaurant can lead to changes in one's tastes and choices of food. Specifically, technology can lead to value change either (1) by bringing some previously unattainable goal within the realm of choice or (2) by making some values easier to implement than heretofore, that is, by changing the costs associated with realizing them.

Dr. Irene Taviss is exploring the ways in which technological change affects intrinsic sources of tension and potential change in value systems. When technology facilitates implementation of some social ideal and society fails to act upon this new possibility, the conflict between principle and practice is sharpened, thus leading to new tensions. For example, the economic affluence that technology has helped to bring to American society makes possible fuller implementation than heretofore of our traditional values of social and economic equality. Until it is acted upon, that possibility gives rise to the tensions we associate with the rising expectations of the underprivileged and provokes both the activist response of the radical left and the hippie's rejection of society as "hypocritical."

Another example related to the effect of technological change on values is implicit in our concept of democracy. The ideal we associate with the old New England town meeting is that each citizen should have a direct voice in political decisions. Since this has not been possible, we have elected representatives to serve our interests and vote our opinions. Sophisticated computer technology, however, now makes possible rapid and efficient collection and analysis of voter opinion and could eventually provide for "instant voting" by the whole electorate on any issue presented to it via television a few hours before. It thus raises the possibility of instituting a system of direct democracy and gives rise to tensions between those who would be violently opposed to such a prospect and those who are already advocating some system of participatory democracy.

This new technological possibility challenges us to clarify what we mean by democracy. Do we construe it as the will of an undifferentiated majority, as the resultant of transient coalitions of different interest groups representing different value commitments, as the considered judgment of the people's elected representatives, or as by and large the kind of government we actually have in the United States, minus the flaws in it that we would like to correct? By bringing us face to face with such questions, technology has the effect of calling society's bluff and thereby preparing the ground for changes in its values.

In the case where technological change alters the relative costs of implementing different values, it impinges on inherent contradictions in our value system. To pursue the same example, modern technology can enhance the values we associate with democracy. But it can also enhance another American value—that of "secular rationality," as sociologists call it—by facilitating the use of scientific and technical expertise in the process of political decision making. This can in turn further reduce citizen participation in the democratic process. Technology thus has the effect of facing us with contradictions in our own value system and of calling for deliberate attention to their resolution.

The Value Implications of Economic Change

In addition to the relatively direct effects of technology on values, as illustrated above, value change often comes about through the intermediation of some more general social change produced by technology, as in the tension imposed on our individualistic values by the external benefits and costs of technological development that was alluded to in the earlier discussion of the negative effects of technology. Professor Nathan Rosenberg is exploring the closely allied relationship between such values and the need for society to provide what economists call public goods and services.

As a number of economists have shown, such public goods differ from private consumer goods and services in that they are provided on an all-or-none basis and consumed in a joint way, so that more for one consumer does not mean less for another. The clearing of a swamp or a flood-control project, once completed, benefits everyone in the vicinity. A meteorological forecast, once made, can be transmitted by word of mouth to additional users at no additional cost. Knowledge itself may thus be thought of as the public good par excellence, since the research expenses needed to produce it are incurred only once, unlike consumer goods of which every additional unit adds to the cost of production.

As noted earlier, private profit expectation is an inadequate incentive for the production of such public goods, because their benefit is indiscriminate and not fully appropriate to the firm or individual that might incur the cost of producing them. Individuals are therefore motivated to dissimulate by understating their true preferences for such goods in the hope of shifting their cost to others. This creates a "free-loader" problem, which skews the mechanism of the market. The market therefore provides no effective indication of the optimal amount of such public commodities from the point of view of society as a whole. If society got only as much public health care, flood control, or knowledge as individual profit calculations would generate, it would no doubt get less of all of them than it does now or than it expresses a desire for by collective political action.

This gap between collective preference and individual motivation imposes strains on a value system, such as ours, which is primarily individualistic rather than collective or "societal" in its orientation. That system arose out of a simpler, more rustic, and less affluent time, when both benefits and costs were of a much more private sort than now. It is no longer fully adequate for our society, which industrial technology has made productive enough to allocate significant resources to the purchase of public goods and services, and in which modern transportation and communications as well as the absolute magnitude of technological effects lead to extensive ramifications of individual actions on other people and on the environment.

The response to this changed experience on the part of the public at large generally takes the form of increased government intervention in social and economic affairs to contain or guide these wider ramifications, as noted previously. The result is that the influence of values associated with the free reign of individual enterprise and action tends to be counteracted, thus facilitating a change in values. To be sure, the tradition that ties freedom and liberty to a laissez-faire system of decision making remains very strong, and the changes in social structures and cultural attitudes that can touch it at its foundations are still only on the horizon.

Religion and Values

Much of the unease that our society's emphasis on technology seems to generate among various sectors of society can perhaps be explained in terms of the impact that technology has on religion. The formulations and institutions of religion are not immune to the influences to technological change, for they, too, tend toward an accommodation to changes in the social milieu in which they function. But one way in which religion functions is as an ultimate belief system that provides legitimation, that is, a "meaning" orientation, to moral and social values. This ultimate meaning orientation, according to Professor Harvey Cox, is even more basic to human existence than the value orientation. When the magnitude or rapidity of social change threatens the credibility of that belief system, therefore, and when the changes are moreover seen as largely the results of technological change, the meanings of human existence that we hold most sacred seem to totter and technology emerges as the villain.

Religious change thus provides another mediating mechanism through which technology affects our values. That conditions are ripe for religious change at the present time has been noted by many observers, who are increasingly questioning whether our established religious syntheses and symbol systems are adequate any longer to the religious needs of a scientific and secular society that is changing so fundamentally as to strain traditional notions of eternity. If they are not, how are they likely to change? Professor Cox is addressing himself to this problem with specific attention to the influence of technology in guiding the direction of change.

He notes that religion needs to come to terms with the pluralism of belief systems that is characteristic of the modern world. The generation of knowledge and the use of technology are so much a part of the style and self-image of our own society that men begin to experience themselves, their power, and their relationships to nature and history in terms of open possibility, hope, action, and self-confidence. The symbolism of such traditional religious postures as subservience, fatefulness, destiny, and suprarational faith begin then to seem irrelevant to our actual experience. They lose credibility, and their religious function is weakened. Secular belief systems arise to compete for the allegiance of men: political belief systems, such as communism; or scientific ones, such as modern-day humanism; or such inexplicit, noninstitutionalized belief complexes as are characteristic of agnosticism.

This pluralism poses serious problems for the ultimate legitimation or "meaning" orientation for moral and social values that religion seeks to provide, because it demands a religious synthesis that can integrate the fact of variant perspectives into its own symbol system. Western religions have been notoriously incapable of performing this integrating function and have rather gone the route of schism and condemnation of variance as heresy. The institutions and formulations of historical Christianity in particular, which once served as the foundations of

Western society, carry the added burden of centuries of conflict with scientific world views as these have competed for ascendancy in the same society. This makes it especially difficult for traditional Christianity to accommodate to a living experience so infused by scientific knowledge and attitude as ours and helps explain why its adequacy is coming under serious question at the present time.

Cox notes three major traditions in the Judeo-Christian synthesis and finds them inconsistent in their perceptions of the future: an "apocalyptic" tradition foresees imminent catastrophe and induces a negative evaluation of this world; a "teleological" tradition sees the future as the certain unfolding of a fixed purpose inherent in the universe itself; a "prophetic" tradition, finally, sees the future as an open field of human hope and responsibility and as becoming what man will make of it.[2]

Technology, as noted, creates new possibilities for human choice and action but leaves their disposition uncertain. What its effects will be and what ends it will serve are not inherent in the technology, but depend on what man will do with technology. Technology thus makes possible a future of open-ended options that seems to accord well with the presuppositions of the prophetic tradition. It is in that tradition above others, then, that we may seek the beginnings of a religious synthesis that is both adequate to our time and continuous with what is most relevant in our religious history. But this requires an effort at deliberate religious innovation for which Cox finds insufficient theological ground at the present time. Although it is recognized that religions have changed and developed in the past, conscious innovation in religion has been condemned and is not provided for by the relevant theologies. The main task that technological change poses for theology in the next decades, therefore, is that of deliberate religious innovation and symbol reformulation to take specific account of religious needs in a technological age.

What consequences would such changes in religion have for values? Cox approaches this question in the context of the familiar complaint that, since technology is principally a means, it enhances merely instrumental values at the expense of expressive, consummatory, or somehow more "real" values. The appropriate distinction, however, is not between technological instrumental values and nontechnological expressive values, but among the expressive values that attach to different technologies. The horse-and-buggy was a technology too, after all, and it is not prima facie clear that its charms were different in kind or superior to the sense of power and adventure and the spectacular views that go with jet travel.

Further, technological advance in many instances is a condition for the emergence of new creative or consummatory values. Improved sound boxes in the past and structural steel and motion photography in the present have made possible the artistry of Jascha Heifetz, Frank Lloyd Wright, and Charles Chaplin, which have opened up wholly new ranges of expressive possibility without, moreover, in any way inhibiting a concurrent renewal of interest in medieval instruments and primitive art. If religious innovation can provide a meaning orientation broad enough to accommodate the idea that new technology can be creative of new values, a long step will have been taken toward providing a religious belief system adequate to the realities and needs of a technological age.

Individual Man in a Technological Age

What do technological change and the social and value changes that it brings with it mean for the life of the individual today? It is not clear that their effects are all one-way. For example, we are often told that today's individual is alienated by the vast proliferation of technical expertise and complex bureaucracies, by a feeling of impotence in the face of "the machine," and by a decline in personal privacy. It is probably true that the social pressures placed on individuals today are more complicated and demanding than they were in earlier times. Increased geographical and occupational mobility and the need to function in large organizations place difficult demands on the individual to conform or "adjust." It is also evident that the privacy of many individuals tends to be encroached upon by sophisticated eavesdropping and surveillance devices, by the accumulation of more and more information about individuals by governmental and many private agencies, and by improvements in information-handling technologies such as the proposed institution of centralized statistical data banks. There is little doubt, finally, that the power, authority, influence, and scope of government are greater today than at any time in the history of the United States.

But, as Professor Edward Shils points out in his study on technology and the individual, there is another, equally compelling side of the coin. First, government seems to be more shy and more lacking in confidence today than ever before. Second, while privacy may be declining in the ways indicated above, it also tends to decline in a sense that most individuals are likely to approve. The average man in Victorian times, for example, probably "enjoyed" much more privacy than today. No one much cared what happened to him, and he was free to remain ignorant, starve, fall ill, and die in complete privacy; that was the "golden age of privacy," as Shils puts it. Compulsory universal education, social security legislation, and public health measures—indeed, the very idea of a welfare state—are all antithetical to privacy in this sense, and it is the rare individual today who is loath to see that kind of privacy go.

It is not clear, finally, that technological and social complexity must inevitably lead to reducing the individual to "mass" or "organization" man. Economic productivity and modern means of communication allow the individual to aspire to more than he ever could before. Better and more easily available education not only provides him with skills and with the means to develop his individual potentialities, but also improves his self-image and his sense of value as a human being. This is probably the first age in history in which such high proportions of people have *felt* like individuals; no eighteenth-century English factory worker, so far as we know, had the sense of individual worth that underlies the demands on society of the average resident of the black urban ghetto today. And, as Shils notes, the scope of individual choice and action today are greater than in previous times, all the way from consumer behavior to political or religious allegiance. Even the much-maligned modern organization may in fact "serve as a mediator or buffer between the individual and the full raw impact of technological change," as an earlier study supported by the Harvard Program has concluded.

Recognition that the impact of modern technology on the individual has two faces, both negative and positive, is consistent with the double effect of technological change that was discussed above. It also suggests that appreciation of that impact in detail may not be achieved in terms of old formulas, such as more or less privacy, more or less government, more or less individuality. Professor Shils is therefore attempting to couch his inquiry in terms of the implications of technological change for the balance that every individual must strike between his commitment to private goals and satisfactions and his desires and responsibilities as a public citizen. The citizens of ancient Athens seem to have been largely public beings in this sense, while certain segments of today's hippie population seem to pursue mainly private gratifications. The political requirements of our modern technological society would seem to call for a relatively greater public commitment on the part of individuals than has been the case in the past, and it is by exploring this hypothesis that we may enhance our understanding of what technology does to the individual in present-day society.

ECONOMIC AND POLITICAL ORGANIZATION

The Enlarged Scope of Public Decision Making

When technology brings about social changes which impinge on our existing system of values, it poses for society a number of problems that are ultimately political in nature. The term "political" is used here in the broadest sense: it encompasses all of the decision-making structures and procedures that have to do with the allocation and distribution of wealth and power in society. The political organization of society thus includes not only the formal apparatus of the state but also industrial organizations and other private institutions that play a role in the decision-making process. It is particularly important to attend to the organization of the entire body politic when technological change leads to a blurring of once clear distinctions between the public and private sectors of society and to changes in the roles of its principal institutions.

It was suggested above that the political requirements of our modern technological society call for a relatively greater public commitment on the part of individuals than in previous times. The reason for this, stated most generally, is that technological change has the effect of enhancing the importance of public decision making in society, because technology is continually creating new possibilities for social action as well as new problems that have to be dealt with.

A society that undertakes to foster technology on a large scale, in fact, commits itself to social complexity and to facing and dealing with new problems as a normal feature of political life. Not much is yet known with any precision about the political imperatives inherent in technological change, but one may nevertheless speculate about the reasons why an increasingly technological society seems to be characterized by enlargement of the scope of public decision making.

For one thing, the development and application of technology seems to require large-scale, and hence increasingly complex, social concentrations, whether these be large cities, large corporations, big universities, or big government. In instances where technological advance appears to facilitate reduction of such first-order concentrations, it tends instead to enlarge the relevant *system* of social organization, that is, to lead to increased centralization. Thus, the physical dispersion made possible by transportation and communications technologies, as Meyer and Kain have shown, enlarges the urban complex that must be governed as a unit.

A second characteristic of advanced technology is that its effects cover large distances, in both the geographical and social senses of the term. Both its positive and negative features are more extensive. Horsepowered transportation technology was limited in its speed and capacity, but its nuisance value was also limited, in most cases to the owner and to the occupant of the next farm. The supersonic transport can carry hundreds across long distances in minutes, but its noise and vibration damage must also be suffered willy-nilly by everyone within the limits of a swath 3,000 miles long and several miles wide.

The concatenation of increased density (or enlarged system) and extended technological "distance" means that technological applications have increasingly wider ramifications and that increasingly large concentrations of people and organizations become dependent on technological systems. A striking illustration of this was provided by the widespread effects of the power blackout in the northeasten part of the United States. The result is not only that more and more decisions must be social decisions taken in public ways, as already noted, but that, once made, decisions are likely to have a shorter useful life than heretofore. That is partly because technology is continually altering the spectrum of choices and problems that society faces, and partly because any decision taken is likely to generate a need to take ten more.

These speculations about the effects of technology on public decision making raise the problem of restructuring our decision-making mechanisms—including the system of market incentives—so that the increasing number and importance of social issues that confront us can be resolved equitably and effectively.

Private Firms and Public Goods

Among these issues, as noted earlier, is that created by the shift in the composition of demand in favor of public goods and services—such as education, health, transportation, slum clearance, and recreational facilities—which, it is generally agreed, the market has never provided effectively and in the provision of which government has usually played a role of some significance. This shift in demand raises serious questions about the relationship between technological change and existing decision-making structures in general and about the respective roles of government and business in particular. A project initiated under the direction of Dr. Robin Marris is designed to explore those questions in detail.

In Western industrialized countries, new technological developments generally originated in and are applied through joint stock companies whose shares are widely traded on organized capital markets. Corporations thus play a dominant role in the development of new methods of production, of new methods of satisfying consumer wants, and even of new wants. Most economists appear to accept the thesis originally proposed by [Joseph] Schumpeter that corporations play a key role in the actual process of technological innovation in the economy. Marris himself has recently characterized this role as a perceiving of latent consumer needs and of fostering and regulating the rate at which these are converted into felt wants.[3]

There is no similar agreement about the implications of all this for social policy. J. K. Galbraith, for example, argues that the corporation is motivated by the desire for growth subject to a minimum profit constraint and infers (1) a higher rate of new-want development than would be the case if corporations were motivated principally to maximize profit, (2) a bias in favor of economic activities heavy in "technological content" in contrast to activities requiring sophisticated social organization, and (3) a bias in the economy as a whole in favor of development and satisfaction of private needs to the neglect of public needs and at the cost of a relatively slow rate of innovation in the public sector.

But Galbraith's picture is not generally accepted by economists, and his model of the corporation is not regarded as established economic theory. There is, in fact, no generally accepted economic theory of corporate behavior, as Marris points out, so that discussions about the future of the system of corporate enterprise usually get bogged down in an exchange of unsubstantiated assertions about how the existing system actually operates. What seems needed at this time, then, is less a new program of empirical research than an attempt to synthesize what we know for the purpose of arriving at a more adequate theory of the firm. This is the objective of phase 1 of the Marris project.

On the basis of the resulting theoretical clarification, phase 2 will go on to address such questions as (1) the costs of a policy of economic growth, (2) the incommensurability of individual incentive and public will, (3) the desirable balance between individual and social welfare when the two are inconsistent with each other, (4) changes in the roles of government and industrial institutions in the political organization of American society, and (5) the consequences of those changes for the functions of advertising and competing forms of communication in the process of public education. In particular, attention will be directed to whether existing forms of company organization are adequate for marshaling technology to social purposes by responding to the demand for public goods and services, or whether new productive institutions will be required to serve that end.

We can hope to do no more than raise the level of discussion of such fundamental and difficult questions, of course, but even that could be a service.

The Promise and Problems of Scientific Decision Making

There are two further consequences of the expanding role of public decision making. The first is that the latest information-handling of devices and techniques tends to be utilized in the decision-making process. This is so (1) because public policy can be effective only to the degree that it is based on reliable knowledge about the actual state of the society, and thus requires a strong capability to collect, aggregate, and analyze detailed data about economic activities, social patterns, pop-[...] [p]olitical trends, and (2) because it is recognized increasingly that [...] one area impinge on and have consequences for other policy areas [...] [...] of as unrelated, so that it becomes necessary to base decisions on [...] [soc]iety that sees it as a system and that is capable of signaling as m[...] p[...] of the probable consequences of a contemplated action.

[...] Alan F. Westin points out, reactions to the prospect of more deci-[sion m]aki[ng bas]ed on computerized data banks and scientific management tech-niques ru[n the g]amut of optimism to pessimism mentioned in the opening of this [chapter. Negat]ive reactions take the form of rising political demands for greater [particip]ation in decision making, for more equality among different seg-ments of [the po]pulation, and for greater regard for the dignity of individuals. The increasing dependence of decison making on scientific and technological devices and techniques is seen as posing a threat to these goals, and pressures are generated in opposition to further "rationalization" of decision-making processes. These pressures have the paradoxical effect, however, not of deflecting the supporters of technological decision making from their course, but of spurring them on to renewed effort to save the society before it explodes under planlessness and inadequate administration.

The paradox goes further, and helps to explain much of the social discontent that we are witnessing at the present time. The greater complexity and the more extensive ramifications that technology brings about in society tend to make social processes increasingly circuitous and indirect. The effects of actions are widespread and difficult to keep track of, so that experts and sophisticated techniques are increasingly needed to detect and analyze social events and to formulate policies adequate to the complexity of social issues. The "logic" of modern decision making thus appears to require greater and greater dependence on the collection and analysis of data and on the use of technological devices and scientific techniques. Indeed, many observers would agree that there is an "increasing relegation of questions which used to be matters of political debate to professional cadres of technicians and experts which function almost independently of the democratic political process."[4] In recent times, that process has been most noticeable, perhaps, in the areas of economic policy and national security affairs.

This "logic" of modern decision making, however, runs counter to that element of traditional democratic theory that places high value on direct participation in the political processes and generates the kind of discontent referred to above. If it turns out on more careful examination that direct participation is becoming less relevant to a society in which the connections between causes and effects are long and often hidden—which is an increasingly "indirect" society, in other words—elaboration of a new democratic ethos and of new democratic processes more adequate to the realities of modern society will emerge as perhaps the major intellectual and political challenge of our time.

The Need for Institutional Innovation

The challenge is, indeed, already upon us, for the second consequence of the enlarged scope of public decision making is the need to develop new institutional forms and new mechanisms to replace established ones that can no longer deal effectively with the new kinds of problems with which we are increasingly faced. Much of the political ferment of the present time—over the problems of technology assessment, the introduction of statistical data banks, the extension to domestic problems of techniques of analysis developed for the military services, and the modification of the institutions of local government—is evidence of the need for new institutions. It will be recalled that Professor Oettinger's study concludes that innovation is called for in the educational establishment before instructional technology can realize the promise that is potential in it. Our research in the biomedical area has repeatedly confirmed the need for institutional innovation in the medical system, and Marris has noted the evolution that seems called for in our industrial institutions. The Rosenbloom research group, finally, has documented the same need in the urban area and is exploring the form and course that the processes of innovation might take.

Direct intervention by business or government to improve ghetto conditions will tend to be ineffective until local organizations come into existence which enable residents to participate in and control their own situation. Such organizations seem to be a necessary condition for any solution of the ghetto problem that is likely to prove acceptable to black communities. Professors Richard S. Rosenbloom, Paul R. Lawrence, and their associates are therefore engaged in the design of two types of organization suited to the peculiar problems of the modern ghetto. These are (1) a state- or area-wide urban development corporation in which business and government join to channel funds and provide technical assistance to (2) a number of local development corporations, under community control, which can combine social service with sound business management.

Various "ghetto enrichment" strategies are being proposed at the present time, all of which stress the need for institutional innovation of some kind and in many of which creation of one sort or another of community development corporation is a prominent feature. In none of these respects does our approach claim any particular originality. What does seem promising, however, is our effort to design a local development corporation that is at once devoted to social service and built on sound business principles.

These characteristics point to large and powerful organizations that can serve as engines of indigenous ghetto development. They would of course interact with "outside" institutions, not only those at various levels of government, but especially their counterpart state or area urban development corporations. They would not be dependent principally on such outside institutions, however, since they would be engines that, once started, could keep running largely on their own power. In economic terms, the local development corporations would become "customers" of business. In political terms, they would be partners of existing governmental structures. In broader social terms, they could become vehicles for integrating underprivileged urban communities into the mainstream of American society.

The design for the state or area urban development corporation, in Professor Rosenbloom's description, would be a new form of public-private partnership serving to pull together the resources and programs of the business sector, of universities and research institutions, of public agencies, and of community organizations. This corporation could act as a surrogate for the "invisible hand" of the market, able to reward the successes of the local development corporations through command of a pool of unrestricted funds. Since there is no necessary relationship between profitability and social benefit for economic venture in the ghetto, however, success would need to be measured, not in usual profit-and-loss terms, but in terms of such social indicators as employment levels, educational attainment, health statistics, and the like.

The collaborative arrangements we have entered into in New Jersey and in Boston offer us a welcome opportunity to test and develop some of our hypotheses and designs. In both of these programs, our research group is in a position to contribute know-how and advice, based on its understanding of organizational and corporate behavior, and to acquire insight and primary data for research that can prove useful in other contexts. As long ago as our first annual report, we announced the hope and expectation that the Harvard Program could supplement its scholarly production by adding a dimension of action research. New Jersey and Boston are providing us with our first opportunity to realize that objective.

CONCLUSION

As we review what we are learning about the relationship of technological and social change, a number of conclusions begin to emerge. We find, on the one hand, that the creation of new physical possibilities and social options by technology tends toward and appears to require the emergence of new values, new forms of economic activity, and new political organizations. On the other hand, technological change also poses problems of social and psychological displacement.

The two phenomena are not unconnected, nor is the tension between them new; man's technical prowess always seems to run ahead of his ability to deal with and profit from it. In America, especially, we are becoming adept at extracting the new techniques, the physical power, and the economic productivity that are inherent in our knowledge and its associated technologies. Yet we have not fully accepted the fact that our progress in the technical realm does not leave our institutions, values, and political processes unaffected.

Individuals will be fully integrated into society only when we can extract from our knowledge not only its technological potential but also its implications for a system of values and a social, economic, and political organization appropriate to a society in which technology is so prevalent.

NOTES

1. Robin Williams, "Individual and Group Values," *Annals of the American Academy of Political and Social Science* 37 (May 1967): 30.

2. See Harvey Cox, "Tradition and the Future," pts. 1 and 2, in *Christianity and Crisis* 27, nos. 16 and 17 (October 2 and 16, 1968): 218–20 and 227–31.

3. Robin Marris, *The Economic Theory of "Managerial" Capitalism* (Glencoe, Ill.: The Free Press, 1964).

4. Harvey Brooks, "Scientific Concepts and Cultural Change," in *Science and Culture*, ed. G. Holton (Boston: Houghton Mifflin, 1965), p. 71.

9

Technology and Culture in a Developing Country

Kwame Gyekye

Even though the subject of my paper is "Technology and Culture in a Developing Country," it seems appropriate to preface it by examining science itself in the cultural traditions of a developing country, such as Ghana, in view of the fact that the lack of technological advancement, or the ossified state in which the techniques of production found themselves, in the traditional setting of Africa and, in many ways, even in modern Africa, is certainly attributable to the incomprehensible inattention to the search for scientific principles by the traditional technologists. I begin therefore with observations on how science and knowledge fared in the traditional culture of a developing country.

SCIENCE AND OUR CULTURE

In a previous publication I pointed out—indeed I stressed—the empirical orientation of African thought: maintaining that African proverbs, for instance, a number of which bear some philosophical content, addressed—or resulted from reflections on—specific situations, events, or experiences in the lives of the people, and that even such a metaphysical concept as destiny (or fate) was reached inductively, experience being the basis of the reasoning that led to it.[1] Observation and experience constituted a great part of the sources of knowledge in African traditions.[2] The empirical basis of knowledge had immediate practical results in such areas as agriculture and herbal medicine: our ancestors, whose main occupation was farming, knew of the system of rotation of crops; they knew when to allow a piece of land to lie fallow for a while; they had some knowledge of the technology of food processing and preservation; and there is a great deal

Kwame Gyekye, "Technology and Culture in a Developing Country," in *Philosophy and Technology*, ed. Roger Fellows (Cambridge: Cambridge University Press, 1995), pp. 121–41. Copyright © 1995 Kwame Gyekye. Reprinted with permission.

of evidence about their knowledge of the medicinal potencies of herbs and plants—the main source of their health care delivery system long before the introduction of Western medicine. (Even today, there are countless testimonies of people who have received cures from "traditional" healers where the application of Western therapeutics could not cope.)

It has been asserted by several scholars that African life in the traditional setting is intensely religious or spiritual. [John S.] Mbiti opined that "Africans are notoriously religious, and each people has its own religious system with a set of beliefs and practices. Religion permeates into all the departments of life so fully that it is not easy or possible to isolate it."[3] According to him, "in traditional life there are no atheists."[4] [K. A.] Busia observed that Africa's cultural heritage "is intensely and pervasively religious,"[5] and that "in traditional African communities, it was not possible to distinguish between religious and nonreligious areas of life. All life was religious."[6] Many colonial administrators in Africa used to refer to Africans, according to [G.] Parrinder, as "this incurably religious people."[7] Yet, despite the alleged religiosity of the African cultural heritage, the empirical orientation or approach to most of their enterprises was very much to the fore. I strongly suspect that even the African knowledge of God in the traditional setting was, in the context of a nonrevealed religion of traditional Africa, empirically reached.

Now, one would have thought that such a characteristically empirical, epistemic outlook would naturally lead to a profound and extensive interest in science as a theory, that is, in the acquisition of theoretical knowledge of nature, beyond the practical knowledge which they seem to have had of it, albeit not in a highly developed form, and which they utilized to their benefit. But, surprisingly, there is no evidence that such an empirical orientation of thought in traditional African culture led to the creation of the scientific outlook or a deep scientific understanding of nature. It is possible, arguably, to credit people who practiced crafts and pursued such activities as food preservation, food fermentation, herbal therapeutics, etc. (see next section) with some amount of scientific knowledge; after all, the traditional technologies, one would assume, must have had some basis in science. Yet, it does not appear that their practical knowledge of crafts or forms of technologies led to any deep scientific understanding or analysis of the enterprises they were engaged in. Observations made by them may have led to interesting facts about the workings of nature; but those facts needed to be given elaborate and coherent theoretical explanations. Science requires explanations that are generalizable, facts that are disciplined by experiments, and experiments that are repeatable and verifiable elsewhere. But the inability (or, is it lack of interest?) of the users of our culture to engage in sustained investigations and to provide intelligible scientific explanations or analyses of their own observations and experiences stunted the growth of science.

Science begins not only in sustained observations and investigations into natural phenomena, but also in the ascription of causal explanations or analyses to those phenomena. The notion of causality is of course crucial to the pursuit of science. Our cultures appreciated the notion of causality very well. But, for a reason which must be linked to the alleged intense religiosity of the cultures, causality was generally understood in terms of spirit, of mystical power. The consequence of this was that purely scientific or empirical causal explanations, of which the users of our culture were somehow aware, were often not regarded as profound enough to offer complete satisfaction. This led them to give up, but too soon, on the search for empirical causal explanations, even of causal relations between natural phenomena or events, and resort to supernatural causation.

Empirical causation, which asks what and how questions, too quickly gave way to agentive causation which asks who and why questions. Agentive causation led to the postulation of spirits or mystical powers as causal agents; so that a particular metaphysic was at the basis of this sort of agentive causation. According to Mbiti, "The physical and spiritual are but two dimensions of one and the same universe. These dimensions dove-tail into each other to the extent that at times and in places one is apparently more real than, but not exclusive of, the other."[8] It is the lack of distinction between the purely material (natural) and the immaterial (supernatural, spiritual) that led to the postulation of agentive causation in all matters. For, in a conception of a hierarchy of causes, it was easy to identify the spiritual as the agent that causes changes in relations even among empirical phenomena. In view of the critical importance of causality to the development of the science of nature, a culture that was obsessed with supernaturalistic or mystical causal explanations would hardly develop the scientific attitude in the users of that culture, and would, consequently, not attain knowledge of the external world that can empirically be ascertained by others, including future generations.

Yet, the alleged intense religiosity of the African cultural heritage need not have hindered interest in science, that is, in scientific investigations both for their own sake and as sure foundations for the development of technology. Religion and science, even though they perceive reality differently, need not, nevertheless, be incompatible. Thus it is possible for religious persons to acquire scientific knowledge and outlook. But to be able to do so most satisfactorily, one should be able to separate the two, based on the conviction that purely scientific knowledge and understanding of the external world would not detract from one's faith in an ultimate being. A culture may be a religious culture, even an intensely religious culture at that; but, in view of the tremendous importance of science for the progress of many other aspects of the culture, it should be able to render unto Caesar what is Caesar's and unto God what is God's ("Caesar" here referring to the pursuit of the knowledge of the natural world). The inability of our traditional cultures to separate religion from science, as well as the African conception of nature as essentially animated or spirit-filled (leading to the belief that natural objects contained mystical powers to be feared or kept at bay or, when convenient, to be exploited for man's immediate material benefit), was the ground of the agentive causal explanations enamored of the users of our cultures in the traditional setting. Science, as already stated, is based on a profound understanding and exploitation of the important notion of causality, that is, on a deep appreciation of the causal interactions between natural phenomena. But where this is enmeshed with—made inextricable from—supernaturalistic molds and orientations, it, as a purely empirical pursuit, hardly makes progress.

Also, religion, even if it is pursued by a whole society or generation, is still a highly subjective cognitive activity, in that its postulates and conclusions are not immediately accessible to the objective scrutiny or verification by others outside it. Science, on the contrary, is manifestly an objective, impartial enterprise whose conclusions are open to scrutiny by others at any time or place, a scrutiny that may lead to the rejection or amendment or confirmation of those conclusions. Now, the mesh in which both religion and science (or, rather the pursuit of science) found themselves in African traditional cultures, made the relevant objective approach to scientific investigations into nature well-nigh impossible. Moreover, in consequence of this mesh, what could have become scientific knowledge accessible to all others became an esoteric knowledge, a specialized knowledge, accessible only to initiates probably under an oath of secrecy administered by priests and priestesses, traditionally acknowledged as the custodians of the verities and secrets of nature. These custodians, it was, who "knew," and were often consulted on, the causes of frequent low crop yield, lack of adequate rainfall over a long period of time, the occurrence of bush fires, and so on. Knowledge-claims about the operations of nature became not only esoteric but also, if for that reason, personal rather than exoteric and impersonal. This preempted the participative nature of the search for deep and extensive knowledge of the natural world; for, others would not have access to, let alone participate in, the type of knowledge that is regarded as personal and arcane.

Knowledge of the potencies of herbs and other medicinal plants was in the traditional setting probably the most secretive of all. Even if the claims made by African medicine men and women of having discovered cures for deadly diseases could be substantiated scientifically, those claims cannot be pursued for verification since their knowledge-claims were esoteric and personal. The desire to make knowledge of the external world personal has been the characteristic attitude of our traditional healers who claim to possess knowledge of medicinal plants, claims at least some of which can be scientifically investigated. In the past, all such possibly credible claims to knowledge of medicinal plants just evaporated on the death of the traditional healer or priest. And science, including the science of medicine, stagnated.

I think that the personalization of the knowledge of the external world is attributable to the mode of acquiring that knowledge: that mode was simply not based on experiment. And, in the circumstance, the only way one could come by one's knowledge of, say, herbal therapeutics, was most probably through mystical or magical means, a means not subject to public or objective scrutiny and analysis.

The lack of the appropriate attitude to sustained scientific probing, required for both vertical and horizontal advancement of knowledge, appears to have been a characteristic of our African cultural past. One need not have to put this want of the appropriate scientific attitude to the lack of the capacity for science. And I, on my part, would like to make a distinction here: between the having of the intellectual capacity on the one hand, and the having of the proclivity or impulse to exercise that capacity on a sustained basis that would yield appreciable results on the other hand. The impulse for sustained scientific or intellectual probing does not appear to have been nurtured and promoted by our traditional cultures.

It appears in fact that the traditional cultures rather throttled the impulse toward sustained and profound inquiry for reasons that are not fully known or intelligible. One reason, however, may be extracted from the Akan (Ghanaian) maxim, literally translated as: "If you insist on probing deeply into the eye sockets of a dead person, you see a ghost."[9] The translation is of course not clear enough. But what the maxim is saying is that curiosity or deep probing may lead to dreadful consequences (the ghost is something of which most people are apprehensive). The maxim, as [E.] Laing also saw, stunts the "development of the spirit of inquiry, exploration, and adventure."[10] The attitude sanctioned by the maxim would, as Laing pointed out, be "inimical to science";[11] but not only to science, but, I might add, to all kinds of knowledge. My colleague, [Kofi Asare] Opoku,[12] however, explained to me in a conversation that the intention of the maxim is to put an end to a protracted dispute, which might tear a family or lineage apart: a dispute that has been settled, in other words, should not be resuscitated, for the consequences of the resuscitation would not be good for the solidarity of the family. Thus, Opoku would deny that this maxim is to be interpreted as damaging to intellectual or scientific probing. In response to Opoku's interpretation, one would like to raise the following questions: Why should further evidence not be looked for if it would indeed help settle the matter more satisfactorily? Why should further investigation be stopped if it would unravel fresh evidence and lead to what was not previously known? To end a dispute prematurely for the sake of family solidarity to the dissatisfaction of some members of the family certainly does violence to the pursuit of moral or legal knowledge. So, whether in the area of legal, moral, or scientific knowledge, it seems to me that the maxim places a damper on the impulse or proclivity to deep probing, to the pursuit of further knowledge.

The general attitude of the users of the African traditional cultures expressed in oft-used statements as "this is what the ancestors said," "this is what the ancestors did," and similar references to what are regarded as the ancestral habits or modes of thought and action, may be put down to the inexplicable reluctance— or lack of the impulse—to pursue sustained inquiries into the pristine ideas and values of the culture. It is this kind of mind-set, one might add, which often makes the elderly people even in our contemporary (African) societies try to hush and stop children with inquisitive minds from persistently asking certain kinds of questions and, thus, from pursuing intellectual exploration on their own. (I do not have the space to provide evidence to show that our forefathers did not expect later generations to regard their modes of thought and action as sacrosanct and unalterable, and to think and act in the same way they did. So that, if later generations, i.e. their descendants, failed to make changes, amendments, or refinements such as may be required by their own times and situations, that would have to be put down to the intellectual indolence or shallowness of the descendants.)

Finally let me say this: The pursuit of science—the cultivation of rational or theoretical knowledge of the natural world—seems to presuppose an intense desire, at least initially, for knowledge for its own sake, not for the sake of some immediate practical results. It appears that our cultures had very little if any conception of knowledge for its own sake. It had a conception of knowledge that was practically oriented. Such an epistemic conception seems to have had a parallel in the African conception of art. For, it has been said by several scholars[13] that art was conceived in the African traditional setting in functional (or, teleological) terms, that the African aesthetic sense did not find the concept of "art for art's sake" hospitable. Even though I think that the purely aesthetic element of art was not lost sight of, this element does not appear to have been stressed in African art appreciation, as was the functional conception. This practical or functional conception of art, which dwarfed a conception of art for art's sake, must have infected the African conception of knowledge, resulting in the lack of interest in the acquisition of knowledge, including scientific knowledge, *for its own sake*.

It is clear from the foregoing discussion of the attitude of our indigenous African cultures to science that: (i) the cultures did not have a commitment, however spasmodic, to the advancement of the scientific knowledge of the natural world; (ii) they made no attempts, however feeble, to investigate the scientific theories underpinning the technologies they developed, as I will point out in some detail below; (iii) the disposition to pursue sustained inquiries into many areas of their life and thought does not seem to have been fostered by our African cultures; and (iv) the successive generations of the participants in the culture could not, consequent upon (iii), augment the compendium of knowledge that they had inherited from their forefathers, but rather gleefully felt satisfied with it, making it into a hallowed or mummified basis of their own thought and action. In our contemporary world, when sustainable development, a great aspect of which is concerned with the enhancement of the material well-being of human beings, depends on the intelligent and efficient exploitation of the resources of nature—an exploitation that can be effected only through science and its progeny technology—the need to cultivate the appropriate scientific attitudes is an imperative.

Contemporary African culture will have to come to terms with the contemporary scientific attitudes and approaches to looking at things in Africa's own environment, attitudes, and approaches that have been adopted in the wake of the contact with the Western cultural traditions. The governments of African nations have for decades been insisting on the cultivation of science in the schools and universities as an unavoidable basis for technological, and hence industrial, advancement. More places and facilities are made available for those students who are interested in the pursuit of science. Yet, *mirabile dictu*, very many more students register for courses in the humanities and the social sciences than in the mathematical and natural sciences. Has the traditional culture anything to do with this lack of real or adequate or sustained interest in the natural sciences, or not?

TECHNOLOGY AND OUR CULTURE

Like science, technology—which is the application of knowledge or discovery to practical use—is also a feature or product of culture. It develops in the cultural milieu of a people and its career or future is also determined by the characteristics of the culture. Technology is an enterprise that can be said to be common to all human cultures; it can certainly be regarded as among the earliest creations of any human society. This is because the material existence and survival of the human society depend on the ability of man to make at least simple tools and equipment and to develop techniques essential for the production of basic human needs such as food, clothing, shelter, and security. The concern for such needs was naturally more immediate than the pursuit and acquisition of the systematic knowledge of nature—that is, science. Thus, in all human cultures and societies the creation of simple forms of technology antedates science—the rational and systematic pursuit of knowledge and understanding of the natural world, of the processes of nature, based on observation and experiment. The historical and functional priority of technology over science was also a phenomenon even in the cultures of Western societies, historically the home of advanced and sophisticated technology. From antiquity on, and through the Middle Ages into the modern European world, innovative technology showed no traces of the application of consciously scientific principles.[14] Science-based technology was not developed until about the middle of the nineteenth century.[15] Thus, technology was for centuries based on completely empirical knowledge.

The empirical character of African thought in general and of its epistemology in particular was pointed up in the preceding section. The pursuit of empirical knowledge—knowledge based on experience and observation, and generally oriented toward the attainment of practical results—underpinned a great deal of the intellectual enterprise of the traditional setting. (Note that philosophical knowledge was also thought to have a practical orientation.) And so, like other cultures of the world, practical knowledge and the pursuit of sheer material well-being and survival led the cultures of Africa to develop technologies and techniques, simple in their forms, as would be the case in premodern times. Basic craftsmanship emerged: farming implements such as the cutlass, hoe, and axe were made by the blacksmith; the goldsmith produced the bracelet, necklace, and rings (including the earring): "African coppersmiths have for centuries produced wire to make bracelets and ornaments—archaeologists have found the draw-plates and other wire-making tools."[16] There were carpenters, wood-carvers, potters, and cloth-weavers, all of whom evolved techniques for achieving results. Food production, processing, and preservation techniques were developed, and so were techniques for extracting medicinal potencies from plants, herbs, and roots. A number of these technical activities in time burgeoned into industries.

There was, needless to say, a great respect and appreciation for technology because of what it could offer the people by way of its products. The need for, and the appreciation of, technology should have translated into real desire for innovation and improvement on existing technological products and techniques. There is, however, not much evidence to support the view that there were attempts to innovate technologies and refine techniques received from previous generations. There were no doubts whatsoever about the potencies of traditional medicines extracted from plants and herbs—the basis of the health care delivery system in the traditional setting and, to a very great extent, in much of rural Africa today. Yet, there were—and are—enormous problems about both the nature of the diagnosis and the appropriate or reliable dosage, problems which do not seem to have been grappled with. Diagnosis requires systematic analysis of cause and effect, an approach which would not be fully exploited in a system, like the one evolved by our cultures, which would often explain the causes of illness, as they would many other natural occurrences, in agentive (i.e. supernatural, mystical) terms, as I explained in the previous section. Such a causal approach to coping with disease would hardly dispose a people toward the search for effective diagnostic technologies.

Traditional healers were often not short on prescriptive capabilities: they were capable in a number of cases of prescribing therapies often found to be efficacious. But their methods here generated two problems: one was the preparation of the medicine to be administered to a patient; the other was the dosage—the quantity of the medicine for a specific illness. Having convinced himself of the appropriate therapeutic for a particular disease—a therapeutic which would often consist of a concoction, the next step for the herbal healer was to decide on the proportion (quantity) of each herbal ingredient for the concoction. Second, a decision had to be taken on the appropriate and effective dosage for a particular illness. Both steps obviously required exact measurement of quantity. The failure to provide exact measurement would affect the efficacy of the concoction as well as the therapeutic effect of the dosage; in the case of the latter, there was the possibility of underdosage or over-dosage. Yet the need for exact measurement does not seem to have been valued and pursued by our cultures, a cultural defect which in fact is still taking its toll also in the maintenance of machines by our mechanics of today. [Kwasi] Wiredu mentions the case of a Ghanaian mechanic who, in working on engine maintenance, would resort to the use of his sense of sight rather than of a feeler gauge in adjusting the contact breaker point in the distributor of a car.[17] The mechanic, by refusing to use a feeler gauge and such other technical aids, of course fails to achieve the required precision measurement.

When it is realized that the habit or attitude of the mechanic was not peculiar to him but that it is a habit of a number of mechanics in our environments, it can be said that the development of that habit is a function of the culture. If one considers that precision measurement is basic not only for the proper maintenance of machines, but also for the quality of manufactured products of all kinds, one can appreciate the seriousness of the damage to the growth of technology caused by our cultures' failure to promote the value of precision measurement.

Even though it is true to say that historically technology was for centuries applied without resort to scientific principles, it is also conceivable that this fact must have slowed down the advancement of technology. It deprived technology of a necessary scientific base. The making of simple tools and equipment may not require or rest on the knowledge of scientific principles; but not so the pursuit of most other technological enterprises and methods. It cannot be doubted that the preparation of medicinal concoctions by traditional African herbal healers and their prescriptive dosages, for instance, must have been greatly hampered by the failure to attend to the appropriate scientific testing of the potencies of the various herbs and the amounts of each (herb or plant) required in a particular concoction. Theoretical knowledge should have been pursued to complement their practical knowledge.

Food technology, practiced in the traditional setting mainly by African women, was a vibrant activity, even though the scientific aspect of it was not attended to. According to [S.] Sefa-Dede, who has done an enormous amount of research in traditional food technology in Ghana, "The scientific principles behind the various unit operations may be the same as found in modern food technologies, but the mode of application may be different."[18] The techniques traditionally deployed in food preservation undoubtedly involve the application of principles of science: physics, chemistry, and biology, which the users of those techniques may not have been aware of. The techniques of preserving food all over Africa include drying, smoking, salting, and fermenting. The drying technique is aimed at killing bacteria and other decay-causing microorganisms and thus preserving food intact for a long time; smoking serves as a chemical preservative; and so does salting which draws moisture and microorganisms from foods; fermentation of food causes considerable reduction of acidity levels and so creates conditions that prevent microbial multiplication.[19] It is thus clear that there are scientific principles underlying these methods.

Let us take the case of a woman in the central region of Ghana underlying whose practice of food technology is clearly a knowledge of some principles of physics, chemistry, and metallurgy. The woman in question is a processor of *fante kenkey*, a fermented cereal dumpling made from maize. Maize dough is fermented for two to four days. A portion of the dough is made into a slurry and cooked into a stiff paste. This is mixed with the remaining portion through a process called aflatization to produce aflata. This is wrapped in dried banana leaves and boiled for three to four hours until it is cooked. Now, this woman is able to solve a problem arising from the technique she uses in processing *fante kenkey*, to the amazement of a modern scientific research team interested in studying traditional food technology.[20] The woman challenged the research team to indicate how they could solve a very practical problem which can arise when one is boiling *fante kenkey* in a 44-gallon drum. This was the problem:

> Imagine that you have loaded a 44-gallon barrel with uncooked *fante kenkey*. You set the system up on the traditional cooking stove, which uses firewood. The fire is lit and the boiling process starts. In the middle of the boiling process, you notice that the barrel has developed a leak at its bottom. The boiling water is gushing into the fire and gradually putting off the fire. What will you do to save the situation?

The possible solutions suggested by the research team were found to be impractical. One solution given by the team was to transfer the product from the leaking barrel into a new one. There are at least two reasons why this could not be done: one was that the *kenkey* will be very hot and difficult to unpack; the process will also be time-consuming; another reason was that another barrel may not be available.

The traditional woman food technologist then provided the solution: Adjust the firewood in the stove to allow increased burning; then collect two or three handfuls of dry palm kernel and throw them into the fire—this will heat up and turn red hot; finally, collect coarse table salt and throw it into the hot kernels. The result will be that the salt will explode and in the process seal the leak at the bottom of the barrel.

According to Sefa-Dede, the solution provided by the woman is based on the sublimation of the salt with the associated explosion. The explosion carries with it particles of salt which fill the opening. It is possible that there is interaction between the sodium chloride in the salt and the iron and other components forming the structure of the barrel. A few questions may arise as one attempts to understand the source of knowledge of the traditional practitioners: Why was dry palm kernel used as a heat exchange medium? What is peculiar about table salt (sodium chloride) in this process? In the case under discussion, it can certainly be said that the woman has some knowledge about the thermal properties of palm kernels. (It is possible that there is traditional knowledge about the excellent heat properties of palm kernels. For, traditional metal smelters, blacksmiths, and goldsmiths are known to use palm kernels for heating and melting various metals.) The woman, it can also be said, does have added knowledge of some chemistry and metallurgy. Even though it is clear that the ideas and solutions which the woman was able to come up with are rooted in basic and applied scientific principles, she cannot, like most other traditional technology practitioners, explain and articulate those principles. But not only that: they must have thought that the whys and hows did not matter: it was enough to have found practical ways to solve practical problems of human survival.

Thus, the pursuit of the principles would not have been of great concern to the users of traditional technologies, concerned, as they were, about reaping immediate practical results from their activities. The result was that there was no real understanding of the scientific processes involved in the technologies they found so useful. Yet, the concern for investigating and understanding those principles would most probably have led to innovation and improvement of the technologies. It can therefore be said that the weak scientific base of the traditional technology stunted its growth, and accounted for the maintenance and continual practice of the same old techniques. The understanding of the principles involved would probably have generated extensive innovative practices and the application of those principles to *other* yet-unknown technological possibilities. It clearly appears to be the case that once some technique or equipment was known to be working, there was no desire or enthusiasm on the part of its creators or users to innovate and improve on its quality, to make it work better or more efficiently, to build other—and more—efficient tools. Was this sort of complacency, or the feeling of having reached a cul-de-sac or of having come to the end of one's intellectual or technological tether, a reflection on the levels of capability that could be attained by our cultures?

APPROACHES TO DEVELOPING A MODERN TECHNOLOGY

It can hardly be denied that technology, along with science, has historically been among the central pillars—as well as the engines—of modernity. It is equally undeniable that the modern world is increasingly becoming a technological world: technology is, by all indications, going to become the distinguishing feature of global culture in the coming decades. Africa will have to participate significantly in the cultivation and promotion of this aspect of human culture, if it is to benefit from it fully. But the extensive and sustained understanding and acquisition of modern technology insistently require adequate cultivation of science and the scientific outlook. The acquisition of scientific and technological outlook will in turn require a new mental orientation on the part of the African people, a new and sustained interest in science to provide a firm base for technology, a new intellectual attitude to the external world uncluttered by superstition, mysticism, and other forms of irrationality; the alleged spirituality of the African world, which was allowed in many ways to impede sustained inquiries into the world of nature, will have to come to terms with the physical world of science. Knowledge of medicinal plants, for instance, qua scientific knowledge, must be rescued from the quagmire of mysticism and brought in to the glare of publicity, and its language made exoteric and accessible to many others.

The need for sustained interest in science is important for at least two reasons: to provide an enduring base for a real technological take-off at a time in the history of the world when the dynamic connections between science and technology have increasingly been recognized and made the basis of equal attention to both: technology has become science-based, while science has become technology-directed. The second reason, a corollary of the first, is that it is the application of science to technology that will help improve traditional technologies.

Ideally, technology, as a cultural product, should take its rise from the culture of a people, if it is to be directly accessible to a large section of the population and its nuances fully appreciated by them. For this reason, one approach to creating modern technology in a developing country is to upgrade or improve existing traditional technologies whose developments, as I have already indicated, seem to have been stunted in the traditional setting because of their very weak science base. Let us recall the case of the woman food technologist referred to in the previous section. She was able to find practical ways of solving problems that emerged in the course of utilizing some technology by resorting to ideas and solutions which are obviously rooted in basic science, but without the benefit of the knowledge of chemistry, physics, engineering, or metallurgy. From the technology she used, questions that arise would include the following: Why were dry palm kernels used as a heat exchange medium? What is peculiar about table salt (sodium chloride) in this process? Yet for most traditional technology practitioners the whys and how did not often matter, so long as some concrete results can be achieved through the use of a particular existing technology. But the why and how questions of course do matter a great deal. Improving traditional technologies will require not only looking for answers to such questions, but also searching for areas or activities to which the application of existing technologies (having been improved) can be extended.

Traditional technologies have certain characteristics which could—and must—be featured in the approach to modern technology in a developing country. Traditional technologies are usually simple, not highly specialized technologies: this fact makes for the involvement of large numbers of the people in the application or use of the technologies, as well as in their development; but it also promotes indigenous technological awareness. The materials that are used are locally available and the processes are effective. (In the case of the woman's food technology the materials in question, namely, palm kernels and table salt, are household items which are readily available.) Traditional technologies are developed to meet material or economic needs—to deal with specific problems of material survival. They can thus be immediately seen both as having direct connections with societal problems and as appropriate to meeting certain basic or specific needs. If the technologies that will be created by a developing country in the modern world feature some of the characteristics of the traditional technologies, they will have greater relevance and impact on the social and economic life of the people.

The improvement of traditional technologies is contingent on at least two factors. One is the existence or availability of autonomous, indigenous technological capacities. These capacities would need to be considerably developed. The development of capacities in this connection is not simply a matter of acquiring skills or techniques but, perhaps more importantly, of understanding, and being able to apply, the relevant scientific principles. It might be assumed that the ability to acquire skills presupposes the appreciation of scientific principles; such an assumption, however, would be false. One could acquire skills without understanding the relevant underpinning scientific principles. The situation of the woman food technologist is a clear case in point. However, the lack of understanding of the relevant scientific principles will impede the improvement exercise itself. The other factor relates to the need for change in certain cultural habits and attitudes on the part of artisans, technicians, and other practitioners of traditional technologies. Practitioners of traditional technologies will have to be weaned away from certain traditional attitudes and be prepared to learn and apply new or improved techniques and practices. Some old, traditional habits, such as the habit, referred to in the previous section, of the automatic use of the senses in matters of precision measurements, will have to be abandoned; adaptation to new—and generally more effective—ways of practicing technology, such as resorting to technical aids in precision measurements, will need to be pursued. It is the cultivation of appropriate attitudes to improved—or modern—technology and the development of indigenous technological capacity that will provide the suitable cultural and intellectual receptacle for the modern technologies that may be transferred from the technologically advanced industrial countries of the world to a developing country.

Now, the transfer of technology from the technologically developed world is a vital approach to bringing sophisticated technology to a developing country. It could also be an important basis for developing, in time, an indigenous technological capacity and the generation of fairly advanced indigenous technologies. But all this will depend on how the whole complex matter of technology transfer is tackled. If the idea is not executed well enough—if it is bungled—it may lead to complacency and passivity on the part of the recipients, reduce them to permanent technological dependency, and involve them in technological pursuits that may not be immediately appropriate to their objectives of social and economic development. On the other hand, an adroit approach to technology transfer by its recipients will, as I said, be a sure basis for a real technological take-off for a developing country.

Transfer of technology involves taking some techniques and practices developed in some technologically advanced country to some developing country. The assumption or anticipation is that the local people, i.e. the technicians or technologists in the developing country, will be able to acquire the techniques transferred to them. Acquiring techniques theoretically means being able to learn, understand, analyze, and explain the whys and hows of those techniques and thus, finally, being able to replicate and design them off the local technologist's own bat. It is also anticipated that the local technologist, who is the beneficiary of the transferred technology, will be able not only to adapt the received technology to suit the needs and circumstances of the developing country, but also to build on it and, if the creative capacity is available, to use it as an inspiration to create new technologies appropriate to the development requirements and objectives of the developing country.

The assumptions and anticipations underlying the transfer of technology of course presuppose the existence, locally, of an autonomous technological capacity which can competently deal with, i.e. disentangle, the intricacies of the transferred technology. In the event of the nonexistence of an adequate indigenous technological capacity, the intentions in transferring technology will hardly be achieved. There is some kind of paradox here: autonomous, indigenous technological capacities are expected to be developed *through* dealing with transferred technologies (this is certainly the ultimate goal of technology transfer); yet, the ability to deal effectively with transferred technologies requires or presupposes the existence of indigenous technological capacities adequate for the purpose. However, the paradox can be resolved if we assume that the indigenous technological capacities will exist, albeit of a minimum kind, which would, therefore, need to be nurtured, developed, and augmented to some level of sophistication required in operating a modern technology. The assumptions also presuppose that the transferred technology, developed in a specific cultural milieu different in a number of ways from that of a developing country, is easily adaptable to the social and cultural environment of the developing country. This presupposition may not be wholly true. However, despite the problems that may be said to be attendant to the transfer of technology, technology transfer is, as I said, an important medium for generating a more efficient modern technology in a developing country.

Now, technology is of course developed within a culture; it is thus an aspect—a product—of culture. Technology transfer, then, is certainly an aspect of the whole phenomenon of cultural borrowing or appropriation which follows on the encounters between cultures. There appears, however, to be a difference between transfer of technology to a developing country and the normal appropriation by a culture of an alien cultural product. The difference arises because of the way the notion of technology transfer is conceived and executed. It can be admitted that what is anticipated in technology transfer is primarily *knowledge* of techniques, methods and materials all of which are relevant to matters of industrial production. But knowledge is acquired through the active participation of the recipient; it is not transferred on to a passive agent or receptacle. In the absence of adequate and extensive knowledge and understanding of the relevant scientific principles, the attitudes of the recipients of transferred technology will only be passive, not responsive in any significant way to the niceties of the new cultural products being introduced to them. In the circumstance, that which is transferred will most probably remain a thin veneer, hardly affecting their scientific or technological outlook and orientation. Machines and equipment can be transferred to passive recipients who may be able to use them for a while; but the acquisition of knowledge (or, understanding) of techniques—which is surely involved in the proper meaning of technology—has to be prosecuted *actively*, that is, through the active exercise of the intellects of the recipients.

In an ideal situation of cultural borrowing, an element or product of the cultural tradition of one people is accepted and taken possession of by another people. The alien cultural product is not simply "transferred" to the recipients. Rather, goaded by their own appreciation of the significance of the product, they would seek it, acquire it, and appropriate it, i.e. make it their own; this means that they would participate actively and purposefully in the acquisition of the product. To the extent, (i) that what is called technology transfer is an aspect of the phenomenon of cultural borrowing, and (ii) that the people to whom some technology is transferred are, thus, expected to understand and take possession of it through active and purposeful participation in its acquisiton, "transfer of technology" is, in my view, a misnomer. For, what is transferred may not be acquired, appropriated, or assimilated.

For the same reasons, Ali Mazrui's biological metaphor of "technology transplant" will not do either. In Mazrui's view, "there has been a considerable amount of technology transfer to the Third World in the last thirty years—but very little technology transplant. Especially in Africa very little of what has been transferred has in fact been successfully transplanted."[21] To the extent, (i) that this biological or medical metaphor clearly involves passivism on the part of the recipient (i.e. the patient), who thus has no choice in actively deciding the "quality" of the foreign body tissue to be sewn onto his own body, and (ii) that there is no knowing whether the physical constitution of the recipient will accept or reject the new body tissue, the biological perception of acquiring the technological products of other cultures is very misleading. On a further ground, the biological metaphor will not do: the body onto which a foreign body tissue is to be transplanted is in a diseased condition which makes it impossible for it to react in a wholly positive manner to its new "addition" and to take advantage of it; even if we assume, analogically, that the society that is badly in need of the technological products of other cultures is technologically or epistemically "diseased," the fact would still remain that, in the case of the human society, the members of the society would, guided by their needs, be in a position not only to decide on which technological products of foreign origin they would want to acquire, but also to participate actively and positively in the appropriation of those products.

Thus, neither technology transfer nor technology transplant is a fruitful way of perceiving—and pursuing—the acquisition of technology from other cultures; neither has been a real feature or method in the phenomenon of cultural borrowing. Our historical knowledge of how the results of cultural encounters occur seems to suggest the conviction that what is needed is, not the transfer or transplant of technology, but the *appropriation* of technology—a perception or method which features the active, adroit, and purposeful initiative and participation of the recipients in the pursuit and acquisition of a technology of foreign production.

It must also be noted that just as in cultural borrowing there are surely some principles or criteria that guide the borrowers in their selection of products from the alien—i.e., the encountered—culture, so, in the appropriation of technology some principles or criteria would need to be established to guide the choice of the products of technology created in one cultural environment for use in a different environment. Technology can transform human society in numerous ways. For this reason, a developing country will have to consider technology rather as an instrument for the realization of *basic* human needs than as an end—as merely a way of demonstrating human power or ingenuity. The word "basic" is important here and is used advisedly: to point up the need for technology to be concerned fundamentally and essentially with such human needs as food, shelter, clothing, and health. The pursuit and satisfaction of these basic needs should guide the choice and appropriation of technology. Thus, what ought to be chosen is the technology that will be applied to industry, food and agriculture, water, health, housing, road and transportation, and other most relevant activities that make ordinary life bearable. On this showing, military and space exploration technologies, for instance, may not be needed by a developing country. However, as a developing country comes to be increasingly shaped by technology, certain aspects of technology will become a specialized knowledge; it will *then* become necessary to create a leaven of experts to deal with the highly specialized aspects of those technologies.

The adaptability of technological products to local circumstances and objectives must be an important criterion in the appropriation and development of technology.

Finally, the fundamental, most cherished values of a culture will also constitute a criterion in the choice of technology. Technology, I said, can transform human society. This social transformation will involve changes not only in our ways and patterns of living, but also in our values. But human beings will have to decide whether the (new) values spewed out by technology are the kinds of values we need and would want to cherish. Technology emerges in, and is fashioned by, a culture; thus, right from the outset, technology is driven or directed by human purposes, values, and goals. And, if this historical relation between technology and values is maintained, what will be produced for us by technology will (have to) be in consonance with those purposes, values, and goals. Technology was made by man, and not man for technology. This means that human beings should be the center of the focus of the technological enterprise. Technology and humanism (i.e. concern for human welfare) are—and should not be— antithetical concepts; technology and industrialism should be able to coexist with the concern for the interests and welfare of the people in the technological society. So, it should be possible for an African people to embark on the "technologicalization" of their society without losing the humanist essence of their culture. The value of concern for human well-being is a fundamental, intrinsic, and self-justifying value which should be cordoned off against any technological subversion of it. In this connection, let me refer to the views expressed by Kenneth Kaunda in the following quotation:

> I am deeply concerned that this high valuation of Man and respect for human dignity which is a legacy of our (African) tradition should not be lost in the new Africa. However "modern" and "advanced" in a Western sense the new nations of Africa may become, we are fiercely determined that this humanism will not be obscured. African society has always been Man-centred. We intend that it will remain so.[22]

I support the view that the humanist essence of African culture—an essence that is basically moral[23]—ought to be maintained and cherished in the attempt to create modernity in Africa. It must be realized that technology alone cannot solve all the deep-rooted social problems such as poverty, exploitation, economic inequalities, and oppression in human societies *unless* it is underpinned and guided by some basic moral values; in the absence of the strict application of those values, technology can in fact create other problems, including environmental problems. Social transformation, which is an outstanding goal of the comprehensive use of technology, cannot be achieved unless technology moves along under the aegis of basic human values. Technology is a human value, of course. And because it is basic to the fulfillment of the material welfare of human beings, there is a tendency to privilege it over other human values. But to do so would be a mistake. The reason is that technology is obviously an instrument value, not an intrinsic value to be pursued for its own sake. As an instrument in the whole quest for human fulfillment, its use ought to be guided by other—perhaps intrinsic and ultimate—human values, in order to realize its maximum relevance to humanity.

In considering technology's aim of fulfilling the material needs of humans, the pursuit of the humanist and social ethic of the traditional African society can be of considerable relevance because of the impact this ethic can have on the distributive patterns in respect of the economic goods that will result from the application of technology: in this way, extensive and genuine social—and in the sequel, political—transformation of the African society can be ensured, and the maximum impact of technology on society achieved.

NOTES

1. Kwame Gyekye, *An Essay on African Philosophical Thought* (Cambridge: Cambridge University Press, 1987), pp. 16–18, 106–107.

2. It is instructive to note that the Ewe word for "knowledge" is *nunya*, a word which actually means "thing observed." This clearly means that observation or experience was regarded as the source of knowledge in Ewe thought: see N. K. Dzobo, "Knowledge and Truth: Ewe and Akan Conceptions," in *Person and Community: Ghanaian Philosoph-*

ical Studies, eds. Kwame Gyekye and Kwasi Wiredu (Washington, D.C.: The Council for Research in Values and Philosophy, 1992), pp. 74ff. The empirical character of African thought generally can most probably not be doubted.

3. John S. Mbiti, *African Religions and Philosophy* (New York: Doubleday and Company, 1970), p. 1.

4. Ibid., p. 38.

5. K. A. Busia, *Africa in Search of Democracy* (New York: Praeger, 1967), p. 1.

6. Ibid., p. 7.

7. G. Parrinder, *African Traditional Religion* (New York: Harper & Row, 1962), p. 9.

8. Mbiti, *African Religions and Philosophy*, p. 74.

9. The Akan version is: *we feefee efun n'aniwa ase a, wohu saman.*

10. E. Laing, *Science and Society in Ghana*, the J. B. Danquah Memorial Lectures (Ghana Academy of Arts and Sciences, 1990), p. 21.

11. Ibid., p. 21.

12. Kofi Asare Opoku of the Institute of African Studies, University of Ghana, Legon.

13. Robert W. July, for example, says: "Art for art's sake had no place in traditional African society" and that it was "essentially functional." See his *An African Voice: The Role of The Humanities in African Independence* (Durham: Duke University Press, 1987), p. 49; also Claude Wauthier, *The Literature and Thought of Modern Africa* (London: Heinemann, 1978), pp. 173–74.

14. Lynn White, *Medieval Religion and Technology: Collected Essays* (Berkeley: University of California Press, 1978), p. 127.

15. Lord Todd, *Problems of the Technological Society*, The Aggrey-Fraser-Guggisberg Memorial Lectures (Published for the University of Ghana by the Ghana Publishing Corporation, Accra, 1973), p. 8.

16. Arnold Pacey, *The Culture of Technology* (Oxford: Basil Blackwell, 1983), p. 145.

17. Kwasi Wiredu, *Philosophy and an African Culture* (Cambridge: Cambridge University Press, 1980), p. 15.

18. S. Sefa-Dede, "Traditional Food Technology," in *Encyclopedia of Food Science, Food Technology, and Nutrition*, eds. R. Macrae, R. Robinson and M. Sadler (New York: Academy Press, 1993), p. 4600.

19. S. Sefa-Dede, ibid.; also, "Harnessing Food Technology for Development," in *Harnessing Traditional Food Technology for Development*, eds. S. Sefa-Dede and R. Orraca-Tetteh (Department of Nutrition and Food Science, University of Ghana, Legon, 1989); Esi Colecraft, "Traditional Food Preservation: An Overview," *African Technology Forum* 6, no. 1, (February/March 1993): 15–17.

20. The encounter was between this traditional woman food technologist and research scientists and students from the Department of Nutrition and Food Science of the University of Ghana headed by Professor S. Sefa-Dede. The account of the encounter presented here was given to me by Sefa-Dede both orally and in writing, and I am greatly indebted to him.

21. Ali A. Mazrui, "Africa between Ideology and Technology: Two Frustrated Forces of Change," in *African Independence: The First Twenty-Five Years*, eds. Gwendolen M. Carter and Patrick O'Meara (Bloomington: Indiana University Press, 1985), pp. 281–82.

22. Kenneth Kaunda, *A Humanist in Africa* (Nashville: Abingdon, 1966), p. 28.

23. Gyekye, *An Essay on African Philosophical Thought*, pp. 143–46.

PART 4
THE AUTONOMY OF TECHNOLOGY AND ITS PHILOSOPHICAL CRITICS

INTRODUCTION

The essays in this part concern the idea of autonomous technology, the idea that technological development follows its own path without the rational control of human beings or human society. To speak this way of autonomous technology conjures up the image of a Frankenstein monster with a will of its own. No one in this part, however, intends to attribute will or volition to technology. More precisely, the advocates of the idea of autonomous technology argue that there is no will, no human purpose driving (at least some) technological development. As we saw in part 1, Schell argued that pure science was autonomous in this sense. If technology is indeed autonomous, there are no effective "political actors" in control of technological change.

It is a common practice to label the advocates of the idea of autonomous technology—Jacques Ellul, Hans Jonas, and Langdon Winner—as antitechnology, but this characterization requires some qualification. Even Ellul, who is regarded as the paradigmatic antitechnologist, is not opposed to technology per se, but sees tragic consequences in the modern technological system that are a result of the autonomy of technology. Traditional techniques were limited to the satisfaction of specific human needs, but it is a profound mistake to think of modern technology in this way. It is not technology in the narrow sense that these authors see as dangerous, but more broadly the modern "system." They do, however, call into question the orthodox faith in inevitable progress, and in this sense could be thought of as opposed to technology.

Jacques Ellul, in the selection "The Autonomy of Technique" (also taken from his book, *The Technological Society*), takes a very hard line in regard to the autonomy of technique. As we saw in Ellul's essay in part 2, the problem is not the machine, but the extension of machinelike thinking and procedures. Perhaps this can best be understood if we recall Pacey's concept of technology-practice.

157

According to Ellul, Pacey's "technical aspect" is no mere aspect but a determinant of the whole of technology-practice.

Although in principle the cultural and organizational aspects could interact and modify the technical, in practice they have become ineffective in contemporary society. While Ellul claims that technique evolves independently of economics, politics, and the social situation, his point can be better put by saying that these cultural and organizational factors can no longer be distinguished from technique. Traditional cultural values and ways of organizing our lives have all been swept away and transformed by the imperatives of technological "progress." Technique has created a new civilization.

Although persons participate in technological development, they turn out to be co-conspirators on Ellul's view, since their traditional values have been replaced by the worship of "progress." Human beings themselves become a means to the increasingly efficient perfection of means. Thus it happens that technique evolves according to its own internal laws, free from any decisive interference from external factors. Every next step in technological development is inevitably determined by every preceding step, beyond the intervention or desire of any single human agency. In this sense, it is not that technique is literally autonomous but that it builds its own momentum. Hiroshima was destroyed by the atomic bomb, not because it was necessary to win the war, but because it was the necessary next step in the development of atomic energy.

Of course new technologies create new problems, but this only fuels the demand for further development to find "technical fixes" for these problems in a never-ending circle. This is an all-or-nothing situation. Once a society opts for technological development, that is its last option. Traditional ideologies (democracy, capitalism, or communism) become irrelevant in the technological society, and political distinctions between societies become blurred. Mikhail Gorbachev's choice of technological development sounded the death knell of traditional communism in his country. Although Ellul denies being a determinist, he sees no way to prevent "a worldwide totalitarian dictatorship which will allow technique its full scope."

In the second reading, "Toward a Philosophy of Technology," Hans Jonas is primarily concerned with revealing some important, and much neglected, philosophical issues embedded in technology. The task is immense, but we can begin by selecting the most obvious aspects of this "focal fact of modern life" under the categories of (1) formal dynamics, (2) material works, and (3) ethics.

Although Jonas is more circumspect than Ellul, he clearly agrees with the central tenets of the autonomous technology thesis. Viewed from an abstract, formal perspective, modern technology is clearly not a possession or state—a mere tool—but a restless, dynamic process. This force does not aim at an equilibrium, but is constantly disequilibrating. Following its own "laws of motion," each new stage in the evolution of technology gives us the next developments. There is a technological imperative at work here; "can" becomes "must" in our technological society. Thus, it seems that "technology is destiny."

Having described the traits of modern technology, Jonas analyzes what he considers to be the major forces driving technological development: pressures of competition, population growth and scarcity of natural resources, the idea of progress, and the need for coordination and control. Although the competitive pressures for profit inhere in capitalism, the rest operate independently of the economic system. The technological system is out of control.

Jonas's discussion of the material works of technology makes the point that technological objects are not mere means but radical transformations of our way of life, creating a life that is qualitatively different. The revolution in biology may even, for example, make it possible to engineer a human being. Surely, then, we need to redirect our most serious philosophical thinking in grappling with these unprecedented possibilities.

The point in describing our situation with regard to technology is to do something about it. In contrast to the impression Ellul gives, Jonas sees no "hard" determinism in the process of technology. The first step in gaining control, in opposing the enormous inertial forces at work, is to become aware of how the "system" works. But to do something requires social and political action, and to be effective we may have to employ those "techniques" we despise: propaganda, persuasion, indoctrination, and manipulation. However, "the best hope of man rests in his most troublesome gift: the spontaneity of human acting which confounds all prediction."

Langdon Winner, in "Reverse Adaptation and Control," does not argue that technology in general is autonomous, but that autonomy is characteristic of some technological complexes. While the conventional wisdom continues to think of technology as a means to preestablished human ends, modern, large-scale systems frequently reverse the means-ends distinction. Rather than a linear view of technological development in which desired ends drive the quest for means, we find a circularity here in which the means provided by technology drive the quest for ends as means to further means. On this model, ends are invented ex post facto for public relations as a means to keep the system going. While all specific technologies are purposive in their original design, the complexity of the system sometimes produces means for which some special purpose has to be designed if there is to be "progress."

In supporting this view, Winner describes in some detail five ways in which, to quote philosopher Herbert Marcuse, the technological society's "sweeping rationality, which propels efficiency and growth, is itself irrational." Curiously enough, Winner claims that he "is not arguing that there is anything inherently wrong with this." However, it is clear that if our social and political ideals are to guide technological development, this increasingly prevalent tendency of large-scale systems needs to be understood.

The last three essays in this part offer refinements or criticisms of the thesis of autonomous technology. In "Artifacts, Neutrality, and the Ambiguity of 'Use,'" Russell Woodruff closely examines one of the central philosophical issues concerning the thesis of technological autonomy: the alleged neutrality of technological objects. Critics of the idea of autonomous technology and those advocates of technological expansion (such as the authors in part 3 and Samuel Florman in the essay following this one) contend that technological artifacts are in themselves value-neutral. The use of these technological tools or instruments confers value upon them. Since human beings or human institutions control the use and development of technology, technology is not autonomous, but rather guided by human goals and values. Woodruff challenges this claim—and thereby indirectly supports the autonomous technology thesis—by showing that the concept of "use" has several different and ambiguous meanings, many of which are clearly not value-neutral. The word "use" has at least three different senses—act, purpose, and method—and within each of these senses there are ways in which the use of technology is not neutral. For example, the idea of "purpose" can be applied to the use of artifacts in two ways, as the *goal* of some humans using the technological artifact or as the *function* of the artifact itself. Human goals are clearly not value-neutral. Moreover, Woodruff demonstrates that if we focus on the function of a technological object, we are constrained in the *method* of using it. To function properly, an artifact must be used in the proper way. This proscribed method of use also tends to vitiate any sense of the neutrality of the technological artifact. Although Woodruff makes no claims that his argument is important as a defense of the autonomous technology thesis—indeed, he does not even mention the idea of autonomous technology—his essay is a useful analysis of a key theoretical issue in the debate over the value of technology and the control of technological development.

In his essay, "In Praise of Technology," Samuel Florman decries attempts to alter the current direction of technology. Florman recognizes that we face serious social problems brought on in part by technology. The orthodox belief that technology is necessarily good is a myth. However, Florman argues that critics of technological development such as Ellul have simply substituted a new myth, namely that technology has escaped human control and become autonomous. This belief of Ellul, Florman argues, is so absurd that it hardly deserves serious refutation. The fundamental mistake, he thinks, is the reification of technology, that is, treating the abstract notion of technology as if it were a thing existing independently of its human context. There is no such thing as technology in the abstract; there exist only specific technologies. Some observers have become so obsessed with the unanticipated and undesirable consequences of modern technology that they have forgotten the intended, and central, results that do satisfy people's desires.

Florman argues that although the critics of technology assume a high moral purpose, there is a strong totalitarian bias in their thinking. The critics denigrate the intelligence of ordinary citizens when they claim that technology forces men to work, to consume, and to live in cities. The fact is that persons freely choose their style of life. They want to raise their material standard of living and are quite willing to work hard in order to achieve this. Modern technology has provided the opportunity for people to live better while at the same time expanding their political freedoms. To think that life was more humane, more satisfying, or more free in preindustrial society is pure, unfounded romanticism.

The problems we confront, Florman claims, are not caused by technology (or, by his own profession, engineering) but by indomitable human nature. "The vast majority of people in the world want to move forward, whatever the consequences." The basic problem, Florman thinks, is obvious: there are simply too many people wanting too many things. To address the real dangers, we must reject the negative, vacuous, and mystifying doctrines of the advocates of the idea of autonomous technology. Progress is not automatic, but with a good dose of common sense, courage, and citizen activism, the promise of technology may still be realized.

Finally, in "The Autonomy of Technology," Joseph Pitt more directly attacks the autonomous technology thesis in an essay that combines a philosophical argument with a historical example. Pitt builds on the criticism offered by Florman that the central mistake of Ellul and his followers is the reification of technology—thinking that Technology with a capital T is some kind of living thing with a mind of its own. There is an illusion of an autonomous technology, thinks Pitt, but this is only the result of the "lack of absolute predictability" in the outcomes of technological developments and new inventions. Critics of technology tend to forget that at each stage of technological development there is a good reason for using a particular tool or instrument, a good reason to "make concessions" to the technological development thereby altering human behavior.

Pitt's major example, however, is the development of the telescope and its use by Galileo in the science of astronomy to help prove the Copernican theory of the heliocentric universe. His central claim is that the development of science and technology are interrelated. Science reveals what kind of technological instruments are possible and what kind are needed; in turn, technological developments make possible new discoveries that alter scientific theories. On this view, once science and a particular technological artifact are seen to be interrelated, then the development of neither can be autonomous. The technological instrument—in this case, the telescope—did not force Galileo to challenge the Church over the truth of Aristotle's view of the universe. The telescope that Galileo built simply gave him an instrument, a weapon, to use in his battle. No technology forces humans to use it. For Pitt, the advocates of autonomous technology are really afraid that they cannot control the society and the individuals who create and use technology. Human beings are the real problem, not the technological artifacts—"the tools by themselves do nothing."

As we move along into the first decade of the twenty-first century, it is fascinating to read Ellul, writing from the vantage point of the early 1960s. At the end of the selection printed here, he lists some of the sanguine predictions by his scientific contemporaries for the year 2000: the elimination of disease and universal hunger, an end to fuel shortages, and routine trips to the moon. Needless to say, science seems to have failed to deliver on many of its promises. Worsening global pollution, the *Challenger* disaster, and the proliferation of AIDS indicate that the "perfect world" the savants envisioned is much further off. What do the failures of contemporary science and technology mean for the thesis of autonomous technology? What do these failures mean for the hegemony of science and the "inevitability" of technological progress?

10

The Autonomy of Technique

Jacques Ellul

The primary aspect of autonomy is perfectly expressed by Frederick Winslow Taylor, a leading technician. He takes, as his point of departure, the view that the industrial plant is a whole in itself, a "closed organism," an end in itself. Giedion adds: "What is fabricated in this plant and what is the goal of its labor— these are questions outside its design." The complete separation of the goal from the mechanism, the limitation of the problem to the means, and the refusal to interfere in any way with efficiency; all this is clearly expressed by Taylor and lies at the basis of technical autonomy.

Autonomy is the essential condition for the development of technique, as Ernst Kohn-Bramstedt's study of the police clearly indicates. The police must be independent if they are to become efficient. They must form a closed, autonomous organization in order to operate by the most direct and efficient means and not be shackled by subsidiary considerations. And in this autonomy, they must be self-confident in respect to the law. It matters little whether police action is legal, if it is efficient. The rules obeyed by a technical organization are no longer rules of justice or injustice. They are "laws" in a purely technical sense. As far as the police are concerned, the highest stage is reached when the legislature legalizes their independence of the legislature itself and recognizes the primacy of technical laws. This is the opinion of Best, a leading German specialist in police matters.

The autonomy of technique must be examined in different perspectives on the basis of the different spheres in relation to which it has this characteristic. First, technique is autonomous with respect to economics and politics. We have already seen that, at the present, neither economic nor political evolution condi-

From *The Technological Society*, by Jacques Ellul, translated by John Wilkinson (1964), published by Jonathan Cape. Reprinted by permission of the The Random House Group Ltd.

tions technical progress.* Its progress is likewise independent of the social situation. The converse is actually the case, a point I shall develop at length. Technique elicits and conditions social, political, and economic change. It is the prime mover of all the rest, in spite of any appearance to the contrary and in spite of human pride, which pretends that man's philosophical theories are still determining influences and man's political regimes decisive factors in technical evolution. External necessities no longer determine technique. Technique's own internal necessities are determinative. Technique has become a reality in itself, self-sufficient, with its special laws and its own determinations.

Let us not deceive ourselves on this point. Suppose that the state, for example, intervenes in a technical domain. Either it intervenes for sentimental, theoretical, or intellectual reasons, and the effect of its intervention will be negative or nil; or it intervenes for reasons of political technique, and we have the combined effect of two techniques. There is no other possibility. The historical experience of the last years shows this fully.

To go one step further, technical autonomy is apparent in respect to morality and spiritual values. Technique tolerates no judgment from without and accepts no limitation. It is by virtue of technique rather than science that the great principle has become established: *chacun chez soi.*† Morality judges moral problems; as far as technical problems are concerned, it has nothing to say. Only technical criteria are relevant. Technique, in sitting in judgment on itself, is clearly freed from this principal obstacle to human action. (Whether the obstacle is valid is not the question here. For the moment we merely record that it is an obstacle.) Thus, technique theoretically and systematically assures to itself that liberty which it has been able to win practically. Since it has put itself beyond good and evil, it need fear no limitation whatever. It was long claimed that technique was neutral. Today this is no longer a useful distinction. The power and autonomy of technique are so well secured that it, in its turn, has become the judge of what is moral, the creator of a new morality. Thus, it plays the role of creator of a new civilization as well. This morality—internal to technique—is assured of not having to suffer from technique. In any case, in respect to traditional morality, technique affirms itself as an independent power. Man alone is subject, it would seem, to moral judgment. We no longer live in that primitive epoch in which things were good or bad in themselves. Technique in itself is neither, and can therefore do what it will. It is truly autonomous.

However, technique cannot assert its autonomy in respect to physical or biological laws. Instead, it puts them to work; it seeks to dominate them.

Giedion, in his probing study of mechanization and the manufacture of bread, shows that "wherever mechanization encounters a living substance, bacterial or animal, the organic substance determines the laws." For this reason, the

---

*See Part 2, chapter 4.
† "Each [is master] in his own house." (Ed.)

mechanization of bakeries was a failure. More subdivisions, intervals, and precautions of various kinds were required in the mechanized bakery than in the nonmechanized bakery. The size of the machines did not save time; it merely gave work to larger numbers of people. Giedion shows how the attempt was made to change the nature of the bread in order to adapt it to mechanical manipulations. In the last resort, the ultimate success of mechanization turned on the transformation of human taste. Whenever technique collides with a natural obstacle, it tends to get around it either by replacing the living organism by a machine, or by modifying the organism so that it no longer presents any specifically organic reaction.

The same phenomenon is evident in yet another area in which technical autonomy asserts itself: the relations between techniques and man. We have already seen, in connection with technical self-augmentation, that technique pursues its own course more and more independently of man. This means that man participates less and less actively in technical creation, which, by the automatic combination of prior elements, becomes a kind of fate. Man is reduced to the level of a catalyst. Better still, he resembles a slug inserted into a slot machine: he starts the operation without participating in it.

But this autonomy with respect to man goes much further. To the degree that technique must attain its result with mathematical precision, it has for its object the elimination of all human variability and elasticity. It is a commonplace to say that the machine replaces the human being. But it replaces him to a greater degree than has been believed.

Industrial technique will soon succeed in completely replacing the effort of the worker, and it would do so even sooner if capitalism were not an obstacle. The worker, no longer needed to guide or move the machine to action, will be required merely to watch it and to repair it when it breaks down. He will not participate in the work any more than a boxer's manager participates in a prize fight. This is no dream. The automated factory has already been realized for a great number of operations, and it is realizable for a far greater number. Examples multiply from day to day in all areas. Man indicates how this automation and its attendant exclusion of men operates in business offices; for example, in the case of the so-called tabulating machine.* The machine itself interprets the data, the elementary bits of information fed into it. It arranges them in texts and distinct numbers. It adds them together and classifies the results in groups and subgroups, and so on. We have here an administrative circuit accomplished by a single, self-controlled machine. It is scarcely necessary to dwell on the astounding growth of automation in the last ten years. The multiple applications of the automatic assembly line, of automatic control of production operations (so-called cybernetics) are well known. Another case in point is the automatic pilot. Until recently the automatic pilot was used only in rectilinear flight; the finer operations were carried out by the living pilot. As early as 1952 the automatic pilot effected the

---

*We might substitute for Ellul's tabulating machine the personal computer. (Ed.)

operations of take-off and landing for certain supersonic aircraft. The same kind of feat is performed by automatic direction finders in anti-aircraft defense. Man's role is limited to inspection. This automation results from the development servomechanisms which act as substitutes for human beings in more and more subtle operations by virtue of their "feedback" capacity.

This progressive elimination of man from the circuit must inexorably continue. Is the elimination of man so unavoidably necessary? Certainly! Freeing man from toil is in itself an ideal. Beyond this, every intervention of man, however educated or used to machinery he may be, is a source of error and unpredictability. The combination of man and technique is a happy one only if man has no responsibility. Otherwise, he is ceaselessly tempted to make unpredictable choices and is susceptible to emotional motivations which invalidate the mathematical precision of the machinery. He is also susceptible to fatigue and discouragement. All this disturbs the forward thrust of technique.

Man must have nothing decisive to perform in the course of technical operations; after all, he is the source of error. Political technique is still troubled by certain unpredictable phenomena, in spite of all the precision of the apparatus and the skill of those involved. (But this technique is still in its childhood.) In human reactions, howsoever well calculated they may be, a "coefficient of elasticity" causes imprecision, and imprecision is intolerable to technique. As far as possible, this source of error must be eliminated. Eliminate the individual, and excellent results ensue. Any technical man who is aware of this fact is forced to support the opinions voiced by Robert Jungk, which can be summed up thus: "The individual is a brake on progress." Or: "Considered from the modern technical point of view, man is a useless appendage." For instance, ten percent of all telephone calls are wrong numbers, due to human error. An excellent use by man of so perfect an apparatus!

Now that statistical operations are carried out by perforated-card machines instead of human beings, they have become exact. Machines no longer perform merely gross operations. They perform a whole complex of subtle ones as well. And before long—what with the electronic brain—they will attain an intellectual power of which man is incapable.

Thus, the "great changing of the guard" is occurring much more extensively than Jacques Duboin envisaged some decades ago. Gaston Bouthoul, a leading sociologist of the phenomena of war, concludes that war breaks out in a social group when there is a "plethora of young men surpassing the indispensable tasks of the economy." When for one reason or another these men are not employed, they become ready for war. It is the multiplication of men who are excluded from working which provokes war. We ought at least to bear this in mind when we boast of the continual decrease in human participation in technical operations.

However, there are spheres in which it is impossible to eliminate human influence. The autonomy of technique then develops in another direction. Technique is not, for example, autonomous in respect to clock time. Machines, like abstract technical laws, are subject to the law of speed, and coordination presupposes time adjustment. In his description of the assembly line, Giedion writes: "Extremely precise time tables guide the automatic cooperation of the instruments, which, like the atoms in a planetary system, consist of separate units but gravitate with respect to each other in obedience to their inherent laws." This image shows in a remarkable way how technique became simultaneously independent of man and obedient to the chronometer. Technique obeys its own specific laws, as every machine obeys laws. Each element of the technical complex follows certain laws determined by its relations with the other elements, and these laws are internal to the system and in no way influenced by external factors. It is not a question of causing the human being to disappear, but of making him capitulate, of inducing him to accommodate himself to techniques and not to experience personal feelings and reactions.

No technique is possible when men are free. When technique enters into the realm of social life, it collides ceaselessly with the human being to the degree that the combination of man and technique is unavoidable, and that technical action necessarily results in a determined result. Technique requires predictability and, no less, exactness of prediction. It is necessary, then, that technique prevail over the human being. For technique, this is a matter of life or death. Technique must reduce man to a technical animal, the king of the slaves of technique. Human caprice crumbles before this necessity; there can be no human autonomy in the face of technical autonomy. The individual must be fashioned by techniques, either negatively (by the techniques of understanding man) or positively (by the adaptation of man to the technical framework), in order to wipe out the blots his personal determination introduces into the perfect design of the organization.

But it is requisite that man have certain precise inner characteristics. An extreme example is the atomic worker or the jet pilot. He must be of calm temperament, and even temper, he must be phlegmatic, he must not have too much initiative, and he must be devoid of egotism. The ideal jet pilot is already along in years (perhaps thirty-five) and has a settled direction in life. He flies his jet in the way a good civil servant goes to his office. Human joys and sorrows are fetters on technical aptitude. Jungk cites the case of a test pilot who had to abandon his profession because "his wife behaved in such a way as to lessen his capacity to fly. Every day, when he returned home, he found her shedding tears of joy. Having become in this way accident conscious, he dreaded catastrophe when he had to face a delicate situation." The individual who is a servant of technique must be completely unconscious of himself. Without this quality, his reflexes and his inclinations are not properly adapted to technique.

Moreover, the physiological condition of the individual must answer to technical demands. Jungk gives an impressive picture of the experiments in training and control that jet pilots have to undergo. The pilot is whirled on centrifuges until he "blacks out" (in order to measure his toleration of acceleration). There are catapults, ultrasonic chambers, etc., in which the candidate is forced to undergo unheard-of tortures in order to determine whether he has adequate resistance and whether he is capable of piloting the new machines. That the human organism is, technically speaking, an imperfect one is demonstrated by the experiments. The sufferings the individual endures in these "laboratories" are considered to be due to "biological weaknesses," which must be eliminated. New experiments have pushed even further to determine the reactions of "space pilots" and to prepare these heroes for their roles of tomorrow. This has given birth to new sciences, biometry for example; their one aim is to create the new man, the man adapted to technical functions.

It will be objected that these examples are extreme. This is certainly the case, but to a greater or lesser degree the same problem exists everywhere. And the more technique evolves, the more extreme its character becomes. The object of all the modern "human sciences" is to find answers to these problems.

The enormous effort required to put this technical civilization into motion supposes that all individual effort is directed toward this goal alone and that all social forces are mobilized to attain the mathematically perfect structure of the edifice. ("Mathematically" does not mean "rigidly." The perfect technique is the most adaptable and, consequently, the most plastic one. True technique will know how to maintain the illusion of liberty, choice, and individuality; but these will have been carefully calculated so that they will be integrated into the mathematical reality merely as appearances!) Henceforth it will be wrong for a man to escape this universal effort. It will be inadmissible for any part of the individual not to be integrated in the drive toward technicization; it will be inadmissible that any man even aspire to escape this necessity of the whole society. The individual will no longer be able, materially or spiritually, to disengage himself from society. Materially, he will not be able to release himself because the technical means are so numerous that they invade his whole life and make it impossible for him to escape the collective phenomena. There is no longer an uninhabited place, or any other geographical locale, for the would-be solitary. It is no longer possible to refuse entrance into a community to a highway, a high-tension line, or a dam. It is vain to aspire to live alone when one is obliged to participate in all collective phenomena and to use all the collective's tools, without which it is impossible to earn a bare subsistence. Nothing is gratis any longer in our society; and to live on charity is less and less possible. "Social advantages" are for the workers alone, not for "useless mouths." The solitary is a useless mouth and will have no ration card—up to the day he is transported to a penal colony. (An attempt was made to institute this procedure during the French Revolution, with deportations to Cayenne.)

Spiritually, it will be impossible for the individual to disassociate himself from society. This is due not to the existence of spiritual techniques which have increasing force in our society, but rather to our situation. We are constrained to be "engaged," as the existentialists say, with technique. Positively or negatively, our spiritual attitude is constantly urged, if not determined, by this situation. Only bestiality, because it is unconscious, would seem to escape this situation, and it is itself only a product of the machine.

Every conscious being today is walking the narrow ridge of a decision with regard to technique. He who maintains that he can escape it is either a hypocrite or unconscious. The autonomy of technique forbids the man of today to choose his destiny. Doubtless, someone will ask if it has not always been the case that social conditions, environment, manorial oppression, and the family conditioned man's fate. The answer is, of course, yes. But there is no common denominator between the suppression of ration cards in an authoritarian state and the family pressure of two centuries ago. In the past, when an individual entered into conflict with society, he led a harsh and miserable life that required a vigor which either hardened or broke him. Today the concentration camp and death await him; technique cannot tolerate aberrant activities.

Because of the autonomy of technique, modern man cannot choose his means any more than his ends. In spite of variability and flexibility according to place and circumstance (which are characteristic of technique) there is still only a single employable technique in the given place and time in which an individual is situated.

At this point, we must consider the major consequences of the autonomy of technique. This will bring us to the climax of this analysis.

Technical autonomy explains the "specific weight" with which technique is endowed. It is not a kind of neutral matter, with no direction, quality, or structure. It is a power endowed with its own peculiar force. It refracts in its own specific sense the wills which make use of it and the ends proposed for it. Indeed, independently of the objectives that man pretends to assign to any given technical means, that means always conceals in itself a finality which cannot be evaded. And if there is a competition between this intrinsic finality and an extrinsic end proposed by man, it is always the intrinsic finality which carries the day. If the technique in question is not exactly adapted to a proposed human end, and if an individual pretends that he is adapting the technique to this end, it is generally quickly evident that it is the end which is being modified, not the technique. Of course, this statement must be qualified by what has already been said concerning the endless refinement of techniques and their adaptation. But this adaptation is effected with reference to the techniques concerned and to the conditions of their applicability. It does not depend on external ends. Perrot has demonstrated this in the case of judicial techniques, and Giedion in the case of mechanical techniques. Concerning the overall problem of the relation between the ends and the means, I take the liberty of referring to my own work, *Présence au monde moderne.*

Once again we are faced with a choice of "all or nothing." If we make use of technique, we must accept the specificity and autonomy of its ends, and the totality of its rules. Our own desires and aspirations can change nothing.

The second consequence of technical autonomy is that it renders technique at once sacrilegious and sacred. (*Sacrilegious* is not used here in the theological but in the sociological sense.) Sociologists have recognized that the world in which man lives is for him not only a material but also a spiritual world; that forces act in it which are unknown and perhaps unknowable; that there are phenomena in it which man interprets as magical; that there are relations and correspondences between things and beings in which material connections are of little consequence. This whole area is mysterious. Mystery (but not in the Catholic sense) is an element of man's life. Jung has shown that it is catastrophic to make superficially clear what is hidden in man's innermost depths. Man must make allowance for a background, a great deep above which lie his reason and his clear consciousness.

The characteristics we have examined permit me to assert with confidence that there is no common denominator between the technique of today and that of yesterday. Today we are dealing with an utterly different phenomenon. Those who claim to deduce from man's technical situation in past centuries his situation in this one show that they have grasped nothing of the technical phenomenon. These deductions prove that all their reasonings are without foundation and all their analogies are astigmatic.

The celebrated formula of Alain has been invalidated: "Tools, instruments of necessity, instruments that neither lie nor cheat, tools with which necessity can be subjugated by obeying her, without the help of false laws; tools that make it possible to conquer by obeying." This formula is true of the tool which puts man squarely in contact with a reality that will bear no excuses, in contact with matter to be mastered, and the only way to use it is to obey it. Obedience to the plow and the plane was indeed the only means of dominating earth and wood. But the formula is not true for our techniques. He who serves these techniques enters another realm of necessity. This new necessity is not natural necessity; natural necessity, in fact, no longer exists. It is technique's necessity, which becomes the more constraining the more nature's necessity fades and disappears. It cannot be escaped or mastered. The tool was not false. But technique causes us to penetrate into the innermost realm of falsehood, showing us all the while the noble face of objectivity of result. In this innermost recess, man is no longer able to recognize himself because of the instruments he employs.

The tool enables man to conquer. But, man, dost thou not know there is no more victory which is thy victory? The victory of our days belongs to the tool. The tool alone has the power and carries off the victory. Man bestows on himself the laurel crown, after the example of Napoleon III, who stayed in Paris to plan the strategy of the Crimean War and claimed the bay leaves of the victor.

But this delusion cannot last much longer. The individual obeys and no longer has victory which is his own. He cannot have access even to his apparent triumphs except by becoming himself the object of technique and the offspring of the mating of man and machine. All his accounts are falsified. Alain's definition no longer corresponds to anything in the modern world. In writing this, I have, of course, omitted innumerable facets of our world. There are still artisans, petty tradesmen, butchers, domestics, and small agricultural landowners. But theirs are the faces of yesterday, the more or less hardy survivals of our past. Our world is not made of these static residues of history, and I have attempted to consider only moving forces. In the complexity of the present world, residues do exist, but they have no future and are consequently disappearing . . . .

A LOOK AT THE YEAR 2000

In 1960 the weekly *Express* of Paris published a series of extracts from texts by American and Russian scientists concerning society in the year 2000. As long as such visions were purely a literary concern of science-fiction writers and sensational journalists, it was possible to smile at them.* Now we have like works from Nobel Prize winners, members of the Academy of Sciences of Moscow, and other scientific notables whose qualifications are beyond dispute. The visions of these gentlemen put science fiction in the shade. By the year 2000, voyages to the moon will be commonplace; so will inhabited artificial satellites. All food will be completely synthetic. The world's population will have increased fourfold but will have been stabilized. Sea water and ordinary rocks will yield all the necessary metals. Disease, as well as famine, will have been eliminated; and there will be universal hygienic inspection and control. The problems of energy production will have been completely resolved. Serious scientists, it must be repeated, are the source of these predictions, which hitherto were found only in philosophic utopias.

The most remarkable predictions concern the transformation of educational methods and the problem of human reproduction. Knowledge will be accumulated in "electronic banks" and transmitted directly to the human nervous system by means of coded electronic messages. There will no longer be any need of reading or learning mountains of useless information; everything will be received and registered according to the needs of the moment. There will be no need of attention or effort. What is needed will pass directly from the machine to the brain without going through consciousness.

In the domain of genetics, natural reproduction will be forbidden. A stable population will be necessary, and it will consist of the highest human types. Artificial insemination will be employed. This, according to Muller, will "permit the introduction into a carrier uterus of an ovum fertilized *in vitro*, ovum and sperm . . . having been taken from persons representing the masculine ideal and the feminine ideal, respectively. The reproductive cells in question will preferably be those of persons dead long enough that a true perspective of their lives and works, free of all personal prejudice, can be seen. Such cells will be taken from cell

---

*Some excellent works, such as Robert Jungk's *Le fusur a déja commencé,* were included in this classification.

banks and will represent the most precious genetic heritage of humanity. . . . The method will have to be applied universally. If the people of a single country were to apply it intelligently and intensively . . . they would quickly attain a practically invincible level of superiority. . . ." Here is a future Huxley never dreamed of.

Perhaps, instead of marveling or being shocked, we ought to reflect a little. A question no one ever asks when confronted with the scientific wonders of the future concerns the interim period. Consider, for example, the problems of automation, which will become acute in a very short time. How, socially, politically, morally, and humanly, shall we contrive to get there? How are the prodigious economic problems, for example, of unemployment, to be solved? And, in Muller's more distant utopia, how shall we force humanity to refrain from begetting children naturally? How shall we force them to submit to constant and rigorous hygienic controls? How shall man be persuaded to accept a radical transformation of his traditional modes of nutrition? How and where shall we relocate a billion and a half persons who today make their livings from agriculture and who, in the promised ultrarapid conversion of the next forty years, will become completely useless as cultivators of the soil? How shall we distribute such numbers of people equably over the surface of the earth, particularly if the promised fourfold increase in population materializes? How will we handle the control and occupation of outer space in order to provide a stable *modus vivendi?* How shall national boundaries be made to disappear? (One of the last two would be a necessity.) There are many other "hows," but they are conveniently left unformulated. When we reflect on the serious although relatively minor problems that were provoked by the industrial exploitation of coal and electricity, when we reflect that after a hundred and fifty years these problems are still not satisfactorily resolved, we are entitled to ask whether there are any solutions to the infinitely more complex "hows" of the next forty years. In fact, there is one and only one means to their solution, a world-wide totalitarian dictatorship which will allow technique its full scope and at the same time resolve the concomitant difficulties. It is not difficult to understand why the scientists and worshippers of technology prefer not to dwell on this solution, but rather to leap nimbly across the dull and uninteresting intermediary period and land squarely in the golden age. We might indeed ask ourselves if we will succeed in getting through the transition period at all, or if the blood and the suffering required are not perhaps too high a price to pay for this golden age.

If we take a hard, unromantic look at the golden age itself, we are struck with the incredible naïveté of these scientists. They say, for example, that they will be able to shape and reshape at will human emotions, desires, and thoughts and arrive scientifically at certain efficient, preestablished collective decisions. They claim they will be in a position to develop certain collective desires, to constitute certain homogeneous social units out of aggregates of individuals, to forbid men to raise their children, and even to persuade them to renounce having any. At the same time, they speak of assuring the triumph of freedom and of the necessity of avoiding dictatorship at any price.* They seem incapable of grasping the contradiction involved, or of understanding that what they are proposing, even after the intermediary period, is in fact the harshest of dictatorships. In comparison, Hitler's was a trifling affair. That it is to be a dictatorship of test tubes rather than of hobnailed boots will not make it any less a dictatorship.

When our savants characterize their golden age in any but scientific terms, they emit a quantity of down-at-the-heel platitudes that would gladden the heart of the pettiest politician. Let's take a few samples. "To render human nature nobler, more beautiful, and more harmonious." What on earth can this mean? What criteria, what content, do they propose? Not many, I fear, would be able to reply. "To assure the triumph of peace, liberty, and reason." Fine words with no substance behind them. "To eliminate cultural lag." What culture? And would the culture they have in mind be able to subsist in this harsh social organization? "To conquer outer space." For what purpose? The conquest of space seems to be an end in itself, which dispenses with any need for reflection.

We are forced to conclude that our scientists are incapable of any but the emptiest platitudes when they stray from their specialties. It makes one think back on the collection of mediocrities accumulated by Einstein when he spoke of God, the state, peace, and the meaning of life. It is clear that Einstein, extraordinary mathematical genius that he was, was no Pascal; he knew nothing of political or human reality, or, in fact, anything at all outside his mathematical reach. The banality of Einstein's remarks in matters outside his specialty is as astonishing as his genius within it. It seems as though the specialized application of all one's faculties in a particular area inhibits the consideration of things in general. Even J. Robert Oppenheimer,† who seems receptive to a general culture, is not outside this judgment. His political and social declarations, for example, scarcely go beyond the level of those of the man in the street. And the opinions of the scientists quoted by *l'Express* are not even on the level of Einstein or Oppenheimer. Their pomposities, in fact, do not rise to the level of the average. They are vague generalities inherited from the nineteenth century, and the fact that they represent the furthest limits of thought of our scientific worthies must be symptomatic of arrested development or of a mental block. Particularly disquieting is the gap between the enormous power they wield and their critical ability, which must be estimated as null. To wield power well entails a certain faculty of criticism, discrimination, judgment, and option. It is impossible to have confidence in men who apparently lack these faculties. Yet it is apparently our fate to be facing a "golden age" in the power of the sorcerers who are totally blind to the meaning of the human adventure. When they speak of preserving the seed of outstanding men, whom, pray, do they mean to be the judges? It is clear, alas, that they propose to sit in judgment themselves. It is hardly likely that they will deem a Rim-

---

*The material here and below is cited from actual texts.

†J. Robert Oppenheimer (1904–1967). Physicist and director of the Manhattan Project that developed the atomic bomb. (Ed)

baud or a Nietszche worthy of posterity. When they announce that they will conserve the genetic mutations which appear to them most favorable, and that they propose to modify the very germ cells in order to produce such and such traits; and when we consider the mediocrity of the scientists themselves outside the confines of their specialties, we can only shudder at the thought of what they will esteem most "favorable."

None of our wise men ever pose the question of the end of all their marvels. The "wherefore" is resolutely passed by. The response that would occur to our contemporaries is: for the sake of happiness. Unfortunately, there is no longer any question of that. One of our best-known specialists in diseases of the nervous system writes: "We will be able to modify man's emotions, desires, and thoughts, as we have already done in a rudimentary way with tranquillizers." It will be possible, says our specialist, to produce a conviction or an impression of happiness without any real basis for it. Our man of the golden age, therefore, will be capable of happiness amid the worst privations. Why, then, promise us extraordinary comforts, hygiene, knowledge, and nourishment if, by simply manipulating our nervous systems, we can be happy without them? The last meager motive we could possibly ascribe to the technical adventure thus vanishes into thin air through the very existence of technique itself.

But what good is it to pose questions of motives? of Why? All that must be the work of some miserable intellectual who balks at technical progress. The attitude of the scientists, at any rate, is clear. Technique exists because it is technique. The golden age will be because it will be. Any other answer is superfluous.

REFERENCES

Bouthol, Gaston. *La Guerre [War]*. Paris: Presses Universitaires de France, 1953.

Duboin, Jacques. *La grande relève des hommes par la machine [The Great Replacement of Man by the Machine]*. Paris: Les Editions Nouvelles, 1932.

Ellul, Jacques. *Présence au monde moderne [Presence in the Modern Word]*. Geneva: Roulet, 1948.

Giedion, Siegfried. *Mechanization Takes Command*. New York: Oxford University Press, 1948.

Jung, Carl Gustav. *Modern Man in Search of a Soul*. New York: Harcourt Brace. 1956.

Jungk, Robert. *Die Zukunft hat schon begonnen: Amerikas Allmacht und Ohnmacht*. Stuttgart: Scherz and Goverts, 1952. [Translated as *Tomorrow Is Already Here: Scenes from a Man Made World*. London: R. Hart-Davis, 1954.]

Kohn-Bramstedt, Ernst. *Dictatorship and Political Police: The Technique of Control by Fear*. London: K. Paul, Trench, Trubner, 1945.

Taylor, Frederick Winslow. *The Principles of Scientific Management, 1911*. Reprint New York: Norton, 1947, 1967.

Toward a Philosophy of Technology

Hans Jonas

Are there philosophical aspects to technology? Of course there are, as there are to all things of importance in human endeavor and destiny. Modern technology touches on almost everything vital to man's existence—material, mental, and spiritual. Indeed, what of man is *not* involved? The way he lives his life and looks at objects, his intercourse with the world and with his peers, his powers and modes of action, kinds of goals, states and changes of society, objectives and forms of politics (including warfare no less than welfare), the sense and quality of life, even man's fate and that of his environment: all these are involved in the technological enterprise as it extends in magnitude and depth. The mere enumeration suggests a staggering host of potentially philosophic themes.

To put it bluntly: if there is a philosophy of science, language, history and art; if there is social, political, and moral philosophy; philosophy of thought and of action, of reason and passion, of decision and value—all facets of the inclusive philosophy of man—how then could there not be a philosophy of technology, the focal fact of modern life? And at that a philosophy so spacious that it can house portions from all the other branches of philosophy? It is almost a truism, but at the same time so immense a proposition that its challenge staggers the mind. Economy and modesty require that we select, for a beginning, the most obvious from the multitude of aspects that invite philosophical attention.

The old but useful distinction of "form" and "matter" allows us to distinguish between these two major themes: (1) the *formal dynamics* of technology as a continuing collective enterprise, which advances by its own "laws of motion"; and (2) the *substantive content* of technology in terms of the things it puts into human use, the powers it confers, the novel objectives it opens up or dictates, and the altered manner of human action by which these objectives are realized.

From *The Hastings Center Report* 9, no. 1 (1979): 34–93. Reproduced by permission. © The Hastings Center.

The first theme considers technology as an abstract whole of movement; the second considers its concrete uses and their impact on our world and our lives. The formal approach will try to grasp the pervasive "process properties" by which modern technology propels itself—through our agency, to be sure—into ever-succeeding and superceding novelty. The material approach will look at the species of novelties themselves, their taxonomy, as it were, and try to make out how the world furnished with them looks. A third, overarching theme is the *moral* side of technology as a burden on human responsibility, especially its long-term effects on the global condition of man and environment. This—my own main preoccupation over the past years—will only be touched upon.

THE FORMAL DYNAMICS OF TECHNOLOGY

First some observations about technology's form as an abstract whole of movement. We are concerned with characteristics of *modern* technology and therefore ask first what distinguishes it *formally* from all previous technology. One major distinction is that modern technology is an enterprise and process, whereas earlier technology was a possession and a state. If we roughly describe technology as comprising the use of artificial implements for the business of life, together with their original invention, improvement, and occasional additions, such a tranquil description will do for most of technology through mankind's career (with which it is coeval), but not for modern technology. In the past, generally speaking, a given inventory of tools and procedures used to be fairly constant, tending toward a mutually adjusting, stable equilibrium of ends and means, which—once established—represented for lengthy periods an unchallenged optimum of technical competence.

To be sure, revolutions occurred, but more by accident than by design. The agricultural revolution, the metallurgical revolution that led from the neolithic to the iron age, the rise of cities, and such developments, *happened* rather than were consciously created. Their pace was so slow that only in the time-contraction of historical retrospect do they appear to be "revolutions" (with the misleading connotation that their contemporaries experienced them as such). Even where the change was sudden, as with the introduction first of the chariot, then of armed horsemen into wartime—a violent, if short-lived, revolution indeed—the innovation did not originate from within the military art of the advanced societies that it affected, but was thrust on it from outside by the (much less civilized) peoples of Central Asia. Instead of spreading through the technological universe of their time, other technical breakthroughs, like Phoenician purple-dying, Byzantine "greek fire," Chinese porcelain and silk, and Dumascene steel-tempering, remained jealously guarded monopolies of the inventor communities. Still others, like the hydraulic and steam playthings of Alexandrian mechanics, or compass and gunpowder of the Chinese, passed unnoticed in their serious technological potentials.[1]

On the whole (not counting rare upheavals), the great classical civilizations had comparatively early reached a point of technological saturation—the aforementioned "optimum" in equilibrium of means with acknowledged needs and goals—and had little cause later to go beyond it. From there on, convention reigned supreme. From pottery to monumental architecture, from food growing to shipbuilding, from textiles to engines of war, from time measuring to stargazing: tools, techniques, and objectives remained essentially the same over long times; improvements were sporadic and unplanned. Progress therefore—if it occurred at all*—was by inconspicuous increments to a universally high level that still excites our admiration and, in historical fact, was more liable to regression than to surpassing. The former at least was the more noted phenomenon, deplored by the epigones with a nostalgic remembrance of a better past (as in the declining Roman world). More important, there was, even in the best and most vigorous times, no proclaimed *idea* of a future of *constant progress* in the arts. Most important, there was never a deliberate method of going about it like "research," the willingness to undergo the risks of trying unorthodox paths, exchanging information widely about the experience, and so on. Least of all was there a "natural science" as a growing body of theory to guide such semitheoretical, prepractical activities, plus their social institutionalization. In routines as well as panoply of instruments, accomplished as they were for the purposes they served, the "arts" seemed as settled as those purposes themselves.†

Traits of Modern Technology

The exact opposite of this picture holds for modern technology, and this is its first philosophical aspect. Let us begin with some manifest traits.

1. Every new step in whatever direction of whatever technological field tends *not* to approach an equilibrium or saturation point in the process of fitting means to ends (nor is it meant to), but, on the contrary, to give rise, if successful, to further steps in all kinds of direction and with a fluidity of the ends themselves. "Tends to" becomes a compelling "is bound to" with any major or important step (this almost being its criterion); and the innovators themselves expect, beyond the accomplishment, each time, of their immediate task, the constant future repetition of their inventive activity.

---

*Progress did, in fact, occur even at the heights of classical civilizations. The Roman arch and vault, for example, were distinct engineering advances over the horizontal establature and flat ceiling of Greek (and Egyptian) architecture, permitting spanning feats and thereby construction objectives not contemplated before (stone bridges, aqueducts, the vast baths, and other public halls of Imperial Rome). But materials, tools, and techniques were still the same, the role of human labor and crafts remained unaltered, stonecutting and brickbaking went on as before. An existing technology was enlarged in its scope of performance, but none of its means or even goals made obsolete.

†One meaning of "classical" is that those civilizations had somehow implicitly "defined" themselves and neither encouraged nor even allowed to pass beyond their innate terms. The—more or less—achieved "equilibrium" was their very pride.

2. Every technical innovation is sure to spread quickly through the technological world community, as also do theoretical discoveries in the sciences. The spreading is in terms of knowledge and of practical adoption, the first (and its speed) guaranteed by the universal intercommunication that is itself part of the technological complex, the second enforced by the pressure of competition.

3. The relation of means to ends is not unilinear but circular. Familiar ends of longstanding may find better satisfaction by new technologies whose genesis they had inspired. But equally—and increasingly typical—new technologies may suggest, create, even impose new ends, never before conceived, simply by offering their feasibility. (Who had ever wished to have in his living room the Philharmonic orchestra, or open heart surgery, or a helicopter defoliating a Vietnam forest? or to drink his coffee from a disposable plastic cup? or to have artificial insemination, test-tube babies, and host pregnancies? or to see clones of himself and others walking about?) Technology thus adds to the very objectives of human desires, including objectives for technology itself. The last point indicates the dialectics or circularity of the case: once incorporated into the socioeconomic demand diet, ends first gratuitously (perhaps accidentally) generated by technological invention become necessities of life and set technology the task of further perfecting the means of realizing them.

4. Progress, therefore, is not just an ideological gloss on modern technology, and not at all a mere option offered by it, but an inherent drive which acts willy-nilly in the formal automatics of its *modus operandi* as it interacts with society. "Progress" is here not a value term but purely descriptive. We may resent the fact and despise its fruits and yet must go along with it, for—short of a stop by the fiat of total political power, or by a sustained general strike of its clients or some internal collapse of their societies, or by self-destruction through its works (the last, alas, not the least likely of these)—the juggernaut moves on relentlessly, spawning its always mutated progeny by coping with the challenges and lures of the now. But while not a value term, "progress" here is not a neutral term either, for which we could simply substitute "change." For it is in the nature of the case, or a law of the series, that a later stage is always, in terms of technology itself, *superior* to the preceding stage.* Thus we have here a case of the entropy-defying sort (organic evolution is another), where the internal motion of a system, left to itself and not interfered with, leads to ever "higher," not "lower" states of itself. Such at least is the present evidence.† If Napoleon once said, "Politics is destiny," we may well say today, "Technology is destiny."

---

*This only seems to be but is not a value statement, as the reflection on, for example, an ever more destructive atom bomb shows.

†There may conceivably be internal degenerative factors—such as the overloading of finite information-processing capacity—that may bring the (exponential) movement to a halt or even make the system fall apart. We don't know yet.

These points go some way to explicate the initial statement that modern technology, unlike traditional, is an enterprise and not a possession, a process and not a state, a dynamic thrust and not a set of implements and skills. And they already adumbrate certain "laws of motion" for this restless phenomenon. What we have described, let us remember, were formal traits which as yet say little about the contents of the enterprise. We ask two questions of this descriptive picture: *why* is this so, that is, what *causes* the restlessness of modern technology: what is the nature of the thrust? And, what is the philosophical import of the facts so explained?

The Nature of Restless Technology

As we would expect in such a complex phenomenon, the motive forces are many, and some causal hints appeared already in the descriptive account. We have mentioned *pressure of competition*—for profit, but also for power, security, and so forth—as one perpetual mover in the universal appropriation of technical improvements. It is equally operative in their origination, that is, in the process of invention itself, nowadays dependent on constant outside subsidy and even goal-setting: potent interests see to both. War, or the threat of it, has proved an especially powerful agent. The less dramatic, but no less compelling, everyday agents are legion. To keep one's head above the water is their common principle (somewhat paradoxical, in view of an abundance already far surpassing what former ages would have lived with happily ever after). Of pressures other than the competitive ones, we must mention those of population growth and of impending exhaustion of natural resources. Since both phenomena are themselves already by-products of technology (the first by way of medical improvements, the second by the voracity of industry), they offer a good example of the more general truth that to a considerable extent technology itself begets the problems which it is then called upon to overcome by a new forward jump. (The Green Revolution and the development of synthetic substitute materials or of alternate sources of energy comes under this heading.) These compulsive pressures for progress, then, would operate even for a technology in a noncompetitive, for example, a socialist setting.

A motive force more autonomous and spontaneous than these almost mechanical pushes with their "sink or swim" imperative would be the pull of the quasi-utopian *vision* of an ever better life, whether vulgarly conceived or nobly, once technology has proved the open-ended capacity for procuring the conditions for it: perceived possibility whetting the appetite ("the American dream," "the revolution of rising expectations"). This less palpable factor is more difficult to appraise, but its playing a role is undeniable. Its deliberate fostering and manipulation by the dream merchants of the industrial-mercantile complex is yet another matter and somewhat taints the spontaneity of the motive, as it also degrades the quality of the dream. It is also moot to what extent the vision itself is *post hoc* rather than *ante hoc*, that is, instilled by the dazzling feats of a technological process already underway and thus more a response to than a motor of it.

Groping in these obscure regions of motivation, one may as well descend, for an explanation of the dynamism as such, into the Spenglerian* mystery of a "Faustian soul" innate in Western culture, that drives it, nonrationally, to infinite novelty and unplumbed possibilities for their own sake; or into the Heideggerian† depths of a fateful, metaphysical decision of the will for boundless power over the world of things—a decision equally peculiar to the Western mind: speculative intuitions which do strike a resonance in us, but are beyond proof and disproof.

Surfacing once more, we may also look at the very sober, functional facts of industrialism as such, of production and distribution, output maximization, managerial and labor aspects, which even apart from competitive pressure provide their own incentives for technical progress. Similar observations apply to the requirements of *rule* or control in the vast and populous states of our time, those giant territorial superorganisms which for their very cohesion depend on advanced technology (for example, on information, communication, and transportation, not to speak of weaponry) and thus have a stake in its promotion: the more so, the more centralized they are. This holds for socialist systems no less than for free-market societies. May we conclude from this that even a communist world state, freed from external rivals as well as from internal free-market competition, might still have to push technology ahead for purposes of control on this colossal scale? Marxism, in any case, has its own inbuilt commitment to technological progress beyond necessity. But even disregarding all dynamics of these conjectural kinds, the most monolithic case imaginable would, at any rate, still be exposed to those noncompetitive, natural pressures like population growth and dwindling resources that beset industrialism as such. Thus, it seems, the compulsive element of technological progress may not be bound to its original breeding ground, the capitalist system. Perhaps the odds for an eventual stabilization look somewhat better in a socialist system, provided it is worldwide—and possibly totalitarian in the bargain. As it is, the pluralism we are thankful for ensures the constancy of compulsive advance.

We could go on unraveling the causal skein and would be sure to find many more strands. But none nor all of them, much as they explain, would go to the heart of the matter. For all of them have one premise in common without which they could not operate for long: the premise that there *can* be indefinite progress because there *is* always something new and better to find. The, by no means obvious, givenness of this objective condition is also the pragmatic conviction of the performers in the technological drama; but without its being true, the conviction would help as little as the dream of the alchemists. Unlike theirs, it is backed up by an impressive record of past successes, and for many this is sufficient ground for their belief. (Perhaps holding or not holding it does not even greatly

---

*Oswald Spengler (1880–1936). German writer, whose best known work is *The Decline of the West*, in which he predicted the breakdown of Western civilization. (Ed.)

† Martin Heidegger (1889–1976). German existentialist philosopher. (Ed.)

matter.) What makes it more than a sanguine belief, however, is an underlying and well-grounded, theoretical view of the nature of things and of human cognition, according to which they do not set a limit to novelty of discovery and invention, indeed, that they of themselves will at each point offer another opening for the as yet unknown and undone. The corollary conviction, then, is that a technology tailored to a nature and to a knowledge of this indefinite potential ensures its indefinitely continued conversion into the practical powers, each step of it begetting the next, with never a cutoff from internal exhaustion of possibilities.

Only habituation dulls our wonder at this wholly unprecedented belief in virtual "infinity." And by all our present comprehension of reality, the belief is most likely true—at least enough of it to keep the road for innovative technology in the wake of advancing science open for a long time ahead. Unless we understand this ontologic-epistemological premise, we have not understood the inmost agent of technological dynamics, on which the working of all the adventitious causal factors is contingent in the long run.

Let us remember that the virtual infinitude of advance we here seek to explain is in essence different from the always avowed perfectibility of every human accomplishment. Even the undisputed master of his craft always had to admit as possible that he might be surpassed in skill or tools or materials; and no excellence of product ever foreclosed that it might still be better, just as today's champion runner must know that his time may one day be beaten. But these are improvements within a given genus, not different in kind from what went before, and they must accrue in diminishing fractions. Clearly, the phenomenon of an exponentially growing *general* innovation is qualitatively different.

Science as a Source of Restlessness

The answer lies in the interaction of *science* and *technology* that is the hallmark of modern progress, and thus ultimately in the kind of nature which modern science progressively discloses. For it is here, in the movement of *knowledge*, where relevant novelty first and constantly occurs. This is itself a novelty. To Newtonian physics, nature appeared simple, almost crude, running its show with a few kinds of basic entities and forces by a few universal laws, and the application of those well-known laws to an ever greater variety of composite phenomena promised ever widening knowledge indeed, but no real surprises. Since the mid-nineteenth century, this minimalistic and somehow finished picture of nature has changed with breathtaking acceleration. In a reciprocal interplay with the growing subtlety of exploration (instrumental and conceptual), nature itself stands forth as ever more subtle. The progress of probing makes the object grow richer in modes of operation, not sparer as classical mechanics had expected. And instead of narrowing the margin of the still-undiscovered, science now surprises itself with unlocking dimension after dimension of new depths. The very essence of matter has turned from a blunt, irreducible ultimate to an always reopened challenge for further penetration. No one can say whether this will go on forever, but a suspicion of intrinsic infinity in the very being of things obtrudes itself and therewith an anticipation of unending inquiry of the sort where succeeding steps will not find the same old story again (Descartes's "matter in motion"), but always add new twists to it. If then the art of technology is correlative to the knowledge of nature, technology too acquires from this source that potential of infinity for its innovative advance.

But it is not just that indefinite scientific progress offers the *option* of indefinite technological progress, to be exercised or not as other interests see fit. Rather the cognitive process itself moves by interaction with the technological, and in the most internally vital sense: for its own *theoretical* purpose, science must generate an increasingly sophisticated and physically formidable technology as its tool. What it finds with this help initiates new departures in the practical sphere, and the latter as a whole, that is, technology at work, provides with its experiences a large-scale laboratory for science again, a breeding ground for new questions, and so on in an unending cycle. In brief, a mutual feedback operates between science and technology; each requires and propels the other; and as matters now stand, they can only live together or must die together. For the dynamics of technology, with which we are here concerned, this means that (all external promptings apart) an agent of restlessness is implanted in it by its functionally integral bond with science. As long, therefore, as the cognitive impulse lasts, technology is sure to move ahead with it. The cognitive impulse, in its turn, culturally vulnerable in itself, liable to lag or to grow conservative with a treasured canon—that theoretical eros itself no longer lives on the delicate appetite for truth alone, but is spurred on by its hardier offspring, technology, which communicates to it impulsions from the broadest arena of struggling, insistent life. Intellectual curiosity is seconded by interminably self-renewing practical aim.

I am conscious of the conjectural character of some of these thoughts. The revolutions in science over the last fifty years or so are a fact, and so are the revolutionary style they imparted to technology and the reciprocity between the two concurrent streams (nuclear physics is a good example). But whether these scientific revolutions, which hold primacy in the whole syndrome, will be typical for science henceforth—something like a law of motion for its future—or represent only a singular phase in its longer run, is unsure. To the extent, then, that our forecast of incessant novelty for technology was predicated on a guess concerning the future of science, even concerning the nature of things, it is hypothetical, as such extrapolations are bound to be. But even if the recent past did not usher in a state of permanent revolution for science, and the life of theory settles down again to a more sedate pace, the scope for technological innovation will not easily shrink; and what may no longer be a revolution in science, may still revolutionize our lives in its practical impact through technology. "Infinity" being too large a word anyway, let us say that present signs of potential and of incentives point to an indefinite perpetuation and fertility of the technological momentum.

The Philosophical Implications

It remains to draw philosophical conclusions from our findings, at least to pin-point aspects of philosophical interest. Some preceding remarks have already been straying into philosophy of science in the technical sense. Of broader issues, two will be ample to provide food for further thought beyond the limitations of this [chapter]. One concerns the status of knowledge in the human scheme, the other the status of technology itself as a human goal, or its tendency to become that from being a means, in a dialectical inversion of the means-end order itself.

Concerning knowledge, it is obvious that the time-honored division of theory and practice has vanished for both sides. The thirst for pure knowledge may persist undiminished, but the involvement of knowing at the heights with doing in the lowlands of life, mediated by technology, has become inextricable; and the aristocratic self-sufficiency of knowing for its own (and the knower's) sake has gone. Nobility has been exchanged for utility. With the possible exception of philosophy, which still can do with paper and pen and tossing thoughts around among peers, all knowledge has become thus tainted, or elevated if you will, whether utility is intended or not. The technological syndrome, in other words, has brought about a thorough *socializing* of the theoretical realm, enlisting it in the service of common need. What used to be the freest of human choices, an extravagance snatched from the pressure of the world—the esoteric life of thought—has become part of the great public play of necessities and a prime necessity in the action of the play.* Remotest abstraction has become enmeshed with nearest concreteness. What this pragmatic functionalization of the once highest indulgence in impractical pursuits portends for the image of man, for the restructuring of a hallowed hierarchy of values, for the idea of "wisdom," and so on, is surely a subject for philosophical pondering.

Concerning technology itself, its actual role in modern life (as distinct from the purely instrumental definition of technology as such) has made the relation of means and ends equivocal all the way up from the daily living to the very vocation of man. There could be no question in former technology that its role was that of humble servant—pride of workmanship and aesthetic embellishment of the useful notwithstanding. The Promethean enterprise of modern technology speaks a different language. The word "enterprise" gives the clue, and its unendingness another. We have mentioned that the effect of its innovations is disequilibrating rather than equilibrating with respect to the balance of wants and supply, always

---

*There is a paradoxical side effect to this change of roles. That very science which forfeited its place in the domain of leisure to become a busy toiler in the field of common needs, creates by its tools a growing domain of leisure for the masses, who reap this with the other fruits of technology as an additional (and no less novel) article of forced consumption. Hence leisure, from a privilege of the few, has become a problem for the many to cope with. Science, not idle, provides for the needs of this idleness too: no small part of technology is spent on filling the leisure-time gap which technology itself has made a fact of life.

breeding its own new wants. This in itself compels the constant attention of the best minds, engaging the full capital of human ingenuity for meeting challenge after challenge and seizing the new chances. It is psychologically natural for that degree of engagement to be invested with the dignity of dominant purpose. Not only does technology dominate our lives in fact, it nourishes also a belief in its being of predominant worth. The sheer grandeur of the enterprise and its seeming infinity inspire enthusiasm and fire ambition. Thus, in addition to spawning new ends (worthy or frivolous) from the mere invention of means, technology as a grand venture tends to establish *itself* as the transcendent end. At least the suggestion is there and casts its spell on the modern mind. At its most modest, it means elevating *homo faber* to the essential aspect of man; at its most extravagant, it means elevating *power* to the position of his dominant and interminable goal. To become ever more masters of the world, to advance from power to power, even if only collectively and perhaps no longer by choice, *can* now be seen to be the chief vocation of mankind. Surely, this again poses philosophical questions that may well lead unto the uncertain grounds of metaphysics or of faith.

I here break off, arbitrarily, the formal account of the technological movement in general, which as yet has told us little of what the enterprise is about. To this subject I now turn, that is, to the new kinds of powers and objectives that technology opens to modern man and the consequently altered quality of human action itself.

THE MATERIAL WORKS OF TECHNOLOGY

Technology is a species of power, and we can ask questions about how and on what object any power is exercised. Adopting Aristotle's rule in *De anima* that for understanding a faculty one should begin with its objects, we start from them too—"objects" meaning both the visible *things* technology generates and puts into human use, and the *objectives* they serve. The objects of modern technology are first everything that had always been an object of human artifice and labor: food, clothing, shelter, implements, transportation—all the material necessities and comforts of life. The technological intervention changed at first not the product but its production, in speed, ease, and quantity. However, this is true only of the very first stage of the industrial revolution with which large-scale scientific technology began. For example, the cloth for the steam-driven looms of Lancashire remained the same. Even then, one significant new product was added to the traditional list—the machines themselves, which required an entire new industry with further subsidiary industries to build them. These novel entities, machines—at first capital goods only, not consumer goods—had from the beginning their own impact on man's symbiosis with nature by being consumers themselves. For example: steam-powered water pumps facilitated coal mining, required in turn extra coal for firing their boilers, more coal for the foundries and forges that made those boilers, more for the mining of the requisite iron ore, more for its transportation to the foundries, more—both coal and iron—for the rails and locomotives made in these same foundries, more for the conveyance of the foundries' product to the pitheads and return, and finally more for the distribution of the more abundant coal to the users outside this cycle, among which were increasingly still more machines spawned by the increased availability of coal. Lest it be forgotten over this long chain, we have been speaking of James Watt's modest steam engine for pumping water out of mine shafts. This syndrome of self-proliferation—by no means a linear chain but an intricate web of reciprocity—has been part of modern technology ever since. To generalize, technology exponentially increases man's drain on nature's resources (of substances and of energy), not only through the multiplication of the final goods for consumption, but also, and perhaps more so, through the production and operation of its own mechanical means. And with these means—machines—it introduced a new category of goods, not for consumption, added to the furniture of our world. That is, among the objects of technology a prominent class is that of technological apparatus itself.

Soon other features also changed the initial picture of a merely mechanized production of familiar commodities. The final products reaching the consumer ceased to be the same, even if still serving the same age-old needs; new needs, or desires, were added by commodities of entirely new kinds which changed the habits of life. Of such commodities, machines themselves became increasingly part of the consumer's daily life to be used directly by himself, as an article not of production but of consumption. My survey can be brief as the facts are familiar.

New Kinds of Commodities

When I said that the cloth of the mechanized looms of Lancashire remained the same, everyone will have thought of today's synthetic fiber textiles for which the statement surely no longer holds. This is fairly recent, but the general phenomenon starts much earlier in the synthetic dyes and fertilizers with which the chemical industry—the first to be wholly a fruit of science—began. The original rationale of these technological feats was substitution of artificial for natural materials (for reasons of scarcity or cost), with as nearly as possible the same properties for effective use. But we need only think of plastics to realize that art progressed from substitutes to the creation of really new substances with properties not found in any natural one, raw or processed, thereby also initiating uses not thought of before and giving rise to new classes of objects to serve them. In chemical (molecular) engineering, man does more than in mechanical (molar) engineering which constructs machinery from natural materials; his intervention is deeper, redesigning the infra-patterns of nature, making substances to specification by arbitrary disposition of molecules. And this, be it noted, is done deductively from the bottom, from the thoroughly analyzed last elements, that is, in a real *via compositiva* after the completed *via resolutiva*, very different from the long-known empirical practice of coaxing substances into new properties, as in metal alloys from the bronze age on. Artificiality or creative engineering with abstract construction invades the heart of matter. This, in molecular biology, points to further, awesome potentialities.

With the sophistication of molecular alchemy we are ahead of our story. Even in straightforward hardware engineering, right in the first blush of the mechanical revolution, the objects of use that came out of the factories did not really remain the same, even where the objectives did. Take the old objective of travel. Railroads and ocean liners are relevantly different from the stage coach and from the sailing ship, not merely in construction and efficiency but in the very feel of the user, making travel a different experience altogether, something one may do for its own sake. Airplanes, finally, leave behind any similarity with former conveyances, except the purpose of getting from here to there, with no experience of what lies in between. And these instrumental objects occupy a prominent, even obtrusive place in our world, far beyond anything wagons and boats ever did. Also they are constantly subject to improvement of design, with obsolescence rather than wear determining their life span.

Or take the oldest, most static of artifacts: human habitation. The multistoried office building of steel, concrete, and glass is a qualitatively different entity from the wood, brick, and stone structures of old. With all that goes into it besides the structures as such—the plumbing and wiring, the elevators, the lighting, heating, and cooling systems—it embodies the end products of a whole spectrum of technologies and far-flung industries, where only at the remote sources human hands still meet with primary materials, no longer recognizable in the final result. The ultimate customer inhabiting the product is ensconced in a shell of thoroughly derivative artifacts (perhaps relieved by a nice piece of driftwood). This transformation into utter artificiality is generally, and increasingly, the effect of technology on the human environment, down to the items of daily use. Only in agriculture has the product so far escaped this transformation by the changed modes of its production. We still eat the meat and rice of our ancestors.*

Then, speaking of the commodities that technology injects into private use, there are machines themselves, those very devices of its own running, originally confined to the economic sphere. This unprecedented novum in the records of individual living started late in the nineteenth century and has since grown to a pervading mass phenomenon in the Western world. The prime example, of

---

*Not so, objects my colleague Robert Heilbroner in a letter to me: "I'm sorry to tell you that meat and rice are both *profoundly* influenced by technology. Not even they are left untouched." Correct, but they are at least generically the same (their really profound changes lie far back in the original breeding of domesticated strains from wild ones—as in the case of all cereal plants under cultivation). I am speaking here of an order of transformation in which the results bear no resemblance to the natural materials at their source, nor to any naturally occurring state of them.

course, is the automobile, but we must add to it the whole gamut of household appliances—refrigerators, washers, dryers, vacuum cleaners—by now more common in the lifestyle of the general population than running water or central heating were one hundred years ago. Add lawn mowers and other power tools for home and garden; we are mechanized in our daily chores and recreations (including the toys of our children) with every expectation that new gadgets will continue to arrive.

These paraphernalia are machines in the precise sense that they perform work and consume energy, and their moving parts are of the familiar magnitudes of our perpetual world. But an additional and profoundly different category of technical apparatus was dropped into the lap of the private citizen, not labor-saving and work-performing, partly not even utilitarian, but—with minimal energy input—catering to the senses and the mind: telephone, radio, television, tape recorders, calculators, record players—all the domestic terminals of the electronics industry, the latest arrival on the technological scene. Not only by their insubstantial, mind-addressed output, also by the subvisible, not literally "mechanical" physics of their functioning do these devices differ in kind from all the macroscopic, bodily moving machinery of the classical type. Before inspecting this momentous turn from power engineering, the hallmark of the first industrial revolution, to communication engineering, which almost amounts to a second industrial-technological revolution, we must take a look at its natural base: electricity.

In the march of technology to ever greater artificiality, abstraction, and subtlety, the unlocking of electricity marks a decisive step. Here is a universal force of nature which yet does not naturally appear to man (except in lightning). It is not a datum of uncontrived experience. Its very "appearance" had to wait for science, which contrived the experience for it. Here, then, a technology depended on science for the mere providing of its "object," the entity itself it would deal with—the first case where theory alone, not ordinary experience, wholly preceded practice (repeated later in the case of nuclear energy). And what sort of entity! Heat and steam are familiar objects of sensuous experience, their force bodily displayed in nature; the matter of chemistry is still the concrete, corporeal stuff mankind had always known. But electricity is an abstract object, disembodied, immaterial, unseen; in its usable form, it is entirely an artifact, generated in a subtle transformation from grosser forms of energy (ultimately from heat via motion). Its theory indeed had to be essentially complete before utilization could begin.

Revolutionary as electrical technology was in itself, its purpose was at first the by now conventional one of the industrial revolution in general: to supply motive power for the propulsion of machines. Its advantages lay in the unique versatility of the new force, the ease of its transmission, transformation, and distribution—an unsubstantial commodity, no bulk, no weight, instantaneously delivered at the point of consumption. Nothing like it had ever existed before in man's traffic with matter, space, and time. It made possible the spread of mechanization to every home; this alone was a tremendous boost to the technological tide, at the same time hooking private lines into centralized public networks and thus making them dependent on the functioning of a total system as never before, in fact, for every moment. Remember, you cannot hoard electricity as you can coal and oil, or flour and sugar for that matter.

But something much more unorthodox was to follow. As we all know, the discovery of the universe of electromagnetics caused a revolution in theoretical physics that is still underway. Without it, there would be no relativity theory, no quantum mechanics, no nuclear and subnuclear physics. It also caused a revolution in technology beyond what it contributed, as we noted, to its classical program. The revolution consisted in the passage from electrical to electronic technology which signifies a new level of abstraction in means and ends. It is the difference between power and communication engineering. Its object, the most impalpable of all, is information. Cognitive instruments had been known before—sextant, compass, clock, telescope, microscope, thermometer, all of them for information and not for work. At one time, they were called "philosophical" or "metaphysical" instruments. By the same general criterion, amusing as it may seem, the new electronic information devices, too, could be classed as "philosophical instruments." But those earlier cognitive devices, except the clock, were inert and passive, not generating information actively, as the new instrumentalities do.

Theoretically as well as practically, electronics signifies a genuinely new phase of the scientific-technological revolution. Compared with the sophistication of its theory as well as the delicacy of its apparatus, everything which came before seems crude, almost natural. To appreciate the point, take the man-made satellites now in orbit. In one sense, they are indeed an imitation of celestial mechanics— Newton's laws finally verified by cosmic experiment: astronomy, for millennia the most purely contemplative of the physical sciences, turned into a practical art! Yet, amazing as it is, the astronomic imitation, with all the unleashing of forces and the finesse of techniques that went into it, is the least interesting aspect of those entities. In that respect, they still fall within the terms and feats of classical mechanics (except for the remote-control course corrections).

Their true interest lies in the instruments they carry through the void of space and in what these do, their measuring, recording, analyzing, computing, their receiving, processing, and transmitting abstract information and even images over cosmic distances. There is nothing in all nature which even remotely foreshadows the kind of things that now ride the heavenly spheres. Man's imitative practical astronomy merely provides the vehicle for something else with which he sovereignly passes beyond all the models and usages of known nature.* That

---

*Note also that in radio technology, the medium of action is nothing material, like wires conducting currents, but the entirely immaterial electromagnetic "field," i.e., space itself. The symbolic picture of "waves" is the last remaining link to the forms of our perceptual world.

the advent of man portended, in its inner secret of mind and will, a cosmic event was known to religion and philosophy: now it manifests itself as such by fact of things and acts in the visible universe. Electronics indeed creates a range of objects imitating nothing and progressively added to by pure invention.

And no less invented are the ends they serve. Power engineering and chemistry for the most part still answered to the natural needs of man: for food, clothing, shelter, locomotion, and so forth. Communication engineering answers to needs of information and control solely created by the civilization that made this technology possible and, once started, imperative. The novelty of the means continues to engender no less novel ends—both becoming as necessary to the functioning of the civilization that spawned them as they would have been pointless for any former one. The world they help to constitute and which needs computers for its very running is no longer nature supplemented, imitated, improved, transformed, the original habitat made more habitable. In the pervasive mentalization of physical relationships, it is a *transnature* of human making, but with this inherent paradox: that it threatens the obsolescence of man himself, as increasing automation ousts him from the places of work where he formerly proved his humanhood. And there is a further threat: its strain on nature herself may reach a breaking point.

The Last Stage of the Revolution

That sentence would make a good dramatic ending. But it is not the end of the story. There may be in the offing another, conceivably the last, stage of the technological revolution, after the mechanical, chemical, electrical, electronic stages we have surveyed, and the nuclear we omitted. All these were based on physics and had to do with what man can put to his use. What about biology? And what about the user himself? Are we, perhaps, on the verge of a technology, based on biological knowledge and wielding an engineering art which, this time, has man himself for its object? This has become a theoretical possibility with the advent of molecular biology and its understanding of genetic programming; and it has been rendered morally possible by the metaphysical neutralizing of man. But the latter, while giving us the license to do as we wish, at the same time denies us the guidance for knowing what to wish. Since the same evolutionary doctrine of which genetics is a cornerstone has deprived us of a valid image of man, the actual techniques, when they are ready, may find us strangely unready for their responsible use. The antiessentialism of prevailing theory, which knows only of *de facto* outcomes of evolutionary accident and of no valid essences that would give sanction to them, surrenders our being to a freedom without norms. Thus the technological call of the new microbiology is the twofold one of physical feasibility and metaphysical admissibility. Assuming the genetic mechanism to be completely analyzed and its script finally decoded, we can set about rewriting the text. Biologists vary in their estimates of how close we are to the capability; few seem to doubt the right to use it. Judging by the rhetoric of its prophets, the idea of taking our evolution into our own hands is intoxicating even to many scientists.

In any case, the idea of making over man is no longer fantastic, nor interdicted by an inviolable taboo. If and when *that* revolution occurs, if technological power is really going to tinker with the elemental keys on which life will have to play its melody in generations of men to come (perhaps the only such melody in the universe), then a reflection on what is humanly desirable and what should determine the choice—a reflection, in short, on the image of man, becomes an imperative more urgent than any ever inflicted on the understanding of mortal man. Philosophy, it must be confessed, is sadly unprepared for this, its first cosmic task.

TOWARD AN ETHICS OF TECHNOLOGY

The last topic has moved naturally from the descriptive and analytic plane, on which the objects of technology are displayed for inspection, onto the evaluative plane where their ethical challenge poses itself for decision. The particular case forced the transition so directly because there the (as yet hypothetical) technological object was man directly. But once removed, man is involved in all the other objects of technology, as these singly and jointly remake the worldly frame of his life, in both the narrower and the wider of its senses: that of the artificial frame of civilization in which social man leads his life proximately, and that of the natural terrestrial environment in which this artifact is embedded and on which it ultimately depends.

Again, because of the magnitude of technological effects on both these vital environments in their totality, both the quality of human life and its very preservation in the future are at stake in the rampage of technology. In short, certainly the "image" of man, and possibly the survival of the species (or of much of it), are in jeopardy. This would summon man's duty to his cause even if the jeopardy were not of his own making. But it is, and, in addition to his ageless obligation to meet the threat of things, he bears for the first time the responsibility of prime agent in the threatening disposition of things. Hence nothing is more natural than the passage from the objects to the ethics of technology, from the things made to the duties of their makers and users.

A similar experience of inevitable passage from analysis of fact to ethical significance, let us remember, befell us toward the end of the first section. As in the case of the matter, so also in the case of the form of the technological dynamics, the image of man appeared at stake. In view of the quasi-automatic compulsion of those dynamics, with their perspective of indefinite progression, every existential and moral question that the objects of technology raise assumes the curiously eschatological quality with which we are becoming familiar from the extrapolating guesses of futurology. But apart from thus raising all challenges of present particular matter to the higher powers of future exponential magnifi-cation, the despotic dynamics of the technological movement as such, sweeping its captive movers along in its breathless momentum, poses its own questions to man's axiological conception of himself. Thus, form and matter of technology alike enter into the dimension of ethics.

The questions raised for ethics by the objects of technology are defined by the major areas of their impact and thus fall into such fields of knowledge as ecology (with all its biospheric subdivisions of land, sea, and air), demography, economics, biomedical and behavioral sciences (even the psychology of mind pollution by television), and so forth. Not even a sketch of the substantive prob-lems, let alone of ethical policies for dealing with them, can here be attempted. Clearly, for a normative rationale of the latter, ethical theory must plumb the very foundations of value, obligation, and the human good.

The same holds of the different kind of questions raised for ethics by the sheer fact of the formal dynamics of technology. But here, a question of another order is added to the straightforward ethical questions of both kinds, subjecting any resolution of them to a pragmatic proviso of harrowing uncertainty. Given the mastery of the creation over its creators, which yet does not abrogate their responsibility nor silence their vital interest, what are the chances and what are the means of gaining *control* of the process, so that the results of any ethical (or even purely prudential) insights can be translated into effective action? How in short can man's freedom prevail against the determinism he has created for him-self? On this most clouded question, whereby hangs not only the effectuality or futility of the ethical search which the facts invite (assuming it to be blessed with *theoretical* success!), but perhaps the future of mankind itself, I will make a few concluding, but—alas—inconclusive, remarks. They are intended to touch on the whole ethical enterprise.

Problematic Preconditions of an Effective Ethics

First, a look at the novel state of determinism. Prima facie, it would seem that the greater and more varied powers bequeathed by technology have expanded the range of choices and hence increased human freedom. For economics, for example, the argument has been made[2] that the uniform compulsion which scarcity and subsistence previously imposed on economic behavior with a virtual denial of alternatives (and hence—conjoined with the universal "maximization" motive of capitalist market competition—gave classical economics at least the appearance of a deterministic "science") has given way to a latitude of indeter-minacy. The plenty and powers provided by industrial technology allow a plu-ralism of choosable alternatives (hence disallow scientific protection). We are not here concerned with the status of economics as a science. But as to the altered state of things alleged in the argument, I submit that the change means rather that one, relatively homogeneous determinism (thus relatively easy to formalize into a law) has been supplanted by another, more complex, multifarious determinism, namely, that exercised by the human artifact itself upon its creator and user. We, abstractly speaking the possessors of those powers, are concretely subject to their emancipated dynamics and the sheer momentum of our own multitude, the vehicle of those dynamics.

I have spoken elsewhere[3] of the "new realm of necessity" set up, like a second nature, by the feedbacks of our achievements. The almighty we, or Man personified is, alas, an abstraction. *Man* may have become more powerful; *men* very probably the opposite, enmeshed as they are in more dependencies than ever before. What ideal Man can do is not the same as what real men permit or dictate to be done. And here I am thinking not only of the immanent dynamism, almost automatism, of the impersonal technological complex I have invoked so far, but also of the pathology of its client society. Its compulsions, I fear, are at least as great as were those of unconquered nature. Talk of the blind forces of nature! Are those of the sorcerer's creation less blind? They differ indeed in the serial shape of their causality: the action of nature's forces is cyclical, with periodical recurrence of the same, while that of the technological forces is linear, progressive, cumulative, thus replacing the curse of constant toil with the threat of maturing crisis and possible catastrophe. Apart from this significant vector difference, I seriously wonder whether the tyranny of fate has not become greater, the latitude of spontaneity smaller; and whether man has not actually been weakened in his decision-making capacity by his accretion of collective strength.

However, in speaking, as I have just done, of "his" decision-making capacity, I have been guilty of the same abstraction I had earlier criticized in the use of the term "man." Actually, the subject of the statement was no real or representative individual but Hobbes' "Artificial Man," "that great Leviathan, called a Common-Wealth," or the "large horse" to which Socrates likened the city, "which because of its great size tends to be sluggish and needs stirring by a gadfly." Now, the chances of there being such gadflies among the numbers of the commonwealth are today no worse nor better than they have ever been, and in fact they are around and stinging in our field of concern. In that respect, the free spontaneity of personal insight, judgment, and responsible action by speech can be trusted as an ineradicable (if also incalculable) endowment of humanity, and smallness of number is in itself no impediment to shaking public complacency. The problem, however, is not so much complacency or apathy as the counterforces of active, and anything but complacent, interests and the complicity with them of all of us in our daily consumer existence. These interests themselves are factors in the determinism which technology has set upon the space of its sway. The question, then, is that of the possible chances of unselfish insight in the arena of (by nature) selfish *power*, and more particularly: of one long-range, interloping insight against the short-range goals of many incumbent powers. Is there hope that wisdom itself can become power? This renews the thorny old subject of Plato's philosopher-king and—with that inclusion of realism which the utopian Plato did not lack—or the role of myth, not knowledge, in the education of the guardians. Applied to our topic: the *knowledge* of objective dangers and of values endangered, as well as of the technical remedies, is beginning to be there and to be disseminated; but to make it prevail in the marketplace is a matter less of the rational dissemination of truth than of public relations techniques, persuasion, indoctrination, and manipulation, also of unholy alliances, perhaps even conspiracy. The philosopher's descent into the cave may well have to go all the way to "if you can't lick them, join them."

That is so not merely because of the active resistance of special interests but because of the optical illusion of the near and the far which condemns the long-range views to impotence against the enticement and threats of the nearby: it is this incurable shortsightedness of animal-human nature more than ill will that makes it difficult to move even those who have no special axe to grind, but still are in countless ways, as we all are, beneficiaries of the untamed system and so have something dear in the present to lose with the inevitable cost of its taming. The taskmaster, I fear, will have to be actual pain beginning to strike, when the far has moved close to the skin and has vulgar optics on its side. Even then, one may resort to palliatives of the hour. In any event, one should try as much as one can to forestall the advent of emergency with its high tax of suffering or, at the least, prepare for it. This is where the scientist can redeem his role in the technological estate.

The incipient knowledge about technological danger trends must be developed, coordinated, systematized, and the full force of computer-aided projection techniques deployed to determine priorities of action, so as to inform preventive efforts wherever they can be elicited, to minimize the necessary sacrifices, and at the worst to preplan the saving measures which the terror of beginning calamity will eventually make people willing to accept. Even now, hardly a decade after the first stirrings of "environmental" consciousness, much of the requisite knowledge, plus the rational persuasion, is available inside and outside academia for any well-meaning powerholder to draw upon. To this, we—the growing band of concerned intellectuals—ought persistently to contribute our bit of competence and passion.

But the real problem is to get the well-meaning into power and have that power as little as possible beholden to the interests which the technological colossus generates on its path. It is the problem of the philosopher-king compounded by the greater magnitude and complexity (also sophistication) of the forces to contend with. Ethically, it becomes a problem of playing the game by its impure rules. For the servant of truth to join in it means to sacrifice some of his time-honored role: he may have to turn apostle or agitator or political operator. This raises moral questions beyond those which technology itself poses, that of sanctioning immoral means for a surpassing end, of giving unto Caesar so as to promote what is not Caesar's. It is the grave question of moral casuistry, or of Dostoevsky's Grand Inquisitor, or of regarding cherished liberties as no longer affordable luxuries (which may well bring the anxious friend of mankind into odious political company)—questions one excusably hesitates to touch but in the further rule of things may not be permitted to evade.

What is, prior to joining the fray, the role of philosophy, that is, of a philosophically grounded ethical knowledge, in all this? The somber note of the last remarks responded to the quasi-apocalyptic prospects of the technological tide, where stark issues of planetary survival loom ahead. There, no philosophical ethics is needed to tell us that disaster must be averted. Mainly, this is the case of the ecological dangers. But there are other, noncatastrophic things afoot in technology, where not the existence but the image of man is at stake. They are with us now and will accompany us and be joined by others at every new turn technology may take. Mainly, they are in the biomedical, behavioral, and social fields. They lack the stark simplicity of the survival issue, and there is none of the (at least declaratory) unanimity on them which the specter of extreme crisis commands. It is here where a philosophical ethics or theory of values has its task. Whether its voice will be listened to in the dispute on policies is not for it to ask; perhaps it cannot even muster an authoritative voice with which to speak—a house divided, as philosophy is. But the philosopher must try for normative knowledge, and if his labors fall predictably short of producing a compelling axiomatics, at least his clarifications can counteract rashness and make people pause for a thoughtful view.

Where not existence but "quality" of life is in question, there is room for honest dissent on goals, time for theory to ponder them, and freedom from the tyranny of the lifeboat situation. Here, philosophy can have its try and its say. Not so on the extremity of the survival issue. The philosopher, to be sure, will also strive for a theoretical grounding of the very proposition that there ought to be men on earth, and that present generations are obligated to the existence of future ones. But such esoteric, ultimate validation of the perpetuity imperative for the species—whether obtainable or not to the satisfaction of reason—is happily not needed for consensus in the face of ultimate threat. Agreement in favor of life is pretheoretical, instinctive, and universal. Averting disaster takes precedence over everything else, including pursuit of the good, and suspends otherwise inviolable prohibitions and rules. All moral standards for individual or group behavior, even demands for individual sacrifice of life, are premised on the continued existence of human life. As I have said elsewhere,[4] "No rules can be devised for the waiving of rules in extremities. As with the famous shipwreck examples of ethical theory, the less said about it, the better."

Never before was there cause for considering the contingency that all mankind may find itself in a lifeboat, but this is exactly what we face when the viability of the planet is at stake. Once the situation becomes desperate, then what there is to do for salvaging it must be done, so that there be life—which "then," after the storm has been weathered, can again be adorned by ethical conduct. The moral inference to be drawn from this lurid eventuality of a moral pause is that we must never allow a lifeboat situation for humanity to arise.[5] One part of the ethics of technology is precisely to guard the space in which any ethics can operate. For the rest, it must grapple with the cross-currents of value in the complexity of life.

A final word on the question of determinism versus freedom which our presentation of the technological syndrome has raised. The best hope of man rests in his most troublesome gift: the spontaneity of human acting which confounds all prediction. As the late Hannah Arendt never tired of stressing: the continuing arrival of newborn individuals in the world assures ever-new beginnings. We should expect to be surprised and to see our predictions come to naught. But those predictions themselves, with their warning voice, can have a vital share in provoking and informing the spontaneity that is going to confound them.

NOTES

1. But as serious an actuality as the Chinese plough "wandered" slowly westward with little traces of its route and finally caused a major, highly beneficial revolution in medieval European agriculture, which almost no one deemed worth recording when it happened (cf. Paul Leser, *Entstehung und Verbreitung des Pfluges* [Münster, 1931: reprint The International Secretariate for Research on the History of Agricultural Implements, Brede-Lingby, Denmark, 1971]).

2. I here loosely refer to Adolph Lowe, "The Normative Roots of Economic Values," in *Human Values and Economic Policy*, ed. Sidney Hook, (New York: New York University Press, 1967), and more, perhaps, to the many discussions I had with Lowe over the years. For my side of the argument, see "Economic Knowledge and the Critique of Goals," in *Economic Means and Social Ends*, ed. R. I. Heilbroner (Englewood Cliffs, N.J.: Prentice-Hall, 1969), reprinted in Hans Jonas, *Philosophical Essays* (Englewood Cliffs, N.J.: Prentice-Hall, 1969, 1974).

3. "The Practical Uses of Theory," *Social Research* 26 (1959), reprinted in Hans Jonas, *The Phenomenon of Life: Toward a Philosophical Biology* (New York: Harper and Row, 1966). The reference is to pp. 209–10 in the latter edition.

4. "Philosophical Reflections on Experimenting with Human Subjects," in *Experimentation with Human Subjects*, ed. Paul A. Freund (New York: George Braziller, 1970), reprinted in Hans Jonas, *Philosophical Essays*. The reference is to pp. 124–25 in the latter edition.

5. For a comprehensive view of the demands which such a situation or even its approach would make on our social and political values, see Geoffrey Vickers, *Freedom in a Rocking Boat* (London: Allen Lane, 1970).

Reverse Adaptation and Control

Langdon Winner

The process which I call *reverse adaptation* is the key to the critical interpretation of how ends are developed for large-scale systems and for the activities of the technological society as a whole. Here the conception of autonomous technology as the rule of a self-generating, self-perpetuating, self-programming mechanism achieves its sharpest definition. The basic hypothesis is this: *that beyond a certain level of technological development, the rule of freely articulated, strongly asserted purposes is a luxury that can no longer be permitted.* I want now to state the logic of this position.

Of interest to the theory are technological systems or networks of a highly advanced development—systems characterized by large size, concentration, extension, and the complex interconnection of a great number of artificial and human parts. Such conditions of size and interconnectivity mark a new "state" in the history of technical means. Components that were developed and operated separately are now linked together to form organized wholes. The resulting networks represent a quantum jump over the power and performance capabilities of smaller, more segmental systems. In this regard, the genius of the twentieth century consists in the final connecting of technological elements taken from centuries of discovery and invention.

Characteristic also of this new stage of development is the interdependence of the major functioning components. Services supplied by one part are crucial to the successful working of other parts and to the system as a whole. This situation has both an internal and external dimension. Within the boundaries of any specific system, the mutual dependencies are tightly arranged and controlled. But internally well-integrated systems are also in many cases dependent upon each other. Through relationships of varying degrees of certainty and solidity, the systems

197

establish meta-networks, which supply "inputs" or receive "outputs" according to the purposes at hand. One need only consider the relationships among the major functional components—systems of manufacturing, energy, communications, food supply, transportation—to see the pulse beat of the technological society.

Large-scale systems can succeed in their ambitious range of activities only through an extension of *control*. Interdependence is a productive relationship only when accompanied by the ability to guarantee its outcome. But if a system must depend on elements it does not control, it faces a continuing uncertainty and the prospect of disruption. For this reason, highly organized technologies of the modern age have a tendency to enlarge their boundaries so that variables which were previously external become working parts of the system's internal structure.

The name usually given to the process of thought and action that leads to the extension of control is *planning,* which means much more than the sort of planning done by individuals in everyday life. Planning in this context is a formalized technique designed to make new connections with a high degree of certainty and manipulability. Clear intention, foresight, and calculation combine with the best available means of action. In some typical passages from [Jacques] Ellul we read:

> The more complex manufacturing operations become, the more necessary it is to take adequate precautions and to use foresight. It is not possible to launch modern industrial processes lightly. They involve too much capital, labor, and social and political modifications. Detailed forecasting is necessary.[1]

> Planning permits us to do more quickly and more completely whatever appears desirable. Planning in modern society is *the* technical method.[2] In the complexity of economic phenomena arising from techniques, how could one justify refusal to employ a trenchant weapon that simplifies and resolves all contradictions, orders incoherences, and rationalizes the excesses of production and consumption?[3]

Size, complexity, and costliness in technological systems combine to make planning—intelligent anticipation plus control—a virtual necessity. This is more than just convenience. Planning is crucial to the coherence of the technological order at a particular stage in its development.

One can ask, What would be the consequences of an inability to plan or to control the span of interdependencies? A reasonable answer would be that many specific kinds of enterprise known to us would fail. The system could not complete its tasks or achieve its purposes. Another consequence, more drastic, might be that the disturbance and disorientation would eventually ruin the internal structure of the system. The whole organized web of connections would collapse.

More important than either of these are the implications for the technological ensemble of the civilization as a whole. Ultimately, if large-scale, complex, interconnected, interdependent systems could not successfully plan, technological apraxia would become endemic. Society would certainly move to a different sort of technological development. This is clear enough to those who read Jacques Ellul, Lewis Mumford, Herbert Marcuse, Paul Goodman, or Ivan Illich and experience horror at the critique of social existence founded on large-scale systems. What if the critics were taken seriously? What if the necessary operating conditions of such systems were tampered with? Surely society would move "backward."[4]

This idea of moving "backward" is a fascinating one. At work here is a quaint, two-dimensional, roadlike image that almost everyone (including this writer) falls into as easily as sneezing. One moves, it seems forward (positive) or backward (negative). Never does one move upward and to the right or off into the distance at, say, a thirty-four degree angle. No; it is forward or backward in a straight line. What is understood, furthermore, is that forward means larger, more complex, based on the latest scientific knowledge and the centralized control of an increasingly greater range of variables. Hence it is clear that not to plan, not to control the circumstances of large-scale systems, is to risk a kind of ghastly cultural regression. This lends extra urgency to these measures and extra vehemence toward any criticism of the world that they produce. Surely, it is believed (and this is no exaggeration), the critics would have us *back* in the stone age.[5]

Now, everything I have said so far presupposes that large-scale technological systems are at the outset based on independent ends or purposes. It makes sense to say that technologies "serve" this or that end or need or to say that they are "used" to achieve a preconceived purpose or set of purposes. Nothing argued here seeks to deny this. In the original design, all technologies are *purposive*.

But within the portrait of advanced technics just sketched, this situation is cast in a considerably different light. Under the logic that takes one from size, interconnection, and interdependence to control and planning, it can happen that such things as ends, needs, and purposes come to be dysfunctional to a system. In some cases, the originally established end of a system may turn out to be a restraint upon the system's ability to grow or to operate properly. Strongly enforced, the original purpose may serve as a troublesome obstacle to the elaboration of the network toward a higher level of development. In other instances, the whole process that leads to the establishment of ends for the system may become an unacceptable source of uncertainty, interference, and instability. Formerly a guide to action, the end-setting process is now a threat. If the system must depend on a source that is truly independent in its ability to enforce new ends, then it faces the perils of dependency.

In instances of this kind, a system may well find it necessary to junk the whole end-means logic and take a different course. It may decide to take direct action to extend its control over the ends themselves. After all, when strongly asserted needs, purposes, or goals begin to pose a risk to the system's effective operation, why not choose transcendence? Why not treat the ends as an "input" like any other, include them in the plan, and tailor them to the system's *own needs?* Obviously *this* is the "one best way."

At this point the idea of rationality in technological thinking once again begins to wobble, for if one takes rationality to mean the accommodation of means to ends, then surely reverse-adapted systems represent the most flagrant violation of rationality. If, on the other hand, one understands rationality to be the effective, logical ordering of technological parts, then systems which seek to control their own ends are the very epitome of the rational process. It is this contrast that enables Herbert Marcuse to conclude that the technological society's "sweeping rationality, which propels efficiency and growth, is itself irrational."[6] How one feels about this depends on which model of rationality one wishes to follow. The elephant can do his dance if you ask him to, but he sometimes crushes a beautiful maiden during the performance.

Let us briefly examine some of the patterns reverse adaptation can take. Remember that as I use the term *system* here I am referring to large sociotechnical aggregates with human beings fully present, acting, and thinking. The behavior suggested, however, is meant as an attribute of the aggregate. Later I shall ask whether a change in the identities or ideologies of those "in control" is likely to make any difference.

1. *The system controls markets relevant to its operations.* One institution through which technologies are sometimes thought to be regulated is the market. If all went according to the ideal of classical economics, the market ought to provide the individual and social collectivity a powerful influence over the products and services that technological systems offer. Independent agents acting through the market should have a great deal to say about what is produced, how much, and at what price.

In point of fact, however, there are many ways in which large-scale systems circumvent the market, ways that have become the rule rather than the exception in much of industrial production. In J. K. Galbraith's version of what has become a mundane story: "If, with advancing technological and associated specialization, the market becomes increasingly unreliable, industrial planning will become increasingly impossible unless the market also gives way to planning. Much of what the firm regards as planning consists in minimizing or getting rid of market influences."[7]

Galbraith mentions three common procedures through which this is accomplished. The first is vertical integration, in which the market is superseded. "The planning unit takes over the source of supply or the outlet; a transaction that is subject to bargaining over prices and amounts is thus replaced with a transfer within the planning unit."[8]

A second means is market control, which "consists in reducing or eliminating the independence of action of those to whom the planning unit sells or from whom it buys."[9] Such control, Galbraith asserts, is a function of size. Large systems are able to determine the price they ask or pay in transactions with smaller organizations. To some extent, they are also able to control the amount sold.[10]

A third means suspends the market through contract. Here the systems agree in advance on amounts and prices to prevail in exchanges over a long period of time. The most stable and desirable of these involve contracts with the state.

Galbraith illustrates each of these with examples from General Motors, U.S. Steel, General Electric, and others, examples I shall not repeat. The matter is now part of modern folklore. That the market is an effective means for controlling large-scale systems is known to be a nostalgic, offbeat, or fantastic utopian proposal with little to do with reality.

2. *The system controls or strongly influences the political processes that ostensibly regulate its output and operating conditions.* Other possible sources of independent control are the institutions of politics proper. According to the model, clear-minded voters, legislators, executives, judges, and administrators make choices, which they impose upon the activities of technological systems. By establishing wise goals, rules, and limits for all such systems, the public benefit is ensured.

But the technological system, the servant of politics, may itself decide to find political cures for its own problems. Why be a passive tool? Why remain strictly dependent on political institutions? Such systems may act directly to influence legislation, elections, and the content of law. They may employ their enormous size and power to tailor political environments to suit their own efficient workings.

One need only review the historical success of the railroads, oil companies, food and drug producers, and public utilities in controlling the political agencies that supposedly determine what they do and how. A major accomplishment of political science is to document exactly how this occurs. The inevitable findings are now repeated as Ralph Nader retraces the footsteps of Grant McConnell. It is apparently still a shock to discover that "regulatory" commissions are dominated by the entities they regulate.[11]

The consequences of this condition are well known. In matters of safety, price, and quality of goods and services, the rules laid down reflect the needs of the system rather than some vital, independent, and forceful expression of public interest. It is not so much that the political process is always subverted in this way: the point is that occurrences of this sort happen often enough to be considered normal. No one is surprised to find Standard Oil spending millions to fight antipollution legislation. There is little more than weariness in our discovery that the Food and Drug Administration regularly allows corporations of the food industry to introduce untested and possibly unsafe additives into mass-processed foods.[12] I am not speaking here of the influence of private organizations only. Indeed, the best examples in this genre come from public agencies able to write their own tickets, for example, the Army Corps of Engineers.

3. *The system seeks a "mission" to match its technological capabilities.* It sometimes happens that the original purpose of a megatechnical organization is

accomplished or in some other way exhausted. The original, finite goals may have been reached or its products become outmoded by the passage of time. In the reasonable, traditional model of technological employment one might expect that in such cases the "tool" would be retired or altered to suit some new function determined by society at large.

But this is an unacceptable predicament. The system with its massive commitments of manpower and physical resources may not wish to steal gracefully into oblivion. Unlike the fabled Alexander, therefore, it does not weep for new worlds to conquer. It sets about creating them. Fearing imminent extinction, the system returns to the political arena in an attempt to set new goals for itself, new reasons for social support. Here a different kind of technological invention occurs. The system suggests a new project, a new mission, or a new variety of apparatus, which, according to its own way of seeing, is absolutely vital to the body politic. It places all of its influence into an effort to convince persons in the political sphere of this new need. A hypothesis suggests itself: if the system is deemed important to society as a whole, and if the new purpose is crucial to the survival of the system, then that purpose will be supported regardless of its objective value to the society.

Examples of this phenomenon are familiar in contemporary political experience and are frequently the subject of heated debate. The National Aeronautics and Space Administration [NASA] faces the problem of finding new justifications for its existence as a network of "big technology." NASA has successfully flown men to the moon. Now what? Many interesting new projects have been proposed by the agency: the space shuttle, the VSTOL [vertical short takeoff and landing] aircraft system, explorations to Mars, Venus, and Jupiter, asteroid space colonies. But whatever the end put forward, the fundamental argument is always the same: the aerospace "team" should not be dismantled, the great organization of men, technique, and equipment should not be permitted to fall to pieces. Give the system something to do. Anything. In the early 1960s resources were sought to fly the astronauts to unknown reaches of space; now funds are solicited to fly businessmen from Daly City to Lake Tahoe or the president to a space station for lunch.

Similar instances can be found in the recent histories of the ABM [antiballistic missile] project, the SST [supersonic transport], Boeing, General Dynamics, and Lockheed.[13] The nation may not need a particular new fighter plane, transport, bomber, or missile system. But the aerospace firms certainly need the contracts. And Los Angeles, Seattle, Houston, and other cities certainly need the aircraft companies. Therefore the nation needs the aircraft. The connections of the system to society as a whole give added punch to the effort to have reverse-adapted technological ends embraced as the most revered of national goals.

4. *The system propagates or manipulates the needs it also serves.* But even if one grants a certain degree of interference in the market and political processes, is it not true that the basic human *needs* are still autonomous? The institutional-

ized means to their satisfaction may have been sidetracked or corrupted, but the original needs and desires still exist as vital and independent phenomena. After all, persons in society do need food, shelter, clothing, health, and access to the amenities of modern life. Given the integrity of these needs, it is still possible to establish legitimate ends for all technical means.

The theory of technological politics finds such views totally misleading, for reverse adaptation does not stop with deliberate interference in political and economic institutions. It also includes control of the needs in society at large. Megatechnical systems do not sit idly by while the whims of public taste move toward some specifically desired product or service. Instead they have numerous means available to bring about that most fortunate of circumstances in which the social need and what the system is best able to produce coincide in a perfect one-to-one match. All the knowledge of the behavioral sciences and all of the tools of refined psychological technique are put to work on this effort. Through the right kinds of advertising, product design, and promotion, through the creation of a highly energized, carefully manipulated universe of symbols, man as consumer is mobilized to want and seek actively the goods and services that the instruments of technology are able to provide at that moment. "Roughly speaking," Ellul observes, "the problem here is to modify human needs in accordance with the requirements of planning."[14]

If the system were truly dependent upon a society with autonomous needs, if it were somehow forced to take a purely responsive attitude, then the whole arrangement of modern technology would be considerably different. In all likelihood there would be fewer such systems and with less highly developed structures. Autonomous needs are in this sense an invitation to apraxia. Adequate steps must be taken to insure that wants and needs of the right sort arrive at the correct time in predetermined quantity. In Ellul's words: "If man does not have certain needs, they must be created. The important concern is not the psychic and mental structure of the human being but the uninterrupted flow of any and all goods which invention allows the economy to produce. Whence the measureless trituration of the human soul, the true issue of which is propaganda. And propaganda, reduced to advertising, relates happiness and a meaningful life to consumption."[15]

The point here raises an important issue, which neither the apologists nor the critics of the technological society have addressed very well. Assumed in most of the writing is the continued growth of human wants and needs in response to the appearance of new technological achievements. With each new invention or innovation it becomes possible to awaken and satisfy an appetite latent in the human constitution. Potentially there is no limit to this. A want or need will arise to meet any breakthrough. But precisely how this occurs is never fully elucidated. Apparently the human being is by nature a creature of infinite appetite.

But even thinkers who believe this to be true are sometimes sobered by its implications. What if all the wrong needs are awakened? Marxists grapple with this dilemma in their analyses of "false consciousness" and "commodity fetishism," trying to explain how the proletariat should have taken such a serious interest in the debased consumer goods and status symbols of bourgeois society. Much of the neo-Marxian criticism of the Frankfurt school—the writings of Theodor Adorno, Max Horkheimer, Marcuse, Jürgen Habermas, and others—focuses on the corruption of Marx's vision of human fulfillment in technological societies.[16] Persons in such societies certainly do lead lives of great material abundance, as predicted by Marx's theory. But the quality of their desire and of their relationships to material things is certainly not what the philosopher had in mind.

Many of Ellul's lamentations, similarly, come from his conclusion that man, or at least modern man, is indeed infinitely malleable and appetitive and, therefore, an easy mark. There is nothing that a well-managed sales campaign cannot convince him to crave with all his heart. Manipulated by "propaganda," the sum total of all psychological and mass media techniques, man wildly pursues a burgeoning glut of consumer products of highly questionable worth.[17]

It is incorrect, however, to say that needs of this sort are false. Persons who express the needs undoubtedly have them. To those persons, they are as real as any other needs ever experienced. "False" is not a response to someone who says he absolutely must have an automatic garage door opener, extra-dry deodorant for more protection, or air-flow torsion-bar suspension. No; the position of the theory is not that such needs are false but rather that they are not autonomous. A need becomes a need in substantial part because a megatechnical system external to the person needed that need to be needed.

A possible objection here is that human needs are always some variant of what is available at the time and generally desired by the society. Personal needs do not exist independent of the social environment and specific state of technics in which they occur. That is undoubtedly correct. Nevertheless, it is clear that the degree of conscious, rational, well-planned stimulation and manipulation of need is now much greater than in any previous historical period. Systems in the technological order are able to engender and give direction to highly specific needs, which in the aggregate constitute much of the "demand" for products and services. The combined impact of such manipulation produces a climate of generalized, intense needfulnees bordering on mass hysteria, which keeps the populace permanently mobilized for its necessary tasks of consumption.

True, other cultures at other times have been as effective in suppressing need for religious or purely practical reasons. But this fact merely gives additional focus to the peculiar turn that a culture based on high-tech systems has taken.

5. *The system discovers or creates a crisis to justify its own further expansion.* One way in which large systems measure their own vitality is on the scale of growth. If a system is growing, it is maintaining its full structure, replacing worn-out parts, and expanding into new areas of activity. Thinking on this subject has become highly specialized, but the basic maxim is still simple: healthy things grow.

There are times, however, when a system may find that its growth has slowed or even stopped. Even worse, it may discover that its rationale for growth has eroded. Public need for the goods or services the system provides may have leveled off; social and political support for expansion may have withered. In such cases the system has, from its own point of view, failed in its very success.

But the system is not helpless in this predicament. It does control its own internal structure, and it has command of a great deal of information about its role in society. With a little care it can manipulate either its own structure or the relevant information to create the appearance of a public "crisis" surrounding its activities. This is not to say that the system lies or deceives. It may, however, read and publicize its own condition and the condition of its environment very selectively. From the carefully selected portrait may come an image of a new and urgent social need.

Two scenarios of this sort have become familiar in recent years: the threat and the shortage. Under the psychology of the threat, the system finds an external and usually very nebulous enemy whose existence demands the utmost in technological preparation. Foreign military powers and crime in the streets have been traditional favorites. Statistics are cited to demonstrate that the enemy is well armed and busy. Society, on the other hand, is asleep at the switch, woefully bereft of tools and staff. The only logical conclusion is that the relevant system must, therefore, be given the means necessary to meet the threat as soon as possible.

The Department of Defense, to cite one noteworthy example, keeps several such plot lines in various stages of preparation at all times. If public or congressional interest in new projects or impressive hardware fails, the latest "intelligence" is readily available to show that a "gap" has appeared in precisely the area in question. This practice works best when defense systems on two continents are able to justify their growth in terms of each other's activities. Here one can see firsthand one of nature's rarities: the perfect circle.

If well orchestrated, the shortage can be equally impressive. Here the system surveys the data on its own operations and environment and announces that a crucial resource, product, or service is in dangerously short supply. Adequate steps must be taken to forestall a crisis for the whole society. The system must be encouraged to expand and to extend its sphere of control.

It may well be that there is a demonstrable shortage. What is important, however, is that the system may command a virtual monopoly of information concerning the situation and can use this monopoly for self-justification. Persons and groups outside usually do not have access to or interest in the information necessary to scrutinize the "need" in a critical way. This allows the system to define the terms of the "shortage" in its own best way. Outsiders are able to say, "Yes, I see; there is a shortage." But they are usually not prepared to ask: What is its nature? What are the full circumstances? What alternatives are available? Thus the only response is, "Do what is necessary." A number of "shortages" of this kind have been well publicized of late. There are now "crises" in natural gas,

petroleum, and electricity, which a guileless public is discovering from predictable sources of information. In this instance, as in others, the almost
inevitable outcome may well be "crisis-system growth supported by a huge
public investment," all with a dubious relation to any clearly demonstrated need.

Some of the more interesting cases of reverse adaptation combine the above
strategies in various ways. Numbers 1 and 2 as well as 2 and 3 could be anticipated as successful pairings. A particularly ironic case is that which brings
together numbers 4 and 5, the propagation of need and the discovery of shortage.
This is presently a popular strategy with power companies who spend millions
advertising power-consuming luxury appliances while at the same time trum-
peting the dire perils of the "energy crisis." Such cases might be called double
reverse adaptations.

I am not saying that the patterns noted are universal in the behavior of
megatechnical systems. It may also happen that the traditionally expected
sequence of relating ends to means does occur, or some mix may take place. The
hypothesis of the theory of technological politics is that as large-scale systems
come to dominate various areas of modern social life, reverse adaptation will
become an increasingly important way of determining what is done and how.

I am not arguing that there is anything inherently wrong with this. My point
is that such behavior violates the models of technical practice we normally
employ. To the extent to which we employ tool-use and ends-means conceptions,
our experience will be out of sync with our expectations.

NOTES

1. Jacques Ellul. *The Technological Society,* trans. John Wilkinson (New York:
Alfred A. Knopf, 1964), p. 166.

2. Ibid., p. 184.

3. Ibid., p. 177.

4. A sample of this response is found in Melvin Kranzberg, "Historical Aspects of
Technology Assessment," *Technology Assessment Hearings Before the Subcommittee on
Science, Research, and Development of the U.S. House of Representatives* (Washington,
D.C.: U.S. Government Printing Office. 1970). After a brief survey of the ideas of Ellul,
Mumford, and Marcuse, Kranzberg concludes: "While such wholesale indictments may
stimulate nihilistic revolutionary movements, they really tell us very little about what can
be done to guide and direct technological innovation along socially beneficial lines" (ibid.,
p. 385).

5. Another interesting side to the "forward"-"backward" view counsels that the forward direction is ineluctable. In his statement to a U.S. Senate hearing in 1970, Harvey
Brooks took care to deny the proposition that "technological progress" is a "largely
autonomous development." But he went on to say, "While this pessimistic view of technology is not without evidence to support it, I believe it represents only a partial truth. Furthermore, it is an essentially sentimental and irrational view, because man in fact has *no*

choice but to push forward with his technology. The world is already *irrevocably committed to a technological culture* [emphasis added]." Reprinted ibid., p. 331.

6. Herbert Marcuse, *Negations: Essays in Critical Theory,* trans. Jeremy J. Shapiro (Boston: Beacon Press, 1969), p. xiii.

7. John Kenneth Galbraith, *The New Industrial State* (New York: The New American Library, 1968), p. 37.

8. Ibid., p. 39.

9. Ibid.

10. Ibid., p. 41.

11. Compare Grant McConnell, *Private Power and American Democracy* (New York: Alfred A. Knopf, 1966), and the reports of the Ralph Nader Study groups: James S. Turner, *The Chemical Feast: The Ralph Nader Study Group Report on the Food and Drug Administration* (New York: Grossman Publishers, 1970); Robert Fellmeth, *The Interstate Commerce Commission: The Ralph Nader Study Group Report on the Interstate Commerce Commission and Transportation* (New York: Grossman Publishers, 1970).

12. Gene Marine and Judy Van Allen, *Food Pollution* (New York: Holt, Rinehart & Winston, 1972).

13. See Clark R. Mollenhoff, *The Pentagon: Politics, Profits, and Plunder* (New York: Pinnacle Books, 1972): Murray Weidenbaum, "Arms and the American Economy: A Domestic Convergence Hypothesis," *Quarterly Review of Economics and Business* 8 (spring 1968); Ralph Lapp, *The Weapons Culture* (Baltimore: Penguin Books, 1968); Seymour Melman, *Pentagon Capitalism: The Political Economy of War* (New York: McGraw-Hill, 1970).

14. Ellul, *Technological Society,* p. 225.

15. Ibid., p. 221.

16. See Theodor Adorno, *Minima Moralia,* trans. E. F. N. Jephcott (London: NLB, 1974); Max Horkheimer, *Critical Theory,* trans. Matthew J. O'Connell et al. (New York: Herder and Herder, 1972); Jürgen Habermas, *Legitimation Crisis,* trans. Thomas McCarthy (Boston: Beacon Press, 1975). An interesting, polemical review of the progress of the Frankfurt school is given in Göran Therborn's article, "A Critique of the Frankfurt School," *New Left Review* 63 (September–October 1970): 65–96.

17. "Propaganda is a set of methods employed by an organized group that wants to bring about the active or passive participation in its actions of a mass of individuals, psychologically unified through psychological manipulations and incorporated in an organization." Jacques Ellul, *Propaganda,* trans. Konrad Kellen (New York: Alfred A. Knopf, 1967), p. 61.

Artifacts, Neutrality, and the Ambiguity of "Use"

Russell Woodruff

The main purpose of this paper is to call attention to a neglected argument concerning the neutrality of artifacts. Most arguments about neutrality, whether supportive or critical, have as their subject the relation between artifacts and the purposes they serve. At the same time, a significant argument focusing on the methods of using artifacts generally goes unnoticed. This neglect compromises our understanding of the extent and limits of the neutrality of artifacts. As will be shown, when we ask if artifacts are neutral with respect to purposes, the answer, crudely put, is "sometimes," whereas when we ask if artifacts are neutral with respect to method, the answer is "never."

Failure to attend to the issue of method may come from overlooking the ambiguity of the term "use." There are at least three different senses of "use" that apply to our interactions with artifacts: act, purpose, and method. The secondary purpose of this paper is to call attention to these different senses so that we may avoid confusing or conflating them, and so that we are sensitive to the need for clarity regarding what precisely is being claimed when someone says something like "it's not the tool, it's how it's used," or "the tool determines its use."

The subject of this paper is not technology as a whole. Broadly conceived, technology includes at least knowledge, methods of inquiry, communities and organizations, as well as objects. But claims about neutrality typically are about technological objects (artifacts), and since this paper examines such claims, it will adopt the same focus.

Russell Woodruff, "Artifacts, Neutrality, and the Ambiguity of 'Use,'" *Research in Philosophy and Technology* 16 (1997): 119–27. Copyright © 1997 Russell Woodruff. Reprinted with permission of the author.

THE ACT OF USING

One kind of claim used to support the idea that artifacts are neutral is that the fabrication of an artifact does not necessitate any future event, specifically the event of using that artifact to achieve some purpose. Emmanuel Mesthene writes, "There is nothing in the nature or fact of a new tool, of course, that requires its use."[1] "The tools by themselves do nothing," states Joseph Pitt. The idea is that we can always choose not to use an artifact, and so avoid having our actions determined or influenced by it. While in principle this is correct, in practice the possibility of this rejection depends on a number of matters. I can think of three: (a) whether there exist alternate means to serve the purpose that using the artifact would serve; (b) if not, whether it is practically possible to reject the purpose served by the functioning of the artifact; and (c) whether the particular individual confronting this situation has the freedom or power to reject using the artifact.

It is true that the fabrication of an artifact does not necessitate any future event, specifically the use of that artifact. But we must remember the real-life context of fabrication. Humans purposefully fabricate artifacts in order to extend their capabilities to manipulate material and information. Ultimately motivated by the fact that humans find their needs insufficiently met by their unmediated interaction with nature, humans are in a position where they *must* make and use artifacts in order to live. So while it is true that there is no causal necessity whereby the making of a particular artifact "pushes" the using of it, there is a practical necessity that "pulls" us to use artifacts, and thus to make them. Although we may sometimes be free to choose not to use a particular artifact, we are never free to choose to use no artifacts.

The idea that artifacts are causally passive may remind us that "ultimately" it is "we" who are in control. But saying that we can decide if we want to interact with particular artifacts says nothing about the structure of those human-artifact interactions that we do undertake, and, since we must undertake some such interactions, this argument for neutrality fails us precisely when we need to understand if and how artifacts determine our interactions with them.

USES OF ARTIFACTS

Other claims about the neutrality of artifacts address the question whether or to what extent artifacts have different uses, including uses with contrary moral value. Here "use" refers to what an artifact is used *for* (as when we say, "it has many uses"), or in other words, what *purposes* the artifact can serve. The basic question here concerns the existence of a range of purposes that can be served by an artifact. A second kind of question concerns the locus of the decision as to which of these alternate purposes the artifact will in fact serve. The second kind of question depends on the first in that we can inquire into the locus of decision about which purpose will be served only if there is a decision to be made; if there were only one purpose that a particular artifact could serve, such an inquiry would be pointless. For this reason, the analysis here will center on the first kind of question.

The claim that artifacts can serve multiple purposes, particularly purposes with contrary values, seems to be used as the central pillar of attempts to support artifact neutralism. For example, Emmanuel Mesthene writes:

> We can explore the heavens with [technology], or destroy the world. We can cure disease, or poison entire populations. We can free enslaved millions, or enslave millions more. Technology spells only possibility, and is in that respect neutral.[2]

Mario Bunge describes the neutrality of artifacts thus:

> Most industrial products are morally neutral, in the sense that they can be used for good or for evil. A knife may be used for cutting loaves of bread or throats; a powerful drug to cure or to kill.[3]

As indicated, this kind of claim is distinct from, although often accompanied by, a second kind of claim regarding the locus of decision as to which uses (purposes) an artifact will serve. Mesthene writes:

> Technology . . . creates new possibilities for human choice and action, but leaves their disposition uncertain. What its effects will be and what ends it will serve are not inherent in the technology but depend on what man will do with technology.[4]

Joseph Pitt states the same position in this way:

> It is not the machine that is frightening, but what some men will do with the machine; or, given the machine, what we fail to do by way of assessment and planning. It may be only a slogan, but there is a ring of truth to: "Guns don't kill, people do." There is no problem about the autonomy of technology. The problem is man.[5]

Opposing neutralism on its own terms, some authors challenge the idea that artifacts can serve multiple purposes. For example, Friedrich Rapp argues that the separation of artifact from purpose is dubious, especially in the case of highly specialized artifacts. Very often, Rapp claims, such artifacts have only one possible application.

> Technological processes and objects are certainly not always *factually* neutral. One will never, for example, be able to find a peaceful application of highly specialized military weapons. The range of applicability of a technological system is inversely proportional to its degree of specialization; the more specialized the

design, the narrower its applicable latitude. While a hammer is a relatively versatile tool, modern technological devices and systems are as a rule, tailored to a very specific purpose, which, as a result, they very effectively fulfill. Within the context of a given technological potential a decision for or against selection of a specific goal is possible. Once the choice of goal is made and the particular technological system constructed, there is typically not much leeway as to its possible applications.[6]

In order to assess claims about the range of purposes that artifacts can serve, we need to get clear on what is being referred to by "purpose" in such claims. In particular, we need to distinguish the purposes of people from the purposes of things. Both people and things have purposes, but it seems that they have them in different ways. When I have a purpose, what I have is an intentional state reflecting some *goal* that I want to bring about. When an artifact has a purpose, what it has is a *function* that it more or less adequately performs. This distinction can be contended. It can be said that artifacts in some sense have intentions, or that "goal" and "function" name the same thing. I do not think they name exactly the same thing. One difference is that humans *want* to achieve their purposes, while artifacts cannot *want* anything. But nothing here hinges on whether the purposes of humans are of the same nature as those of artifacts. All that needs be granted is that both humans and artifacts have purposes in some sense of that term.

In this regard, consider the knife mentioned by Bunge in his claim about artifact neutrality. It has a purpose, that is, a function: to cut things. In turn, humans can use the knife for alternate purposes (goals): to kill people, make sandwiches, carve wooden figurines. Note that at the level of artifact function, the neutralist claim about an artifact having a range of purposes is false. The knife has only one purpose (function); indeed, even among knives there is widespread specialization of function. The knife for cutting bread has quite different features than the knife for carving wooden figurines. So too, with the hammer mentioned earlier by Rapp. If he were talking about functions, Rapp's claim that "a hammer is a relatively versatile tool" would be puzzling. The carpenter's hammer fits its two functions (driving and pulling nails) quite well, and no others. Indeed, in comparison with other kinds of hammers, it is even ill-fitted to other "hammerly" functions such as pounding sheet metal into a desired shape. In this sense, there is "not much leeway" to the "possible applications" of hammers and bombers alike.

Consider some other examples of artifacts, their functions, and the purposes they can serve:

- a B-1 bomber flies off and drops bombs, in order to damage buildings, infrastructure, and people;
- a numerically controlled machine tool shapes complex metal parts, in order to construct airplanes, compressors, dynamos, and so forth;

- a wool topcoat shields one's body from the cold, in order to make it easier to get around in cold weather;
- a washing machine cleans clothes, in order to enhance our appearance and eliminate odors;
- a canoe transports people on water, in order that they may engage in fishing, exercise, trading, sightseeing, or piracy;
- a bulletproof vest shields one's torso from injury by bullets, in order to survive a shooting; and
- an electrical power network generates and distributes electricity, in order to illuminate homes, offices, shops, factories, hospitals, schools, prisons, and in order to power computers, air conditioners, refrigerators, vacuum cleaners, electric chairs, respirators, and so on.

Each of these artifacts has a narrowly ascribed set of functions, as does every other artifact ever created. This being the case, claims about artifacts serving a range of purposes obviously make sense only as claims about the human purposes that can be served if and when artifacts are used in the performance of their functions. Examining artifacts with respect to human purposes we are confronted, on the one hand, by some with quite limited ranges of servable purposes. Washing machines, bulletproof vests, B-1 bombers, and wool topcoats are good examples of single-purpose artifacts. On the other hand, hammers, canoes, numerically controlled machine tools, and electrical power networks are all multipurpose artifacts.

Rapp contends that the matter of how many different purposes can be served by the performance of an artifact's function is contingent upon the complexity or modernity of the artifact in question. This seems questionable. While it is true that the more specialized something is the narrower its range of purpose (and that is probably analytically true), I see no reason to identify modern or complex tools and systems with specialized tools and systems. On the one hand, there seems to be more or less as broad a range of purposes for electric power grids and numerically controlled machine tools as there is for hammers, perhaps more so. On the other hand, the range of purposes seems more or less equally narrow for B-1 bombers, bulletproof vests, wool topcoats, and washing machines. Even some very simple and ancient stone tools are quite specialized.

Rapp's argument is successful as a criticism of a universal generalization that all artifacts are flexible with respect to their purposes. In some cases—bombers, bulletproof vests, and topcoats—there is only one purpose served by using the artifact to perform its function. In such cases, contrary to neutralism, the purposes served are "inherent in the technology." But it must be granted that there are many artifacts—canoes, hammers, machine tools, electric power grids, and so on—that can serve more than one purpose when used to perform their functions. Nor can modern or complex be equated with single-purpose, nor premodern or simple with multipurpose. The question of the neutrality of artifacts in terms of their flexibility with respect to purposes must be addressed on a case-by-case basis. In the case of functions, however, artifacts are never neutral. Every artifact has a specific set of functions built into its designed features.

METHODS OF USING

In addition to "act," and "purpose" (the latter in turn divisible into "function" and "goal"), there is another sense of "use" sometimes employed when talking about artifact use. This third sense of "use" refers to the method of operating the artifact. This is the sense we mean when we ask *"How* is it used?" An artifact's use in this sense consists of the form of a series of operations by humans. A canoe's use, in this sense, consists in getting in, sitting down, grasping the paddle, and moving it through the water.

As seen in the section on the act of using, it is not necessarily the case that a particular artifact must be used. But it is the case that, *if* it is used, the using must take a particular form. This points to the sense of "use" as method of operation. Artifacts are designed and made by humans to be functional, in order that we might more easily and effectively achieve many of the large number of ends we pursue. In the context of pursuing such ends we get artifacts to perform their functions only by following the rules built into the material forms of the artifacts. While we determine in our fabrication of artifacts what functions they will perform and how, once artifacts are created it is we, as would-be users, who must conform to them, and not them to us. All artifacts, even simple ones like screwdrivers and canoes, impose methods of operation on all who would use them. Whether you are a trader or a pirate there is, roughly speaking, only one way to operate a canoe. The artifact dictates a method of operation and the person who would successfully use it, for *any* purpose, must see that her using adheres to the dictated method.

Attending to use as method of operation reveals a significant limitation of neutralism. While many artifacts can serve diverse purposes, and no artifact necessitates its use, every artifact imposes a method of operation on would-be users. This imposition has the form of a hypothetical imperative: If you want to achieve some purpose with this artifact, your action must take the prescribed forms. The method is built in, and we must adhere to it.

The method of operation of an artifact is the "how" of using. It consists of the forms of physical and mental operations as determined by the material structure of the artifact in the context of a given purpose. These forms of operation are distinct from, although not necessarily separate from, features of technological activity such as duration of operations ("how long"), which is determined by the purpose of the action (one is done when the purpose is achieved), as well as the "where," "when," "who," "why," and "how many times" of the activity, which are determined by decisions of task management within the social context of the action, and by more basic decisions about the overarching purposes of the technological activity and of the social organization in general.

The determination by an artifact of the "how" of technological action presents itself to would-be users as a set of physical and mental requirements that users must meet in order to get the artifact to perform its function. For example, tools impose the basic physical requirement that one has the strength to be able to move them in accordance with their method of operation. Sledgehammers and wooden extension ladders cannot be used by those who cannot lift them.

Strength requirements generally recede in importance as we move from hand tools to machines. With the latter, the physical force required for operation is most often built into the machine. (One exception that comes to mind is the class of newer supersonic fighter airplanes. Putting these sophisticated machines through aerial combat operations requires substantial strength to overcome the gravitational forces exerted upon the pilot's body.)

Artifacts also require that we perceive and attend to various aspects of our interaction with them. When I drive a nail with a hammer, I must pay attention to the position of the nail relative to the wood, the position of my free hand relative to the whole operation, the space available for swinging the hammer, the trajectory of the hammer as it strikes the nail, and the sound and feel of the nail as it is being driven. When I use a nail gun to perform the same function, the attentional requirements change, and in this case generally are reduced in number. Since nail and gun are combined in one unit, I no longer have to attend to their relation. Since the operation—the actual driving—is performed by means of the tool's internal processes, I no longer have to attend to the trajectory of tool and arm. The attentional demands reduce to attending to the position of the gun relative to the wood, and to the position of my free hand and other body parts. This last requirement is heightened with this new tool due to the power it brings to bear upon the operation. While one is quite unlikely to accidentally drive a nail through one's hand with a hammer or cuts one's leg off with a hand saw, such accidents are much more likely possibilities with nail guns and chain saws.

In addition to perceiving and attending to features of artifact operation, there is the operation itself. To get an artifact to perform its function something must be done, and, unless the doing is to be left to chance, its execution requires *skill*, the "ability to use one's knowledge effectively and readily in execution or performance."[7] Thus, in addition to physical and attentional demands, artifacts impose skill requirements.

It may be objected that some artifacts perform their functions when humans merely "push a button," and that, thus, no skill requirements are imposed on their users. How much skill is required to turn on an amplifier or an air conditioner? Many artifacts can be operated by "unskilled" people.

This objection fails to notice the difference between saying that an artifact imposes low skill requirements and saying that an artifact does not impose skill requirements. Just as a low temperature is still a temperature, a low skill requirement is still a skill requirement. While some artifacts may not require any skills specific to their operation, and thus their operation may be labeled "unskilled," these same artifacts require basic human skills. To perform the "unskilled" work of pushing a button requires that we recognize the button as a switch, that we comprehend the meaning of mathematical or linguistic symbols, and that we can judge when to push the button. Skills held in common with the majority of adult humans are no less skills than those specialized skills that are specific to tools or tasks. The fact that we take for granted the possession of common skills does not mean they are not required. While we may devalue both the kinds of performances that "anyone can do" and the artifacts that require only these kinds of performances (and to an extent such devaluing is justified), this should not blind us to the fact that such artifacts nonetheless impose skill requirements.

Focusing on use as method of operation reveals an important way in which artifacts are not value neutral. Given that artifacts impose a method of operation requiring the utilization of skills, we need only make the additional point that skillful activity can be satisfying in itself, apart from purposes served, objects made, knowledge gained, and payments or praise earned. Many empirical studies provide evidence that use of skill is intrinsically satisfying.[8] In the context of hypothetical imperatives of practical activity, an artifact determines what sensory, manual, and mental capacities must be brought to bear by humans in order to operate it. To the extent that the use of these capacities is directly beneficial or detrimental to the agent, the artifact is a causal condition of something valuable, and is so independently of the purposes to which it is put, independently of the further consequences of its use, and independently of human intentions regarding its use and the consequences of its use. Thus artifacts are not value-neutral.

On this line of reasoning about use as method, artifacts matter as more than storehouses of possibilities. Their materiality imposes forms of acting upon humans, forms that can make favorable and unfavorable differences in people's lives. We must pay attention to the ways artifacts directly affect the human components of our technological systems. This is not merely a matter of providing experiences with certain "feels." It is also a matter of whether we will reduce or enhance our opportunities to enrich our lives through the development of certain capacities that are part of what it is to be human.

NOTES

1. Emmanuel Mesthene, "How Technology Will Shape the Future," *Science* (12 July 1968): 135.

2. Emmanuel Mesthene, "Technology and Wisdom," in *Technology and Social Change* (Indianapolis: Bobbs-Merrill, 1967), p. 59.

3. Mario Bunge, *Treatise on Basic Philosophy*, vol. 7 (Boston: Reidel, 1984), p. 310.

4. Emmanuel Mesthene, *Technological Change: Its Impact on Man and Society* (New York: New American Library, 1970), p. 60.

5. Joseph Pitt, "The Autonomy of Technology," in *Technology and Responsibility,* ed. P. T. Durbin (Boston: Reidel, 1987), p. 113.

6. Friedrich Rapp, *Analytical Philosophy of Technology* (Boston: Reidel, 1978), p. 54–55.

7. *Webster's New Collegiate Dictionary,* 1980 ed., s.v. "skill."

8. Robert Cooper, "Task Characteristics and Intrinsic Motivation," *Human Relations* 26, no. 3 (1973): 387–413; S. R. Helpingstine, T. C. Head, and Peter F. Sorenson, "Job Characteristics, Job Satisfaction, Motivation, and Satisfaction with Growth: A Study of Industrial Engineers," *Psychological Reports* 49 (1981): 381–82; P. Humphrys and Gordon E. O'Brien, "The Relationship between Skill Utilization, Professional Orientation, and Job Satisfaction for Pharmacists," *Journal of Occupational Psychology* 59 (1986): 315–26; E. E. Lawler and D. T. Hall, "Relationship of Job Characteristics to Job Involvement, Satisfaction, and Intrinsic Motivation," *Journal of Applied Psychology* 54 (1970): 259–86.

In Praise of Technology

Samuel C. Florman

generation ago most people believed, without doubt or qualification, in the beneficial effects of technological progress. Books were written hailing the coming of an age in which machines would do all the onerous work, and life would become increasingly utopian.

Today there is a growing belief that technology has escaped from human control and is making our lives intolerable. Thus do we dart from one false myth to another, ever impressed by glib and simple-minded prophets.

Hostility to technology has become such a familiar staple of our reading fare that rarely do we stop to consider how this new doctrine has so quickly and firmly gained its hold upon us. I believe that critical scrutiny of this strange and dangerous phenomenon is very much overdue. The founding father of the contemporary anti-technological movement is Jacques Ellul, whose book, *The Technological Society*, was published in France in 1954, and in the United States ten years later. When it appeared here, Thomas Merton, writing in *Commonweal*, called it "one of the most important books of this mid-century." In *Book Week* it was labeled "an essay that will likely rank among the most important, as well as tragic, of our time."

Ellul's thesis is that "technique" has become a Frankenstein monster that cannot be controlled. By technique he means not just the use of machines, but all deliberate and rational behavior, all efficiency and organization. Man created technique in prehistoric times out of sheer necessity, but then the *bourgeoisie* developed it in order to make money, and the masses were converted because of their interest in comfort. The search for efficiency has become an end in itself, dominating man and destroying the quality of his life.

The second prominent figure to unfurl the banner of anti-technology was Lewis Mumford. His conversion was particularly significant since for many years he had been known and respected as the leading historian of technology. His massive *Myth of the Machine* appeared in 1967 (Part I: *Technics and Human Development*) and in 1970 (Part II: *The Pentagon of Power*). Each volume in turn was given front-page coverage in the *New York Times Sunday Book Review*. On the first page of *Book World* a reviewer wrote, "Hereafter it will be difficult indeed to take seriously any discussion of our industrial ills which does not draw heavily upon this wise and mighty work." The reviewer was Theodore Roszak, who, as we shall see, was soon to take his place in the movement.

The next important convert was René Dubos, a respected research biologist and author. In *So Human an Animal*, published in 1968, Dubos started with the biologist's view that man is an animal whose basic nature was formed during the course of his evolution, both physical and social. This basic nature, molded in forests and fields, is not suited to life in a technological world. Man's ability to adapt to almost any environment has been his downfall, and little by little he has accommodated himself to the physical and psychic horrors of modern life. Man must choose a different path, said Dubos, or he is doomed. This concern for the individual, living human being was just what was needed to flesh out the abstract theories of Ellul and the historical analyses of Mumford. *So Human an Animal* was awarded the Pulitzer Prize, and quickly became an important article of faith in the anti-technology crusade.

In 1970 everybody was talking about Charles A. Reich's *Greening of America*. In paperback it sold more than a million copies within a year. Reich, a law professor at Yale, spoke out on behalf of the youthful counterculture and its dedication to a liberating consciousness-raising. Theodore Roszak's *Where The Wasteland Ends* appeared in 1972 and carried Reich's theme just a little further, into the realm of primitive spiritualism. Roszak, like Reich, is a college professor. Unlike *The Greening of America*, his work did not capture a mass audience. But it seemed to bring to a logical climax the anti-technological movement started by Ellul. As the reviewer in *Time* magazine said, "he has brilliantly summed up once and for all the New Arcadian criticism of what he calls 'postindustrial society.'"

There have been many other contributors to the anti-technological movement, but I think that these five—Ellul, Mumford, Dubos, Reich, and Roszak—have been pivotal.

They are united in their hatred and fear of technology, and surprisingly unanimous in their treatment of several key themes:

1. Technology is a "thing" or a force that has escaped from human control and is spoiling our lives.
2. Technology forces man to do work that is tedious and degrading.
3. Technology forces man to consume things that he does not really desire.
4. Technology creates an elite class of technocrats, and so disenfranchises the masses.
5. Technology cripples man by cutting him off from the natural world in which he evolved.

6. Technology provides man with technical diversions which destroy his existential sense of his own being.

The anti-technologists repeatedly contrast our abysmal technocracy with three cultures that they consider preferable: the primitive tribe, the peasant community, and medieval society.

Recognizing that we cannot return to earlier times, the anti-technologists nevertheless would have us attempt to recapture the satisfactions of these vanished cultures. In order to do this what is required is nothing less than *a change in the nature of man.* The anti-technologists would probably argue that the change they seek is really a return to man's *true* nature. But a change from man's present nature is clearly their fondest hope.

In the often-repeated story, Samuel Johnson and James Boswell stood talking about Berkeley's theory of the nonexistence of matter. Boswell observed that although he was satisfied that the theory was false, it was impossible to refute it. "I never shall forget," Boswell tells us, "the alacrity with which Johnson answered, striking his foot with mighty force against a large stone, till he rebounded from it—'I refute it *thus.*'"

The ideas of the anti-technologists arouse in me a mood of exasperation similar to Dr. Johnson's. Their ideas are so obviously false, and yet so persuasive and widely accepted, that I fear for the common sense of us all.

The impulse to refute this doctrine with a Johnsonian kick is diminished by the fear of appearing simplistic. So much has been written about technology by so many profound thinkers that the nonprofessional cannot help but be intimidated. Unfortunately for those who would dispute them, the anti-technologists are masters of prose and intellectual finesse. To make things worse, they display an aesthetic and moral concern that makes the defender of technology appear like something of a philistine. To make things worse yet, many defenders of technology are indeed philistines of the first order.

Yet the effort must be made. If the anti-technological argument is allowed to stand, the engineer is hard pressed to justify his existence. More important, the implications for society, should anti-technology prevail, are most disquieting. For, at the very core of anti-technology, hidden under a veneer of aesthetic sensibility and ethical concern, lies a yearning for a totalitarian society.

The first anti-technological dogma to be confronted is the treatment of technology as something that has escaped from human control. It is understandable that sometimes anxiety and frustration can make us feel this way. But sober thought reveals that technology is not an independent force, much less a thing, but merely one of the types of activities in which people engage. Furthermore, it is an activity in which people engage because they choose to do so. The choice may sometimes be foolish or unconsidered. The choice may be forced upon some members of society by others. But this is very different from the concept of technology *itself* misleading or enslaving the populace.

Philosopher Daniel Callahan has stated the case with calm clarity:

> At the very outset we have to do away with a false and misleading dualism, one
> which abstracts man on the one hand and technology on the other, as if the two
> were quite separate kinds of realities. I believe that there is no dualism inherent
> here. Man is by nature a technological animal; to be human is to be technolog-
> ical. If I am correct in that judgment, then there is no room for a dualism at all.
> Instead, we should recognize that when we speak of technology, this is another
> way of speaking about man himself in one of his manifestations.

Although to me Callahan's statement makes irrefutable good sense, and Ellul's
concept of technology as being a thing-in-itself makes absolutely no sense, I recognize that this does not put an end to the matter, any more than Samuel Johnson
settled the question of the nature of reality by kicking a stone.

It cannot be denied that, in the face of the excruciatingly complex problems
with which we live, it seems ingenuous to say that men invent and manufacture
things because they want to, or because others want them to and reward them
accordingly. When men have engaged in technological activities, these activities
appear to have had *consequences*, not only physical but also intellectual, psychological, and cultural. Thus, it can be argued, technology is *deterministic*. It
causes other things to happen. Someone invents the automobile, for example, and
it changes the way people think as well as the way they act. It changes their living
patterns, their values, and their expectations in ways that were not anticipated
when the automobile was first introduced. Some of the changes appear to be not
only unanticipated but undesired. Nobody wanted traffic jams, accidents, and
pollution. Therefore, technological advance seems to be independent of human
direction. Observers of the social scene become so chagrined and frustrated by
this turn of events—and its thousand equivalents—that they turn away from the
old commonsense explanations, and become entranced by the demonology of the
anti-technologists.

In addition to confounding rational discourse, the demonology outlook of the
anti-technologists discounts completely the integrity and intelligence of the ordinary person. Indeed, pity and disdain for the individual citizen is an essential
aspect of anti-technology. It is central to the next two dogmas, which hold that
technology forces man to do tedious and degrading work, and then forces him to
consume things that he does not really desire.

Is it ingenuous, again, to say that people work, not to feed some monstrous
technological machine, but, as since time immemorial, to feed themselves? We
all have ambivalent feelings toward work, engineers as well as anti-technologists.
We try to avoid it, and yet we seem to require it for our emotional well-being.
This dichotomy is as old as civilization. A few wealthy people are bored because
they are not required to work, and a lot of ordinary people grumble because they
have to work hard.

The anti-technologists romanticize the work of earlier times in an attempt to make it seem more appealing than work in a technological age. But their idyllic descriptions of peasant life do not ring true. Agricultural work, for all its appeal to the intellectual in his armchair, is brutalizing in its demands. Factory and office work is not a bed of roses either. But given their choice, most people seem to prefer to escape from the drudgery of the farm. This fact fails to impress the anti-technologists, who prefer their sensibilities to the choices of real people.

As for the technological society forcing people to consume things that they do not want, how can we respond to this canard? Like the boy who said, "Look, the emperor has no clothes," one might observe that the consumers who buy cars and electric can openers could, if they chose, buy oboes and oil paints, sailboats and hiking boots, chess sets and Mozart records. Or, if they have no personal "increasing wants," in Mumford's phrase, could they not help purchase a kidney machine which would save their neighbor's life? If people are vulgar, foolish, and selfish in their choice of purchases, is it not the worst sort of cop-out to blame this on "the economy," "society," or "the suave technocracy"? Indeed, would not a man prefer being called vulgar to being told he has no will with which to make choices of his own?

Which brings us to the next tenet of anti-technology, the belief that a technocratic elite is taking over control of society. Such a view at least avoids the logical absurdity of a demon technology compelling people to act against their own interests. It does not violate our common sense to be told that certain people are taking advantage of other people. But is it logical to claim that exploitation increases as a result of the growth of technology?

Upon reflection, this claim appears to be absolutely without foundation. When camel caravans traveled across the deserts, there were a few merchant entrepreneurs and many disenfranchised camel drivers. From earliest historical times, peasants have been abused and exploited by the nobility. Bankers, merchants, landowners, kings, and assorted plunderers have had it good at the expense of the masses in practically every large social group that has ever been (not just in certain groups like pyramid-building Egypt, as Mumlord contends). Perhaps in small tribes there was less exploitation than that which developed in large and complex cultures, and surely technology played a role in that transition. But since the dim, distant time of that initial transition, it simply is not true that advances in technology have been helpful to the Establishment in increasing its power over the masses.

In fact, the evidence is all the other way. In technologically advanced societies, there is more freedom for the average citizen than there was in earlier ages. There has been continuing apprehension that new technological achievements *might* make it possible for governments to tyrannize the citizenry with Big Brother techniques. But, in spite of all the newest electronic gadgetry, governments are scarcely able to prevent the antisocial actions of criminals, much less control every act of every citizen. Hijacking, technically ingenious robberies, computer-aided

embezzlements, and the like, are evidence that the outlaw is able to turn technology to his own advantage, often more adroitly than the government. The FBI
has admitted that young revolutionaries are almost impossible to find once they
go "underground." The rebellious individual is more than holding his own.

Exploitation continues to exist. That is a fact of life. But the anti-technologists are in error when they say that it has increased in extent or intensity because
of technology. In spite of their extravagant statements, they cannot help but recognize that they are mistaken, statistically, at least. Reich is wrong when he says
that "decisions are made by experts, specialists, and professionals safely insu-
lated from the feelings of the people." (Witness changes in opinion, and then in
legislation, concerning abortion, divorce, and pornography.) Those who were
slaves are now free. Those who were disenfranchised can now vote. Rigid class
structures are giving way to frenetic mobility. The barons and abbots and merchant princes who treated their fellow humans like animals, and convinced them
that they would get their reward in heaven, would be incredulous to hear the anti-
technologists theorize about how technology has brought about an increase in
exploitation. We need only look at the underdeveloped nations of our present era
to see that exploitation is not proportionate to technological advance. If anything,
the proportion is inverse.

Next we must confront the charge that technology is cutting man off from
his natural habitat, with catastrophic consequences. It is important to point out
that if we are less in touch with nature than we were—and this can hardly be dis-
puted—then the reason does not lie exclusively with technology. Technology
could be used to put people in very close touch with nature, if that is what they
want. Wealthy people could have comfortable abodes in the wilderness, could
live among birds in the highest jungle treetops, or even commune with fish in the
ocean depths. But they seem to prefer penthouse apartments in New York and
villas on the crowded hills above Cannes. Poorer people could stay on their farms
on the plains of Iowa, or in their small towns in the hills of New Hampshire, if
they were willing to live the spare and simple life. But many of them seem to tire
of the loneliness and the hard physical labor that goes with rusticity, and succumb
to the allure of the cities.

It is Roszak's lament that "the malaise of a Chekhov play" has settled upon
daily life. He ignores the fact that the famous Chekhov malaise stems in no small
measure from living in the country. "Yes, old man," shouts Dr. Astrov at Uncle
Vanya, "in the whole district there were only two decent, well-educated men: you
and I. And in some ten years the common round of the trivial life here has
swamped us, and has poisoned our life with its putrid vapors, and made us just
as despicable as all the rest." There is tedium in the countryside, and sometimes
squalor.

Nevertheless, I personally enjoy being in the countryside or in the woods and
so feel a certain sympathy for the anti-technologists' views on this subject. But I
can see no evidence that frequent contact with nature is *essential* to human well-

being, as the anti-technologists assert. Even if the human species owes much of its complexity to the diversity of the natural environment, why must man continue to commune with the landscapes in which he evolved? Millions of people, in ages past as well as present, have lived out their lives in city environs, with very little if any contact with "nature." Have they lived lives inherently inferior because of this? Who would be presumptuous enough to make such a statement?

The next target of the anti-technologists is Everyman at play. It is particularly important to anti-technology that popular hobbies and pastimes be discredited, for leisure is one of the benefits generally assumed to follow in the wake of technological advances. The theme of modern man at leisure spurs the anti-technologists to derision.

In their consideration of recreation activities, the anti-technologists disdain to take into account anything that an actual participant might feel. For even when the ordinary man considers himself happy—at a ball game or a vacation camp, watching television or listening to a jukebox, playing with a pinball machine or eating hot dogs—we are told that he is only being fooled into *thinking* that he is happy.

It is strategically convenient for the anti-technologists to discount the expressed feelings of the average citizen. It then follows that (1) those satisfactions which are attributed to technology are illusory, and (2) those dissatisfactions which are the fault of the individual can be blamed on technology, since the individual's choices are made under some form of hypnosis. It is a can't-lose proposition.

Under these ground rules, how can we argue the question of what constitutes the good life? The anti-technologists have every right to be gloomy, and have a bounden duty to express their doubts about the direction our lives are taking. But their persistent disregard of the average person's sentiments is a crucial weakness in their argument—particularly when they ask us to consider the "real" satisfactions that they claim ordinary people experienced in other cultures of other times.

It is difficult not to be seduced by the anti-technologists' idyllic elegies for past cultures. We all are moved to reverie by talk of an arcadian golden age. But when we awaken from this reverie, we realize that the anti-technologists have diverted us with half-truths and distortions. The harmony which the anti-technologists see in primitive life, anthropologists find in only certain tribes. Others display the very anxiety and hostility that anti-technologists blame on technology— as why should they not, being almost totally vulnerable to every passing hazard of nature, beast, disease, and human enemy? As for the peasant, was he "foot-free," "sustained by physical work," with a capacity for a "nonmaterial existence"? Did he crack jokes with every passerby? Or was he brutal and brutalized, materialistic and suspicious, stoning errant women and hiding gold in his mattress? And the Middle Ages, that dimly remembered time of "moral judgment," "equilibrium," and "common aspirations." Was it not also a time of pestilence, brigandage, and public tortures? "The chroniclers themselves," admits a noted admirer of the period (J. Huizinga), tell us "of covetousness, of cruelty, of cool calculation, of well-understood self-interest. . . ." The callous brutality, the unrelievable pain, the ever-present threat of untimely death for oneself (and worse, for one's children) are the realities with which our ancestors lived and of which the anti-technologists seem totally oblivious.

It is not my intention to assert that, because we live longer and in greater physical comfort than our forebears, life today is better than it ever was. It is this sort of chamber of commerce banality that has driven so many intellectuals into the arms of the anti-technological movement. Nobody is satisfied that we are living in the best of all possible worlds.

Part of the problem is the same as it has always been. Men are imperfect, and nature is often unkind, so that unhappiness, uncertainty, and pain are perpetually present. From the beginning of recorded time we find evidence of despair, melancholy, and ennui. We find also an abundance of greed, treachery, vulgarity, and stupidity. Absorbed as we are in our own problems, we tend to forget how replete history is with wars, feuds, plagues, fires, massacres, tortures, slavery, the wasting of cities, and the destruction of libraries. As for ecology, over huge portions of the earth men have made pastures out of forests, and then deserts out of pastures. In every generation prophets, poets, and politicians have considered their contemporary situation uniquely distressing, and have looked about for something—or someone—to blame. The anti-technologists follow in this tradition, and, in the light of history, their condemnation of technology can be seen to be just about as valid as the Counter Reformation's condemnation of witchcraft.

But it will not do to say *plus ça change plus c'est la même chose,** and let it go at that. We do have some problems that are unique in degree if not in kind, and in our society a vague, generalized discontent appears to be more widespread than it was just a generation ago. *Something* is wrong, but what?

Our contemporary problem is distressingly obvious. We have too many people wanting too many things. This is not caused by technology; it is a consequence of the type of creature that man is. There are a few people holding back, like those who are willing to do without disposable bottles, a few people turning back, like the young men and women moving to the counterculture communes, and many people who have not gotten started because of crushing poverty and ignorance. But the vast majority of people in the world want to move forward, whatever the consequences. Not that they are lemmings. They are wary of revolution and anarchy. They are increasingly disturbed by crowding and pollution. Many of them recognize that "progress" is not necessarily taking them from worse to better. But whatever their caution and misgivings, they are pressing on with a determination that is awesome to behold.

Our blundering, pragmatic democracy may be doomed to fail. The increasing demands of the masses may overwhelm us, despite all our resilience and ingenuity. In such an event we will have no choice but to change. The Chinese have shown us that a different way of life is possible. However, we must not

---

*Editors' note: "The more things change, the more they are the same."

deceive ourselves into thinking that we can undergo such a change, or maintain such a society, without the most bloody upheavals and repressions.

We are all frightened and unsure of ourselves, in need of good counsel. But where we require clear thinking and courage, the anti-technologists offer us fantasies and despair. Where we need an increase in mutual respect, they exhibit hatred for the powerful and contempt for the weak. The times demand more citizen activism, but they tend to recommend an aloof disengagement. We surely could use a sense of humor, but they are in the grip of an unrelenting dolefulness. Nevertheless, the anti-technologists have managed to gain a reputation for kindly wisdom.

This reputation is not entirely undeserved, since they do have many inspiring and interesting things to say. Their sentiments about nature, work, art, spirituality, and many of the good things in life, are generally splendid and difficult to quarrel with. Their ecological concerns are praiseworthy, and their cries of alarm have served some useful purpose. In sum, the anti-technologists are good men, and they mean well.

But, frightened and dismayed by the unfolding of the human drama in our time, yearning for simple solutions where there can be none, and refusing to acknowledge that the true source of our problems is nothing other than the irrepressible human will, they have deluded themselves with the doctrine of anti-technology. It is a hollow doctrine, the increasing popularity of which adds the dangers inherent in self-deception to all of the other dangers we already face.

The Autonomy of Technology

Joseph C. Pitt

It might seem that it is but one step from the view that technology is ideologically neutral to the view that technology is autonomous. If a tool or system can contribute to the decision-making process by forcing changes in values, then surely, it might be suggested, the system itself becomes an independent actor in the process. Maybe so, but probably not. But the view that technology is autonomous is a popular one. Consider what Jacques Ellul has to say on the subject:

- Technique is autonomous with respect to economics and politics

- Technique elicits and conditions social, political and economic change. It is the prime mover of all the rest, in spite of any appearance to the contrary and in spite of human pride, which pretends that man's philosophical theories are still determining influences and man's political regimes are decisive factors in technical evolution.[1]

Ellul may be right about the role philosophical theories and political regimes play in technical evolution, but his claims also sound somewhat exaggerated. More important, the kind of claim he makes for the autonomy of technology makes it sound as if it were unfalsifiable, especially given assertions such as "in spite of any appearance to the contrary."

Unfortunately, claims like Ellul's have become commonplace. They amount to treating technology as a kind of "thing," and in so doing they reify it, attributing causal powers to it and endowing it with a mind and intentions of its own. In addition to the fact that it is empirically false that *Technology* has these characteristics, reifying Technology moves the discussion, and hence any hope of philosophical progress, down blind alleys. The profit in treating Technology in this way, to the extent there is any, is only negative. It lies in removing the responsibility from human shoulders for the way in which we make our way around in the world. Now we can blame all the terrible things that happen to us on Technology! It is only after the first moves have been made toward reifying Technology that we hear about such things as the "threat" of technology taking over our lives. Likewise, reification leads to misleading talk about technology being the handmaiden to science, or some variant on that theme. In other words, reification makes talk about autonomy possible. But, I will argue, it is a major mistake to think there is any *useful* sense in which we could conceive of technology as autonomous.[2]

It is important to stress the "useful" here. It is no doubt possible to contrive outrageous examples to show there is something called autonomous technology. But before we allow misdirected philosophical analysis to take us into the world of science fiction, we can at least take the time to understand what is really going on. Technology, even understood in its more popular-culture sense as new gadgets and electronics, among other things, is such an integral part of our society and culture that unless we ferret out the ways in which these devices are actually embedded in our lives, we may fall victim to a kind of intellectual hysteria that makes successful dealings with the real world impossible. The first step to take if we are to avoid this danger is to clarify the kinds of issues that can reasonably be addressed. To a large degree this means separating the significant from the trivial.

TRIVIAL AUTONOMY

There are at least two cases of talk about the autonomy of technology that are non-starters. That is, if these popular topics of discussion are considered carefully, they easily can be shown to be irrelevant to serious consideration of the issue, since the kind of autonomy they address is trivial.

In the first case, some version of the following account is given of what it means for technology to be autonomous: technology is autonomous when the inventor of a technology, once the technology is made available, loses control over his or her invention. The development of the digital computer can be used as such an example. Once computers entered the public domain, it was impossible for anyone to call them back. The rapid increase in their sophistication and the all-pervasiveness of their employment in society made it impossible to avoid them once they entered the marketplace. Surely, the story goes, this is a case of autonomous technology.

Well, yes and no. Yes, it is autonomous, if by that is meant only that the inventor alone can no longer control the development of the technology. But this is a trivial sense of "autonomy," since it is true of all aspects of our society. Once in the public domain, each item is beyond the control of its inventor in some sense or other. But that does not make the item autonomous. Its further development is a direct function of how people employ it and extend it. To the extent that people are necessarily involved in that process, the invention cannot be autonomous. Rather than being conceived of as an independent agent that acts on its own, the invention is seized opportunistically as a means to an end. It is used, changed, augmented, or discarded, depending on the goals of the agents. That these various uses were not envisioned or intended by its inventor does not make the invention autonomous in any interesting sense.

The second trivial case of autonomous technology concerns the consequences of innovation. Here it might be claimed, for instance, that because the inventor of a device or system failed to see the consequences of employing it in a certain way, the item has a life of its own and is autonomous. Thus it would appear on this scenario that the use of nuclear plants to generate electricity is evidence for the autonomy of nuclear energy, since this use was not foreseen by Einstein in his famous letter to President Roosevelt informing him of the wartime potential for nuclear energy. This, too, is an incorrect conclusion. The fact of the matter is that no one can foresee all the consequences of any act. That fact, however, does not entail that once some action is taken, the consequences of that action are autonomous. That the full consequences of introducing large-scale manufacturing techniques for the production of automobiles were not anticipated by Henry Ford does not mean that those consequences were due to the automobile or to the processes, economic, social, and engineering, that produced it.

The key to understanding this second point lies in realizing that once an invention or innovation leaves the hands of its inventor, it also leaves behind the circumstances in which the actions of only one person can affect its development and employment. Once it enters the public domain, its diffusion generally will be the result of community decisions; and as we noted these are the kinds of decisions that are the results of compromises. That there is no logical order to the patterns these decisions take should come as no surprise. Compromise is a function of a variety of factors, and it is impossible to tell in advance which of them will be persuasive in any given situation. Furthermore, it may be that *it is this lack of absolute predictability with respect to the outcome of community decisions that itself produces the illusion of the autonomy of technology.* But the fact that the role an innovation acquires in a society is a function of complicated community decisions, which decisions are at best compromises (at worst they are the result of collusion and corruption, which themselves involve compromise), does not entail that the innovation is autonomous. *Quite the contrary.* Given the kind of buffeting and manipulation this process involves, it would appear that it would be anything but autonomous!

Thus arguments from the eventual lack of control of the inventor and the failure to foresee all the consequences fail to secure the case for the "autonomy of technology." But there are also other arguments we need to consider.

THE PROCESS OF TECHNOLOGY

Well-intentioned writers and critics have commented on various aspects of technology which they see as raising the possibility of a serious sense of autonomous technology and, along with it, the specter of apocalypse. One of the best examples of the kind of worry expressed by these authors can be found in John McDermott's essay review of Emmanuel Mesthene's *Technological Change*, "Technology: The Opiate of the Intellectuals."[3] In that review McDermott speaks of a kind of momentum certain devices or systems acquire, thereby providing the appearance of autonomy.

Consider the following McDermottian scenario. A growing retail company located in Fairbanks has just hired a fancy up-to-date accountant with an MBA to manage the financial records of the company, which records are currently in a condition closely resembling chaos. Our accountant is a bright young urban professional. Given the size of the company and its projected growth, she argues persuasively that in the long run it will be cheaper and more efficient to buy a couple of computers than to hire additional staff and to continue handling the books in the traditional way, with ledgers entered by hand, etc. She produces a report showing the projected costs of people versus machines, calculating only for the long run the cost of benefits and retirement for the people and maintenance for the machines. She wins her case and the computers are purchased. But once the computers are introduced, air conditioning is not far behind, because the computers need a cool environment to function optimally. But, our fictional tale continues, air conditioning simply can't be added on to the current structure housing the company offices. Either we redesign the old building to handle air flow and pressure, or we look for a new one. Finally, our storyteller says with a knowing look, the president of the company is totally confused and dismayed and yells: "How did we get into this fix? The old building is perfectly good, we really don't need air conditioning in Alaska; since we introduced those machines, things have gotten out of control!"

This is a typical story—one often told and perhaps even representing a situation often experienced. But just because such stories are told, and some people may interpret their experiences in this fashion, it doesn't follow that they have lost control to some autonomous technology that has taken over their company. What the tale allows us to see is that despite the fact that machines play a prominent role in the unfolding sequence of events, the major overlooked fact is that people often tend to forget the reasons for which they introduced a certain kind of tool or procedure. Instead of taking time to assess critically the impact of making further accommodations to the tools, possibly even concluding that it may be time to reexamine the whole situation, people often simply "go with the flow" and take what appears to be the course of least resistance. Still, from the fact that people sometimes tend to react to the circumstances of a situation in certain ways, perhaps accommodating a new procedure at first, rather than either replacing it either with another or eliminating it altogether, it does not follow that the procedure is autonomous.

A basic point we sometimes tend to forget is that *there is no getting rid of tools, written large*. Humanity making its way around in the world is humanity using tools of wide variety and complexity, e.g., hammers, automobiles, governments, electricity. The tools we invent to help us survive and go beyond are essential—perhaps even to the concept of humanity. It isn't as if we can remove tools altogether and continue without them. When we introduce an implement or a complex system, it is to help us achieve a goal. If we find that the device produces results or side effects in conflict with other goals and/or values, we may replace it or modify it. Whichever we choose, devices, tools, and systems remain with us; they are part of how we go about making our way in the world. What McDermott overlooked (when he spoke of how technologies become so ingrained in our procedures that in accommodating the requirements of the technology we lose our independence of action) was that it is the *perception, or lack of it, that people have of the usefulness of a new product that determines the extent to which they are willing to make concessions in its direction*. They may also lose sight of the goal that first guided their actions and, therefore, may react blindly to the circumstance with which they are now faced. But that is not to say that the product has "taken over." For nothing *in principle* rules out later modifications and, if necessary, replacements. What is required is that the individuals involved keep their objectives in mind and be strong enough to act in their own best interests.

COMMON SENSE

Phrased as I have put it, technology conceived of as humanity at work represents the results of the systematic application of common sense; common sense is how people first gain experience and then knowledge by acting on that experience. Nor should this result come as a surprise. Since, if we acknowledge that the concept of a tool lies at the commonsense heart of technology, and if we accept the rather obvious point that not all tools are physical tools, i.e., that there are conceptual tools, social tools, economic tools, etc., then it is not difficult to agree that knowledge is a tool, and if knowledge is constantly being updated, the tool is constantly being honed. In other words, if science produces knowledge, then the knowledge science produces is constantly being upgraded and changed by virtue of the impact of various other tools on the efforts of science to discover more and more about the world. Or to put it differently, quite aside from the resolution of the question of the independence or interdependence of technology and science, if science produces knowledge, and if that knowledge is sometimes used to develop tools that are used in the world, then what those tools produce should generate a form of knowledge that ought to have a bearing in turn on the original knowledge that produced the tools. In addition, it follows that what we do and how we do it is also constantly changing in the face of these developments, and that is as it should be. The bottom line is that, on this account, once a relation between a science and some tool or procedure is established, neither can lay further claims to autonomy—the interdependence is an essential aspect of the process of science itself.[4] But this point of view cannot be established only by *a priori* argument. We need to look at what actually goes on; and I have selected a historical case study to illustrate my points. This is not to say that the analysis of one historical example will settle the issue, but it should help clarify some matters.

Indeed, the case I want to look at, Galileo and the telescope, ought to help exhibit just the issues relevant to sorting out some of the confusions surrounding the interrelations between the development of science and the use of tools and systems of tools. Furthermore, there is a punch line. The general thesis, as already expressed, is that science and technology—where "technology" should now be read as tools, techniques, and systems of tools and techniques—where they interact at all, are mutually nurturing. There is also a caveat, to wit, in point of fact some technologies are science-independent, e.g., the roads of Rome. This is not to say that they are autonomous, since those technologies were responses to needs and goals also; just not the needs and goals of some scientific theory. And some science generates no technology, e.g., Aristotelian biology. The punch line is this: once that is said, something of a paradox emerges. For the history of science is the history of failed theories. But the failure of theory most often does not force a discarding of whatever technology that theory generated or was involved with, nor does the failure of the theory force the abandoning of the technology if a technology was responsible for that theory. To oversimplify: sciences come and go, but their technologies remain. But oversimplification is what got us into trouble at the start, so a more accurate claim would be: scientific theories come and go, but some technologies with which they are in one way or another associated remain. It is also the case that some technologies associated with specific scientific descriptions disappear when they are replaced or superseded by new techniques.

But there is one sense in which the transient character of scientific theories becomes somewhat problematic. That is, if, as I put it earlier, technology is an integral part of science and partially responsible for changing the science, then the failure of the particular theories could be construed as a failure of the technology involved as well. This may in fact be true. But we should also emphasize that technology is seen as a process of policy formation, implement-system implementation, assessment, and updating, which process functions at a variety of levels and with varying degrees of significance for technologies further up and down the line; e.g., the initial failure of the Hubble to produce clear pictures of the heavens did not spell disaster for the entire project. Goal-achieving activities are nested within one another and, as we shall see, as a matter of historical and physical accident, the nesting will have different degrees of importance depending on the case. Thus placing the blame for a failed scientific theory on its associated technology once again oversimplifies the situation.

GALILEO AND THE TELESCOPE

To illustrate some of the notions introduced here, let us turn to an examination of the development of the telescope by Galileo and its effect on some of the theoretical problems he faced in his efforts to show that Copernicus's theory was worthy of serious scientific consideration. As we shall see, the story is not a simple one, and the issue takes on an increasing degree of complexity as the tale proceeds.

To begin with, we need to be perfectly clear that Galileo did not begin his work on the telescope in order to prove anything about Copernicus. The full story of how Galileo came to construct his first telescope is clearly and succinctly put forth by [Stillman] Drake in his *Galileo at Work*. There, quoting from a number of Galileo's letters and published works, Drake makes it clear that Galileo was first drawn to the idea of constructing a telescope out of financial need. To summarize the account: in July 1609 Galileo was in poor health and, as always, if not nearly broke at least bothered by his lack of money. Having heard of the telescope, Galileo claims to have thought out the principles on which it worked by himself, "my basis being the theory of refraction." Drake acknowledges that there was no theory of refraction at the time, but excuses Galileo's claim on the grounds that this was not the first time that Galileo arrived at a correct result by reasoning from false premises. (Historians of the logic of discovery, take note.) Once having reconstructed the telescope, Galileo writes: "Now having known how useful this would be for maritime as well as land affairs, and seeing it desired by the Venetian government, I resolved on the 25th of this month [August] to appear in the College and make a free gift of it to his Lordship."[6] The result of this gift was the offer of a lifetime appointment with a nice salary increase from 520 to 1000 florins per year. What was unclear at the time, and later became the source of major annoyance on Galileo's part, was that along with the stipend came the provision that there was also to be no further increase for life. So he reinitiated his efforts, eventually successful, to return to Florence.

Now there are some problems here that need not delay us, but they ought to be mentioned in passing. How Galileo managed to reconstruct the telescope from just having heard reports of its existence in Holland remains something of a mystery. Galileo provides us with his own account of the reasoning he followed; but, as Drake notes, his description has been ridiculed by historians because, despite the fact that the telescope he constructed worked, he did not quite think it through correctly. Nevertheless, Drake's observation, that "the historical question of discovery (or in this case, rediscovery) relates to results, not to rigorous logic," seems to the point.[7] Despite the fact that a telescope using two convex lenses can be made to exceed the power of one using a convex and a concave lens, the truth of the matter is that Galileo's telescope worked. On the other hand, this point about faulty reasoning leading to good results seems to tie into the paradoxical way in which technologies (thought of as artifacts of varying degrees of complexity and abstractness) emerge and remain with us. But more of this later.

We can now turn to the question of the impact of the telescope on Galileo's work. As he reports it, Galileo first turned his original eight-power telescope toward the moon in the presence of Cosimo, the Grand Duke of Florence. He and Cosimo apparently discussed the mountainous nature of the moon, and shortly after his return to Padua in late 1609, Galileo built a twenty-power telescope, apparently to confirm his original observations of the moon. He did so and then wrote to the Grand Duke's secretary to announce his results. So far then, Galileo has constructed the telescope for profit and is continuing to use it to advance his own position by courting Cosimo.

Galileo, never retiring about his work, continued to use the telescope and to make his new discoveries known through letters to close friends. Consequently, he also began to attract attention. But others such as Clavius now also had access to telescopes. That meant Galileo had to put his results before the public in order to establish his priority of discovery. Therefore, in March 1610 Galileo published *The Starry Messenger*, reporting his lunar observations as well as accounts of the Medicean stars and the hitherto unobserved density of the heavens. At this point controversy enters the picture. These reports of Galileo essentially challenge one of the fundamental assumptions of the Aristotelian theory of the nature of the heavenly sphere: its perfection and immutability. While the rotation of the Medicean stars around Jupiter can be shown to be compatible with both the Copernican and the Tychonian mathematical astronomies, it conflicts with the philosophical and metaphysical view that demands that the planets be carried about a stationary earth embedded in crystalline spheres. And to be clear about the way the battle lines were drawn, remember that Galileo's major opposition came primarily from the philosophers, not from the proto-scientists and other astronomers of his time.

The consequences of Galileo's telescopic observations were more far-reaching than even Copernicus's mathematical model. For the problems Copernicus set were problems in astronomical physics and, as such, had to do with meeting the observational restraints represented by detailed records of celestial activity. Galileo's results, however, and his further arguments concerning the lack of an absolute break between terrestrial and celestial phenomena, maintaining as they did the similarities between the moon and the earth, etc., forced the philosophers to the wall. It was the philosophers' theories that were being challenged when the immutability of the heavens was confronted with the Medicean stars, the phases of Venus, sunspots, and new comets. One might conclude, then, that this represented something akin to a radical Kuhnian paradigm switch.

Much has been written about the extent to which Kuhn's paradigm shifts and their purported likeness to Gestalt switches actually commit someone who experiences one to seeing a new and completely different world. But to see mountains on the moon in a universe in which celestial bodies are supposed to be perfectly smooth comes pretty close to making sense of what this extreme interpretation of Kuhn might mean. Prior to the introduction of the telescope, observations of the heavens, aside from providing inspiration for poets and lovers, were limited to supporting efforts to plot the movements of the planets against the rotation of the heavenly sphere. Furthermore, metaphysical considerations derived from Aristotle interfered with the conceptual possibility of learning much more, given the absence of alternatives. The one universally accepted tool that was employed in astronomical calculation was geometry, and its use was not predicated on any claims of realism for the mathematical models that were developed, another point derived from Aristotelian methodology. The acceptable problem for mathematical astronomy was to plot the relative positions of various celestial phenomena, not to try to explain them. Nor were astronomers expected to astound the world with new revelations about the population of the heavens, since that was assumed to be fixed and perfect. So whatever else astronomers were to do, it was not to discover new facts; there were not supposed to be any.

But the telescope revealed new facts. And for Galileo this meant that some way had to be found to accommodate them. Furthermore, to make the new telescopic findings acceptable, Galileo had to do more than merely let people look and see for themselves. The strategy he adopted was to link the telescopic data to something already secure in the minds of the community: geometry. This, however, was not as simple as it sounds. He had to build a case for extending geometry as a tool for physics, thereby releasing it from the restrictions under which it labored when used only as a modeling device for descriptive astronomy. In other words, Galileo had to advance the case of Archimedean mechanics. To this end he was forced to do two different things: (1) emphasize rigor in proof—extolling the virtues of geometry and decrying the lack of demonstrations by his opposition; and (2) de-emphasize the appeal to causes in providing explanations of physical phenomena (since abandoning the Aristotelian universe entailed abandoning the metaphysics of causes and teleology—without which the physics was empty).

GEOMETRY AS A TECHNOLOGY

This is not the place to detail the actual way in which Galileo employed geometry to radicalize the notions of proof, explanation, and evidence.[8] Suffice it to say that he did and that it met with mixed success. The general maneuver was to begin by considering a problem of terrestrial physics, proceed to "draw a little picture," analyze the picture using the principles of Euclidean geometry, and (1) interpret the geometric proof in terrestrial terms, just as a logical positivist would interpret an axiomatic system *via* a "neutral" observation language, and then (2) extend the terrestrial interpretation to celestial phenomena. This is how he proceeded with his account of mountains on the moon, namely by establishing an analogy with terrestrial mountains. This process took place in stages. He first subjected the terrestrial phenomena to geometric analysis and then he extended that analysis to the features of the moon. Not all of Galileo's efforts at explanation using this method succeeded, e.g., his account of the tides. Nevertheless, the central role of geometry cannot be denied.

While Galileo used geometry for most of his career, it was not until he was forced to support publicly his more novel observations and hypotheses that we find in his writings the beginnings of what was eventually to become a very sophisticated methodological process. This procedure is most clearly evident in his last two works, the *Dialogue on the Two Chief World Systems* and his *Discourses on Two New Sciences*. But in the end the *geometric method* as employed by Galileo, or to put it more specifically, Galilean science, dies with Galileo. No one significant carried on his research program using his methods. Whatever impetus he gives to mathematics in science, his mathematics, geometry, very quickly gives way to Newton's calculus and the mathematics of the modern era.

Galileo's use of geometry was as much the employing of a technology conceived of as a tool/technique as was his use of the telescope. Furthermore, it represents the first major step toward the mathematization of what today we would call science. This much is commonplace. The challenging part comes in two sections. (1) The telescope was a new technology, whose introduction for primarily nonscientific reasons, i.e., money, was in fact science-independent, i.e., its invention by the Dutch was theory-independent. (The inventor, Hans Lipperhey, was a lens grinder; the invention was apparently the result of simply fooling around with a couple of lenses, the basic properties of which were known through Lipperhey's daily experience.) In many ways, the use of this new technology by Galileo can be held responsible for the extension of the *geometric method* as a radical method of supporting knowledge claims. (2) Geometry was also theory-independent. But, unlike the telescope, geometry was a very old technology. It was called upon to rescue, as it were, the new technology. It was a very different kind of technology from the telescope, being a method for providing justifications, i.e., proofs, of abstract conclusions regarding spatial relations, not a physical thing. Furthermore, despite the fact that this old technology was required to establish the viability of the new, the old was soon to become obsolete with respect to the justificatory role it was to play in science. That it was to be replaced also had nothing to do with any significant relation between the telescope and the development of the theory Newton outlined in his *Principia*. In other words, the telescope itself had little direct bearing on the development of the calculus, and yet it was the calculus that superseded geometry (but did not completely eliminate it) as the mathematical basis for scientific proof.

TECHNOLOGY AND THE DYNAMICS OF CHANGE: AUTONOMY SOCIALIZED

If we try to sort it all out, the results are uncomfortable for standard views of technology and the growth of knowledge. The two technologies remain, the two sciences have been replaced. Furthermore, in one of the truly nice bits of irony that history reveals, one of the superceded technologies, geometry, after being replaced by a different kind of mathematical system for justificatory functions, experiences a resurrection in the nineteenth century and ends up playing a crucial role (but not a justificatory role) in the development of yet another physics, having been modified and expanded in the process.

Where is the autonomy here? Both Galileo's physics and the telescope, while capable of being viewed as independent products of one man's creative energy, can also be seen performing an intricate *pas de deux* of motivation and justification when the process of inquiry is examined. It is getting difficult to determine which view ought to take priority. A resolution of the problem might be found if we stop looking at the history and examine the concept of "autonomy" itself.

If we define "autonomous" as "free from influence in both its development and its use," then technology cannot be autonomous since it is inherently something used to accomplish specific goals. But what happens if we try to define "technology" so as to allow technology to have an impact on us as well as on our environment? Are we then committed to the view that, given a technology in use, there emerges from its use a self-propagating process outside the control of humankind? If (1) technology, is a product, and (2) we do not add some additional properties to technology beyond its being a thing we manipulate, then (3) there is no reason why we should even begin to think of technology as not within our control.

In other words, we can talk of Galileo being forced to employ geometry and to develop novel methods of justification in order to defend his telescopic discoveries, for what sense does "forced" carry here? The telescope did not with logical necessity precipitate him headlong into battle. Much of what Galileo did to defend his claims and insure his priority of discovery was the product of his flamboyant personality. This was a man who loved fights and being in the public eye. How these features of Galileo's personality can be factored into the tool so as to make it appear that the tool itself is responsible for the action of the man is beyond serious consideration. Given the tool, we can plot its history. What that history amounts to is how it is used. How it is used is a complicated process, for it can entail more than intentional application of a device. "Use" may also mean "rely on," and it may be the case that what we rely on we take for granted, never giving thought to the cost. But this does not thereby entail that, in the absence of human deliberation, the tool by default acquires intentionality and, along with it, control of human affairs.

An alternative would be to endorse the idea that both the telescope and geometry used Galileo. This suggests a science fiction scenario in which as soon as any technology is used by a person, it "takes over" that individual. In the case of populations adopting constitutions that establish governments, all freedom of human action is lost since the government "takes over." Surely this amounts to a *reductio*. For the tool used to adopt government is reason. Is reason, too, going to be something sufficiently alien that we should fear it? The image really does become Mephistophelian enough that we ought to worry about the extent to which we have lost touch with reality.

Further, the existence of a technology does not entail that it will be used. We all know people who refuse to use computers today, not because they cannot, but simply because they feel more comfortable with the old technology of pen and paper. Surely we do not want to say that these individuals are controlled by pencil and paper. The decision to employ a certain means to an end requires thought, information, a determination of the nature and desirability of the end, assessment of the long- and short-term costs and benefits, as well as constant updating of the database. What if, in his declining years, our pen-and-pencil advocate changes his mind and opts for the computer, having decided that time is running out and he has too many things to finish by hand? Do we really want to say that the machine won out over man? Surely not; the man initiated the process that led to the machine, so why not include him in that process?

We are at the point where, in closing, we might ask: Why are we so quick to point to the machines and wag our finger? Well, the long and the short of it is that those who fear reified technology really fear men. It is not the machine that is frightening, but what some men will do with the machine; or, given the machine, what we fail to do by way of assessment and planning. It may be only a slogan, but there is a ring of truth to: "Guns don't kill, people do." There is no problem about the autonomy of technology. Pogo was right: "We have met the enemy and he is us."[9] The tools by themselves do nothing. That, I propose, is the only significant sense of autonomy you can find for technology.

NOTES

1. Jacques Ellul, *The Technological Society* (New York: Vintage Books, 1964), p. 133

2. For examples of this "style" of philosophizing about technology, see Ellul, *The Technological Society* and Langdon Winner, *Autonomous Technology: Technics-Out-of-Control as a Theme in Political Thought* (Boston: Kluwer Academic Publishers, 1977).

3. John McDermott, "Technology: The Opiate of the Intellectuals," *New York Review of Books*, July 31, 1969, 25–35.

4. See John Dewey, *The Quest for Certainty* (New York: Minton, Balch, 1929) for the development of a similar argument.

5. As quoted in Stillman Drake, *Galileo at Work* (Chicago: University of Chicago Press, 1978), p. 139.

6. As quoted in Drake, *Galileo at Work*, p. 141, Galileo's letter to his brother-in-law Bendetto Landucci.

7. Drake, *Galileo at Work*, p. 140.

8. I have worked on the topic. See my "Galileo: Causation and the Use of Geometry," in *New Perspectives on Galileo*, ed. R.E. Butts and J. C. Pitt (Dordrecht: D. Reidel,

1978); "The Role of Inductive Generalizations in Sellar's Theory of Explanation," *Theory and Decision* 13 (1982): 345–56; "The Character of Galilean Evidence," *PSA* (1986): 125–34; and *Galileo, Human Knowledge, and the Book of Nature: Method Replaces Metaphysics* (Dordrecht: Kluwer, 1991). Also see Ernan McMullin, *Galileo, Man of Science* (New York: Basic Books, 1968); William Shea, *Galileo's Intellectual Revolution* (London: Macmillan, 1972); and William A. Wallace, *Galileo's Logic of Discovery and Proof: The Background Content and Use of His Appropriated Treatises on Aristotle's Posterior Analytics* (Boston: Kluwer Academic Publishers, 1992); among others.

9. Quoted in Walt Kelly, *Outrageously Pogo,* eds. Mrs. Walt Kelly and Bill Crouch Jr. (New York: Simon and Schuster, 1985), p. 114.

PART 5

DEMYSTIFYING AUTONOMOUS TECHNOLOGY THROUGH THE HISTORY OF TECHNOLOGY

INTRODUCTION

Modern technology appears to many to be a powerful, mysterious force radically and continuously transforming our world. The thesis of autonomous technology does not attribute an evil intent to technology, but does claim that it develops in a mechanistic way free from any rational human purpose. The next step in technological development is determined by the development before it, and this process in turn is the cause of much social change. As biological evolution seems random and purposeless, so does the evolution of technology.

The authors in this part, while not denying that the notion of autonomous technology illuminates an important factor in social change, argue that this is not the whole story. Technology is *a* cause, but not *the* cause. Clearly, some technological developments have escaped *humane* control, although this is not an inevitable consequence. In this sense, the defenders of modern technology are right. Still, it is too facile to claim that the value of technology simply depends on how we choose to use it. Technologies are not ethically neutral; they are designed by human beings to serve other human beings' ends. The democratic control of technology demands an understanding of the multiple, interacting, dynamic causes and reasons behind the growth and deployment of technology. An adequate philosophy of technology thus requires the help of social scientists to uncover those points in the causal nexus where human values can, and do, exert themselves.

In the first reading, "Is Technology Autonomous?" Michael Goldhaber summarizes in his criticisms of the autonomous technology thesis the kinds of arguments offered by Jacques Ellul. A close examination of these arguments shows, however, that "technological progress is always guided by values and interests that come from outside technology." While technological inventions cannot be unmade—Jonathan Schell is correct about the irreversibility of technology—this fact need not predetermine the future. The construction of nuclear arsenals was not the inevitable "next step" in technological development but a reflection of the fact that the destructive and deterrent power of nuclear energy has been more highly valued than its constructive, life-sustaining applications.

Technology is not value-free, nor, argues Goldhaber, is science. Even pure science is largely a social construction. While the scientist pursues knowledge, the ends being sought are dependent on his interests, and these in turn are affected by what research can get funded. Scientific theories are never "proved" by experiment, for proof is always subject to interpretation. Referring to philosopher of science Imre Lakatos, Goldhaber argues that rival explanatory systems operate much like the stock market. Choices are made on the basis of which system is likely to grow the fastest. This is not a mere metaphor, for the desire for profits fuels the direction of much scientific research. Science and technology are deeply interrelated in today's society, and to understand the direction both take we must also understand the various ideological commitments that drive them.

Science and technology could be used to serve the interests of humanity as a whole. "Unfortunately," says Goldhaber, "the motives of profit, international competitiveness, national expansion, and perpetuating power of the already powerful tend to prevail in the institutions that currently set the direction of innovation. Deep changes in these institutions are sorely needed."

In the second essay in this part, Robert Heilbroner asks, "Do machines make history?" Was Karl Marx right when he said, "The hand mill gives you society with the feudal lord; the steam mill, society with the industrial capitalist"? The answer, Heilbroner believes, is yes in the sense that there is a necessary sequence in the development of technology, and clearly machines do affect the socioeconomic order in important ways. Although this cannot be proven beyond doubt, Heilbroner offers what he considers some convincing lines of argument.

Technology, however, is not a sufficient explanation of historical development. A variety of social factors are necessary to explain the explosive growth of technology, beginning with the Industrial Revolution. While the steam mill did not by itself give us industrial capitalism, it helped make this possible. However, Heilbroner does believe that technology has a distinct role to play in the composition and organization of the labor force. If Marx had "written that the steam mill gives you society with the industrial *manager*, he would have been closer to the truth."

Fundamentally, it is the laissez-faire ideology of capitalism and socialism which is committed to maximizing production that, according to Heilbroner, give the impression that technology is autonomous. Thus, so-called technological determinism is a problem unique to a certain historical period "in which the forces of technical change have been unleashed, but when the agencies for control or guidance of technology are still rudimentary."* It seems likely, he thinks,

---

*Events occurring since Heilbroner's published warning (1967) have shown him to be prophetic. Only recently have we become aware of the incredibly reckless manner in which the government plants that are producing weapons-grade uranium have stored their radioactive wastes. In some cases they have

that in our own day the pace of technological change will accelerate in an uncontrolled fashion unless agencies of public control are greatly strengthened.

In his essay, "Social Choice in Machine Design," David F. Noble argues that we cannot strengthen the public, democratic agencies to control technological development until we go beyond, that is, see through, the prevailing myth of technological determinism. Combined with the myth of inevitable "progress," this doctrine blinds us to the fact that there are choices to be made in regard to technology. Technology does not "give us" capitalism, nor (against Heilbroner) a hierarchical bureaucratic structure of social relations. Technological development is neither unilinear nor autonomous, but a social process which reflects the dominant ideology of those who design technology.

This abstract notion of technological development as autonomous bespeaks a profound ignorance of the actual process that brings specific technologies into being. To refute the notion of technological determinism, Noble reconstructs the social history of a specific technology—the automation of machine tools, that is, of the machines that are used to make machines. Although the push for automation may be explained by the desire for increased efficiency, the question is, why this specific technology? Why numerically controlled machine tools rather than record-playback? There was a choice to be made here, and, on Noble's account, management opted for the development of NC (numerical control) tools, not in the interest of efficiency, but because such tools promised management increased control over the worker. Such "rationalization" of the modes of production is, of course, irrational from the workers' point of view, leading as it does to a tragic deskilling of the worker and a subsequent reduction in his autonomy. This was not an inevitable next step in some autonomous process, but a deliberate choice on the part of management. The challenge for labor is to reject the claims made in the name of some autonomous, mystical technological "progress" and not merely to react to new developments, but to demand control of the design of new technologies. The larger end, for Noble, is "the eclipse of the capitalist system as a whole."

Thomas P. Hughes also argues, in "Technological Momentum," that the deterministic inevitability of autonomous technology needs to be modified in light of historical facts. Hughes argues for a position that he claims is midway between technological determinism and social constructivism. Technology both controls human development and is controlled by it. He evokes a concept he terms "technological momentum," which suggests that the causative power of technological systems to effect changes in human organizations gains force through time. "Evolving technological systems are time-dependent." Once a technological system begins to be developed—Hughes uses the example of the development of electric light and power systems—momentum to continue the process builds up because of the time and energy invested in the acquisition of new skills and knowl-

---

simply vented their wastes into the atmosphere—this despite the fact that these facilities were "supervised" by the Atomic Energy Commission (AEC) and, since 1947, by the Department of Energy (DOE). Clearly, the health and safety of American citizens was a low priority in developing our nuclear arsenal.

edge, the invention of specialized machines, the building of large physical structures, and the establishment of an organizational bureaucracy. Indeed, industrial and government policymakers use the notion of technological momentum to ensure that their favorite projects are continued. Again, consider the amount of money and time that has been spent on the Star Wars missile defense shield. Even though there are many scientific critics of the proposal, the project continues because of its political and technological momentum. The lesson that Hughes wants to impart is that the time to control new technologies and projects is at the beginning, before their momentum becomes fully developed.

Finally, in "The Ruination of the Tomato," Mark Kramer's piece of investigative journalism, we see a case study of one of the ways Langdon Winner's thesis of reverse-adaptation works. Kramer asks, "Why did modern agriculture have to take the taste (of the tomato) away?" The answer is not to be understood in terms of a conspiracy. No single person, or group, wanted to deprive us of tasty tomatoes. The result was the unintended consequence of a highly integrated technological system. Although individual actors in the business have purposes—efficiency, control, profits, wages, and so on—no one willed a tasteless tomato. There are causes for the tomatoes we find in the supermarket, but, in an important sense, no reasons. Although this may seem a trivial matter, may the pattern not be generalized? Consider the arms race, or the environment, or politics. Why, for instance, do our political candidates sound so vapid and noncommittal, even though the electronic media that present their messages have grown so sophisticated in such a short space of time? Are these examples of an autonomous technological system outside of our control? Or are they examples of some kind of technological momentum that we can alter through our collective political will? In part 6 we will begin to examine the possibility of the political control of technological development.

Is Technology Autonomous?

Michael Goldhaber

Technology is a human activity but it often seems to be a force of nature. There are a number of reasons for this.

1. Technology as a system always seems to improve on existing technology. Thus, if past technology has led to higher productivity, future technology, it is commonly assumed, will increase productivity still further.

2. Technological progress is based on competition among firms and nations. Thus, technology as a whole seems to work very much like the arms race: we fear the Japanese will beat us to the "Fifth Generation Computer," so we devote resources to the same end. Whatever advance we do make, whoever is behind will try to copy and build on to remain competitive. So we will have to forge ahead even faster.

3. Decisions about the future direction of technology seem to be made according to ideas in the air in the technological community, rather than by purely individual choices. Thus, for example, as it became possible to produce integrated circuits on a single silicon chip, seemingly obvious economic considerations led to producing a general purpose circuit, i.e., a digital computer processor, rather than specially designing a separate circuit for each different function. Likewise, the goal of higher speed, higher density circuits was also obvious to many in the field.

4. On the other hand, inventions of great influence come from surprising sources, including individuals working on their own (e.g., Chester Carlson, inventor of what became the Xerox machine). Since these inventions obviously cannot be anticipated, there is apparently no way for the direction of technology to be controlled.

From *Reinventing Technology*, "Is Technology Anonymous?" by Michael Goldhaber (New York and London: The Institute for Policy Studies, Taylor & Francis/Routledge Publishers, 1986), pp. 12–24.

5.   Technology builds on science, and scientists are supposedly led to their discoveries by nothing other than the nature of what they already know and their experimental and observational capabilities.

6.   The results of prior technology are everywhere, often are incomprehensible to the lay person and act in unexpected ways, in this way resembling natural processes or living things. For example, without special instruments, it is as impossible to take apart a digital watch to see how it works as it would be to do the same with a housefly.

7.   Finally, products of technology are now more feared than are natural forces. No one anticipates a natural disaster on the order of the "nuclear winter" that might result from nuclear war. (Indeed, theorized catastrophic collisions of the earth with asteroids, such as the one that is hypothesized to have caused the extinction of the dinosaurs, could now probably be prevented by utilizing existing space technology).

Do all these reasons hold? Is technology like a train leaving the station that we can hop onto or get left behind, but whose destination is beyond our control? If so, then it would be correct to speak of "sunrise" industries; it would be correct for politicians simply to urge improving the conditions for technology. At best, a more nuanced political approach to technology would only involve ameliorating its negative consequences.

The answer is that technology is not autonomous; the apparent relentless, natural forward motion of the field is in reality anything but that. Closer examination of the arguments advanced above (or of similar arguments elsewhere) will reveal a very different conclusion: technological progress is always guided by values and interests that come from outside technology. Let us proceed once more through the list above to see what has been left out of account.

1.   The first five items concern technological innovation as a human process. Each of these points leaves unstated that there is a complex social—and therefore ultimately political—process by which technologists, scientists, industrialists, etc., decide on such questions as what constitutes an "improvement." A technological development is always planned for some social setting in which the improvements involved make sense. But what makes sense is always a matter of social consensus, in which a combination of cultural, economic, and ideological elements are socially evaluated to arrive at a set of priorities. For instance, in automobile design there are many possibilities of what might be considered improvements under different circumstances: cars could be faster, safer, use less gas per mile, be bigger to hold more passengers in greater comfort, be smaller to be more easily parked, require less maintenance, or wear out faster (planned obsolescence).

Simply put, what constitutes an improvement is always a question of values and dominant interests, and it is never a purely technological issue.

Even though technologists often seek numerical measures of performance by which to gauge improvements, what they choose to try to measure depends upon nontechnological factors. If pleasing working conditions were valued more highly than efficiency, for instance, it would be possible to arrive at a variety of numerically measurable quantities, which at least partially or indirectly would correlate with this concept. Technological progress could then be measured according to the new parameters.

If technologists were to operate without some guiding set of values, what would count as improvement would constantly change according to the particular measure that happened to catch their momentary attention. Even if they were to persist with some single measure, it would be more likely to be one of the huge number of socially meaningless possibilities than anything that could be called serious. For example, an auto designer might take as a criterion of perfection how close the weight of the brand new car was to exactly two tons; there would be no limit to improvement, for there would always be another decimal place of accuracy to consider. The effort might make for a real challenge to technical skills, but there would be no noticeable benefit to anyone else involved. Needless to say, no corporation or government agency would be likely to support the pursuit of such a goal—unless, mystified by technology, it thought it was doing something else. Technological frivolity of this kind does take place, but only most rarely and accidentally could it have any important consequences. Technology is significant precisely when the goals it adopts are related to values and interests; then the question becomes what values and whose interests are being served.

The range of values that can motivate technological development is wide: from authoritarianism, racism, and destruction—witness Zyklon B poison gas in the Nazi concentration camps, machine guns, and the South African automated pass system; through corporate power, profitability and hierarchy—consider the assembly line, centralized data banks, large office buildings, and containerized shipping; to human equality (at least roughly)—as in mass transit or improvements in nutrition. Encouraging curiosity, extending life, facilitating playfulness, or enlarging democracy have all been goals as well. The values that matter most are those of the institutions and individuals with the greatest power to determine the direction of technology in our society. Overall, at present, these are the values of corporations and the goals of certain government agencies. But, as is especially evident in the latter case, these values and goals can be changed by different political choices.

2. Competition among nations or even between firms is not implicit in technology, and there is no reason technology has to be shaped accordingly. What is true is that technologists have commonly used the potential for competition, and the political weight commonly accorded it, as a basis for urging

support for technological development. They have been fairly successful with this tactic, and, as a result, they have been expected to deliver in terms of advantages over the competition. But it does not therefore follow that such competition is either wise or unavoidable. In fact there are many instances of technological cooperation—for example, making air travel safe—that demonstrate it is possible for nations and corporations to break free of competitive patterns when it is widely acknowledged that they are harmful. The will to do so on a larger scale is a political question.

3. The foregoing should suggest that ideas "in the air" are there for more than purely technological reasons. A bread with triple the current calorie content may be technologically feasible; indeed, it is quite possible that a development in baking technology would point in that direction. But since our culture as a whole would accord little value to such an innovation, it would never be a serious focus for technologists. Clothing that would dissolve in the rain, a pill that would mimic the effects of hay fever, or assembly-line processes so designed that the sounds emanating from them provide renditions of all nine Beethoven symphonies may all be quite feasible, and might offer interesting challenges to technologists. None of them are "in the air," because the context of power and values in which technology is embedded makes these seem nonsensical at the same time as it finds sense in cruise missiles, space stations, and increasing productivity in an era of high unemployment.

4. Individual inventors, or even groups of inventors, can undoubtedly depart from dominant values. But for their inventions to have any significant impact they must be accepted by large corporations, venture capitalists, or government bureaucracies, then by some institution involved with distribution, and finally by some group of users or consumers. An inventor like Chester Carlson may have had difficulty finding a corporation willing to invest in developing his electrostatic photocopying (Xerox) process, but there was nothing remarkably eccentric in his awareness that others besides himself frequently could make use of copies of business letters and other documents. In the 1950s, few individual inventors would have worried about problems of energy conservation; likewise, an inventor of today would probably have little motivation to work on an idea of specific benefit to welfare mothers.

5. Since technical feasibility does not in itself determine the direction of technology, it is obvious that new scientific results do not either. The next section takes a closer look at science itself; though stemming from more complex motives than technology, science too turns out not to be autonomous.

6. This and the next point concern past technological developments already in place. Admittedly, the inventions cannot be unmade, but there is no reason they need determine the future. To a large extent, for example, the unintelligibility of present technology is a deliberate choice, for reasons ranging from

laziness to the wish to extract high replacement and repair charges ("no user-serviceable parts inside"). It would be possible to put more effort into making each technological project comprehensible to its users; alternatively, new products could be designed with intelligibility as a goal. For—unlike nature—the design of modern products began with conscious understanding, and that can be made accessible.

7. As far as the dangers from nuclear war or other direct technological dangers are concerned, these demonstrate the power of technology but not its inevitability. The reason that weapons are so highly developed is a consequence of the fact that destructive power is both highly valued and easier to enunciate than the more complex set of values associated with sustaining and improving human life. Technologists are complicit: many willingly help perpetuate the arms race because they benefit from it in terms of jobs and status. But it is a national political choice that has made it easier to be assured recognition and employment in weapons development than in projects related to other values.

THE PLACE OF SCIENCE

In policy as well as practice, science and technology are interlinked. A political approach to technology has to deal with science as well, in ways that take into account both the linkages and the important distinction* between the two forms of activity.

Like technology, science is a characteristically modern kind of knowledge. Even more than technology, it is intrinsically open, in that for a scientific result to be considered valid it must be reproducible. Reproducibility implies a set of published procedures that any other skilled scientist or group of scientists can use to duplicate the experiment or observation, regardless of any personal beliefs, virtues, conditions of birth, or particular location. All that is needed is some apparatus describable in numerical terms, often built with readily available parts, and

---

*For readers approaching this subject for the first time, the following summary may be of help.

Very loosely, the distinction between science and technology is that if the immediate aim of technology is to achieve practical ends, then the immediate aim of science is increased understanding, or the accumulation of knowledge, especially the knowledge of nature. The term "research" may—again, loosely—be taken to describe science as an activity: "development" is the same for technology. Science or research are both often further subdivided into "basic" (or "pure") and "applied." This distinction has to do with the closeness with which the knowledge likely to be gained is consciously connected to specific possible technological applications. The discovery of nuclear fission involved basic research, but once the practical possibilities were recognized, subsequent experiments to aid in the construction of the atomic bomb were applied, even though an outside observer would have had difficulty seeing much difference in the laboratory procedures involved. Although for administrative purposes, the definitions may be made sharper, basic and applied efforts in fact shade into one another. Finally, it is, of course, unwise to read any moral connotation into the term "pure."

itself understandable according to scientific explanation. Within a particular explanatory and conceptual framework, two very different-looking experimental setups may be said to lead to the same result.

What counts as a scientific explanation changes as theories change, but it characteristically involves natural objects and forces that lack intention, volition, or symbolic meaning—i.e., there are no gods, spirits, angels, demons, ghosts, portents, etc. Science assumes and reinforces a desacralized world view, one that is amenable to commercial development and bureaucratic management.

Technology is necessary for science—in effect, it is the source of scientific apparatus and at least some scientific procedures; and science is necessary for technology, in that it constantly offers new reproducible situations that can be converted into industrial scale forms, and in that through scientific explanation technologists are able to understand how to approach the goals that interest them. For example, the physics of Isaac Newton provides the basis for calculating the dynamics of the orbits of communications satellites, without which they could not be launched. The more "basic" or "pure" the science, the more its explanations are likely to tie together diverse phenomena, and the wider its potential use by technologists is likely to be.

Just as technologists are paid to innovate, scientists are paid to make new discoveries. (Again, scientists are not the only discoverers; poets, psychoanalysts, and investigative reporters all make discoveries different from science.) As a community, scientists can be said to be always seeking to discover the "natural laws" that lie behind appearances; since appearances include the currently accepted laws, in effect scientists, whether they recognize it or not, are always seeking to undermine the apparent limitations on actions those current laws suggest. In asking the question, "What is there?" scientists are really asking how does such and such work, with the implicit goal of understanding how it can be made to work differently. When Newton was trying to discover the physical laws underlying planetary motions, he was implicitly asking how to go about changing those orbits. Implicit in the study of hormones affecting human sexuality is the possibility of changing the nature of that sexuality (for instance, in sex-change operations). Thus, scientific exploration suggests and promotes new technologies (perhaps only for the far future); it may seem that this happens without any specific values or interests other than curiosity, but, again, closer examination suggests that is not so.

To indicate the deep way that values enter science, I shall adapt and summarize the most thoroughgoing account I know of—that of the late philosopher of science Imre Lakatos—on how scientists decide between rival theories. Scientific theories can never be proved or disproved by experiment, since the connection between theory and experiment is always open to interpretation, and interpretations can always be modified or elaborated to account for any disagreements or to explain away agreements. How then are better theories selected?

Lakatos suggests that the scientific community functions like the stock market, in that scientists choose between rival explanatory systems on the basis of which ones are likely to undergo the fastest growth. That is, each explanatory system suggests some new concepts that relate to new kinds of experiments and observations; these in turn help suggest further elaborations and modifications of the explanatory system and thus lead to still further experiments. The systems in which the payoff of interesting new concepts and interesting new experiments is likely to occur fastest have an obvious attraction for scientists since these will help not only in furthering their careers but in placing them closer to heretofore undisclosed knowledge of nature and to novel technological possibilities.

The catch in this explanation is the word "interesting." What is interesting remains a human question answerable differently by different people and at different times. Since one very strong limitation of experimentation is what experiments can be funded, and what social purposes are likely to command the development of new technologies, what is scientifically interesting, and therefore the character of the explanatory system that is likely to survive, will be influenced by who has political or economic power and to what ends that power is exercised.

The choice of explanatory system in which to pursue knowledge influences what experiments are done, and how they are to be interpreted. Thus, the very nature of scientific knowledge and the set of known facts depends in a very complex way on the power structure and values of the society that is — in the final analysis — doing the asking. Since the choice among explanatory systems appears to be a decision about what is true and what is false, truth turns out to be highly, if indirectly, dependent on the larger society of the day.

Furthermore, it is of course not coincidental that the model of science turns out to resemble the stock market. Ideologies of economic growth and of the growth of scientific knowledge evolved at the same time and continually reinforced each other. Rather than splitting apart, these two systems are actually tending toward each other, as exemplified in the recent introduction of genetic engineering stocks on the market. Profitability and scientific truth easily become intertwined. Just as, at the leading edge of technology, companies compete with one another to be the first to produce a new type of commodity (e.g., the first "256,000 bit random access memory chip"—better known as the "256K RAM"—or the first human insulin genetically engineered into bacteria), so scientists compete with each other to be the first to discover something anticipated by the current explanatory system (e.g., the first hormones isolated in mammalian brains or the "top" quark). The sheer joy of being first is common to both such enterprises; so is the increased likelihood of obtaining not only recognition but funding as a result of demonstrating speed that can be related to fast growth—in the one case of profits, and in the other of the explanatory system itself.

THE OVERALL DIRECTION OF
SCIENCE AND TECHNOLOGY

So far in this chapter, technology and science might each appear to be a single seamless unit. The reality of course is far more complex. Some technological efforts focus on highly specific problems such as how to decontaminate the Three Mile Island reactor. The values and interests involved are usually easy to discern. But these specific projects normally make use of other technologies that are less specific in scope. In the Three Mile Island case these would include technologies ranging from structural engineering to video transmission to chelate chemistry to radiation detection, among many others. Such technologies in turn involve others, such as the metallurgy of steel for the structure of a crane, semiconductor electronics for a video camera, and so on. These more general purpose technologies are in turn closely connected to applied science; at an even further remove, various basic sciences would be involved.

The further we go from specific applications, the more difficult it usually is to ferret out the values and interests that underlie the activity. If we look at some particular laboratory, or at one person, or even one special field, the discernible motivations may be quite idiosyncratic. There may well be an engineer somewhere working on improving the color resolution of television screens because he or she would like to see the face of a certain performer better; there may be chemists who just love the smells associated with working on certain compounds; there are computer scientists convinced that computers will help make all people equal. These individual motivations are not to be dismissed as necessarily irrelevant, but they also do not normally determine the overall direction of any particular field, much less the direction of science and technology as a whole.

Roughly speaking, each different subspeciality in science and technology may be viewed as a service to all the other subspecialities where it is applied. It takes on the sum total of the values and interests they serve, approximately in proportion to the degree of demand each application places on the particular subspeciality in question. At each level in this process we must include the institutional interests of the members of the subspeciality and any related bureaucracy. Especially for the sciences, we must also include what might be termed ideological applications— thus the field of ecology may serve the values of the environmental movement, and the field of sociobiology to some extent serves to support sexist ideas.

As each field then builds on past foundations, it continues in directions suggested by the values and interests it has been serving—that is, it expands more in directions helpful to those values than in other directions. In ways both subtle and direct, these values come to imbue the thinking of members of the subspeciality: thus, the value of increasing efficiency and raising productivity becomes central to the thinking of industrial engineers; computer scientists working with them will have these same values reinforced, and so on. The values and interests will also be embodied in the procedures, processes, and product designs emanating from each subspeciality.

When a technology or a science now is applied in a new way, the values and interests that have shaped it before will influence its suitability for the new application, and may even limit the ways the values and interests directly related to that application are served. For example, for most of their four-decade history, computer systems were developed to serve the needs of large, more or less bureaucratic organizations. Now the programs available for personal computers are being written as if the individual user will operate as a scaled-down version of a bureaucracy (with programs for "database management," "word processing," and balance sheets). This emphasis influences not only who can use such systems but how the users will come to view themselves.

Technology and science together both amplify and help perpetuate dominant values. To the extent that these values accord with the broad interests of humanity as a whole, with the needs of the downtrodden and ill-served, that is all to the good. Unfortunately, the motives of profit, international competitiveness, national expansion, and perpetuating the power of the already powerful tend to prevail in the institutions that currently set the direction of innovation. Deep changes in these institutions are sorely needed.

REFERENCE

Lakatos, Imre. *The Methodology of Scientific Research Programs.* Cambridge: Cambridge University Press, 1978.

1 7

Do Machines Make History?

Robert Heilbroner

The hand mill gives you society with the feudal lord; the steam mill, society with the industrial capitalist.

—Karl Marx, *The Poverty of Philosophy*

That machines make history in some sense—that the level of technology has a direct bearing on the human drama—is of course obvious. That they do not make all of history, however that word is defined, is equally clear. The challenge, then, is to see if one can say something systematic about the matter, to see whether one can order the problem so that it becomes intellectually manageable.

To do so calls at the very beginning for a careful specification of our task. There are a number of important ways in which machines make history that will not concern us here. For example, one can study the impact of technology on the *political* course of history, evidenced most strikingly by the central role played by the technology of war. Or one can study the effect of machines on the *social* attitudes that underlie historical evolution: one thinks of the effect of radio or television on political behavior. Or one can study technology as one of the factors shaping the changeful content of life from one epoch to another: when we speak of "life" in the Middle Ages or today we define an existence much of whose texture and substance is intimately connected with the prevailing technological order.

None of these problems will form the focus of this [chapter]. Instead, I propose to examine the impact of technology on history in another area—an area defined by the famous quotation from Marx that stands beneath our title. The question we are interested in, then, concerns the effect of technology in determining the nature of the *socioeconomic* order. In its simplest terms the question is: Did medieval technology bring about feudalism? Is industrial technology the necessary

From Robert Heilbroner, "Do Machines Make History?" *Technology and Culture* 8 (1967): 335–45. © Society for the History of Technology. Reprinted with permission of the Johns Hopkins University Press.

and sufficient condition for capitalism? Or, by extension, will the technology of the computer and the atom constitute the ineluctable cause of a new social order?

Even in this restricted sense, our inquiry promises to be broad and sprawling. Hence, I shall not try to attack it head-on, but to examine it in two stages:

1. If we make the assumption that the hand mill does "give" us feudalism and the steam mill capitalism, this places technological change in the position of a prime mover of social history. Can we then explain the "laws of motion" of technology itself? Or to put the question less grandly, can we explain why technology evolves in the sequence it does?
2. Again, taking the Marxian paradigm at face value, exactly what do we mean when we assert that the hand mill "gives us" society with the feudal lord? Precisely how does the mode of production affect the superstructure of social relationships?

These questions will enable us to test the empirical content—or at least to see if there *is* an empirical content—in the idea of technological determinism. I do not think it will come as a surprise if I announce now that we will find *some* content, and a great deal of missing evidence, in our investigation. What will remain then will be to see if we can place the salvageable elements of the theory in historical perspective—to see, in a word, if we can explain technological determinism historically as well as explain history by technological determinism.

I

We begin with a very difficult question hardly rendered easier by the fact that there exist, to the best of my knowledge, no empirical studies on which to base our speculations. It is the question of whether there is a fixed sequence to technological development and therefore a necessitous path over which technologically developing societies must travel.

I believe there is such a sequence—that the steam mill follows the hand mill not by chance but because it is the next "stage" in a technical conquest of nature that follows one and only one grand avenue of advance. To put it differently, I believe that it is impossible to proceed to the age of the steam mill until one has passed through the age of the hand mill, and that in turn one cannot move to the age of the hydroelectric plant before one has mastered the steam mill, nor to the nuclear power age until one has lived through that of electricity.

Before I attempt to justify so sweeping an assertion, let me make a few reservations. To begin with, I am fully conscious that not all societies are interested in developing a technology of production or in channeling to it the same quota of social energy. I am very much aware of the different pressures that different societies exert on the direction in which technology unfolds. Lastly, I am not unmindful of the difference between the discovery of a given machine and its application as a technology—for example, the invention of a steam engine (the aeolipile) by Hero of Alexandria long before its incorporation into a steam mill. All these problems, to which we will return in our last section, refer however to the way in which technology makes its peace with the social, political, and economic institutions of the society in which it appears. They do not directly affect the contention that there exists a determinate sequence of productive technology for those societies that are interested in originating and applying such a technology.

What evidence do we have for such a view? I would put forward three suggestive pieces of evidence:

1. The simultaneity of invention

The phenomenon of simultaneous discovery is well known.[1] From our view, it argues that the process of discovery takes place along a well-defined frontier of knowledge, rather than in grab-bag fashion. Admittedly, the concept of "simultaneity" is impressionistic,[2] but the related phenomenon of technological "clustering" again suggests that technical evolution follows a sequential and determinate rather than random course.[3]

2. The absence of technological leaps

All inventions and innovations, by definition, represent an advance of the art beyond existing base lines. Yet, most advances, particularly in retrospect, appear essentially incremental, evolutionary. If nature makes no sudden leaps, neither, it would appear, does technology. To make my point by exaggeration, we do not find experiments in electricity in the year *1500*, or attempts to extract power from the atom in the year *1700*. On the whole, the development of the technology of production presents a fairly smooth and continuous profile rather than one of jagged peaks and discontinuities.

3. The predictability of technology

There is a long history of technological prediction, some of it ludicrous and some not.[4] What is interesting is that the development of technical progress has always seemed *intrinsically* predictable. This does not mean that we can lay down future timetables of technical discovery, nor does it rule out the possibility of surprises. Yet I venture to state that many scientists would be willing to make *general* predictions as to the nature of technological capability twenty-five or even fifty years ahead. This too suggests that technology follows a developmental sequence rather than arriving in a more chancy fashion.

I am aware, needless to say, that these bits of evidence do not constitute anything like a "proof" of my hypothesis. At best they establish the grounds on which a prima facie case of plausibility may be rested. But I should like now to strengthen these grounds by suggesting two deeper-seated reasons why technology *should* display a "structured" history.

The first of these is that a major constraint always operates on the technological capacity of an age, the constraint of its accumulated stock of available knowledge. The application of this knowledge may lag behind its reach; the technology of the hand mill, for example, was by no means at the frontier of medieval technical knowledge, but technical realization can hardly precede what men generally know (although experiment may incrementally advance both technology and knowledge concurrently). Particularly from the mid–nineteenth century to the present do we sense the loosening constraints on technology stemming from succesively yielding barriers of scientific knowledge—loosening constraints that result in the successive arrival of the electrical, chemical, aeronautical, electronic, nuclear, and space stages of technology.[5]

The gradual expansion of knowledge is not, however, the only order-bestowing constraint on the development of technology. A second controlling factor is the material competence of the age, its level of technical expertise. To make a steam engine, for example, requires not only some knowledge of the elastic properties of steam but the ability to cast iron cylinders of considerable dimensions with tolerable accuracy. It is one thing to produce a single steam machine as an expensive toy, such as the machine depicted by Hero, and another to produce a machine that will produce power economically and effectively. The difficulties experienced by Watt and Boulton in achieving a fit of piston to cylinder illustrate the problems of creating a technology, in contrast with a single machine.

Yet until a metal-working technology was established—indeed, until an embryonic machine-tool industry had taken root—an industrial technology was impossible to create. Furthermore, the competence required to create such a technology does not reside alone in the ability or inability to make a particular machine (one thinks of Babbage's ill-fated calculator as an example of a machine born too soon), but in the ability of many industries to change their products or processes to "fit" a change in one key product or process.

The necessary requirement of technological congruence[6] gives us an additional cause of sequencing. For the ability of many industries to cooperate in producing the equipment needed for a "higher" stage of technology depends not alone on knowledge or sheer skill but on the division of labor and the specialization of industry. And this in turn hinges to a considerable degree on the sheer size of the stock of capital itself. Thus the slow and painful accumulation of capital, from which springs the gradual diversification of industrial function, becomes an independent regulator of the reach of technical capability.

In making this general case for a determinate pattern of technological evolution—at least insofar as that technology is concerned with production—I do not want to claim too much. I am well aware that reasoning about technical sequences is easily faulted as *post hoc ergo propter hoc*. Hence, let me leave this phase of my inquiry by suggesting no more than that the idea of a roughly ordered progression of productive technology seems logical enough to warrant further empirical investigation. To put it as concretely as possible, I do not think it is just by happenstance that the steam mill follows, and does not precede, the hand mill, nor is it mere fantasy in our own day when we speak of the coming of the automatic factory. In the future as in the past, the development of the technology of production seems bounded by the constraints of knowledge and capability and thus, in principle at least, open to prediction as a determinable force of the historic process.

II

The second proposition to be investigated is no less difficult than the first. It relates, we will recall, to the explicit statement that a given technology imposes certain social and political characteristics upon the society in which it is found. It is true that, as Marx wrote in *The German Ideology*, "A certain mode of production, or industrial stage, is always combined with a certain mode of cooperation, or social stage,"[7] or as he put it in the sentence immediately preceding our hand mill, steam mill paradigm, "In acquiring new productive forces men change their mode of production, and in changing their mode of production they change their way of living—they change all their social relations."

As before, we must set aside for the moment certain "cultural" aspects of the question. But if we restrict ourselves to the functional relationships directly connected with the process of production itself, I think we can indeed state that the technology of a society imposes a determinate pattern of social relations on that society.

We can, as a matter of fact, distinguish at least two such modes of influence:

1. The composition of the labor force

In order to function, a given technology must be attended by a labor force of a particular kind. Thus, the hand mill (if we take this as referring to late medieval technology in general) required a work force composed of skilled or semiskilled craftsmen, who were free to practice their occupations at home or in a small atelier, at times and seasons that varied considerably. By way of contrast, the steam mill—that is, the technology of the nineteenth century—required a work force composed of semiskilled or unskilled operatives who could work only at the factory site and only at the strict time schedule enforced by turning the machinery on or off. Again, the technology of the electronic age has steadily required a higher proportion of skilled attendants; and the coming technology of automation will still further change the needed mix of skills and the locale of work, and may as well drastically lessen the requirements of labor time itself.

2. The hierarchical organization of work

Different technological apparatuses not only require different labor forces but different orders of supervision and coordination. The internal organization of the eighteenth-century handicraft unit, with its typical man-master relationship, presents a social configuration of a wholly different kind from that of the nineteenth-century factory with its men-manager confrontation, and this in turn differs from the internal social structure of the continuous-flow, semi-automated plant of the present. As the intricacy of the production process increases, a much more complex system of internal controls is required to maintain the system in working order.

Does this add up to the proposition that the steam mill gives us society with the industrial capitalist? Certainly the class characteristics of a particular society are strongly implied in its functional organization. Yet it would seem wise to be very cautious before relating political effects exclusively to functional economic causes. The Soviet Union, for example, proclaims itself to be a socialist society although its technical base resembles that of old-fashioned capitalism. Had Marx written that the steam mill gives you society with the industrial *manager*, he would have been closer to the truth.

What is less easy to decide is the degree to which the technological infrastructure is responsible for some of the sociological features of society. Is anomie, for instance, a disease of capitalism or of all industrial societies? Is the organization man a creature of monopoly capital or of all bureaucratic industry wherever found? The questions tempt us to look into the problem of the impact of technology on the existential quality of life, an area we have ruled out of bounds for this [chapter]. Suffice it to say that the similar technologies of Russia and America are indeed giving rise to similar social phenomena of this sort.

As with the first portion of our inquiry, it seems advisable to end this section on a note of caution. There is a danger, in discussing the structure of the labor force or the nature of intrafirm organization, of assigning the sole causal efficacy to the visible presence of machinery and of overlooking the invisible influence of other factors at work. [S. Colum] Gilfillan, for instance, writes, "engineers have committed such blunders as saying the typewriter brought women to work in offices, and with the typesetting machine made possible the great modern newspapers, forgetting that in Japan there are women office workers and great modern newspapers getting practically no help from typewriters and typesetting machines."[8] In addition, even where technology seems unquestionably to play the critical role, an independent "social" element unavoidably enters the scene in the *design* of technology, which must take into account such facts as the level of education of the work force or its relative price. In this way the machine will reflect, as much as mold, the social relationship of work.

These caveats urge us to practice what William James called a "soft determinism" with regard to the influence of the machine on social relations. Nevertheless, I would say that our cautions qualify rather than invalidate the thesis that the prevailing level of technology imposes itself powerfully on the structural organization of the productive side of society. A foreknowledge of the shape of the technical core of society fifty years hence may not allow us to describe the political attributes of that society, and may perhaps only hint at its sociological character, but assuredly it presents us with a profile of requirements, both in labor skills and in supervisory needs, that differ considerably from those of today. We cannot say whether the society of the computer will give us the latter-day capitalist or the commissar, but it seems beyond question that it will give us the technician and the bureaucrat.

III

Frequently, during our efforts thus far to demonstrate what is valid and useful in the concept of technological determinism, we have been forced to defer certain aspects of the problem until later. It is time now to turn up the rug and to examine what has been swept under it. Let us try to systematize our qualifications and objections to the basic Marxian paradigm:

1. Technological progress is itself a social activity

A theory of technological determinism must contend with the fact that the very activity of invention and innovation is an attribute of some societies and not of others. The Kalahari bushmen or the tribesmen of New Guinea, for instance, have persisted in a neolithic technology to the present day; the Arabs reached a high degree of technical proficiency in the past and have since suffered a decline; the classical Chinese developed technical expertise in some fields while unaccountably neglecting it in the area of production. What factors serve to encourage or discourage this technical thrust is a problem about which we know extremely little at the present moment.[9]

2. The course of technological advance is responsive to social reform

Whether technology advances in the area of war, the arts, agriculture, or industry depends in part on the rewards, inducements, and incentives offered by society. In this way the direction of technological advance is partially the result of social policy. For example, the system of interchangeable parts, first introduced into France and then independently into England, failed to take root in either country for lack of government interest or market stimulus. Its success in America is attributable mainly to government support and to its appeal in a society without guild traditions and with high labor costs.[10] The general *level* of technology may follow an independently determined sequential path, but its areas of application certainly reflect social influences.

3. Technological change must be compatible with existing social conditions

An advance in technology not only must be congruent with the surrounding technology but must also be compatible with the existing economic and other institutions of society. For example, labor-saving machinery will not find ready acceptance in a society where labor is abundant and cheap as a factor of production. Nor would a mass production technique recommend itself to a society that did not have mass market. Indeed, the presence of slave labor seems generally to inhibit the use of machinery and the presence of expensive labor to accelerate it.[11]

These reflections on the social forces bearing on technical progress tempt us to throw aside the whole notion of technological determinism as false or misleading.[12] Yet, to relegate technology from an undeserved position of *primum mobile* in history to that of a mediating factor, both acted upon by and acting on the body of society, is not to write off its influence but only to specify its mode of operation with greater precision. Similarly, to admit we understand very little of the cultural factors that give rise to technology does not depreciate its role but focuses our attention on that period of history when technology is clearly a major historic force, namely Western society since 1700.

IV

What is the mediating role played by technology within modern Western society? When we ask this much more modest question, the interaction of society and technology begins to clarify itself for us:

1. The rise of capitalism provided a major stimulus for the development of a technology of production

Not until the emergence of a market system organized around the principle of private property did there also emerge an institution capable of systematically guiding the inventive and innovative abilities of society to the problem of facilitating production. Hence the environment of the eighteenth and nineteenth centuries provided both a novel and an extremely effective encouragement for the development of an *industrial* technology. In addition, the slowly opening political and social framework of late mercantilist society gave rise to social aspirations for which the new technology offered the best chance of realization. It was not only the steam mill that gave us the industrial capitalist but the rising inventor-manufacturer who gave us the steam mill.

2. *The expansion of technology within the market system took on a new "automatic" aspect*

Under the burgeoning market system not alone the initiation of technical improvement but its subsequent adoption and repercussion through the economy was largely governed by market considerations. As a result, both the rise and the proliferation of technology assumed the attributes of an impersonal diffuse "force" bearing on social and economic life. This was all the more pronounced because the political control needed to buffer its disruptive consequences was seriously inhibited by the prevailing laissez-faire ideology.

3. *The rise of science gave a new impetus to technology*

The period of early capitalism roughly coincided with and provided a congenial setting for the development of an independent source of technological encouragement—the rise of the self-conscious activity of science. The steady expansion of scientific research, dedicated to the exploration of nature's secrets and to their harnessing for social use, provided an increasingly important stimulus for technological advance from the middle of the nineteenth century. Indeed, as the twentieth century has progressed, science has become a major historical force in its own right and is now the indispensable precondition for an effective technology.

It is for these reasons that technology takes on a special significance in the context of capitalism—or, for that matter, of a socialism based on maximizing production or minimizing costs. For in these societies, both the continuous appearance of technical advance and its diffusion throughout the society assume the attributes of autonomous process, "mysteriously" generated by society and thrust upon its members in a manner as indifferent as it is imperious. This is why, I think, the problem of technological determinism—of how machines make history—comes to us with such insistence despite the ease with which we can disprove its more extreme contentions.

Technological determinism is thus peculiarly a problem of a certain historic epoch—specifically that of high capitalism and low socialism—*in which the forces of technical change have been unleashed, but when the agencies for the control or guidance of technology are still rudimentary.*

The point has relevance for the future. The surrender of society to the free play of market forces is now on the wane, but its subservience to the impetus of the scientific ethos is on the rise. The prospect before us is assuredly that of an undiminished and very likely accelerated pace of technical change. From what we can foretell about the direction of this technological advance and the structural alterations it implies, the pressures in the future will be toward a society marked by a much greater degree of organization and deliberate control. What other political, social, and existential changes the age of the computer will also bring we do not know. What seems certain, however, is that the problem of technological determinism—that is, of the impact of machines on history—will remain germane until there is forged a degree of public control over technology far greater than anything that now exists.

NOTES

1. See Robert K. Merton, "Singletons and Multiples in Scientific Discovery: A Chapter in the Sociology of Science," *Proceedings of the American Philosophical Society* 105 (October 1961): 470–86.

2. See John Jewkes, David Sawers, and Richard Stillerman, *The Sources of Invention* (New York: Norton, 1960 [paperback edition]), p. 227, for a skeptical view.

3. "One can count 21 basically different means of flying, at least eight basic methods of geophysical prospecting, four ways to make uranium explosive; . . . 20 or 30 ways to control birth. . . . If each of these separate inventions were autonomous, i.e., without cause, how could one account for their arriving in these functional groups?" S. C. Gilfallan, "Social Implications of Technological Advance," *Current Sociology* 1 (1952): 197. See also Jacob Schmookler, "Economic Sources of Inventive Activity," *Journal of Economic History* (March 1962): 1–20; and Richard Nelson, "The Economics of Invention: A Survey of the Literature," *Journal of Business* 32 (April 1959): 101–19.

4. Jewkes et al., *The Sources of Invention*, present a catalogue of chastening mistakes (p. 230f.). On the other hand, for a sober predictive effort, see Francis Bella, "The 1960s: A Forecast of Technology," *Fortune* 59 (January 1959): 74–78; and Daniel Bell, "The Study of the Future," *Public Interest* 1 (fall 1965): 119–30. Modern attempts at prediction project likely avenues of scientific advance or technological function rather than the feasibility of specific machines.

5. To be sure, the inquiry now regresses one step and forces us to ask whether there are inherent stages for the expansion of knowledge, at least insofar as it applies to nature. This is a very uncertain question. But having already risked so much, I will hazard the suggestion that the roughly parallel sequential development of scientific understanding in those few cultures that have cultivated it (mainly, classical Greece, China, the high Arabian culture, and the West since the Renaissance) makes such a hypothesis impossible, provided that one looks to broad outlines and not to inner detail.

6. The phrase is Richard LaPiere's in *Social Change* (New York: McGraw-Hill, 1965), p. 263f.

7. Karl Marx and Friedrich Engels, *The German Ideology* (London: Lawrence and Wisehart, 1942), p. 18.

8. Gilfillan, "Social Implications of Technological Advance," p. 202.

9. An interesting attempt to find a line of social causation is found in E. Hagen, *The Theory of Social Change* (Homewood, Ill.: Dorsey Press, 1962).

10. See K. R. Gilbert, "Machine Tools," in *A History of Technology*, eds. Charles Singer et al, (New York: Oxford University Press, 1958), IV, chap. 14.

11. See LaPiere, *Social Change*, p. 284; also H. J. Habbakuk, *British and American Technology in the 19th Century* (Cambridge: Cambridge University Press, 1962), passim.

12. As, for example, in A. Hansen, "The Technological Determination of History," *Quarterly Journal of Economics* (1921): 76–83.

Social Choice in Machine Design
The Case of Automatically Controlled Machine Tools, and a Challenge for Labor

David F. Noble

BEYOND TECHNOLOGICAL DETERMINISM

The hard-sounding authority of the phrase *technological development* belies its ambiguity. However popular as an explanatory or justificatory device, the notion is but a convenient catch-all, signifying nothing and everything. With reference to human labor, our focus here, the term is commonly used, in a descriptive mode, to suggest the grand evolution of the material artifacts of "work" that set human beings off from the animal world. Alternatively, in a more modern setting, it is used to refer to an integral aspect of the capitalist process of accumulation, wherein costs are minimized and productivity is enhanced through the profitable substitution of machinery for labor. In a normative and teleological mode, the phrase is used to connote "progress," the steady upswing of human society characterized by freedom from want and toil and by greater human dignity, individual creativity, and autonomy. Or, in a less sanguine view, the same process is seen as the dehumanization of people through the progressive rationalization of labor, either as a routine aspect of capitalism (and thus the managerial quest for control, class hegemony, and intensified exploitation) or as an independent process in itself, called industrialization or modernization. The variations are endless.

As can readily be seen, each use of the term is pregnant with intent as well as content. More important for present purposes, all uses tend to have something in common: they reflect the habit of thought that will be referred to here as "technological determinism." However much technological development is understood to be mediated by (unwitting) human agency or tied, in its use, to an economic or social process like capital accumulation, it itself is nevertheless commonly comprehended as an essentially autonomous, unilinear, and causal process.

David F. Noble, "Social Choice in Machine Design," *Politics & Society* 8, nos. 3–4, pp. 247–312. Copyright © 1978 by Sage Publications. Reprinted by permission of Sage Publications and Corwin Press, Inc.

Technological determinism actually embraces two basic, interrelated notions: first, that technological development itself is a "given" that is self-generating and follows a single course; and second, that this process has effects "outside" it, usually referred to as "social impacts." In the first instance, technological development is seen to be essentially independent of social setting; however much it may be embedded in a particular time and place, its form reflects less these historical particulars than the extrapolation of its own immanent dynamic. In the second instance, technological cause is understood to be irreducible and to have singular effects and not others; moreover, the relation between cause and effect is seen to be automatic and inevitable (necessary)—and is more often assumed than demonstrated.[1] Of course, few would claim to be consistent technological determinists, in the precise philosophical meaning of the term; we are talking rather about a prevalent tendency, a subtle although commonplace habit of thought. Like all other determinisms, this habit denies a realm of human freedom; it views history not as a domain of possibilities but as a sequence of necessities.

Why such a tendency should exist is a subject for historical inquiry in itself. Perhaps foremost among the reasons is the common alienation from, and thus ignorance of, technology, on the part of those people who so often write about its supposed impacts. This is not anyone's fault, of course, but rather the result of the professional monopolization of technical knowledge, the industrial monopolization of professionals, and the general specialization and fragmentation of intellect. Thus, for many observers, technology remains mysterious, alien, and impenetrable, simply a given. And this appearance itself is reinforced by the market context in which technology often develops, where everything appears to happen by itself, automatically. But there are reasons why observers remain content with such a superficial view.

For apologists of the status quo, clearly, such a notion of automaticity absolves everyone of blame and responsibility and ratifies things as they are: they could not be any other way given the ineluctable snatch of technological progress. For lazy revolutionaries, who proclaim liberation through technology, and prophets of doom, who forecast ultimate disaster through the same medium, such notions offer justifications for inactivity. And, for the vast majority of us, numbed into a passive complicity in "progress" by the consumption of goods and by daily chants echoing the slogan of the 1933 World's fair—Science Finds, Industry Applies, Man Conforms—such conceptions provide a convenient, albeit often unhappy, excuse for resignation, for avoiding the always difficult task of critically evaluating our circumstances, of exercising our imagination and freedom. Finally, the habit of exaggerating the causal role of technology in history is readily, if subtly, adopted by historians for the simple reason that it makes their work, the writing of history, easier.

But technological determinism is not simply an intellectual shortcoming; it is politically dangerous. Depicting an inexorable, disembodied, and omnipotent historical force that rules out human action and choice, it fosters passivity, quiescence, and resignation, and their correlates, cynicism and despair. Moreover, as a confusion in thought, it generates artificial contradictions that confound effective action. For example, the labor movement's relation to technological development is an unnecessarily contradictory one. Union officials tend readily to welcome rationalization so long as such technological progress guarantees a bigger slice of the pie without jeopardizing jobs and membership. Workers on the shop floor, however—even those whose jobs have been "red circled" (guaranteed)—quite often see such rationalization less as a lightening of their load or as a bigger paycheck than as a loss of control and a tightening of managerial authority over them. Thus, a conflict arises: in general, union spokesmen tend to be "let it" while people on the shop floor tend to be "against it." The "it," however, is taken as a given, in the habit of technological determinism. Rarely, if ever, is the "it" viewed as a range of possibilities, a domain of human choices—at present the reserve of management—that might potentially provide for *both* increased productivity *and* greater shop floor worker control, depending upon the design. The contradictions arise, then, out of a false notion of a fixed technology, a notion symptomatic of technological determinism. (In this case, the notion is in part a reflection of the virtual exclusion of labor from the realm of technological decision-making, which thus leaves it in no position to recognize the extent to which technological development is the special product of managerial choices made to increase productivity through the enhancement of managerial, not worker, control.)

To the extent that social commentators and scholars simply accept this technology as inevitable and necessary, they ratify and legitimate the managerial choices that inform it. However unwitting they may be, their studies are thus partisan acts. The purpose of this [chapter] is to try to overcome the tendency of technological determinism, and the confusion and impotence it generates, by pointing up the possibilities inherent in technological development. It is a deliberately partisan effort, launched from the other side. Before we can ever hope confidently to strive to intervene in this important process, in order to reshape it according to different choices, we must convince ourselves that such choice is possible. And to do this, we must look at the choices that have already been made. Such an investigation, moreover, should make it quite apparent that, first, technological development itself is neither autonomous nor unilinear, and that, second, technological development as a "cause" is neither irreducible nor automatic.

To begin with, we must recognize that technological development is a social process in itself, that it is not only man-made but made of men (and, increasingly, of women, too). Elsewhere, in trying to understand the relationship between modern technology and capitalism, I have examined that social process in some detail, to identify the people who tend most often to give direction and shape to technological development (engineers), to explain how they obtained this important prerogative (professionalization), to describe the social context in which they learn to use it (technical education), to map out the institutional framework within which they exercise it (corporate research organizations, university laboratories, the patent system, a competitive market environment, etc.), and, finally, to articulate a dominant ideology of technical advance (an engineering ideology of profit-making and management).[2] In that study, it became clear that, as Seymour Melman observed, "if one wants to alter our technologies, the place to look is not to molecular structure but to social structure, not to the chemistry of materials but to the rules of man, especially the economic rules of who decides on technology."[3] Technological development, in short, is not an independent force impinging upon society from the outside, according to its own internal logic; rather, it is a social activity in itself, which cannot but reflect the particulars of its setting: the time, the place, the dream and purposes, the relations between people.[4]

But how does the technology reflect the social context, to what extent, and in what specific ways? Where is the imprint of the social activity that gave issue to it? Marx asserted that "instruments of labor not only supply a standard of the degree of development to which human labor has attained, but they are also indicators of the social conditions under which that labor is carried on."[5] How might we go about reading those indicators, and what do they indicate? One way, beyond merely examining the artifact and guessing at its intended use, is to try to recreate the social history of the technology, to reconstruct the choices that became frozen in the hardware. This would then make it possible to examine the ways in which these choices reflect the ideology and thus the social position of the designers. But to look for choices, of course, requires that the assumption be made that technological development is not unilinear, that it could, potentially, follow any number of chosen courses. And the validity of this assumption rests upon the identification, and thus awareness, of possible alternatives.

If it is assumed that technological development is neither autonomous nor unilinear, what are the implications of these assumptions for the causal view of technological development? We discover now that the "impacts" of technology must follow not simply from the technology but from the design choices it embodies, choices that (fully intended or not) define or at least constrain subsequent choices in deployment. In short, technological development, however much it is a cause of social change, is never an irreducible, first cause. It too has been caused, or, more correctly, chosen, as thus indirectly so have its "impacts."

However, and finally, to say that technologies embody human choices and that these choices reflect the intentions (desired impacts) of the designers—intentions that reflect the ideology (often inarticulate and prereflective) and thus social position of the designers—is not to say that these intentions (desired impacts) are automatically fully realized in the simple construction and use of the technology. Far from it. To end the investigation here, by simply extrapolating the impacts from the choices, would be to beg some of the most important questions and confuse mere intentions for the whole of reality (a common and convenient error). In actuality, the "impacts" are always determined subsequent to the introduction of the technology and in ways not altogether consistent with the intentions of designers. That is, if technology does not automatically "cause" a social effect, neither do the choices (intentions) that lie behind it.

The relation between cause and effect is always mediated by a complex and often conflictive social process, and it is this process, and not the technology or the choices that informed the technology, that determines the final outcome. This social process can be broken down into several aspects. Two of these have received considerable attention from industrial engineers and sociologists: the debugging (and possible minor redesign) of the physical technology following trial use, and the psychological, social, and organizational "adaptation" to the technology of those affected by it. A third aspect, often mistaken for (or dismissed as) "irrational fear of change," is the quite rational struggle on the part of those affected by the technology against the choices embodied in it and the impacts implied by them, a struggle that accurately reflects an objective conflict of interest between, say, machine operators and designers. It is here that we find the only point of entry for labor to register its choices in the process of technological development.

The following brief case history of the design, deployment, and subsequent use of a particular production technology—automatically controlled machine tools—is not meant to be a full account. Rather, it is presented in order to illustrate the points made above and, more importantly, to ground them in the historical and the concrete. Theorizing, however vital to our understanding, is never in itself sufficient to comprehend reality (much less to change it). In this social history, we will endeavor to: place the technology in its social setting, identify the designers, and indicate the loci of decision making wherein its shape was determined; identify the choices made in the *design* of the hardware and software systems, by examining selected and abandoned alternatives; explore how these choices reflected the social relations of production (the "horizontal" relations between large and small firms and between firms and the state, and the "vertical" relations between management and labor within the firm); examine the choices made in *deployment* subsequent to design (since in the case of a relatively portable technology like machine tools, deployment—physical and institutional installations—is separable from design, and choices enter in at both stages of development to successively delimit possibilities and contribute to the final outcome); and examine closely the social process subsequent to installation in order to contrast the realities of that experience with the expectations implicit in the original design and deployment choices. Again, the whole purpose of this exercise is to move us beyond the facile and mystifying habit of technological determinism, and the sense of hopelessness that it mirrors and reinforces, by reawakening us to the possibilities.

THE TECHNOLOGY: AUTOMATICALLY CONTROLLED MACHINE TOOLS

The focus here is a particular production technology, numerically controlled contouring machine tools.[6] It appears that the numerical control (NC) revolution in manufacturing, as it has been called, is leading, on one hand, to increased concentration in the metalworking industry and, on the other, to a reorganization of shopfloor activities within the larger firms in the direction of greater managerial control over production and, thus, over the workforce. These changes are most often attributed to the advent of the new technology and are seen to follow logically, inevitably, as a consequence of it. At a Small Business Administration hearing on the impact of NC on small business in 1971, for example, one expert, the president of Data Systems Corporation, intimated that it was inevitable that "we will see some companies die, but I think we will see other companies grow very rapidly."[7] James Childs, president of the Numerical Control Society, was far less sanguine in his assessment of the situation, pointing out that "the technological gap is definitely growing and unless proper action is taken, will continue to grow until the small shop is extinct. Unless we are willing to allow a relatively few large shops to control the market . . . we must support the small shop."[8] Whatever the orientation, there was consensus among those in attendance that the metalworking industry was undergoing a change as a consequence of the new technology, a change that seemed to spell disaster for many of the smaller firms, which, in 1971, constituted 83 percent of the firms in the industry.[9] The most common reason cited for this trend was the inaccessibility of the new technology to the smaller shops, and this was explained in terms of the expense of NC machines (hardware) and the difficulties associated with programming them (software).

If the new technology was seen as having an important effect on the structure of a vital industry, that is, on the relations of production between firms, it was also seen as the cause of significant changes within shops. Indeed, it has been argued that smaller shops could not take advantage of NC because they were unwilling to undertake the social reorganization efforts that the new technology required.[10] Earl Lundgren, a sociologist who has studied the impact of NC on shop reorganization, insists, for example, that the successful employment of the new technology demands extensive training programs for parts programmers, the shift of shop-floor skills from the machinists to the parts programmers and methods personnel, the intensification of centralized control over all aspects of the production process, and a corollary, the detailed coordination of effort throughout the shop.[11] In the course of a site-visit survey of twenty-four NC-using large- and medium-sized firms in 1978, I found much evidence that such a reorganization effort was underway, that in nearly every case management had attempted to transfer skill from the shop floor to the programming office, to tighten up lines of authority, and to extend control over all aspects of production, that is, to meet the social requirements of NC, requirements that signaled profound changes in the relations of production between management and labor.[12]

For the technological determinist, the story is pretty much told: NC leads to concentration in the metalworking industry and to a tightening of managerial control on the shop floor. All that remains to be done is to sit back and watch the inevitable process unfold. But for the critical historian of technology, the problem has merely been defined. The first question to be asked is, Do these consequences really follow from this technological cause? Is the causal connection real? Do the cost and complexity of the new technology really have anything to do with the diffusion of it in the metalworking industry? Might a better explanation be that smaller firms have never even heard of NC because of their distance from the locus of technological change in the industry?[13] Along the same lines, is it really necessary to divide the programming and machine operating functions within the shop? Could programming, like other tooling, be done closer to the floor or by people on the floor? Does the new technology itself make such reorganization impossible?[14] For purposes of brevity, we will here assume that there *is* a sufficient degree of causal connection between the technology and its supposed effects to warrant further investigation.

This leads us to ask the next question: If *this* technology has these significant consequences, why *this* technology? Here I will limit myself to a single aspect of this problem. All would agree that technology affects the social relations of production; here I am turning that around to ask, how have the social relations of production defined the technology? Could it be the case that the new technology, which took shape within the context of these social relations, not only affects them but also reflects them? If the new technology was developed under the auspices of large firms within the metalworking industry, and under the auspices of management within these firms, is it just a coincidence that the technology serves to extend the market control of large firms within the metalworking industry and enhance managerial authority within the shop? Why did NC take the particular *form* that it did, a form that seems to have rendered it inaccessible to small firms, and *why only NC*? Are there other ways of automating machine tools?

Let us begin with the technology. A machine tool (say, a lathe or milling machine) is a machine used to cut away surplus material from a piece of metal to produce a part with the desired shape, size, and finish. Machine tools are really the guts of machine-based industry because they are the means whereby all machinery, including machine tools themselves, are made. Traditionally, the machine tool is operated by a machinist who transmits his skill and his purpose to the machine by means of cranks, levers, and handles. Feedback is achieved through hands, ears, and eyes. Throughout the nineteenth century, technical advances in machining developed by innovative machinists built some intelligence into the machine tools themselves, making them partially "self-acting": automatic feeds, stops, throw-out dogs, mechanical cams. These mechanical devices relieved the machinists of certain manual tasks; the machinist, however, retained control over the operation of the machine. Together with elaborate tooling—fixtures for holding the workpiece in the proper cutting position and jigs for guiding the path of the cutting tool—these design innovations made it possible for less skilled operators to use the machines to cut parts after they had been properly "set up" by more skilled men.[15] Still, the source of the intelligence was the skilled machinist on the floor. Another step in machine design was the advent of tracer technology, perfected during the 1930s and 1940s, which involved the use of patterns, or templates; these were traced by a hydraulic or electronic sensing device, which in turn conveyed the information to a cutting tool, which reproduced the pattern in the workpiece. The template, then, was, in essence, the cam that contained the intelligence for the machine, intelligence that came from a patternmaker. Tracer technology made possible elaborate contour cutting, but it was still only a partial form of automation. You needed different templates for different surfaces on the same workpiece. During the 1940s, with the war-spurred development of a whole host of new sensing and measuring devices as well as precision servomotors, which made possible the accurate control of mechanical motion, people began to think about the complete automation of contour machining.

Automating a machine tool is different from automating, say, automotive manufacturing equipment, which is single-purpose, fixed automation, and cost effective only given a high volume of product and thus high demand. Machine tools are general purpose, versatile machines, used primarily for small-batch, low-volume production of parts. The challenge of automating machine tools, then, is to retain the versatility of the machine. The answer that evolved was the use of some medium—film, lines on paper, magnetic tape, punched cards of tape—that would store the intelligence and control the machine. To change the part produced, you had only to change the tape. The machine remained the same.

The automating of machine tools, then, involves two separate processes. You need machine controls, a means of transmitting information to the machine to make the tables and cutting tool move as desired. And you need a means of getting the information on the medium, the tape, in the first place. The real challenge was the latter. Machine controls were just another step in a known direction, an extension of gunfire control technology developed during the war. The tape preparation was something new. The first viable solution was record playback, a system developed in 1946–47 by General Electric and Gisholt and a few smaller firms.[16] It involved having a machinist make a part while the motions of the machine under his command were recorded on magnetic tape. After the first piece was made, subsequent identical parts could be made automatically by playing back the tape and reproducing the machine motions. John Diebold, the management consultant, among the first to write about such "flexible automation" in his book *Automation*, heralded record playback as "no small achievement . . . it means that automatic operation of machine tools is possible for the job shop—normally the last place in which anyone would expect even partial automation."[17] But record playback enjoyed only a brief existence, for reasons we shall explore. (It was nevertheless immortalized as the inspiration for Kurt Vonnegut's *Player Piano*.[18] Vonnegut was a publicist at G.E. at the time and saw the record-playback lathe he describes in the novel.)

The other solution to the medium-preparation problem was numerical control (NC). Although some trace its history back to the Jacquard loom of 1804, NC was actually of more recent vintage, the brainchild of a defense subcontractor for Bell Aircraft, John Parsons, and engineers at MIT subcontracted by Parsons.[19] Numerical control, first demonstrated in 1952, represented a dramatic departure in machine tool technology.[20] Record playback was, in reality, a multiplier of skill, a means of obtaining repeatability. The intelligence still came from the machinist who made the tape. Numerical control was based upon an entirely different philosophy of manufacturing. Here the specifications for a part, the information contained in the engineering blueprint, is broken down, first, into a mathematical representation of the part, then into a mathematical description of the desired path of the cutting tool, and ultimately, into hundreds or thousands of discrete instructions, translated for economy into a numerical code, which is read by the machine controls. The NC tape, in short, is a means of formally synthesizing the skill of a machinist, circumventing his role as the source of the intelligence of production (in theory). This new approach to machining was heralded by the National Commission on Technology, Automation and Economic Progress, as "probably the most significant development in manufacturing since the introduction of the moving assembly line."[21]

CHOICE IN DESIGN: HORIZONTAL RELATIONS OF PRODUCTION

Numerical control appeared to potential users in the large aircraft contracting firms to be a valuable means of taking the intelligence of production, and thus control over production, off the shop floor. The new technology thus dovetailed nicely with their larger efforts to computerize company operations.[22] At the same time, in the intensely anti-Communist 1950s, NC looked like a solution to security problems, enabling management to remove the blueprints from the floor so that Communists could not get their hands on them. NC did appear to eliminate the need for costly jigs and fixtures and the templates required by tracer machining, and it did make possible the cutting of complex shapes that defied manual methods. But, additionally, and of equal importance, NC seemed to eliminate once and for all the problems of "pacing," to afford management greater control over production by replacing problematic time-study methods with "tape-time" (using the time it takes to run a part tape as the base for calculating rates) and replacing skilled machinists with more tractable "button-pushers," who would simply load and unload the automatic machinery. The intention is the important thing here. If, with hindsight, NC seems to have required organizational changes in the factory, changes that enhanced managerial control, it is because the technology was chosen, in part for just that purpose. This becomes even more clear when one looks at how the new technology was deployed.

CHOICE IN DEPLOYMENT:
MANAGERIAL INTENTIONS

There is no question but that management saw in NC the potential to enhance their authority over production and seized upon it, despite questionable cost effectiveness.[23] Machine tool builders and control manufacturers, of course, also promoted their wares along these lines; well attuned to the needs of their customers, they promised an end to traditional managerial problems. Thus the president of the Landis Machine Company, in a trade journal article entitled "How Can New Machines Cut Costs?" stressed that "with modern automatic controls, the production pace is set by the machine, not by the operator."[24] The advertising copy of the MOOG Machine Company of Buffalo, New York, similarly described how their new machining center "has allowed management to plan and schedule jobs more effectively," while pointing out, benevolently, that "operators are no longer faced with making critical production decisions."[25]

Machine tool manufacturers peddled their wares and the trade journals echoed their pitch.[26] Initially, potential customers believed the hype; they very much wanted to. Earl Lundgren, a sociologist, conducted surveys of user plants in the 1960s and concluded that "the prime interest in each subject company was the transfer of as much planning and control from the shop to the office as possible; ... there was little doubt in all cases that management fully intended to transfer as much planning from the shop floor to the staff offices [methods engineers and programmers] as possible." Moreover, management believed that "under numerical control the operator is no longer required to take part in planning activities."[27]

In my own survey in 1978 of twenty-five plants in the Midwest and New England—including manufacturers of machine tools, farm implements, heavy construction equipment, jet engines and aircraft parts, and specialized industrial machinery—I observed the same phenomenon. Everywhere, management initially believed in the promises of NC promoters and attempted to remove all decision-making from the floor and assign unskilled people to NC machines, to substitute "tape-time" for problematic time studies to set base rates for piecework and measure output quotas, to tighten up authority by concentrating all mental activity in the office, and otherwise to extend detail control over all aspects of the production process.

This is not to say, however, that I drew the same conclusions that Lundgren did in his earlier survey. Characteristically for an industrial sociologist, he viewed such changes as *requirements* of the new technology whereas, in reality, they reflected simply the *possibilities* of the technology that were "seized upon," to use Harry Braverman's phrase, by management to realize particular objectives, social as well as technical. There is nothing inherent in NC technology, for example, that makes it *necessary* to assign programming and machine tending to different people (that is, to management and workers, respectively): the technology merely makes it *possible*. Management philosophy and motives alone make it necessary that the technology be deployed in this way, and this is another form of necessity entirely, namely, the (to their minds, forced) social choices of the powerful.

One illustration of managerial choice in machine deployment is provided by the experience of a large manufacturing firm near Boston. In 1968, owing to low worker morale, turnover, absenteeism, and the general unreliability of programming and machinery, the company faced what it termed a "bottleneck" in its NC lathe section. Plant managers were frantic to figure out a way to achieve the expected output from this expensive equipment. In that prosperous and reform-minded period they decided upon a job enlargement/enrichment experiment wherein machine operators would be organized into groups and their individual tasks would be extended. Although it was the hope of the company that such a reorganization would boost the morale of the men on the floor and motivate them to "optimize the utilization" of the machinery, the union was at first reluctant to cooperate, fearing a speed-up. The company was thus hard-pressed to secure union support for their program and instituted a bonus for all participants. At one of the earliest meetings between management and union representatives on the new program, the company spokesman for the plan began his discussion of the job-enlargement issue with the question (and thinly veiled threat), "Should we make the hourly people button-pushers or responsible people?" (Given the new technology, management believed it now had the choice.)[28]

A second illustration of the managerial imperative behind technological determinism is provided in an interview I had with shop managers of a plant in Connecticut. Here, as elsewhere, much of the NC programming is relatively simple—the technology is used to machine simple parts formerly produced on conventional machinery (as opposed to complex three- to five-axis contour cutting such as of airfoils, etc.). Observing this, I asked the two men why the operators couldn't do their own programming. At first, they dismissed the suggestion as ridiculous, arguing that the operators would have to know how to set feeds and speeds, that is, be "industrial engineers." I countered this by pointing out that the same people probably set the feeds and speeds on conventional machinery, routinely making adjustments on the process sheet provided by the methods engineers in order to make out. They nodded. Then they said that the operators could not understand the programming language. This time I pointed out that the operators could often be seen reading the mylar tape—twice-removed information describing the machining being done—in order to know what was coming (especially to anticipate programming errors that could mess things up). Again, they nodded. Finally, they looked at each other and smiled, and one of them leaned over and confided, "we don't want them to." Here is the reality behind technological determinism.[29]

REALITY ON THE SHOP FLOOR

In the discussions above on choice in design and deployment, we have endeavored to show why it is a mistake to view technological development as an autonomous or unilinear process. The evolution of any technology reflects the social relations that gave birth to it and follows from the human choices that inform it. But, as we argued at the outset, it would be an error to assume that we can now simply substitute one form of determinism for another, to suppose that, in having exposed the choices, we can now simply deduce the rest of reality (rather than just describe a part of it). For it would be as much a serious flaw in reasoning to extrapolate reality from the intentions that underlie the technology[30] as from the technology itself. Intentions are never so automatically realized, no more than desire is identical to satisfaction.

"In the conflict between employer and employed," John G. Brooks observed in 1903, "the 'storm center' is largely at this point where science and invention are applied to industry."[31] It is here, in the storm center, that the reality of NC is hammered out; it is here, in the storm center, where those who choose the technology finally come face to face with those who do not. It is their first confrontation with what philosophers call "other minds."

The introduction of NC control machinery has not been altogether uneventful, especially in plants where the machinists' union (such as the International Association of Machinists, the United Auto Workers, the United Electrical Workers, or the International Union of Electrical Workers) have had a long history. Work stoppages and strikes over rates for the new machinery were common in the 1960s. At General Electric, for example, there were strikes at several plants and one during the winter of 1965 shut down the entire Lynn, Massachusetts, plant for a month. Here we will focus less upon these overt expressions of conflict than upon more subtle indications that management dreams have tended to remain dreams: the use of "tape time" to replace time-studies; the deskilling of machine operators; and the question of control over production.

Early dreams of the use of tape time to set base rates and determine output proved fanciful. The machines could not produce parts to specifications without the repeated manual intervention of the operator, in the early machines to make tool offset adjustments and to correct for program errors and even now to insure a good finish to tolerance (especially with rough castings). At one large factory in New England, for example, where NC has been in use for nearly a quarter of a century, the machines are often used almost like manual machines. There is, in reality, a spectrum of manual intervention requirements depending upon the machine, the product, and the like, but the mythology that the tape ran the machine without the need for people was dispelled pretty early on—and along with it went the hope of using the tape to measure performance.

The deskilling of machine operators has, on the whole, not even taken place, for two reasons. First, as was mentioned, the assigning of labor grades and thus rates for the new machinery was, and is, a hotly contested issue, yet to be finally resolved one way or the other. Second (and the reason why skill levels on numerical control equipment are as high in nonunion shops as in union shops), any determination of skill requirements must take into account the actual degree of automaticity of the machinery and, as we have seen, this was not up to expectations. That is, management has had to have people on their machines who know what they are doing simply because the machines do *not* run by themselves, and they are very expensive. While it is true that many manufacturers initially tried to use unskilled people on the new equipment, they rather quickly saw their error and upgraded the classification (although in some places they merely gave the skilled operators a premium while retaining the lower formal classification, presumably in the hope that some day the skill requirements would actually drop to match the classification). It has been argued, moreover, that, in some ways, NC requires greater skill—as was suggested, by the United Electrical Workers in their UE *Guide to Automation* of 1960—because of the need to oversee the complex and rapid machining of parts, to constantly watch for "nonevents." Whether this is true or not is certainly debatable. One thing is not, though. The intelligence of production in machine shops has not yet been built into the machine or taken off the floor.

And this brings us, once again, to the question of control. In theory, the programmer prepares the tape (and thus sets feeds and speeds, thereby determining the rate of production), proofs it out on the machine, and then turns the show over to the operator who from then on simply presses start and stop buttons and loads and unloads the machine (using standard fixtures). This rarely happens in reality as was pointed out above. Machining to tolerances more often than not requires close attention to the details of the operation and frequent manual interventions through manual feed and speed overrides. This aspect of the technology, of course, reintroduces the control problem for management. Just as in the conventional shop, where operators are able to modify the settings specified on the work sheet (prepared by the methods engineer) in order to restrict output or otherwise "make out" (by running the machine harder), so in the NC shop, the operators are able to adjust feeds and speeds, for similar purposes.

Thus, if you walk into a shop you will often find feedrate override dials set uniformly at, say, 70 percent or 80 percent of tape-determined feedrate. In some places, this is called the "70 percent syndrome"; everywhere it is known as pacing, the collective restriction of output on the floor. To combat it, management sometimes programs the machines at 130 percent; sometimes they actually lock the overrides altogether to keep the operators out of the "planning process." But this, of course, gets management into serious trouble since interventions are required in many cases to get parts out the door.

To what extent the considerable amount of human intervention required for machining (and repairs) is attributable to the inherent unreliability of the technology itself is hard to assess. But it is certainly true that the technology develops shortcomings once it is placed on the shop floor, whether they were there in the original designs or not. Machines often do not do the job they are supposed to, and downtime is still excessive. Whether this is due to operator "adjustments" (or failures to adjust) or technical defects is hard to determine, but sabotage is an acknowledged problem. "I don't care how many computers you have they'll still have a thousand ways to beat you," lamented one manager in a Connecticut plant. "When you put a guy on an NC machine, he gets tempermental," another manager in Rhode Island complained. "And then, through a process of osmosis, the machine gets temperamental."

On the shop floor, it is not only the choices of management that have effects. The same antagonistic social relations that, in their reflection in the minds of designers, gave issue to the new technology, now subvert it in reality. This contradiction of capitalist production, of a Divided Humanity, presents itself most clearly to management as a problem of "worker motivation," and management's acceptance of this challenge is its own tacit acknowledgment that it does not have control over production, that it is still dependent upon the workforce to turn a profit.

Thus, in evaluating the work of those whose intentions to wrest control over production from the workforce informed the design and deployment of NC, we must take into account an article written by two industrial engineers in 1971. Appearing in *The Manufacturing, Engineering, and Management Journal*, it was entitled "A Case for Wage Incentives in the NC Age." It makes it quite clear that the contradiction of capitalist production has not been eclipsed—computers or no computers.

> Under automation, it is argued, the machine basically controls the manufacturing cycle, and therefore, the worker's role diminishes in importance. The fallacy in this reasoning is that if the operator malingers or fails to service the machine for a variety of reasons, both utilization and subsequent return on investment suffer drastically.
>
> Basic premises underlying the design and development of NC machines aim at providing the capability of machining configurations beyond the scope of conventional machines. Additionally, they "deskill" the operator. Surprisingly, however, the human element continues to be a major factor in the realization of optimum utilization or yield of these machines. This poses a continuing problem for management, because a maximum level of utilization is necessary to assure a satisfactory return on investment.[32]

This same contradiction, faced by every capitalist since the emergence of the capitalist mode of production, was more succinctly, if inadvertently, expressed by another plant manager in Connecticut. With a colleague chiming in, he proudly described the elaborate procedure they had developed whereby every production change, even the most minor, had to be okayed by an industrial engineer. "We want absolutely no decisions made on the floor," he insisted; no operator was to make any change from the process sheets without the written authorization of a supervisor. A moment later, however, looking out onto the floor from his glass-enclosed office, he reflected upon the reliability of the machinery and the expense of parts and equipment and emphasized, with equal conviction, that "we need guys out there who can think."

CONCLUSION: LABOR'S CHALLENGE REVISITED

We have here tried to get beyond the ideology of technological determinism by demystifying the development of a particular technology. The purpose of this exercise, as was indicated at the outset, was to regain confidence in the fact that technology consists of a range of possibilities; if certain human choices guided the course of its development in the past, other choices could steer it in the future. This done, we must turn again to the remaining two tasks: to study the possibilities and clarify new choices and to struggle to get in the position to make them.

It is time, it is imperative, that the labor movement as a whole begin to assume the responsibility for the design and deployment of technology—the organization of our shops, the structure of our jobs, the shape of our lives. For only by becoming self-reliant and assuming this responsibility can we regain the confidence to secure the power to exercise it. This challenge, now faced by all unions throughout industry (due, in part, to the homogenizing effect of management-designed automation, which diminishes the differences between industries and between shops and offices) was clearly, if rather modestly, summarized by an NC machine tool operator in a large shop in Massachusetts:

> The introduction of automation means that our skills are being downgraded and instead of having the prospect of moving up to a more interesting job we now have the prospect of either unemployment or a dead-end job. [But] there are alternatives that unions can explore. We have to establish the position that the fruits of technological change can be divided up—some to the workers, not all to the management as is the case today. We must demand that the machinist rise with the complexity of the machine. Thus, rather than dividing his job up, the machinist should be trained to program and repair his new equipment—a task well within the grasp of most people in the industry.
> Demands such as this strike at the heart of most management prerogative clauses which are in many collective bargaining contracts. Thus, to deal with automation effectively, one has to strike at another prime ingredient of business unionism: the idea of "let the management run the business." The introduction of NC equipment makes it imperative that we fight such ideas.[33]

This, in essence, is the challenge. Overt political and covert shop-floor resistance to the choices of management, which exploit the contradictions inherent in the capitalist mode of production, must be encouraged and enlarged; but these are not in themselves sufficient actions, given the current desperate corporate drive to rationalize industry and fight organized labor. It is no longer enough merely to respond to new methods nor even to anticipate the changes that are "coming down." The goal must be to gain control of, and thereafter redirect, the social process of technological development itself. And this end, of course, requires—as it contributes to—a larger transformation of the social relations of production and, thus, the eclipse of the capitalist system as a whole.

NOTES

1. This assumption is not always correct. See, for example, Stephen Marglin, "What Do Bosses Do? The Origins and Functions of Hierarchy in Capitalist Production," and Katherine Stone, "The Origins of Job Structures in the Steel Industry," *Review of Radical Political Economy* 6 (summer 1974).

2. David F. Noble, *America by Design: Science, Technology, and the Rise of Corporate Capitalism* (New York: Alfred A. Knopf, 1977).

3. Seymour Melman, "The Impact of Economics on Technology," *Journal of Economic Issues* 9, no. 1 (March 1975): 71.

4. It perhaps ought to be pointed out that it is a mistake to try to eclipse technological determinism by simply substituting for it another mechanical, say economic, determinism. Market factors are critical ingredients in the process of technological development, but they are not the only ones, nor, at times, even the most important ones. Sweeping economic explanations for technical advance, however logically consistent, provide only a framework for historical analysis, not a substitute for it. This history, like any other, is a complex of scientific, technical, sociological, economic, political, ideological, and cultural threads, all of which give direction to it. Thus we see the role of the government often rendering market considerations beside the point, fascination with remote control on the part of designers and their sponsors outweighing simple cost considerations, and a fetish for order, predictability, continuous-flow, power, or mathematical elegance having more to do with the outcome than other, more practical economic concerns. In short, here the principle of parsimony can be too avidly applied.

5. Karl Marx, *Capital* (Moscow: Foreign Languages Press, 1959), 1:200.

6. The discussion here is restricted to the first few generations of NC (tape control); it does not include later developments such as Computer Numerical Control (CNC), Direct Numerical Control (DNC), or Flexible Manufacturing Systems (FMS), or any other aspects of Computer Aided Design/Computer Aided Manufacture (CAD/CAM). These are subjects for another article.

7. U.S., Congress, Senate, Hearing before the Subcommittee on Science and Technology of the Select Committee on Small Business, *Introduction to Numerical Control and Its Impact on Small Business*. "Statement of Kenneth Stephanz." 92nd Cong., 1st sess., June 24, 1971, p. 76.

8. Ibid., letter from James Childs, p. 87.

9. Ibid. See statements of John C. Williams, Edward Miller, and Senator David Gambrell. For more information about the diffusion of NC tools, see Clifford Fawcett, "Factors and Issues in the Survival and Growth of the U.S. Machine Tool Industry" (Ph.D. diss., George Washington University, 1976); Jacob Sonny, "Technological Change in the U.S. Machine Tool Industry, 1947–1966" (Ph.D. diss., New School for Social Research,

1971); A. Romeo, "Interindustry Differences in the Diffusion of Innovations" (Ph.D. diss., University of Pennsylvania, 1973); "1976 American Machinist Inventory of Metalworking Equipment," *American Machinist*, December 1976: S. Kurlat, "The Diffusion of NC Machine Tools," Technical Memorandum, EIKONIX Corporation, April 1977; and Jack Rosenberg, "A History of Numerical Control, 1949–1973: The Technical Development, Transfer to Industry, and Assimilation," unpublished manuscript, October 1973.

10. Small Business Administration, *The Impact of NC on Small Business* (Washington, D.C.: Government Printing Office, 1971), p. 129. On organizational impacts of NC, see: Ervin M. Birt, "Organizational and Behavioral Aspects of NC Machine Tool Innovation" (M.S. thesis, Course XV, MIT, 1959); James J. Childs, "Organization of a Numerical Control Operation," *Mechanical Engineering* (May 1959): 61; and Earl F. Lundgren, "Effects of NC on Organizational Structure." *Automation* 76 (January 1969): 44–47.

11. Lundgren, "Effects of NC on Organizational Structure." In his study, Lundgren observed that, among other things, "there was little doubt in all cases that management fully intended to transfer as much planning from the shop floor to the staff offices as possible" and that "under NC the operator is no longer required to take part in planning activities." He concluded that therefore "a change in organization structure to define and formalize the relationships required by the NC technology is strongly indicated." See also L. K. Williams and C. R. Williams, "Impact of NC Equipment on Factory Organization," *California Management Review* 7 (winter 1964): 25–34; R. C. Brewer, "Organizational and Economic Aspects of NC," *Engineer's Digest* 146 (August 1965): 103; William H. Bentley, "Management Aspects of NC," *Automation* 7 (October 1960): 64–70; Earl Lundgren and M. H. Sageger, "Impact of NC on First Line Supervision," *Personnel Journal* 46 (December 1967): 715.

12. This survey included primarily large- and medium-sized plants in the Midwest and New England of manufacturers of machine tools, farm implements, heavy construction equipment, jet engines, aircraft parts, and specialized industrial machinery.

13. On this point, see the letter from James Childs in *Introduction to Numerical Control and Its Impact On Small Business*, p. 87.

14. See, for example, Harry Braverman's discussion of the social possibilities of NC in his *Labor and Monopoly Capital* (New York: Monthly Review Press, 1974), p. 199.

15. The use of jigs and fixtures in metalworking dates back to the early nineteenth century and were the heart of interchangeable parts manufacture, as Merritt Roe Smith has shown in *Harpers Ferry Armory and the New Technology* (Ithaca, N.Y.: Cornell University Press, 1976). But it was not until much later, around the turn of the century, that the "toolmaker" as such became a specialist, distinguished from machinists. The new function was a product of scientific management, which aimed to shift the locus of skill, and control, from the production floor, and the operators, to the toolroom, as a matter of basic shop reorganization. But however much the new tools allowed management to employ less skilled, and thus cheaper, machine operators, they were nevertheless very expensive to manufacture and store, and they lent to manufacture a heavy burden of inflexibility—shortcomings that one Taylorite, Sterling Bunnell, warned about as early as 1914 (cited in David Montgomery, *Fall of the House of Labor: The Workplace, the State, and American Labor Activism* [New York: Cambridge University Press, 1989]). The cost savings that resulted from the use of cheaper labor were thus offset by the expense of tooling. Numerical control, as we will see, was developed, in part, to eliminate the costs and inflexibility of jigs and fixtures and, equally important, to take skill, and the control it implied, off the

floor altogether. Here again, however, the expense of the solution was equal to or greater than the problem. It is interesting to note that in both cases, jigs and fixtures and numerical control, where very expensive new technologies were introduced to make it possible to hire cheaper labor, the tab for the conversion was picked up by the state, from the Ordnance Department in the early nineteenth century, to the Departments of the Army and Navy in World War I, to the Air Force in the second half of the twentieth century.

16. The discussion of the record-playback technology is based upon extensive interviews and correspondence with the engineers who participated in the projects at General Electric (Schenectady) and Gisholt (Madison, Wisconsin). For a brief description of the G.E. technology, see Lawrence R. Peaslee, "Tape-Controlled Machines," *Electrical Manufacturing Magazine*, November 1953, pp. 102–108. Key patents for the G.E. system include Orna W. Livingston, "Record-Reproduce Programming Control System for Electric Motors," patent no. 2,755,422 (July 17, 1956); Lawrence R. Peaslee, "Programming Control System," patent no. 2,937,365 (May 17, 1960); and Lowell L. Holmes, "Magnetic Tape Recording Device," patent no. 2,755,160 (July 17, 1956). Key patents for the Gisholt system: Leif Eric de Neergaard, Frederic W. Olmstead, and Hans Trechsel, "Control System for Machine Tools Utilizing Magnetic Recording," patent no. 3,296,606 (January 3, 1967), and Leif Eric de Neegard, "Method and Means for Recording and Reproducing Displacements," patent no. 2,628,539 (February 17, 1953). For a post-NC view of the advantage of record-playback control, see Veljko Milenkovic, "Single-Channel Programmed Tape Motor Control for Machine Tools," patent no. 3,241,020 (March 15, 1966). Milenkovic's patent was used by the American Machine and Foundry Company (AMF) in the development of their "Versatran" robot. Industrial robots, like "unimates," are record-playback controlled.

17. John Diebold, *Automation* (New York: Van Nostrand, 1952), p. 88.

18. Kurt Vonnegut, *Player Piano* (New York: Delacorte Press), 1952; Kurt Vonnegut, letter to author, February 19, 1977.

19. This is why the Numerical Control Society's award given for significant advancements in the technology is called the Jacquard Award. In 1970, this award went to the developers of the first NC machine at MIT. See "How It All Began," *Metalworking Economics*, June 1970, p. 43. For general historical discussion of such automation, see Diebold, *Automation*; James R. Bright, "The Development of Automation," in *Technology in Western Civilization*, eds. Kranzberg and Pursell, vol. 2, chap. 41; Braverman's chapter on "Machinery," in *Labor and Monopoly Capital*; and Ben Seligman, *Most Notorious Victory: Man in an Age of Automation* (Glencoe, Ill.: The Free Press, 1966), esp. pp. 122–27. For a more detailed history of NC, see Donald P. Hunt, "The Evolution of a Machine Tool Control System and Succeeding Developments" (M.S. thesis, course XV, MIT, 1959) and "A Numerically Controlled Milling Machine: Final Report to the U.S. Air Force in Construction and Initial Operation" (Servomechanisms Laboratory, MIT, July 30, 1952), pt. 1.

20. No effort will be made here to trace the early history of NC, which includes the work of Emanuel Scheyer, Max Schenker, Eric de Neegaard, and Fred Cunningham, among many others. Parsons is generally credited with having initiated the first successful hardware project, the one that, with the help of the Air Force, gave rise to the commercial development of NC.

21. Frank Lynn, Thomas Roseberry, and Victor Babich, "A History of Recent Technological Innovations," in *Technology and the American Economy*, ed. National Commission on Technology, Automation, and Economic Progress (Washington, D.C.: Government Printing Office, 1966), vol. 2, *The Employment Impact of Technological Change*, app., p. 89.

22. Like most large companies that entered the computer age early, the aircraft firms were faced with a number of problems: first, how fully to utilize the computer (after payroll, then what?), and second, how to extract the information for data banks from the personnel who possessed it (whose jobs depended upon their access to or control over information). NC contributed to the solution of both problems: the use of APT absorbed about 50 percent of their computer capacity throughout the 1960s, and NC itself appeared to allow managers to circumvent shop floor personnel considerably.

23. The cost effectiveness of NC depends upon many factors, including training costs, programming costs, and computer costs, beyond mere time saved in actual chip-cutting or reduction in direct labor costs. The MIT staff who conducted the early studies on the economics of NC focused on the savings in cutting time and waxed eloquent about the new revolution. At the same time, however, they warned that the key to the economic viability of NC was a reduction in programming costs. Machine tool company salesmen were not disposed to emphasize these potential drawbacks, though, and numerous users went bankrupt because they believed what they were told. In the early days, however, most users were buffered against such tragedy by state subsidy. Today potential users are somewhat more cautious, and machine tool builders are more restrained in their advertising, tempering their promise of economic success with qualifiers about proper use, the right lot and batch size, sufficient training, and so on.

For the independent investigator, it is extremely difficult to assess the economic viability of such a technology. There are many reasons for this. First, the data are rarely available or accessible. Whatever the motivation—technical fascination, keeping up with competitors, and so on—the purchase of new capital equipment must be justified in economic terms. But justifications are not too difficult to come by if the item is desired enough by the right people. They are self-interested anticipations and thus usually optimistic ones. More importantly, firms rarely conduct postaudits on their purchases, to see if their justifications were warranted. Nobody wants to document his errors, and if the machinery is fixed in its foundation, that is where it will stay, whatever a postaudit reveals; you learn to live with it. The point here is that the economics of capital equipment is not nearly so tidy as economists would sometimes have us believe. The invisible hand has to do quite a bit of sweeping up after the fact.

If the data does exist, it is very difficult to get a hold of. Companies have a proprietary interest in the information and are wary about disclosing it for fear of revealing (and thus jeopardizing) their position vis-à-vis labor unions (wages), competitors (prices), and government (regulations and taxes). Moreover, the data, if it were accessible, is not all tabulated and in a drawer somewhere. It is distributed among departments, with separate budgets; and the costs of one are the hidden costs to the others. Also there is every reason to believe that the data that does exist is self-serving information provided by each operating unit to enhance its position in the firm. And, finally, there is the tricky question of how "viability" is defined in the first place. Sometimes, machines make money for a company whether they are used productively or not.

The purpose of this aside is to emphasize that "bottom line" explanations for complex historical developments, like the introduction of new capital equipment, are never in themselves sufficient, nor necessarily to be trusted. If a company wants to introduce something new, it must justify it in terms of making a profit. This is not to say, however, that profit making was its real motive or that a profit was ever made. In the case of automation, steps are taken less out of careful calculation than on the faith that it is always good to replace labor with capital, a faith kindled deep in the soul of manufacturing engineers and managers—as economist Michael Piore, for example, has shown. See his "The Impact of the Labor Market upon the Design and Selection of Productive Techniques within the Manufacturing Plant," *Quarterly Journal of Economics* 32 (1968). Thus, automation is driven forward, not simply by the profit motive, but by the ideology of automation itself, which reflects the social relations of production.

24. Grayson M. Stickell, "How Can New Machines Cut Costs?" *Tooling and Production*, August 1960, p. 61.

25. *MOOG Hydra-Point News*, Buffalo, New York, 1975.

26. Charles Weiner, former associate editor of *Tooling and Production Magazine* (one of the leading trade journals for the metalworking industry), authored several articles on numerical control in the early 1960s, one of which was based upon a "comprehensive survey of machine tool builders and control manufacturers." According to Weiner, all of his information for these articles came from the manufacturers themselves, and no effort was ever made to make an independent evaluation of NC for the readers of the magazine. Although the editors strove for credibility with production men in the metalworking industry, they were guided primarily, as are the editors of all controlled circulation magazines in the trade press, by the need to secure advertisements. Thus they tended to exaggerate the promise of a new technology, seconding the claims of advertisers, in order to appeal to a particular readership and, by so doing, to elicit further support from the advertisers of the new technology, the manufacturers themselves.

27. Lundgren, "Effects of NC on Organizational Structure."

28. This is not to say that the new technology actually enabled management to reduce people to button-pushers, a centuries-old dream of managers everywhere, and still get quality production out the door; it didn't. The point here is that, with the advent of NC, management began in earnest to think about (and make decisions about) redesigning jobs and reorganizing the production process on the assumption that the new technology gave them this capability.

29. This reality can be seen even more clearly in the case of the latest generation of NC machines, which come equipped with a small computer (hence the name: computer numerical control, CNC). With these machines, it has become possible to store programs right at the machine; more important, it is now possible to edit programs (or create them from scratch) on the shop floor by simply punching information into storage numerically (reminiscent of the old Gisholt Factrol system) or by moving the machine manually to the desired position and feeding the information into storage (a digitized version of record playback). In short, these new machines, created to overcome the programming problems of NC, are ideal for shop floor and operator control over the whole process. However, they are almost never used in this way. It is characteristic of the larger metalworking shops that operators are not permitted to edit the programs, much less to create new ones themselves. Management usually defends this policy by arguing that too many cooks spoil the stew.

30. This is an error that Harry Braverman tended to make in his path-breaking *Labor and Monopoly Capital*. As a result, the potential for class struggle was minimized. One is left merely with a juggernaut on the one side and impotence and despair on the other.

31. Brooks quoted by David Montgomery, *The Fall of the House of Labor*, chap. 4, p. 1.

32. Martin R. Doring and Raymond C. Sailing, "A Case for Wage Incentives in the NC Age," *Manufacturing Engineering and Management* 66, no. 6 (1971): 31.

33. Frank Emspak, "Crisis and Austerity in the Seventies," unpublished manuscript.

Technological Momentum

Thomas P. Hughes

The concepts of technological determinism and social construction provide agendas for fruitful discussion among historians, sociologists, and engineers interested in the nature of technology and technological change. Specialists can engage in a general discourse that subsumes their areas of specialization. In this essay I shall offer an additional concept—technological momentum—that will, I hope, enrich the discussion. Technological momentum offers an alternative to technological determinism and social construction. Those who in the past espoused a technological determinist approach to history offered a needed corrective to the conventional interpretation of history that virtually ignored the role of technology in effecting social change. Those who more recently advocated a social construction approach provided an invaluable corrective to an interpretation of history that encouraged a passive attitude toward an overwhelming technology. Yet both approaches suffer from a failure to encompass the complexity of technological change.

All three concepts present problems of definition. Technological determinism I define simply as the belief that technical forces determine social and cultural changes. Social construction presumes that social and cultural forces determine technical change. A more complex concept than determinism and social construction, technological momentum infers that social development shapes and is shaped by technology. Momentum also is time dependent. Because the focus of this essay is technological momentum, I shall define it in detail by resorting to examples.

"Technology" and "technical" also need working definitions. Proponents of technological determinism and of social construction often use "technology" in a narrow sense to include only physical artifacts and software. By contrast, I use "technical" in referring to physical artifacts and software. By "technology" I usually mean technological or sociotechnical systems, which I shall also define by examples.

Discourses about technological determinism and social construction usually refer to society, a concept exceedingly abstract. Historians are wary of defining society other than by example because they have found that twentieth-century societies seem quite different from twelfth-century ones and that societies differ not only over time but over space as well. Facing these ambiguities, I define the social as the world that is not technical, or that is not hardware or technical software. This world is made up of institutions, values, interest groups, social classes, and political and economic forces. As the reader will learn, I see the social and the technical as interacting within technological systems. Technological system, as I shall explain, includes both the technical and the social. I name the world outside of technological systems that shapes them or is shaped by them the "environment." Even though it may interact with the technological system, the environment is not a part of the system because it is not under the control of the system as are the system's interacting components.

In the course of this essay the reader will discover that I am no technological determinist. I cannot associate myself with such distinguished technological determinists as Karl Marx, Lynn White, and Jacques Ellul. Marx, in moments of simplification, argued that waterwheels ushered in manorialism and that steam engines gave birth to bourgeois factories and society. Lenin added that electrification was the bearer of socialism. White elegantly portrayed the stirrup as the prime mover in a train of cause and effect culminating in the establishment of feudalism. Ellul finds the human-made environment structured by technical systems, as determining in their effects as the natural environment of Charles Darwin. Ellul sees the human-made as steadily displacing the natural—the world becoming a system of artifacts, with humankind, not God, as the artificer.[1]

Nor can I agree entirely with the social constructivists. Wiebe Bijker and Trevor Pinch have made an influential case for social construction in their essay "The Social Construction of Facts and Artifacts."[2] They argue that social, or interest, groups define and give meaning to artifacts. In defining them, the social groups determine the designs of artifacts. They do this by selecting for survival the designs that solve the problems they want solved by the artifacts and that fulfill desires they want fulfilled by the artifacts. Bijker and Pinch emphasize the interpretive flexibility discernible in the evolution of artifacts: they believe that the various meanings given by social groups to, say, the bicycle result in a number of alternative designs of that machine. The various bicycle designs are not fixed; closure does not occur until social groups believe that the problems and desires they associate with the bicycle are solved or fulfilled.

In summary, I find the Bijker-Pinch interpretation tends toward social determinism, and I must reject it on these grounds. The concept of technological momentum avoids the extremism of both technological determinism and social construction by presenting a more complex, flexible, time-dependent, and persuasive explanation of technological change.

TECHNOLOGICAL SYSTEMS

Electric light and power systems provide an instructive example of technological systems. By 1920 they had taken on a messy complexity because of the heterogeneity of their components. In their diversity, their complexity, and their large scale, such mature technological systems resemble the megamachines that Lewis Mumford described in *The Pentagon of Power*.[3] The actor networks of Bruno Latour and Michel Callon[4] also share essential characteristics with technological systems. An electric power system consists of inanimate electrons and animate regulatory boards, both of which, as Latour and Callon suggest, can be intractable if not brought in line or into the actor network.

The Electric Bond and Share Company (EBASCO), an American electric utility holding company of the 1920s, provides an example of a mature technological system. Established in 1905 by the General Electric Company, EBASCO controlled through stock ownership a number of electric utility companies, and through them a number of technical subsystems—namely electric light and power networks, or grids.[5] EBASCO provided financial, management, and engineering construction services for the utility companies. The inventors, engineers, and managers who were the system builders of EBASCO saw to it that the services related synergistically. EBASCO management recommended construction that EBASCO engineering carried out and for which EBASCO arranged financing through sale of stocks or bonds. If the utilities lay in geographical proximity, then EBASCO often physically interconnected them through high-voltage power grids. The General Electric Company founded EBASCO and, while not owning a majority of stock in it, substantially influenced its policies. Through EBASCO General Electric learned of equipment needs in the utility industry and then provided them in accord with specifications defined by EBASCO for the various utilities with which it interacted. Because it interacted with EBASCO, General Electric was a part of the EBASCO system. Even though I have labeled this the EBASCO system, it is not clear that EBASCO solely controlled the system. Control of the complex systems seems to have resulted from a consensus among EBASCO, General Electric, and the utilities in the systems.

Other institutions can also be considered parts of the EBASCO system, but because the interconnections were loose rather than tight[6] these institutions are usually not recognized as such. I refer to the electrical engineering departments in engineering colleges, whose faculty and graduate students conducted research or consulted for EBASCO. I am also inclined to include a few of the various state regulatory authorities as parts of the EBASCO system, if their members were greatly influenced by it. If the regulatory authorities were free of this control, then they should be considered a part of the EBASCO environment, not of the system.

Because it had social institutions as components, the EBASCO system could be labeled a sociotechnical system. Since, however, the system had a technical (hardware and software) core, I prefer to name it a technological system, to distinguish it from social systems without technical cores. This privileging of the technical in a technological system is justified in part by the prominent roles played by engineers, scientists, workers, and technical-minded managers in solving the problems arising during the creation and early history of a system. As a system matures, a bureaucracy of managers and white-collar employees usually plays an increasingly prominent role in maintaining and expanding the system, so that it then becomes more social and less technical.

EBASCO AS A CAUSE AND AN EFFECT

From the point of view of technological—better, technical—determinists, the determined is the world beyond the technical. Technical determinists considering EBASCO as a historical actor would focus on its technical core as a cause with many effects. Instead of seeing EBASCO as a technological system with inter-acting technical and social components, they would see the technical core as causing change in the social components of EBASCO and in society in general. Determinists would focus on the way in which EBASCO's generators, by ener-gizing electric motors on individual production machines, made possible the reorganization of the factory floor in a manner commonly associated with Fordism. Such persons would see street, workplace, and home lighting changing working and leisure hours and affecting the nature of work and play. Determin-ists would also cite electrical appliances in the home as bringing less—and more—work for women,[7] and the layout of EBASCO's power lines as causing demographic changes. Electrical grids such as those presided over by EBASCO brought a new decentralized regionalism, which contrasted with the industrial, urban-centered society of the steam age.[8] One could extend the list of the effects of electrification enormously.

Yet, contrary to the view of the technological determinists, the social con-structivists would find exogenous technical, economic, political, and geograph-ical forces, as well as values, shaping with varying intensity the EBASCO system during its evolution. Social constructivists see the technical core of EBASCO as an effect rather than a cause. They could cite a number of instances of social con-struction. The spread of alternating (polyphase) current after 1900, for instance, greatly affected, even determined, the history of the early utilities that had used direct current, for these had to change their generators and related equipment to alternating current or fail in the face of competition. Not only did such external technical forces shape the technical core of the utilities; economic forces did so as well. With the rapid increase in the United States' population and the concen-tration of industry in cities, the price of real estate increased. Needing to expand their generating capacity, EBASCO and other electric utilities chose to build new turbine-driven power plants outside city centers and to transmit electricity by high-voltage lines back into the cities and throughout the area of supply. Small urban utilities became regional ones and then faced new political or regulatory forces as state governments took over jurisdiction from the cities. Regulations also caused technical changes. As the regional utilities of the EBASCO system expanded, they conformed to geographical realities as they sought cooling water, hydroelectric sites, and mine-mouth locations. Values, too, shaped the history of EBASCO. During the Great Depression, the Roosevelt administration singled out utility holding-company magnates for criticism, blaming the huge losses experienced by stock and bond holders on the irresponsible, even illegal, machinations of some of the holding companies. Partly as a result of this attack, the attitudes of the public toward large-scale private enterprise shifted so that it was relatively easy for the administration to push through Congress the Holding Company Act of 1935, which denied holding companies the right to incorporate utilities that were not physically contiguous.[9]

GATHERING TECHNOLOGICAL MOMENTUM

Neither the proponents of technical determinism nor those of social construction can alone comprehend the complexity of an evolving technological system such as EBASCO. On some occasions EBASCO was a cause; on others it was an effect. The system both shaped and was shaped by society. Furthermore, EBASCO's shaping society is not an example of purely technical determinism, for EBASCO, as we have observed, contained social components. Similarly, social constructivists must acknowledge that social forces in the environment were not shaping simply a technical system, but a technological system, including—as systems invariably do—social components.

The interaction of technological systems and society is not symmetrical over time. Evolving technological systems are time-dependent. As the EBASCO system became larger and more complex, thereby gathering momentum, the system became less shaped by and more the shaper of its environment. By the 1920s the EBASCO system rivaled a large railroad company in its level of capital investment, in its number of customers, and in its influence upon local, state, and federal governments. Hosts of electrical engineers, their professional organizations, and the engineering schools that trained them were committed by economic interests and their special knowledge and skills to the maintenance and growth of the EBASCO system. Countless industries and communities interacted with EBASCO utilities because of shared economic interests. These various human and institutional components added substantial momentum to the EBASCO system. Only a historical event of large proportions could deflect or break the momentum of an EBASCO, the Great Depression being a case in point.

CHARACTERISTICS OF MOMENTUM

Other technological systems reveal further characteristics of technological momentum, such as acquired skill and knowledge, special-purpose machines and processes, enormous physical structures, and organizational bureaucracy. During the late nineteenth century, for instance, mainline railroad engineers in the United States transferred their acquired skill and knowledge to the field of intra-urban transit. Institutions with specific characteristics also contributed to this momentum. Professors in the recently founded engineering schools and engineers who had designed and built the railroads organized and rationalized the experience that had been gathered in preparing roadbeds, laying tracks, building bridges, and digging tunnels for mainline railroads earlier in the century. This engineering science found a place in engineering texts and in the curricula of the engineering schools, thus informing a new generation of engineers who would seek new applications for it.

Late in the nineteenth century, when street congestion in rapidly expanding industrial and commercial cities such as Chicago, Baltimore, New York, and Boston threatened to choke the flow of traffic, extensive subway and elevated railway building began as an antidote. The skill and the knowledge formerly expended on railroad bridges were now applied to elevated railway structures; the know-how once invested in tunnels now found application in subways. A remarkably active period of intra-urban transport construction began about the time when the building of mainline railways reached a plateau, thus facilitating the movement of know-how from one field to the other. Many of the engineers who played leading roles in intra-urban transit between 1890 and 1910 had been mainline railroad builders.[10]

The role of the physical plant in the buildup of technological momentum is revealed in the interwar history of the Badische Anilin und Soda Fabrik (BASF), one of Germany's leading chemical manufacturers and a member of the I.G. Farben group. During World War I, BASF rapidly developed large-scale production facilities to utilize the recently introduced Haber-Bosch technique of nitrogen fixation. It produced the nitrogen compounds for fertilizers and explosives so desperately needed by a blockaded Germany. The high-technology process involved the use of high-temperature, high-pressure, complex catalytic action. Engineers had to design and manufacture extremely costly and complex instrumentation and apparatus. When the blockade and the war were over, the market demand for synthetic nitrogen compounds did not match the large capacity of the high-technology plants built by BASF and other companies during the war. Numerous engineers, scientists, and skilled craftsmen who had designed, constructed, and operated these plants found their research and development knowledge and their construction skills underutilized. Carl Bosch, chairman of the managing board of BASF and one of the inventors of the Haber-Bosch process, had a personal and professional interest in further development and application of high-temperature, high-pressure, catalytic processes. He and other managers, scientists, and engineers at BASF sought additional ways of using the plant and the knowledge created during the war years. They first introduced a high-temperature, high-pressure catalytic process for manufacturing synthetic methanol in the early 1920s. The momentum of the now-generalized process next showed itself in management's decision in the mid 1920s to invest in research and development aimed at using high-temperature, high-pressure catalytic chemistry for the production of synthetic gasoline from coal. This project became the largest investment in research and development by BASF during the Weimar era. When the National Socialists took power, the government contracted for large amounts of the synthetic product. Momentum swept BASF and I.G. Farben into the Nazi system of economic autarky.[11]

When managers pursue economies of scope, they are taking into account the momentum embodied in large physical structures. Muscle Shoals Dam, an artifact of considerable size, offers another example of this aspect of technological momentum. As the loss of merchant ships to submarines accelerated during World War I, the United States also attempted to increase its indigenous supply of nitrogen compounds. Having selected a process requiring copious amounts of electricity, the government had to construct a hydroelectric dam and power station. This was located at Muscle Shoals, Alabama, on the Tennessee River. Before the nitrogen-fixation facilities being built near the dam were completed, the war ended. As in Germany, the supply of synthetic nitrogen compounds then exceeded the demand. The U.S. government was left not only with process facilities but also with a very large dam and power plant.

Muscle Shoals Dam (later named Wilson Dam), like the engineers and managers we have considered, became a solution looking for a problem. How should the power from the dam be used? A number of technological enthusiasts and planners envisioned the dam as the first of a series of hydroelectric projects along the Tennessee River and its tributaries. The poverty of the region spurred them on in an era when electrification was seen as a prime mover of economic development. The problem looking for a solution attracted the attention of an experienced problem solver, Henry Ford, who proposed that an industrial complex based on hydroelectric power be located along seventy-five miles of the waterway that included the Muscle Shoals site. An alliance of public power and private interests with their own plans for the region frustrated his plan. In 1933, however, Muscle Shoals became the original component in a hydroelectric, flood-control, soil-reclamation, and regional development project of enormous scope sponsored by Senator George Norris and the Roosevelt administration and presided over by the Tennessee Valley Authority. The technological momentum of the Muscle Shoals Dam had carried over from World War I to the New Deal. This durable artifact acted over time like a magnetic field, attracting plans and projects suited to its characteristics. Systems of artifacts are not neutral forces; they tend to shape the environment in particular ways.[12]

USING MOMENTUM

System builders today are aware that technological momentum—or whatever they may call it—provides the durability and the propensity for growth that were associated more commonly in the past with the spread of bureaucracy. Immediately after World War II, General Leslie Groves displayed his system-building instincts and his awareness of the critical importance of technological momentum as a means of ensuring the survival of the system for the production of atomic weapons embodied in the wartime Manhattan Project. Between 1945 and 1947, when others were anticipating disarmament, Groves expanded the gaseous-diffusion facilities for separating fissionable uranium at Oak Ridge, Tennessee; persuaded the General Electric Company to operate the reactors for producing plutonium at Hanford, Washington; funded the new Knolls Atomic Power Laboratory at Schenectady, New York; established the Argonne and Brookhaven National Laboratories for fundamental research in nuclear science; and provided research funds for a number of universities. Under his guiding hand, a large-scale production system with great momentum took on new life in peacetime. Some of the leading scientists of the wartime project had confidently expected production to end after the making of a few bombs and the coming of peace.[13]

More recently, proponents of the Strategic Defense Initiative (SDI), organized by the Reagan administration in 1983, have made use of momentum. The political and economic interests and the organizational bureaucracy vested in this system were substantial—as its makers intended. Many of the same industrial contractors, research universities, national laboratories, and government agencies that took part in the construction of intercontinental ballistic missile systems, National Air and Space Administration projects, and atomic weapon systems have been deeply involved in SDI. The names are familiar: Lockheed, General Motors, Boeing, TRW, McDonnell Douglas, General Electric, Rockwell, Teledyn, MIT, Stanford, the University of California's Lawrence Livermore Laboratory, Los Alamos, Hanford, Brookhaven, Argonne, Oak Ridge, NASA, the U.S. Air Force, the U.S. Navy, the CIA, the U.S. Army, and others. Political interests reinforced the institutional momentum. A number of congressmen represent districts that receive SDI contracts, and lobbyists speak for various institutions drawn into the SDI network.[14] Only the demise of the Soviet Union as a military threat allowed counter forces to build up sufficient momentum to blunt the cutting edge of SDI.

CONCLUSION

A technological system can be both a cause and an effect; it can shape or be shaped by society. As they grow larger and more complex, systems tend to be more shaping of society and less shaped by it. Therefore, the momentum of technological systems is a concept that can be located somewhere between the poles of technical determinism and social constructivism. The social constructivists have a key to understanding the behavior of young systems: technical determinists come into their own with the mature ones. Technological momentum, however, provides a more flexible mode of interpretation and one that is in accord with the history of large systems.

What does this interpretation of the history of technological systems offer to those who design and manage systems or to the public that might wish to shape them through a democratic process? It suggests that shaping is easiest before the system has acquired political, economic, and value components. It also follows that a system with great technological momentum can be made to change direction if a variety of its components are subjected to the forces of change.

For instance, the changeover since 1970 by U.S. automobile manufacturers from large to more compact automobiles and to more fuel-efficient and less polluting ones came about as a result of pressure brought on a number of components in the huge automobile production and use system. As a result of the oil embargo of 1973 and the rise of gasoline prices, American consumers turned to imported compact automobiles; this, in turn, brought competitive economic pressure to bear on the Detroit manufacturers. Environmentalists helped persuade the public to support, and politicians to enact, legislation that promoted both anti-pollution technology and gas-mileage standards formerly opposed by American manufacturers. Engineers and designers responded with technical inventions and developments.

On the other hand, the technological momentum of the system of automobile production and use can be observed in recent reactions against major environmental initiatives in the Los Angeles region. The host of institutions and persons dependent politically, economically, and ideologically on the system (including gasoline refiners, automobile manufacturers, trade unions, manufacturers of appliances and small equipment using internal-combustion engines, and devotees of unrestricted automobile usage) rallied to frustrate change.

Because social and technical components interact so thoroughly in technological systems and because the inertia of these systems is so large, they bring to mind the iron-cage metaphor that Max Weber used in describing the organizational bureaucracies that proliferated at the beginning of the twentieth century.[15] Technological systems, however, are bureaucracies reinforced by technical, or physical, infrastructures which give them even greater rigidity and mass than the social bureaucracies that were the subject of Weber's attention. Nevertheless, we must remind ourselves that technological momentum, like physical momentum, is not irresistible.

NOTES

1. Lynn White Jr., *Medieval Technology and Social Change* (Oxford: Clarendon, 1962); Jacques Ellul, *The Technological System* (New York: Continuum, 1980); Karl Marx, *Capital: A Critique of Political Economy*, ed. F. Engels (New York: International Publishers, 1967); *Electric Power Development in the U.S.S.R.*, ed. B. I. Weitz (Moscow: INRA, 1936).

2. The essay is found in *The Social Construction of Technological Systems: New Directions in the Sociology and History of Technology*, ed. W. E. Bijker et al. (Cambridge: MIT Press, 1987).

3. Lewis Mumford. *The Myth of the Machine: II. The Pentagon of Power* (New York: Harcourt Brace Jovanovich, 1970).

4. Bruno Latour, *Science in Action: How to Follow Scientists and Engineers through Society* (Cambridge: Harvard University Press, 1987); Michel Gallon, "Society in the Making: The Study of Technology as a Tool for Sociological Analysis," in *The Social Construction of Technological Systems*.

5. Before 1905, General Electric used the United Electric Securities Company to hold its utility securities and to fund its utility customers who purchased GE equipment. See Thomas P. Hughes, *Networks of Power: Electrification in Western Society, 1880–1930* (Baltimore: Johns Hopkins University Press, 1983), pp. 395–96.

6. The concept of loosely and tightly coupled components in systems is found in Charles Perrow's *Normal Accidents: Living with High Risk Technology* (New York: Basic Books, 1984).

7. Ruth Schwartz Cowan, "The 'Industrial Revolution' in the Home," *Technology and Culture* 17 (1976): 1–23. [See chapter 28 in this volume –Ed.]

8. Lewis Mumford, *The Culture of Cities* (New York: Harcourt Brace Jovanovich, 1970), p. 378.

9. More on EBASCO's history can be found on pp. 392–99 of *Networks of Power*.

10. Thomas Parke Hughes, "A Technological Frontier: The Railway," in *The Railroad and the Space Program*, ed. B. Mazlish (Cambridge: MIT Press, 1965).

11. Thomas Parke Hughes, "Technological Momentum: Hydrogenation in Germany 1900–1933," *Past and Present* (August 1969): 106–32.

12. On Muscle Shoals and the TVA, see Preston J. Hubbard's *Origins of the TVA: The Muscle Shoals Controversy, 1920–1932* (Nashville: Vanderbilt University Press, 1961).

13. Richard G. Hewlett and Oscar E. Anderson Jr., *The New World, 1939–1946* (College Park: Pennsylvania State University Press, 1962), pp. 624–38.

14. Charlene Mires, "The Strategic Defense Initiative" (unpublished essay, History and Sociology of Science Department, University of Pennsylvania, 1990).

15. Max Weber, *The Protestant Ethic and the Spirit of Capitalism*, trans. T. Parsons (London: Unwin-Hyman, 1990), p. 155.

The Ruination of the Tomato

Mark Kramer

It wasn't a conspiracy, it was just good business sense—but why did modern agriculture have to take the taste away?

Sagebrush and lizards rattle and whisper behind me. I stand in the moonlight, the hot desert at my back. It's tomato harvest time, 3 A.M. The moon is almost full and near to setting. Before me stretches the first lush tomato field to be taken this morning. The field is farmed by a company called Tejon Agricultural Partners, and lies three hours northeast of Los Angeles in the middle of the bleak, silvery drylands of California's San Joaquin Valley. Seven hundred sixty-six acres, more than a mile square of tomatoes—a shaggy, vegetable-green rug dappled with murky red dots, 105,708,000 ripe tomatoes lurking in the night. The field is large and absolutely level. It would take an hour and a half to walk around it. Yet, when I raise my eyes past the field to the much vaster valley floor, and to the mountains that loom farther out, the enormous crop is lost in a big flat world.

This harvest happens nearly without people. A hundred million tomatoes grown, irrigated, fed, sprayed, now taken, soon to be cooled, squashed, boiled, barreled, and held at the ready, then canned, shipped, sold, bought, and after being sold and bought a few more times, uncanned and dumped on pizza. And such is the magnitude of the vista, and the dearth of human presence, that it is easy to look elsewhere and put this routine thing out of mind. But that quality—of blandness overlaying a wonderous integration of technology, finances, personnel, and business systems—seems to be what the "future" has in store.

Three large tractors steam up the road toward me, headlights glaring, towing three thin-latticed towers that support floodlights. The tractors drag the towers into place around an assembly field, then hydraulic arms raise them to vertical.

They illuminate a large, sandy work yard where equipment is gathering—fuel trucks, repair trucks, concession trucks, harvesters, tractor-trailers towing big open hoppers. Now small crews of Mexicans, their sunburns tinted light blue in the glare of the three searchlights, climb aboard the harvesters; shadowy drivers mount tractors and trucks. The night fills with the scent of diesel fumes and with the sound of large engines running evenly.

The six harvesting machines drift across the gray-green tomato-leaf sea. After a time, the distant ones come to look like steamboats afloat across a wide bay. The engine sounds are dispersed. A company foreman dashes past, tally sheets in hand. He stops nearby only long enough to deliver a one-liner. "We're knocking them out like Johnny-be-good," he says, punching the air slowly with his right fist. Then he runs off, laughing.

The nearest harvester draws steadily closer, moving in at about the speed of a slow amble, roaring as it comes. Up close, it looks like the aftermath of a collision between a grandstand and a San Francisco tram car. It's two stories high, rolls on wheels that don't seem large enough, astraddle a wide row of jumbled and unstaked tomato vines. It is not streamlined. Gangways, catwalks, gates, conveyors, roofs, and ladders are fastened all over the lumbering rig. As it closes in, its front end snuffles up whole tomato plants as surely as a hungry pig loose in a farmer's garden. Its hind end excretes a steady stream of stems and rejects. Between the ingestion and the elimination, fourteen laborers face each other on long benches. They sit on either side of a conveyor that moves the new harvest rapidly past them. Their hands dart out and back as they sort through the red stream in front of them.

Watching them is like peering into the dining car of a passing train. The folks aboard, though, are not dining but working hard for low wages, culling what is not quite fit for pizza sauce—the "greens," "molds," "mechanicals," and the odd tomato-sized clod of dirt which has gotten past the shakers and screens that tug tomato from vine and dump the harvest onto the conveyor.

The absorbing nature of the work is according to plan. The workers aboard this tiny outpost of a tomato sauce factory are attempting to accomplish a chore at which they cannot possibly succeed, one designed in the near past by some anonymous practitioner of the new craft of *management*. As per cannery contract, each truckload of tomatoes must contain no more than 4 percent green tomatoes, 3 percent tomatoes suffering mechanical damage from the harvester, 1 percent tomatoes that have begun to mold, and .5 percent clods of dirt.

"The whole idea of this thing," a farm executive had explained earlier in the day, "is to get as many tons as you can per hour. Now, the people culling on the machines strive to sort everything that's defective. But to us, that's as bad as them picking out too little. We're getting $40 to $47 a ton for tomatoes—a bad price this year—and each truckload is 50,000 pounds, 25 tons, 1100 bucks a load. If we're allowed 7 or 8 percent defective tomatoes in the load and we don't have 7 or 8 percent defective tomatoes in the load, we're giving away money. And what's worse, we're paying these guys to make the load too good. It's a double loss. Still, you can't say to your guys, 'Hey, leave 4 percent greens and 1 percent molds when you sort the tomatoes on that belt. It's impossible. On most jobs you strive for perfection. They do. But you want to stop them just the right amount short of perfection—because the cannery will penalize you if your load goes over spec. So what you do is run the belt too fast, and sample the percentages in the output from each machine. If the load is too poor, we add another worker. If it's too good, we send someone home."

The workers converse as they ride the machine toward the edge of the desert. Their lips move in an exaggerated manner, but they don't shout. The few workers still needed at harvest time have learned not to fight the machine. They speak under, rather than over, the din of the harvest. They chat, and their hands stay constantly in fast motion.

Until a few years ago, it took a crowd of perhaps six hundred laborers to harvest a crop this size. The six machines want about a hundred workers tonight—a hundred workers for 100 million tomatoes, a million tomatoes per worker in the course of the month it will take to clear the field. The trucks come and go. The harvesters sweep back and forth across the field slowly. Now one stands still in midfield. A big service truck of the sort that tends jet planes drives across the field toward it, dome light flashing. It seems that whatever breaks can be fixed here.

After the first survey, there is nothing new to see. It will be this way for the entire month. Like so many scenes in the new agriculture, the essence of this technological miracle is its productivity, and that is reflected in the very uneventfulness of the event. The miracle is permeated with the air of everydayness. Each detail must have persons behind it—the inventions and techniques signal insights into systems, corporate decisions, labor meetings, contracts, phone calls, handshakes, hidden skills, management guidelines. Yet the operation is smooth-skinned. Almost nothing anyone does here requires manual skills or craft beyond the ability to drive and follow orders. And everyone—top to bottom—has his orders.

The workday mood leaves the gentleman standing next to me in good humor. We'll call him Johnny Riley, and at this harvest time he is still a well-placed official at this farm. He is fiftyish and has a neatly trimmed black beard. His eyebrows and eyelashes match the beard, and his whole face, round, ruddy, and boyish, beams behind heavy, black-framed glasses. He's a glad-hander, a toucher, with double-knit everything, a winning smile that demands acknowledgement, and praise to give out. It is enjoyable to talk with him.

"There are too many people out here on the job with their meters running. We can't afford trouble with tomato prices so low. If something hasn't been planned right, and it costs us extra money to get it straightened out, it's my ass," he says.

The tomato harvester that has been closing for some time, bearing down on our outpost by the edge of the field, is now dangerously near. Behind the monster stretches a mile-and-a-quarter-long row of uprooted stubble, shredded leaves, piles of dirt, and smashed tomatoes. Still Johnny Riley holds his ground. He has to raise his voice to make himself heard.

"I don't like to blow my own horn," he shouts, "but there are secrets to agriculture you just have to find out for yourselves. Here's one case in point. It may seem small to you at first, but profits come from doing the small things right. And one of the things I've found over the years is that a long row is better. Here's why. When you get to the end of a row, the machine here. . . ." Riley gestures up at the harvester, notices our plight, and obligingly leads me to one side. He continues, ". . . the machine here has to turn around before it can go back the other way. And that's when people get off and smoke. Long rows keep them on the job more minutes per hour. You've got less turns with long rows, and the people don't notice this. Especially at night, with lights on, row length is an important tool for people management. Three-fourths of the growers don't realize that. I shouldn't tell you so—it sounds like I'm patting myself on the back—but they don't."

And sure enough, as the harvester climbs off the edge of the tomato field and commences its turn on the sandy work road, the crew members descend from the catwalk, scramble to the ground, and light up cigarettes. Johnny Riley nods knowingly to me, then nods again as a young fellow in a John Deere cap drives out of the darkness in a yellow pickup to join us in the circle of light the harvester has brought with it. It's as if he arrived to meet the harvester—which, it turns out, is what he did do. He is introduced as Buck Klein. Riley seems avuncular and proud as he talks about him.

"He's the field supervisor. Just a few years ago he was delivering material for a fertilizer company. Soon he was their dispatcher, then took orders. He organized the job. He came here to do pesticides, and we've been moving him up." Buck Klein keeps a neutral face for the length of this history, for which I admire him. He is of average height, sturdily built, sports a brush moustache that matches his short, dark blond hair. He wears a western shirt, a belt with a huge buckle that says "Cotton" on it, and cowboy boots. He has come on business.

"We just got a truck back," he says, "all the way from the cannery at Fullerton—three hundred miles of travel and it's back with an unacceptable load. It's got 12 percent mechanical damages, so something's beating on the tomatoes. And this is the machine that's been doing it."

Johnny Riley appears to think for a moment. "We had three loads like that today. Seven percent, 11 percent, and 17 percent mechanicals. You got to take the truck back, get some workers to take out the center of the load and put in some real good tomatoes before you send it back. It ties up workers, and it ties up a truck."

Buck and I join the crew for one lap of harvesting. Then, while the crew members smoke, Buck and a staff mechanic go at the machine with wrenches and screwdrivers. Finally, it is fixed. As we drive off in his truck, Buck talks about the nature of corporate farming. "We have budget sheets for every crop. It's what the management spends their time worrying about, instead of how to make the crops better. It's all high finance. It makes sense, if you think about what they have in it. But I'll tell you something. It's expensive to farm here."

Buck points across the darkness, to the lights of the assembly yard. "Just beyond those lights there's a guy owns a piece—a section of land, and he grows tomatoes there, too. A guy who works with the harvester here, he knows tomatoes pretty well. And he says that guy has a break-even of about 18 tons—18 tons of $40 tomatoes pay his costs, and he's watching every row, growing better than 30 tons to the acre. Our break-even is 24 tons. Why? Because we're so much bigger. They give me more acres than I feel I can watch that closely. The partnership charges 35 bucks an acre management fee, good prices for this and that in the budget. And there is a stack of management people here, where that guy drives his own tractor while he thinks about what to do next. You can't beat him. This is not simple enough here.

"Here, they're so big, and yet they are always looking for a way to cut a dollar out of your budget. Trying to get more and more efficient. It's the workers who they see as the big expense here. They say, okay, management is us, but maybe we can cut out some of those people on the harvesting machines. We can rent these machines from the custom harvester company for $6 a ton bare. We got to pay the workers by the hour even when we're holding up the picking. Twenty workers to a machine some nights and $2.90 a worker is 58 bucks for an hour of down time. You keep moving or send people home.

"Of course this will all be a thing of the past soon. There's a new machine out—Blackwelder makes it—and it's not an experimental model. I mean, it's on the job, at $104,000 and up a shot, and it still pays. It does the same work, only better, with only two workers on it. It's faster, and there's no labor bill. It's an electronic sort. It has a blue belt and little fingers and electric eyes, and when it spots a tomato that isn't right, the little fingers push it out of the way. You just set the amount of greens you want left alone, and it does that, too. We're going to have two of them running later in the harvest, soon as they finish another job."

"What about the workers who have always followed the tomato harvest?" I ask.

"They're in trouble," says Buck, shaking his head. "They'll still be needed, but only toward the end of the harvest. At the beginning, most of what these cullers take away is greens. The electric eye can do that. But at the end of the harvest, most of what they take away is spoiled reds, stuff that gets overripe before we pick it, and they say the machines don't do that as well. That leaves a lot of workers on welfare, or whatever they can get, hanging around waiting for the little bit we need them. They get upset about being sent away. This one guy trying to get his sister on a machine, he's been coming up to me all evening saying things about the other workers. I just ignore it, though. It's all part of the job, I guess."

The trouble in which California farm labor finds itself is old trouble. And yet, just a few years ago, when harvesting of cannery tomatoes was still done by hand, ten times the labor was required on the same acreage to handle a harvest that yielded only a third of what Tejon Agricultural Partners and other growers expect these days. The transformation of the tomato industry has happened in the course of about twenty years.

Much has been written recently about this phenomenon, and with good reason. The change has been dramatic, and is extreme. Tomatoes we remember from the past tasted rich, delicate, and juicy. Tomatoes hauled home in today's grocery bag taste bland, tough, and dry. The new taste is the taste of modern agriculture.

The ruination of the tomato was a complex procedure. It required cooperation from financial, engineering, marketing, scientific, and agricultural parties that used to go their separate ways more and cross paths with less intention. Now larger institutions control the money that consumers spend on tomatoes. It is no more possible to isolate a "cause" for this shift than it is possible to claim that it's the spark plugs that cause a car to run. However, we can at least peer at the intricate machinery that has taken away our tasty tomatoes and given us pale, scientific fruit.

Let us start then, somewhat arbitrarily, with processors of tomatoes, especially with the four canners—Del Monte, Heinz, Campbell, and Libby, McNeill & Libby—that sell 72 percent of the nation's tomato sauce. What has happened to the quality of tomatoes in general follows from developments in the cannery tomato trade.

The increasingly integrated processors have consolidated, shifted, and "reconceptualized" their plants. In the fast world of marketing processed tomatoes, the last thing executives want is to be caught with too many cans of pizza sauce, fancy grade, when the marketplace is starved for commercial catsup. What processors do nowadays is capture the tomatoes and process them until they are clean and dead, but still near enough to the head of the assembly line so they have not yet gone past the squeezer that issues tomato juice on the sluice gate leading to the spaghetti sauce vat, the paste vat, the aspic tank, or the cauldrons of anything in particular. The mashed stuff of tomato products is stored until demand is clear. Then it's processed the rest of the way. The new manufacturing concept is known in the trade as aseptic barreling, and it leads to success by means of procrastination.

The growers supplying the raw materials for these tightly controlled processors have contracted in advance of planting season for the sale of their crops. It's the only way to get in. At the same time, perhaps stimulated by this new guaranteed marketplace—or perhaps stimulating it—these surviving growers of tomatoes have greatly expanded the size of their plantings. The interaction of large growers and large processors has thus crowded many smaller growers out of the marketplace, not because they can't grow tomatoes as cheaply as the big growers (they can) but because they can't provide large enough units of production to attract favorable contracts with any of the few canners in their area.

In turn, the increasing size of tomato-growing operations has encouraged and been encouraged by a number of developments in technology. Harvesters (which may have been the "cause" precipitating the other changes in the system) have in large part replaced persons in the fields. But the new machines became practical only after the development of other technological components—especially new varieties of tomato bred for machine harvesting, and new chemicals that make machine harvesting economical.

What is remarkable about the tomato from the grower's point of view is its rapid increase in popularity. In 1920, each American ate 18.1 pounds of tomato. These days we eat each 50.5 pounds of tomato. Half a million acres of cropland grow tomatoes, yielding nearly 9 million tons, worth over $900 million on the market. Today's California tomato acre yields 24 tons, while the same acre in 1960 yielded 17 tons and in 1940, 8 tons.

The increased consumption of tomatoes reflects changing eating habits in general. Most food we eat nowadays is prepared, at least in part, somewhere other than in the home kitchen, and most of the increased demand for tomatoes is for processed products—catsup, sauce, juice, canned tomatoes, and paste for "homemade" sauce. In the 1920s, tomatoes were grown and canned commercially from coast to coast. Small canneries persisted into the 1950s.

Tomatoes were then a labor-intensive crop, requiring planting, transplanting, staking, pruning. And, important in the tale of changing tomato technology, because tomatoes used to ripen a few at a time, each field required three or four forays by harvesting crews to recover successively ripening fruits. The forces that have changed the very nature of tomato-related genetics, farming practices, labor requirements, business configurations, and buying patterns started with the necessity, built so deeply into the structure of our economic system, for the constant perfection of capital utilization.

Some critics sometimes seem to imply that the new mechanization is a conspiracy fostered by fat cats up top to make their own lives softer. But though there are, surely, greedy conspirators mixed in with the regular folks running tomato farms and tomato factories and tomato research facilities, the impulse for change at each stage of the tomato transformation—from the points of view of those effecting the change—is "the system." The system always pressures participants to *meet the competition.*

Even in the 1920s, more tomatoes were grown commercially for processing than for fresh consumption, by a ratio of about two to one. Today the ratio has increased to about seven to one. Fifty years ago, California accounted for about an eighth of all tomatoes grown in America. Today, California grows about 85 percent of tomatoes. Yet as recently as fifteen years ago, California grew only about half the tomato crop. And fifteen years ago, the mechanical harvester first began to show up in the fields of the larger farms.

Before the harvester came, the average California planting was about 45 acres. Today, plantings exceed 350 acres. Tomato production in California used to be centered in family farms around Merced. It has now shifted to the corporate farms of Kern County, where Tejon Agricultural Partners operates. Of the state's

4,000 or so growers harvesting canning tomatoes in the late sixties, 85 percent have left the business since the mechanical harvester came around. Estimates of the number of part-time picking jobs lost go as high as 35,000.

The introduction of the harvester brought about other changes too. Processors thought that tomatoes ought to have more solid material, ought to be less acid, ought to be smaller. Engineers called for tomatoes that had tougher skins and were oblong so they wouldn't roll back down tilted conveyor belts. Larger growers, more able to substitute capital for labor, wanted more tonnage per acre, resistance to cracking from sudden growth spurts that follow irrigation, leaf shade for the fruit to prevent scalding by the hot sun, determinate plant varieties that grow only so high to keep those vines in rows, out of the flood irrigation ditches.

As geneticists selectively bred for these characteristics, they lost control of others. They bred for thick-walledness, less acidity, more uniform ripening, oblongness, leafiness, and high yield—and they could not also select for flavor. And while the geneticists worked on tomato characteristics, chemists were perfecting an aid of their own. Called ethylene, it is in fact also manufactured by tomato plants themselves. All in good time, it promotes reddening. Sprayed on a field of tomatoes that has reached a certain stage of maturity (about 15 percent of the field's tomatoes must have started to "jell"), the substance causes the plants to start the enzyme activity that induces redness. About half of the time a tomato spends between blossom and ripeness is spent at full size, merely growing red. (Tomatoes in the various stages of this ripening are called, in the trade, immature greens, mature greens, breakers, turnings, pinks, light reds, and reds.) Ethylene cuts this reddening time by a week or more and clears the field for its next use. It recovers investment sooner. Still more important, it complements the genetic work, producing plants with a determined and common ripening time so machines can harvest in a single pass. It guarantees precision for the growers. The large-scale manufacturing system that buys the partnership's tomatoes requires predictable results. On schedule, eight or ten or fourteen days after planes spray, the crop will be red and ready. The gas complements the work of the engineers, too, loosening the heretofore stubborn attachment of fruit and stem. It makes it easier for the new machines to shake the tomatoes free of the vines.

The result of this integrated system of tomato seed and tomato chemicals and tomato hardware and tomato know-how has been, of course, the reformation of tomato business.

According to a publication of the California Agrarian Action Project, a reform-oriented research group located at Davis (some of whose findings are reflected in this [chapter]), the effects of an emerging "low-grade oligopoly" in tomato processing are discoverable. Because of labor savings and increased efficiency of machine harvesting, the retail price of canned tomatoes should have dropped in the five years after the machines came into the field. Instead, it climbed 111 percent, and it did so in a period that saw the overall price of processed fruits and vegetables climb only 76 percent.

There are "social costs" to the reorganization of the tomato-processing industry as well. The concentration of plants concentrates work opportunities formerly not only more plentiful but more dispersed in rural areas. It concentrates problems of herbicide, pesticide, and salinity pollution.

As the new age of cannery tomato production has overpowered earlier systems of production, a kind of flexibility in tomato growing, which once worked strongly to the consumer's advantage, has been lost. The new high-technology tomato system involves substantial investment "up front" for seed, herbicides and pesticides, machinery, water, labor, and for the "management" of growing, marketing, and financing the crop.

In order to reduce the enormous risks that might, in the old system, have fallen to single parties, today's tomato business calls for "jointing" of the tomatoes. Growers nowadays share the burden of planting, raising, harvesting, and marketing—"farming" together with a "joint contractor." The tomatoes grown by Johnny Riley and Buck Klein on land held by Tejon Agricultural Partners were grown under a joint contract with Basic Vegetable Products, Inc., of Vacaville, California. TAP's president at the time, Jack Morgan, was previously executive vice president of Basic Vegetable.

"Jointing" deals are expensive both to set up and to administer. The tomato-growing business situation is becoming so Byzantine that the "per unit cost of production," the cost to a grower of producing a pound of tomatoes, is no longer the sole determinant of who gets to grow America's tomatoes. Once, whoever could sell the most cheaply won the competitive race to market. Today, the cost of doing all business supersedes, for large-scale operations, simple notions such as growing tomatoes inexpensively. Market muscle, tax advantages, clout with financiers, control of supply, all affect the competitive position of TAP as much as does the expense of growing tomatoes.

The consequence of joint contracting for the consumer is a higher-priced tomato. Risks that until recently were undertaken by growers and processors and distributors separately, because they were adversaries, are passed on to consumers now by participants that have allied. Growers are more certain they will recover the cost of production.

Howard Leach, who was president of TAP's parent company, Tejon Ranch, at the time of the tomato harvest, understood very well the economic implications for consumers of joint contracting.

"Productivity lessens," Leach explained to me. "Risk to the producer lessens, which is why we do it. The consumer gets more cost because the processor who puts money in will try to lower supply until it matches the anticipated demand. If you're Hunt-Wesson, you gear up to supply what you forecast that sales will be. You want an assured crop, so you contract for an agreed price. You're locked in, and so is the farming organization. But they are locked into a price they are assured of, and they are big enough to affect the supply."

Under this sort of business condition, the marketplace is fully occupied by giants. It is no place for the little guy with a truckload or two of tomatoes—even if his price is right. Farmers who once planted twenty or thirty acres of cannery tomatoes as a speculative complement to other farming endeavors are for the most part out of the picture, with no place to market their crops and no place to finance their operating expenses. As John Wood, a family farmer turned corporate manager, who currently runs TAP, puts it, "The key thing today is the ability to muscle into the marketplace. These days, it's a vicious fight to do so." And Ray Peterson, the economist and former vice president of Tejon Ranch, sums up the importance of the business side of farming now that the new technology has increased the risk and scale of each venture. "Today," he says, "vegetable farming is more marketing than farming."

The "jointing" of vegetable crops integrates the farming operation with the marketing, processing, and vending operations so closely that it takes teams of lawyers to describe just where one leaves off and another begins. And joint contracting is only one of several sorts of financial and managerial integration with suppliers and marketers that occur in the new tomato scene. Today chemical companies consult as technical experts with farming organizations. Equipment companies consult with farming organizations about what machines will do the jobs that need doing. Operations lease equipment from leasing companies run by banks that also lend them funds to operate. Financial organizations that lend growers vast sums of capital for both development and operations receive in return not merely interest but negotiated rights to oversee some decision-making processes. Agricultural academics sit on agri-business corporate boards.

Today the cannery tomato farmer has all but ceased to exist as a discrete and identifiable being. The organizations and structures that do what farmers once did operate as part and parcel of an economy functioning at a nearly incomprehensible level of integration. So much for the tasty tomato.

PART 6

TECHNOLOGY, ETHICS, AND POLITICS

INTRODUCTION

In the midst of the contested 2000 presidential election in the United States, as the campaign of George W. Bush sought a federal injunction to stop the hand counting of ballots that had been authorized by the Florida Supreme Court, James A. Baker, former U.S. senator and advisor to presidents, also became a philosopher of technology. The issue was whether a recount should commence following verified reports that the punch card ballot technology used in Miami-Dade Country, as well as other counties, had failed and consequently incorrectly counted hundreds if not thousands of potential votes for then Vice President Al Gore as spoiled ballots. Baker, hoping to stem the tide of worries about this technology, weighed in on November 10 with the following insight: "Machines are neither Republicans or Democrats and therefore can never be consciously or even unconsciously biased."

Having read the essays in the previous sections, one may find reason to disagree with Baker. Even if we are now convinced that technology is neither neutral nor ultimately autonomous, serious political, and moral, questions nonetheless have been posed for our consideration—not simply about the use of technology but about its design. While it is a category mistake to attribute intentions of any sort, conscious or unconscious, to machines themselves, it is quite reasonable to claim that they can be biased from design to implementation. It is likely that such biases have both moral and political dimensions. Technologies contain moral and political dimensions at least by virtue of the kinds of values they serve and produce in the societies that adopt them. And when we choose not to use technologies in ways that maximize some political practices over others—such as knowingly adopting faulty vote-counting technologies as we become more and more callous about the prospect that every vote really does count—then we have made a technological choice that has moral and political consequences and belies a view of what we value in our communities. The fundamental philosophical questions are therefore not about technology per se, but about values, and how they manifest themselves in technological systems and artifacts. To cast the issue as an argument between those who are protechnology and those who are antitechnology is to misdirect our attention from the more basic dispute.

Norman Balabanian's "Presumed Neutrality of Technology" starts us off by making the argument for the moral and political content of technology in broad terms. He would agree with David Noble that there is no technological determinism. What Balabanian calls for is a "paradigm shift" in the way we think about technology, that is, a radical change in the role technology plays in forming our fundamental views of how we should live in the world. At first glance, it may not occur to us to question the motto of the 1933 World's Fair, "Science Finds—Industry Applies—Man Conforms." According to Balabanian, the traditional view expressed in this motto uncritically accepts a whole cluster of assumptions about the nature of humans in a technological society driven by a market economic system. As we have seen, Jacques Ellul would think the World's Fair motto correctly, and tragically, describes our situation. The "protechnologists" we have read in part 3 would also endorse this view, but be quite happy with this state of affairs. Both reactions are based upon a serious misunderstanding. There is no technological imperative that operates independently of human motives.

Balabanian hastens to add, however, that to think of "human" motives, to think of humans in the abstract, papers over important moral distinctions, begging significant questions of distributive justice. Who wins and who loses from each technology that is adopted or implemented? New technologies satisfy some but possibly at the expense of others. The free choice opened up by the conveniences afforded to some may restrict the choices of others. Contemporary technology "is intimately tied to matters of political power and social control."

In order to understand the lesson that technological systems can be different than they are now, it helps to have the vantage point of history in order to see how technological systems have been different at various moments in time, depending on who was in power. Paul R. Josephson provides this perspective by showing how technologies were consciously designed, funded, constructed, and implemented to serve particular political ideologies. Providing a detailed description of the use of technology to service the ideologies of Nazi Germany and the former Soviet Union (especially during the reign of Stalin), Josephson helps to expand one of the core themes of Balabanian's contribution: It is not just that capitalism leaves its imprint on technology, but most political ideologies engage in some kind of technological production that reflects their core moral and political beliefs. Choices of what to build and what industries to create are as much a product of ideological aspirations of nation building (or, rather, empire building in Josephson's examples) as scientific progress or technological convenience.

It would be a mistake however to paint a picture depicting helpless citizens always at the mercy of designers, technologists, and technocrats. Since the 1960s, people have increasingly demanded a voice in those decisions that affect their lives. However, one of the biggest hurdles to the democratic control of modern, complex technology is that it seems to require specialized scientific and technical knowledge that ordinary citizens simply do not have. So long as this esoteric knowledge was monopolized by vested interests, the people were essentially powerless. However, as scientists become involved in opposing specific developments, this monopoly of expertise is broken. In addressing this issue, Dorothy Nelkin follows the public controversy over the control of technology in two representative cases, focusing in part on the kind of epistemological issues raised in part 2.

The authority of expert knowledge rests upon assumptions about scientific rationality. As we saw in part 1, Jonathan Schell claimed that pure science is guided by the structure of nature, not by any political motives on the part of the individual scientist. Thus, we believe that scientific knowledge is free from subjective value judgments, for it is publicly verified by a rigorous methodology. This seems to offer a way of taking the politics out of public controversies over technology; the "facts" speak for themselves.

This model of scientific rationality is undermined, however, when the experts disagree. Nelkin does not claim that there is no scientific knowledge, but in the cases she follows it is clear that there are a plethora of facts from which experts may select, as well as considerable uncertainty about what the facts themselves are in any given case. Finding ways to oppose technical advice tends to cancel out the importance of such "knowledge" and reduces the authority of the experts. Nelkin concludes that technical expertise is used by the partisan in a controversy to support what is at bottom a political position. Experts, once they become involved in a policy issue, are "just like everybody else"; they, too, have philosophical and ethical commitments.

Another and more recent example of the reassertion of the control of technology by people is provided by Andrew Feenberg, in a selection from his groundbreaking book *Alternative Modernity: The Technical Turn in Philosophy and Social Theory*. Here, however, the issue is more bottom up—a challenge to the end use of technologies that are distributed and promoted by a state for particular purposes rather than a challenge at the level of planning and administration of a technology. Feenberg's example is the French experience with one of the first versions of computer-mediated communication (CMC), and one of the forerunners of the current World Wide Web, the early videotext system called "Teletel." Introduced in 1981 by the French government, Teletel was for many years the most commercially successful version of CMC technology (Feenberg rightly notes that North Americans will find it odd that this technology was promoted by the government rather than by a private firm). The method of delivery was through the French telephone company, which distributed millions of free terminals ("Minitels") to use the Teletel system, originally under the auspices of the creation of a national telephone directory. Understanding the success and full political import of this system, though, cannot be reduced to an understanding of its intended design or the ends it was supposed to serve for the French state.

Feenberg points out that the initial intention of the creation of Teletel was similar in some ways to the ideological projects Josephon identifies as "Gigantomania" in the Nazi and Stalinist regimes. The point this time, however, was not to glorify the ambitions of an expansionist ideology or to celebrate the glories of state collective plans over markets by creating gigantic monuments to the master race or the working class. Instead, the introduction of the Teletel/Minitel technology was intended to mark the arrival of France into "modernity"—to demonstrate that it was as much a modern power as the United States or Germany, more specifically, as much a power in the new information age. But the grand aspirations of the Teletel system were not the ultimate fruit of the enterprise. Following a lackluster response from users to the new technology, the system was successfully hacked in 1982 and eventually became a mechanism for an early version of e-mail used primarily to connect users interested in personal communication around common interests such as particular sexual tastes and preferences. As Feenberg puts it, "The original plans for Teletel had not quite excluded human communication, but they certainly underestimated its importance relative to the dissemination of data." What had been designed predominantly as a one-way information retrieval system became a "general space for free discussion." Eventually this space was even used to organize student protests against the government, surely not the original intention of this state-sponsored CMC technology. Users can, as Feenberg puts it elsewhere, subvert the original rationality of machines and thus open up new means for achieving alternative political ends.

Feenberg describes the use to which the French people eventually put this state-sponsored communications medium in laudatory terms. Langdon Winner, in "Citizen Virtues in a Technological Order," would certainly agree. Putting the point differently, Winner would argue that there is both an intellectual virtue in seeing politics and technology as forces mutually influential on each other, as well as a political virtue involved in the assertion by citizens of control over the technologically infused political systems that they are bound within. But reviewing the history of philosophy on this point, Winner would identify Feenberg's careful analysis of the French Teletel system as the exception and not the rule. Ancient and modern philosophers alike have either ignored the relationship between technology and politics completely or else come up with reasons to separate the spheres of morality and politics from the more mundane world of artifactual production or planning. Identifying several new projects, for example the Swedish Center for Working Life, which provide openings for citizens to contest and help shape the future of technological policy, Winner sees some hope for overcoming this blind spot in practice if not in theory.

In discussions such as these, we are often tempted to talk generally about the need for attention to "morality" or "politics" without getting into specific forms of politics or morality. While all the authors in this section share a concern for how technology damages the democratic fabric of everyday life or can be used to bolster pernicious ideologies, there are a series of further questions that follow any call for more democracy or less constraints on personal choice. What kind of life do we want to see follow from a renewed attention to the role that technology plays in our moral and political decisions? Who benefits with new technologies in a more robust moral and political context and who loses? Diane P. Michelfelder closes this section by raising the particular issue of gender in the ethics and politics of technology.

Reviewing Feenberg's work on the Teletel/Minitel system, Michelfelder asks the pointed question: If the successful hacking of the Teletel system created more freedom, was that necessarily a kind of freedom that we should seek in the democratization of technology or was it, essentially, banal? Drawing on Albert Borgmann's "device paradigm," and his distinction between focal things and devices (introduced in part 2), Michelfelder argues that mere freedom in and of itself (for example, to find new sex partners) should not be the only goal of a reform of technology. What we need, following Borgmann, is a more substantial sense of freedom—a freedom that enriches our everyday life.

But Michelfelder does not stop there. The problem she finds with Borgmann's distinction is that it too quickly condemns certain kinds of technologies (mere devices, such as the central heating unit as opposed to focal things like the wood-burning stove) without understanding the social networks in which they are bound up. Drawing on feminist ethics (specifically an ethic of care) and using a personally illustrated example of the social networks of care that women have formed around telephones, she argues that while telephones would be categorized as devices in Borgmann's distinction, they share more of the beneficial social attributes of focal things when they are embedded in such sets of relationships. If we are to reform technology then in a more politically and morally responsible manner, we must be attentive to the kinds of ethical frameworks that we want to promote and consider how different kinds of technologies fit into these different moral and political frameworks.

All of the authors in this section agree that we cannot cope with the serious crises we face by simply tinkering with technology. What is required is the democratization of the entire social order, a new paradigm whereby we consciously consider what role technology should play in promoting those moral, political, and social values that we think are most important. For example, is the personal convenience provided by the automobile more important to us than the contributions that a car culture makes to the phenomena of global warming, which in turn will seriously threaten the health and well-being of the generations to come after us? We need not sit by helplessly and accept that our cities will be built for the automobile from now on. There are alternatives, representing different senses of what is important. The question is whether we have the moral courage to take up those alternatives and build our cities differently along with everything else.

ᒿ ।

Presumed Neutrality of Technology

Norman Balabanian

ontemporary technological society is experiencing a profound multidimensional crisis. It was not always so. From at least the time of the Industrial Revolution in England, a general feeling of optimism pervaded Western society. It was commonly believed that the growth of scientific knowledge knew no limits, and that scientific knowledge could always be applied to the problems of society. Since science and technology were so successful in producing marvelous inventions, they could eventually solve any human problem.

This attitude was exemplified at the Chicago *Century of Progress* World's Fair in 1933. The motto of the fair was: Science Finds—Industry Applies—Man Conforms. The guidebook to the fair amplified further: "Science discovers, genius invents, industry applies, and man adapts himself, or is molded by, new things. . . . Individuals, groups, entire races of men fall into step with science and technology."

Such sentiments were not expressed in sorrow, but in obvious satisfaction. The irony that human beings should bend willingly to the dictates of a technological imperative, when for ages they have struggled to be free of human tyrants, seemed to escape the "happy technologists" of that day. A similar outlook survives among technological optimists of today, typified by Simon Ramo in *Century of Mismatch:*

> We must now plan on sharing the earth with machines. . . . But much more important is that we share a way of life with them. . . . We become partners. The machines require for their optimum performance, certain patterns of society. We too have preferred arrangements. But we want what the machines can furnish, and so we must compromise. We must alter the rules of society so that we and they can be compatible.[1]

There is no "compromise" here; it is not that the machine will be constructed to be compatible with human processes, but that humanity must conform to the machine and take on the machine's way of life.

For some time now, and at an increasingly rapid pace, many people have begun to realize that the benefits flowing from science and technology to contemporary society have been purchased at a very high price. The social problems associated with science and technology, which point up the nature of the crisis, are well known. A brief, nonexhaustive taxonomy of crisis-level problems includes:

- *Environmental:* pollution, resource depletion, excess population;
- *Medical/Health:* dangers to health and safety from industrial production processes and from industrial products;
- *Psychological/Emotional:* substitution of machine values for human values; transformation of the nature of work from craftsmanship to meaninglessness, leading to worker alienation; a feeling of citizen powerlessness in the face of a complexity said to be understandable only by an expert elite, and thus, alienation from politics; social malaise exhibited in symptoms of increasing crime, senseless vandalism, anxiety, disharmony and tension, apathy and loss of a feeling of community;
- *Military:* potential nuclear annihilation through MAD (mutual assured deterrence);
- *Social:* increasing centralization, bureaucratization, authoritarianism; diminishing real returns on immense capital requirements.

What do commentators mean when they use the term *technology*? The first image conjured by the term is a machine, a physical object. This is an inadequate conception; like the term *society*, *technology* is an abstract concept. Society is not simply a collection of people but subsumes the interrelationships among them. In the same way, technology is not simply a collection of machines but encompasses relationships among them and their uses. Tools, looms, x-ray machines, nuclear reactors, refrigerators, and automobiles are all elements of technology. But if such objects are all that are comprehended by the term *technology*, the imagination is impoverished indeed. Just as a single word, or collection of words, cannot adequately represent the rich texture of language, so also a single machine, or collection of machines, cannot adequately represent contemporary technology.

There are two serious omissions from the simple-minded notion of technology. One of these is the concept of *know-how*. Accumulated knowledge is as much a part of technology as a machine. In the view of Harvey Brooks, technology is nothing but certain kinds of know-how; technology "is not hardware but knowledge, including the knowledge not only of how to fabricate hardware to predetermined specifications and functions, but also of how to design administrative processes and organizations to carry out specified functions, and to influence human behavior toward specified ends."[2]

A more important omission is the concept of organization, of *system*: the organized structures; the mechanisms of management and control; the processes of production; the specific designs of the overall organizations and systems within which the machines are embedded; the linkages that tie together the physical objects—the "hardware"—with the social institutions—the "software." Technology is not just the computer, for example, but large-scale computer networks linked through telecommunications systems; operating and managing systems; data banks, the knowledge and programs to manipulate them, and the power implicit in their control. Any analysis of contemporary technological society which fails to understand this is deeply flawed.

MODELS AND PREMISES

In going about their everyday lives, people carry in their heads models or paradigms of what the world is like, what society is like. Walter Lippmann noted that individuals create for themselves a *pseudoenvironment*, an internal representation of the world, built up over a lifetime. People's perceptions of events are determined by the images, the preconceptions, the premises that underlie this pseudoenvironment. Most people tend not to question these preconceptions, even when the consequences of acting in accordance with their images of reality lead to anomalies that the paradigm cannot reconcile.

In science, when such anomalies develop, creative thinkers question the old paradigm and its premises and develop a new image of the world based on a new way of looking at things. For example, Newtonian mechanics gave way to relativistic mechanics; the economic system of mercantilism of the fifteenth and sixteenth centuries gave way to the classical capitalist economic system of the eighteenth and nineteenth centuries, which in turn has given way to the neoclassical and Keynesian economic systems. The Keynesian analysis evolved because the world depression of the 1930s was a traumatic anomaly that the previous paradigm could not explain.

In everyday life, on another level, many people are aware of the anomalies between the rhetorically cultivated presuppositions of the social system and the adverse effects of the realities they experience on the receiving end. Such people seriously question the premises on which the social system is founded; and sometimes this questioning leads to a violent "paradigm shift"—through revolution. But, more often than not, their questioning is "contained." Historically, whenever a paradigm shift has been necessary (in science or social structure), it has been resisted by those with an interest in the old paradigm (intellectual, emotional, or financial). It is difficult for people who have invested a lifetime in the service of a particular world view to switch to a different outlook. Even if they intellectually understand the need for it, their previous life leaves them unprepared to carry on comfortably in the new paradigm.

Contemporary technological society is now facing anomalies which the social and economic preconceptions cannot explain. These preconceptions are built into the social structure and culture. But reality cries out for a paradigm shift. It is essential to identify and critically examine the premises—sometimes explicit, but often hidden—that undergird the inadequate contemporary model. Some of the major premises are:

- *Self-Seeking Values:* The preferred behavior mode for human beings is pursuit of one's own self-interest. The foundation of the capitalist ethos is that such self-seeking behavior will lead to social good through the operation of an "invisible hand."
- *Elastic Wants:* Human wants are infinitely elastic; they expand without limit. It is necessary to have continual economic growth in order to satisfy them.
- *Man's Domination of Nature:* The physical world is there for man to subdue, conquer, dominate, subjugate, and exploit. By scientific knowledge, said Descartes, "We may be able to make ourselves masters and possessors of nature."
- *Neutrality of Technology:* Technology is morally and politically neutral; it is a mere tool which can be used for good or evil. If it is not used properly, man is to blame.
- *Freedom of Choice:* Individuals in our free market system have free choice. The root cause of such ills as wasteful consumption, urban congestion, pollution, and the design of inappropriate products lies in the free choice exercised by autonomous individuals.

Each of these undergirding pillars fails to withstand critical examination. They must all be rejected if transformation to a humane society is a goal. The first three refer to the presumed nature of human beings. Although I will devote some brief thoughts to them, the bulk of my analysis will be reserved for the last two.

HUMAN NATURE

The premise or assertion that self-seeking behavior is the preferred mode for humans tends to encourage an aggressive, contentious, noncooperative spirit; it cultivates self-aggrandizement, greed, and envy—looking out for Number One at the expense of the community. Its proponents say this is human nature and cannot be changed. The social structure fostered by this outlook is hierarchical, with individuals engaged in a scramble for status on the ladder of success, elbowing their lonely way to the top.

The capitalist culture and social system require individuals to act in a self-centered, contentious manner. The proper operation of the system *demands* it. How convenient, then, to ascribe this attitude to basic human nature! Writing in 1930, John Maynard Keynes said: "For at least another hundred years we must pretend to ourselves and to everyone that fair is foul and foul is fair; for foul is useful and fair is not. Avarice and usury and precaution must be our gods."[3] Keynes thought that self-seeking should be encouraged because it was useful to the operation of the capitalist system. But he at least recognized that it was not laudable, and that you had to work at it to make avarice and greed appear as gods.

Is the greedy, status-seeking pursuit of self-interest a consequence of unalterable human nature? Demonstrating that it is not merely requires finding counterexamples. Have humans ever acted selflessly? If we could find any such cases (and we need only look to ordinary people in common situations to find many counterexamples), we would have to conclude that the proposition that self-seeking is unalterable human nature is, at the least, not proved. With a bit of further thought, we would have to say that some humans act selflessly, others selfishly; that any human sometimes acts one way, sometimes the other; that individuals often have conflicts within themselves as to whether they should respond to a given situation in a self-seeking manner or in a cooperative, communitarian manner. For each example of greedy, self-centered human behavior one can find an example of altruistic, other-directed, cooperative behavior. Without the cooperative and symbiotic working together of its millions of cells, the human body itself could not function. Far from being by nature selfish, humans can just as validly be assumed to be cooperative by nature. How they actually behave in given situations depends on their socialization, on the reinforcements they obtain for their behavior. If the culture, the social order, continually reinforces them for self-seeking behavior, for personal aggrandizement, the chances are they will act this way more often than not. An outside observer would then notice that most people, most of the time, behave in noncooperative, self-seeking, status-enhancing ways. The observer who concludes that such behavior is intrinsic human nature would be a naïve observer, indeed.

Sometimes the happy technologist will unknowingly concede this. Samuel Florman writes: "Man, for all his angelic qualities, is self-seeking and competitive."[4] There it is: a concession that humans have noble impulses as well as base ones. The real question becomes: which qualities should be cultivated and reinforced? It is not a question of changing human nature to something it is not, but arranging conditions so that people can more often exhibit, and behave in accordance with, their "angelic qualities" rather than their base ones.

Economists postulate that human wants are infinite and insatiable—no sooner has one want been served than another is stimulated; humans are not capable of saying "enough." Florman states that "contemporary man is not content because he *wants* more than he can ever have."[5] This being the case, goes the argument, it is essential to maintain economic growth and increasing levels of consumption. The serious flaw in these assertions is the failure to distinguish between those wants that are basic and absolute needs—such as food, clothing, shelter, sex—which humans will experience independent of the condition of other human beings around them, and those which are social and relative. It is clearly untrue that basic human needs cannot be satisfied. The sight of food is not tempting to a person who has just finished dinner. It is also clear that some basic human needs are not at present satisfied for a substantial fraction of the world's population. To this extent, growth is still needed; not generalized growth, but increase in those areas of production intended to satisfy basic needs of the poor.

The second category of wants may well be insatiable; but what is their nature? These wants are experienced only in a relative sense, only if a feeling of superiority to others is achieved, only if vanity and status are enhanced. These wants require continual comparison with others and feverish activity in pursuit of inequality. They are not ennobling but base. In a sane society, their pursuit would be discouraged. But in a society dominated by the capitalist ethos, they are encouraged and cultivated through high-powered promotion and persuasion. Herbert Marcuse refers to such wants as "false needs,"[6] not to deny that they exist but as a judgment of their worthiness. The flourishing of false needs is a reflection not on human nature but on the values consciously cultivated by the social system. It makes no sense actively to promote ego trips, feelings of vanity and prestige, desires for superiority and status, and then to demand economic growth in order to satisfy these desires—at the expense of the crises now being faced. Furthermore, the effort of satisfying these wants is doomed to failure. If people's satisfactions depend almost entirely on status, ego gratification, and feelings of superiority, then increasing levels of consumption cannot yield increasing satisfaction to society as a whole. The superiority and enhanced status of some implies the inferiority and reduced status of others.

The idea that human beings should have dominion over the earth and all that it contains (now extended to the entire universe) was an early tenet of Western civilization; it even carried biblical sanction. And mankind has not been reticent in carrying out this injunction. But there are at least two fallacies in this outlook.

The first relates to depletion. Clearly, intrusive human activities carried out over a period of time can have, and have had, devastating effects on nature and on human society. In the past, the wholesale destruction of forests, improper cultivation leading to soil erosion, and other similar activities have had major effects on the present conditions of many parts of the earth. They have even partly caused flourishing civilizations to vanish. But the scale, intensity, and quality of current human interventions in nature dwarf those of the past. The disappearance of much of the earth's wealth seems to be imminent. Whatever utility the concept of domination, subjugation, conquest, exploitation, and control of nature by humans may have had in the past, it is now counterproductive. Rather than subjugating nature, human beings must learn to live in harmony with it, to cooperate with its processes. That does not mean that people must lie down supinely before nature, but that their activities should take nature's processes into account. Humans should avoid building their structures over faults in the earth's plate, for example. They should avoid building their towns in flood plains. Cooperating with nature also means understanding its limits and minimizing those activities that put pressure on them or use up its irreplaceable parts.

The second fallacy is the mistaken assumption that nature is one thing and human beings something else. That is false, of course; human beings are as much a part of nature as mountains and birds. If the earth were destroyed, human beings would cease to exist also. Human domination and exploitation of nature implies human domination and exploitation of other humans. Although this proposition was accepted in the past, slavery, imperialism, and other institutions of exploitation and domination should no longer be tolerated. Their rejection also implies the rejection of the parent concept: the subjugation of nature by humans.

NEUTRALITY OF TECHNOLOGY

Although champions of "advanced" technology—happy technologists—may approach their subjects from different perspectives, there is a common refrain to their individual verses that amounts to a litany of technology: technology is just a passive tool whose consequences depend on the use to which it is put; if technology is used harmfully, man is to blame; there are no values embodied in technology; technology plays an entirely passive role with respect to issues of power and control; the prime reasons for introducing innovations in production processes are increased efficiency and productivity; the prime reason for introducing innovations in products is to satisfy human needs. This litany of the happy technologist constitutes an ideology that is a collection of errors, illusions, and mystification presenting an inverted, truncated, distorted reflection of reality. It is also a set of values characteristic of a group; the integrated assertions, theories, and aims that constitute a sociopolitical program. The ideology of technology has purposes quite remote from explaining reality to members of society. It fails to take political power and economic interests into account and thus masks their predominant role. It promotes a model which ascribes to technology an objectivity, a value-neutrality, which technology does not in fact possess. A useful clue to the ideological nature of a statement purporting to be explanatory is the ascribing of action to vague collective nouns and pronouns, as in: "*Man*" is to blame; "*Mind* determines the shape and direction of technology";[7] "If technology is sometimes used for bad ends, *all* bear responsibility . . .";[8] " . . . make technology work as *our* servant";[9] "thus, *we* manufacture millions of products to enhance *our* physical comfort and convenience. . . . But in doing this, *we* overlook the need to plan ahead."[10]

Are we *all* equal in responsibility, or are some more equal than others? Who are the "we" who do the manufacturing? Is that the same "we" who forgot to plan? Doesn't somebody's profit enter the picture at all? Surely some specific minds shape technology, not an abstract "mind." The preceding manner of speaking con-

ceals the existence of specific, powerful corporations whose activities in pursuit
of their interests are major factors in the problems of our contemporary society.

Some insight into the role played by technological ideology can be obtained
by analogy, through an examination of the role of economics in our society. Like
any other science, says John Kenneth Galbraith, the purpose of economics is
understanding; in this case, understanding the economic system—how it works,
the nature of money, etc. But economics also has an instrumental function—to
serve the goals of those with power. It creates images in the minds of people—
thus contributing to their model of society—which are not at all consonant with
the reality of the economic system, at least with the more than 50 percent of the
economy represented by what Galbraith calls the "planning system," the large
corporations and their activities. The instrumental function is to induce people to
behave as if the image were the reality. It is to conceal from people the true nature
of most of the economy as a planned system (not free enterprise), with the planning done by a handful of large corporations in their own interest.[11] Like economics, technology also has two goals. Quite apart from their purpose "to
enhance our physical comfort and convenience," technology and technical
expertise serve an ideological, instrumental function. This function is, again,
image-making and concealment, the covering of political and economic power in
a cloak of technical objectivity. The image is created that decisions and actions
serving the interests of those in power are simply the consequences of objective
facts, carried out for such objective reasons as efficiency.

The ideological assertion is that technological innovations and their introduction
serve the objective goals of efficiency, increase in productivity, and human needs satisfaction. To those holding the images of the dominant paradigm, this seems reasonable. But it does not stand up under close scrutiny. In his study of the development
of the textile industry during the Industrial Revolution in Britain, David Dickson[12]
shows that the rise of the factory system of production, the organization of work in
factories, was largely a managerial necessity, rather than a technological one. It was
done for the purpose of "curbing the insolence and dishonesty of men." The rising
capitalist class made no bones about the introduction of specific machines having as
its major purpose the subduing and disciplining of workers. Speaking of one invention in the textile industry, Andrew Ure, an early champion of the factory system,
says, "This invention confirms the great doctrine already propounded, that when capital enlists science in her service, the refractory hand of labor will always be taught
docility."[13] Technological innovation was not so much determined by concern for
production efficiency as by management's desire to maintain fragmentation of
workers, authoritarian forms of discipline, hierarchical structure, and regimentation.
These same concerns are evident to the present day. After analyzing the demands that
contemporary corporations impose on employees, Richard C. Edwards concludes
that the complex hierarchy of the modern corporation grew not from the demands of
technology but from the desire for greater control of workers.[14]

To a large extent, technological innovations in consumer products arise as a consequence of research and development (R&D) activities. Almost all R&D activities directed toward product development are carried out in the laboratories of large, technology-intensive corporations. The goals of these corporations, simply stated, are survival, growth in sales, and growth in profits.[15] All the activities of corporations—production, marketing, sales, *and* R&D—are carried out to reach these goals. Product innovation, no less than production or marketing, serves corporate purposes and would be carried out independent of social need.

Lack of need would not suffice to thwart the corporation's desire to increase sales and profit; if the market does not exist, it is developed—created, cultivated, and nourished.[16] Once the decision has been made to introduce an innovation in furtherance of corporate objectives, "developing the market" is thrown into high gear. The entire arsenal of persuasion is unleashed to convince potential consumers that the new innovation will not only perform the specific function for which the product is designed, but it will also enhance status, satisfy vanity, increase sexual appeal, etc. The result is to intensify these base impulses, and then to prey upon people's expectations that such impulses can be satisfied by the product. People are thus sold the idea that their self-worth is measured by the goods they possess, in general, and by this specific product, in particular.

Another dimension related to technological innovations appears when one seeks to determine the factors that guide the introduction of an innovation. The paramount consideration, of course, is profit; but why introduce one thing rather than another? On a superficial level, it seems that trial-and-error has something to do with it. In 1978, for example, the processed food industry introduced to the market some ten thousand new products. Of these, about 80 percent failed.[17] Was there a societal need for these products? Was "lower cost to the user" a criterion for developing, producing, and marketing them? In fact, pricing of all the corporation's products must reflect losses from those that failed. Hence, processed food prices generally must rise as a result of these innovations, quite contrary to the ideologically stated reason for product innovation.

The preceding analysis evaporates the claims of technological objectivity and value-neutrality. The nature of a society's technology is intimately related to issues of power and control, and reflects the dominant paradigm in terms of which reality is interpreted. A society in which economic growth is highly valued necessitates a particular kind of technology—one with a high level of innovation, quite independent of social need. Policies leading to economic expansion have to be reflected in the particular form of technology through which this expansion is achieved. Hierarchical forms of social control become reflected in the technology. The presumed neutrality of technology then lends legitimacy to social policies, however repressive.

Like other happy technologists, Ramo over and over again explicitly claims the value-neutrality of technology. But, without realizing it, he makes an amazing concession contradicting this position and acknowledging an instrumental function for technology. In reviewing the space program and justifying having spent large resources on this program, he concludes: "The pattern that we are developing—to be far-sighted, to be bold, to want to pioneer, to be willing to take some risks, to carry with us as a part of our way of life the exploration of the unknown—it is these habits that we cultivated when we carried out the space program."[18] This is a remarkable concession that a specific technological development had an agenda quite unrelated to the primary goal; that the program served the ideological purpose of cultivating certain habits. Can it be denied that less glorious-sounding attitudes than the ones admitted by Ramo—greed, status-seeking, self-aggrandizement, contentiousness—are on the agenda? The general truth cannot be escaped: technology serves an instrumental function.

INDIVIDUAL AUTONOMY

Not only is technology passive and neutral, according to the happy technologist, but whatever evil consequences are associated with the deployment of technology result from autonomous individuals exercising their free choice. Not only that, but people perversely go on using technology even though this fact lends to environmental degradation, depletion of resources, and other unpleasant things they themselves experience. A sampling of their assertions follows:

John Gardner: Everyone lampoons modern technology but no one is prepared to give up his refrigerator.[19]

Samuel Florman: However much we deplore our automobile culture, clearly it has been created by people making choices, not by a runaway technology.[20]

Melvin Kranzberg: That kind of spatial freedom (of the past in rural areas) vanished from the onrush of urbanism: but people apparently want to live together in large agglomerations.[21]

Alvin Weinberg: A social problem exists because many people behave, individually, in a socially unacceptable way. . . . Too many people drive cars in Los Angeles with its curious meteorology, and so Los Angeles suffocates from smog.[22]

Emmanuel Mesthene: The negative effects of technology . . . are traceable less to some mystical autonomy presumed to lie in technology and much more to the autonomy that our political and economic institutions grant to individual decision making.[23]

Simon Ramo: National control is really the only answer, and that can come only if a majority of Americans are willing to accept interference in their freedom to choose automobiles.[24]

These assertions constitute a second litany. To be generous, one would ascribe the origins of this assertion litany not to a deliberate agenda, ideologically promoted, but to a fundamental misunderstanding of the nature of technology. As noted earlier, the systems within which the separate components of technology (the machines) are incorporated are its essential features. Failure to understand that the term *technology* subsumes organized and integrated systems implies a profound misunderstanding of the nature of contemporary technology.

The term *lifestyle* designates the manner in which individuals go about their daily activities at home, at work, at play, etc. In American society there is a general perception that lifestyle is a matter of individual choice, at least for a vast majority. Disregarding economic means for a moment, people think that one can choose to lead a "bohemian" lifestyle or a "straight" one, to wear flashy clothes or sober ones, etc. But is one free to choose to have a refrigerator or not? Is it a simple matter of lifestyle choice or do other institutional arrangements of society impinge with demands of their own? Is it the kind of choice referred to by Justice Holmes when he said: "In its evenhanded majesty, the law forbids rich and poor alike to sleep under a bridge"?

A refrigerator (including freezer) performs several functions. It stores perishable food (a necessity) and cools beer or produces ice for cooling drinks (a comfort or luxury). The latter category is not an essential function. The desirability of cold beer, for example, is culturally or socially induced; other cultures (the British) find warm beer more desirable, so people in those societies do not need a refrigerator to perform this particular function. Consider another society—even a developed European society like France a decade or two ago—in which it is possible for people to purchase their perishable foods on a daily basis in markets or small shops, easily accessible and within walking distance of their homes, even in the largest cities. This option is not available to most contemporary Americans. The supermarket as a social institution, not within walking distance of most people, has its own imperatives. One buys for a week of eating, not for a day, so storage in a refrigerator becomes essential to living. It is a necessity induced by a lifestyle over which individuals have little control—an institutionalized lifestyle. To chide individuals for recalcitrance or perversity for their unwillingness to give up a refrigerator is to profoundly misjudge the nature of contemporary technology and its induced social change. No value judgment about the merits of different lifestyles is implied in this scenario. It is irrelevant to the argument whether or not a supermarket/refrigerator society has advantages over the other. The only question is: Do individuals have autonomy to freely choose one or the other?

A similar and even stronger case can be made concerning the automobile. The question, again, is one of choice. Those who claim the existence of free choice gaze out at society as it exists at a given period of time, with the social structure and state of technology as they exist. Within this framework, they argue, individuals can choose. They can choose to buy this model car or that, this color or that, this upholstery material or that, this option or that. The one fact that belies all of this apparent freedom is that the majority of individuals cannot choose *not* to buy a car—if they also want to participate in the normal life of the society, such as going to work, buying food and clothing, etc.

Many communities in the United States had electric railway systems that served admirably in the first third of this century. It was not the autonomous choice of individuals that killed these existing urban and interurban mass public transportation systems and prevented their expansion and improvement. In more than one case they were purchased by automotive corporations (the major culprit being General Motors) and converted to motor buses, and then allowed to die in order to promote the use of the private automobile.[25] A case in point is Los Angeles. It had an extensive electric streetcar and interurban rail system in the early part of this century. During the 1920s the Pacific Electric Railway operated 1,200 miles of interurban rail service. When the population of the area was only 1 million in 1924, the system carried a volume of 109 million passengers. (By comparison, forty-five years later when the population was eight to nine times greater, public transit buses carried only 190 million passengers annually.) The reason we have smog in Los Angeles, says Weinberg, is because too many people drive cars. A much more accurate reason is because General Motors bought the Pacific Electric Railway system and destroyed it. The single most important cause for not only the smog, but the fact that over half the land area of Los Angeles—including freeways, streets, driveways, parking lots, gas stations, sales rooms, etc.—is dedicated to the automobile, is the power of large corporations.

Simon Ramo decries federal "interference" in setting safety regulations, emission standards, etc., because it would control people's "freedom to choose automobiles." But many government laws and regulations have had, and continue to have, a major impact in instituting, cultivating, and maintaining the U.S. automobile culture. Among these are the investment tax credit that encouraged auto manufacturers to produce cars and the oil depletion allowance [that] encouraged oil companies to produce fuel for the cars. But of far greater significance was the establishment and funding method of the federal interstate highway system in the late fifties. This was one of the most far-reaching acts of interference by government. Yet ideologically committed technologists like Ramo do not cry "control" when the actions of government favor the corporate interests.

People were not asked to debate the merits of different transportation systems and then choose what they favored. Specific corporate interests, in order to promote their own welfare, caused the current state of the U.S. transportation system. The design of the transportation system is not a consequence of 200 million citizens exercising their free choice, but a handful of powerful groups pursuing their own interest. It is not descriptive of reality to say now that people have free choice of their mode of transportation. To set up the social system so that individuals are compelled to buy cars just in order to be active members of society, and then to sneer at them because they are unwilling to give up their cars, is to add insult to injury. It is like blaming the victim for the crime. For most people, most of the time, driving a car is not discretionary; it is mandatory.

Even within the context of a market and the regulation of technological developments by the market, is it possible accurately to describe the current status of specific technologies (e.g., the transportation system) as the consequence of untrammelled individual choice guiding the invisible hand? Market prices can be kept artificially low by transferring some of the costs associated with production or use from the manufacturer and/or user to third parties. This can be done in at least two ways—by subsidies from the government or by failure to account for "external costs" or negative "externalities." Both of these processes have operated extensively to distort price structures. Vast sums have been transferred to corporations in subsidies by the federal government, either directly (through grants and low-interest loans) or indirectly (through the taxing mechanism or having the government assume responsibility for certain components of technological systems, like highways and airports). For example, subsidies have been running at $10 billion annually just to the traditional energy industries; subsidies to the oil corporations alone have totalled more than $75 billion.

Economic activity involves producers and consumers, sellers and buyers. External costs are those undesirable effects upon other people resulting from these activities. Such costs can be tangible—such as cleaning bills and medical bills due to the effects of economic activity on the environment—or intangible, yet real. Inestimably huge external costs, both privately and socially borne, have been transferred to others. Purchasing decisions are influenced by prices that are artificially depressed in such ways. If this circumstance permits a large-scale technological development to take place, which then induces major changes in the way people live, would it be meaningful to assert that the detailed forms of the resulting society are consequences of individual free choice?

A cogent and compelling analysis of these issues is provided by James Carroll:

> [T]echnology often embodies and expresses political value choices that, in their operations and effects, are binding on individuals and groups, whether such choices have been made in political forums or elsewhere . . . technological processes in contemporary society have become the equivalent of a form of law—that is, an authoritative or binding expression of social norms and values from which the individual or group may have no immediate recourse.[26]

Furthermore, says Carroll, there are often no appropriate political processes for "identifying and debating the value choices implicit in what appear to be technical alternatives." Technological processes are technically complex and occur in administrative organizations (either government agencies or corporations) to which citizens have no access. Ordinary people on the outside have neither the opportunity nor the means to identify the value questions, far less to resolve them in a public forum. *De facto*, the locus of the political value choices becomes the technological processes themselves; there is no public debate and issues are posed and resolved in technical terms. Individuals have no autonomous choice in the matter.

Is Kranzberg serious when he offers pure individual choice as the explanation for the specific way that Americans are distributed around the country? Did people leave the Dust Bowl in the thirties because of free choice? Were individuals consulted on the mechanization of the farms which drove farm workers away from the rural areas? The locations of industries will greatly influence where people live; were individuals consulted as to where specific industries should be located? Did the interests of huckstering land developers have nothing to do with luring people to southern California? Was it individual choice or the manipulation of specific corporate interests that caused the specific layout of Los Angeles where it is necessary to drive long distances just to carry on everyday activities?

The answers to all these questions are obvious. The design of cities, the locations of services, places of employment, shopping centers, etc., are all based on the private motor car as the dominant mode of transportation. Individuals in most places have little choice; they cannot decide to buy either at the supermarket or the corner grocery—they must buy at the supermarket because there are no longer many small stores. They cannot choose either to take mass transportation or drive a car—they are compelled to drive a car. They cannot choose to buy unpackaged, unwrapped, unboxed products because such products do not exist in most places. They cannot choose to recycle much of the waste materials (paper, glass, metal) they are compelled to bring into their homes because no recycling facilities are available in most places.

NECESSARY HARMONY

Technology is not a neutral, passive tool devoid of values; it takes the shape of and, in turn, helps to shape, its embedding social system. The ideologically promoted neutral-tool/use-abuse model of technology conceals the issues of economic and political power relationships among different groups in society, and thus serves the instrumental function of legitimating the dominant ideology. In addition, contemporary technology, far from increasing freedom, limits individual autonomy and imposes a style of living about which individuals have little choice.

The needs to which technology is said to respond are induced by the social system. In a social order with different values and goals, the needs would be different and so the nature of technology would be correspondingly different. Thus, the crisis of contemporary society cannot be resolved if the contemporary form of technology remains dominant; it must be replaced by a technology of a different nature. Such technology is intimately tied to matters of political power and social control; changing the technology implies a profound change in the social order.

The crisis is not a crisis *within* technological society which can be overcome by patching up the system, but a crisis of the technological system *itself*. The major question is not *who* is to control the means of production, but *what* the means of production shall be; and *what* shall be produced. It is not *where* to locate the nuclear power plants, but *whether* to have nuclear power at all. It is not merely a question of possibly limiting growth but of radically altering the very nature of technology.[27] Contemporary technology is based on a narrowly conceived economic efficiency, on social control, and on profit. I would characterize the alternate technology needed as *harmonious* and based on different criteria. Harmonious technology would respect ecological values and be in symbiosis with nature. This does not mean that there would be no human intervention in nature, just that such interventions would not be destructive and exploitative, but in harmony with ecological values; consequently, harmonious technology would rely mainly on renewable energy and be minimally consumptive of nonrenewable resources. Harmonious technology would be responsive to direct social needs and would not require a hierarchical, exploitative, and alienating relationship among human beings. It would not oppress people nor treat them as appendages to machines, but would be satisfying to work with. Harmonious technology would value durability and equality of products, decentralization of production, diversity—rather than monoculture—in agriculture, and pluralism in lifestyle and culture.

Accomplishing this transformation requires a new consciousness that sees the interrelations among the physical, biological, and social spheres; collectively they constitute a system of which humanity is a part. A new style of living which is in harmony with the natural world is also needed. A harmonious technology in a harmonious society is the goal. An appropriate motto might be: *Science Discovers—Humanity Decides—Technology Conforms*.

NOTES

1. Simon Ramo, *Century of Mismatch* (New York: David McKay, 1970), p. 120.

2. Harvey Brooks, "The Technology of Zero Growth," *Daedalus* (fall 1983): 139.

3. John Maynard Keynes, *Essays in Persuasion* (London, p. 372; reprint, New York: Norton, 1953).

4. Samuel C. Florman, *The Existential Pleasures of Engineering* (New York: St. Martin's Press, 1976), p. 84.

5. Ibid., p. 75.

6. Herbert Marcuse, *One-Dimensional Man* (Boston: Beacon, 1964), p. 5.

7. Bruce O. Watkins and Roy Meador, *Technology and Human Values* (Ann Arbor: Ann Arbor Science, 1978), p. 55.

8. Ibid., p. 157.

9. Melvin Kranzberg and Carroll Pursell, *Technology in Western Civilization*, vol. 2 (New York: Oxford University Press, 1967), p. 32.

10. Simon Ramo, *Cure for Chaos* (New York: David McKay, 1969), p. 1.

11. John Kenneth Galbraith, *Economics and the Public Purpose* (New York: Houghton Mifflin, 1973).

12. David Dickson, *The Politics of Alternative Technology* (New York: Universe, 1974), pp. 71–83.

13. Andrew Ure, *The Philosophy of Manufacturers* (London, 1835); quoted in Dickson, *The Politics of Alternative Techology*, p. 80.

14. Richard C. Edwards, *Contested Terrain: The Transformation of the Workplace in the 20th Century* (New York: Basic Books, 1979).

15. Galbraith, *Economics and the Public Purpose*.

16. R. M. Hall and F. S. Hill Jr., *Introduction to Engineering* (New York: Prentice Hall, 1975), p. 24.

17. "The Great Consumer Rip-Off," Home Box Office television program, February 28, 1979.

18. Ramo, *Century of Mismatch*, p. 51.

19. John Gardner, Godkin Lecture at Harvard University, reported in the *New York Times*, March 30, 1969, sec. 4, p. 9E.

20. Florman, *The Existential Pleasures of Engineering*, p. 60.

21. Kranzberg and Pursell, *Technology in Western Civilization*, p. 700.

22. Alvin Weinberg, "Can Technology Replace Social Engineering?" *University of Chicago Magazine* 59 (October 1966); reprinted in Albert H. Teich, *Technology and Man's Future* (New York: St. Martin's Press, 1977). [See chapter 7 of this volume.]

23. Emmanuel G Mesthene, *Technological Change: Its Impact on Man and Society* (Cambridge: Harvard University Press, 1970), p. 40.

24. Ramo, *Century of Mismatch*, p. 117.

25. Bradford C. Snell, *American Ground Transport*, 1974. Prepared for the U.S. Senate Subcommittee.

26. James D. Carroll, "Participatory Technology," *Science* 171 (February 19, 1971): 647.

27. Ivan Illich, *Tools for Conviviality* (New York: Harper and Row. 1973).

Technology and Politics in Totalitarian Regimes

Paul R. Josephson

Technologies are symbols of national achievement. They demonstrate the prowess of the nation's scientists and engineers. They are central to national security strategies. They serve foreign policy purposes through technology transfer. They entrance a public who can become intoxicated with the artifact's symbolism and overlook its potential dangers to society (and at other times provoke fear and dislike). We need only think of the space race between the United States and the USSR, or other technological posturing between the two superpowers, to comprehend the importance of technology in securing a regime's legitimacy at home and abroad.

Skyscrapers and apartment housing, subway systems, assembly lines, canals and bridges, hydropower stations, and nuclear reactors have an imposing physical presence. They also have what has been called "display value."[1] Display value includes the social, cultural, and ideological significance of technology. While countries differ in terms of economic and political organization—market or centrally planned economy, single-party or multiparty system, centralized or decentralized decision-making apparatus—the display value of large-scale technologies applies to all.

Yet surely the place of technology in totalitarian regimes differs from that in pluralist regimes. If economic, political, and social concerns shape the practice of biology and physics in totalitarian regimes, then it should come as no surprise that technology, too, has a particular style in those regimes. At first glance technology would seem to be value-neutral, serving the rational ends of achieving a desired outcome in the "one best way." This way means efficiency maximization. Technologies are various devices, techniques, or systems intended to give us control over the natural environment—and also over our political, economic, and social structures. The latter include scientific management for industry, the gath-

Reprinted from Paul R. Josephson, *Totalitarian Science and Technology* (Amherst, N.Y.: Humanity Books, 2000).

ering, collating, and analysis of data for national planning, and so on. Engineers strive for efficiencies in production by optimizing the use of labor and capital inputs. They work toward these ends through planning to set prices, allocating raw materials and market share, and designing shop layouts and material flows.[2]

The "one best way" distinction is crucial, for it implies that given any engineering problem the solution will be universal, based on engineering calculations that employ the scientific method. The "one best way" means that rockets and jets the world over resemble each other because other designs would not fly. All hydroelectric stations, subways, bridges, and skyscrapers share essential materials, structural elements, and components, or they would not stand. The first-glance differences between technologies in different settings reflect, literally, superficial elements: the skin of glass and steel or aluminum and plexiglas of a skyscraper, for example. You could go so far as to say that functional efficiency determines design. Yet technologies are more than components assembled in the "one best way" to create a large system. Economic and social obstacles as much as technical ones must be overcome to ensure successful diffusion of technology. Capital, political and human organizations are vital to technology.[3]

Engineers trained in a given milieu tend to accept the broader cultural values of their system. What are rational means for achieving desired ends in one society may be abhorrent in another. For example, the mass production of consumer goods through the "American system" of interchangeable parts and Fordism (standardized production; a controlled and steady flow of energy and materials in production processes from acquisition to the assembly line; and mass production to lower unit costs) will be crass materialism to conservative German engineers. The factory assembly line symbolized the exploitation of the proletariat to Soviet engineers. On the other hand, Soviet leaders idealized Taylorism (a doctrine of scientific management in industry) and established an officially sponsored materialism. And when the ends are full employment, social welfare, inexpensive housing, or universal health care (and not *simply* the design of a jet engine!), disagreements over the means and ends pour forth.

Take the example of public housing. Recognizing that their expertise could be used to achieve social goals, and responding to housing needs, engineers in Nazi Germany and the Stalinist USSR sought a prominent role in factory organization, housing, and urban planning. In the USSR, a Marxist urban industrial ideology held sway. Urban centers swelled in the 1920s and 1930s as peasants streamed into cities. Stalin encouraged this behavior by cutting capital investment in the countryside to focus on the creation of heavy industry and by forcing the peasantry into hated collected farms. Planners' preferences held sway in this centrally planned economy. Housing had to be built rapidly. Why not use inexpensive, standardized designs based on prefabricated forms that could be assembled rapidly by unskilled and illiterate workers into dwellings?

Soviet housing was proletarian in its minimal space, threadbare appointments, and shared bathrooms. This housing frequently incorporated the "collectivist" ethos in communal kitchens, child care facilities, and rooms for workers' clubs, but these were introduced more often to cut costs than to uphold a proletarian social ideal. The apartments and clubrooms, like the factory itself, also had a political function as the appropriate setting for the Communist Party to employ various media (radio, film, mass publications, meetings) to educate the masses about Stalin's programs. In Nazi Germany, these communal means and ends were rejected out of hand as anathema to the *völkisch* peasant and reviled as "Bolshevist." More appropriate for the German were thatched-roof cottages that showed his organic and blood ties to the soil.

THE TOTALITARIAN MACHINE

Does the machine, the symbol of the engineer, have the same effect on societies everywhere? Do all of the world's engineers employ machines for the same ends? Is the universal goal of the machine such efficiencies as increased output per unit of input, economies of scale through mass production, and speed of output? More to the point, can the engineer make rational, optimalistic choices in a totalitarian system? The answer is often yes, but the path to that answer has been arduous in every system.

Yet the ideological underpinnings of National Socialism and Soviet Marxism differed significantly. Nazism was an antiurban, racially based ideology. According to its myth, Aryan "settlers" whose blood rooted them organically in the German soil created a technology that served *völkisch* needs, not the profit motive of international capitalists. At the same time, German technological achievements—for example, in the chemical and automotive industries—were pioneering efforts that displayed complex elegance. So there was a disjunction between the advanced state of German technology and the Nazi myth, described by historian Jeffrey Herf as "reactionary modernism."[4]

Soviet technologies were intended to reflect the collectivist ethos of serving the basic housing, transport, and food demands of the masses. Simultaneously, they served state goals of economic self-sufficiency and military might. The construction sites were also forums for educating the unskilled workers, not only about technical details but also in the messages of Stalinism. The result was bland, functional designs in which workers' safety and environmental concerns played a secondary role.

In spite of these differences, several features distinguish technologies in totalitarian regimes from those in pluralist regimes. The most obvious is the fact that the state serves as prime mover behind development and diffusion. Whether in Soviet research institutes or Nazi ordnance laboratories, this engenders "big science" approaches to research and development. The absence of market forces and the exclusion of the public from decisions about how or whether to diffuse a technology permits the development of technologies that persist no matter their questionable efficacy or environmental soundness. These characteristics also apply in the nonmarket sectors of democracies, notably defense industries, which are infamous for projects that waste billions of dollars.

A second feature of totalitarian technology is overly centralized administration of research and development. This is not surprising in countries of state socialism, such as the former USSR, where the state owns most of the means of production. But in fascist regimes, too, the persistence of private property is tempered by centrally funded projects that rely on the state for their impetus. Major industrialists prosper in close cooperation with the state, while smaller businesses are subjugated to the "national good." This leads to irrational use of resources, as the cases of the Soviet Magnitogorsk steel combine and Albert Speer's monumental plans to rebuild Berlin will demonstrate. To be sure, decision making about which projects to fund involves give-and-take among engineers, economic planners, and party officials. Naturally, officials stress the interests of the state. Hence engineers in totalitarian regimes tend to be more accountable to the state; those in pluralist regimes find greater autonomy in setting the research agenda. Their professional societies sell expertise, receiving the exclusive right to practice their professions through licenses granted by the state: for example, as medical doctors. In totalitarian regimes, societies, clubs, and associations for architects, scientists, lawyers, and doctors are subjugated to single-party organizations ruled from above.

Third, technologies in totalitarian regimes are characterized by gigantomania: for example, Speer's plans for wide-gauge (four-meter) railroad tracks with two-story-high cars or Stalin's seven "wedding-cake" Moscow skyscrapers and the world speed and distance records that were set in aviation.[5] This scale concerns both physical parameters and the display value of the technology. Gigantomania often results in waste of labor and capital resources, especially in centrally planned economies where the state is the prime mover behind every project. In totalitarian regimes projects seem to take on a life of their own, so important are they for cultural and political ends as opposed to the ends of engineering rationality.

THE TECHNOLOGICAL STYLE OF THE SOVIET UNION

The Soviet Union embraced large-scale technologies with an energy that belied its economic backwardness.[6] Its leaders saw technology as a means to convert a peasant society into a well-oiled machine of workers dedicated to the construction of Communism. They believed large-scale technologies would marshall scarce resources efficiently and provide the appropriate forum for the political and cultural education of a burgeoning working class. Soviet leaders had the utmost confidence in the ability of technology to transform nature and bring freedom to Soviet citizens. Constructivist visions of the Communist future found expression in Lenin's electrification, Stalin's canals and hydropower stations,

Khruschev's atomic energy, and Brezhnev's Siberian river diversion project. There were glorious chapters in the history of large-scale technology in the USSR, including the pioneering conquests of the atom and the cosmos.

In all of these projects, Soviet engineers and Party leaders took an extremely utilitarian view of the importance of science and technology to secure dominion over nature. This view was central to the works of thinkers as far back as Francis Bacon in the seventeenth century.[7] Marxists further believed that natural laws not only existed in nature but could be discovered in human institutions and applied for the betterment of humanity.

The Soviet Union sought modern technology in the West through technology transfer, including "turnkey" factories (supplied ready to work) and other cooperative arrangements, and through espionage. Its leaders were particularly enamored of borrowing American technology, which they considered the most progressive. Most Soviet engineers believed that once technology had been lifted physically and psychologically out of its capitalist environment, it would cease to serve its capitalist masters and work for the good of the Soviet state and proletariat. It is ironic that Ford, General Electric, and other giant American corporations were a model for Soviet planning.

Both engineers and the Communists favored a strong central government. They shared technological goals. When they seized power the Soviets had no blueprint for industrialization or expertise in organizing it, so the technical specialists provided that. Historian Kendall Bailes argues that "the Communist Party supplied the link, largely missing before the revolution, between the masses of the population and the plans and projects of the techno-structure."[8]

The 1920s are often seen as the golden age of Soviet society, uplifted by utopian and constructivist visions for the advent of Communism. For Communist leaders, the view of progress was inextricably tied to technological development, which would be achieved in short order, they believed, through economies of scale, centralized economic planning, mass production, and universal mechanization. This in turn would lead to more rational and equitable distribution of goods and services. Of course, the Soviets encountered great challenges in rebuilding their economy after the ravages of war, revolution, and civil war. It was a long time before modern technology penetrated industry.

Although it entered the economy slowly, technology rapidly became a central aspect of Soviet daily life. There were festivals of machines, symphonies of factory steam whistles. Newly married peasants were conveyed in celebration on a tractor.[9] (The Nazis, too, embraced these public spectacles as a way to gather thousands of believers together to gape at *Wehrmacht* weapons.) In Soviet literature, technology was displayed with utopian fervor. Posters with technological themes supplanted the Russian Orthodox icon. The machine was central to Soviet commercial art. Technology even had an impact on language, as when proud parents named their boys "Tractor" and girls "Electrification" or "Domna" (forge).

The style of Soviet technologies was characterized by an aesthetics based on two concerns. The first was an exaggerated level of interest in mass production, owing both to egalitarian ideological precepts and scarcities of finished goods. The latter contributed to a premature fixing of parameters for many technologies. The second was the gigantomania that grew out of a fascination and commitment to a technology of display. On the surface, some of these characteristics are reminiscent of the Western Bauhaus movement, Fordism, and Taylorism, with their aesthetics based on standardization, rationalization, and mass production of components.[10] But the Soviet characteristics conspired with political forces to create a style noteworthy for bland, functional designs in which safety and comfort played a secondary role, environmental issues were rarely raised, and large-scale systems acquired substantial technological momentum. Moreover, there was a progression of the objects of the transforming visions of large-scale technologies from people to nature itself. First, peasants and enemies of the people were to be transformed into workers and citizens. The metamorphosis of capricious nature into something rationally ordered by technology followed.

STALINISM, HERO PROJECTS, AND GIGANTOMANIA

Administration, organization, and technological hubris were joined in the Stalinist utopia. Stalin's views of the place of technology in society were far more economically determinist and far less subtle than Lenin's. This means that the development of the productive forces was the sine qua non of Stalinism. Centralized political control, organization, and economic management were basic to this endeavor, the means for identifying labor and capital reserves and overcoming any human, natural, or technical obstacles. Such obstacles were labeled as evidence of "wrecking." No scale or tempo seemed impossible; everything qualified as superlative, as the "best" or the "biggest" in the world. The most modern technology would secure a safe haven for "socialism in one country."

The Soviets had come to power without a coherent urban policy, but by the end of 1918, after they nationalized land and then abolished private housing in cities, the state became the single client for large-scale construction projects. This gave planners unprecedented control over the urban environment. Soviet architects of the 1920s debated nearly every urban and architectural issue. Would the Marxist utopia consist of nodal points? Would cities be built in a linear fashion adjacent to power and transportation corridors, a style called "automobile socialism" by its detractors?[11]

The first technology to serve as the flagship of the Stalinist system was the Moscow subway. This was part of the plan to transform the urban landscape in a socialist fashion. Moscow lagged far behind the cities of Europe and the United States in terms of transportation, hot and cold running water, heating, electric supply, and sewers. Socialist reconstruction, as it was called, would result in well-illuminated streets, parks, and transport systems that returned the worker home rested, not exhausted as in the West, and with the appropriate political education.[12] In contrast to the dirty, damp, and dark subways in capitalist systems, the socialist metro would invigorate the workers' spirits with its modern ventilation. The aesthetics of the subway stations attracted attention, too. The architecture, sculpture, and paintings of the ornate marble and granite underground palaces reflected dominant themes of Soviet culture: industrialization, collectivization, literacy, and, later, the victory over the fascists: "Each of these palaces burns with one flame—the flame of fast approaching victorious socialism."[13]

The Moscow metro became the exemplar of future Soviet large-scale technologies. The Party forced the pace of construction against all technological challenges with centralized control, borrowing some techniques from the West but always striving to demonstrate that the Soviet way was better. The method involved serial production at factories of large components, which were transported by rail to site for assembly. Valuable resources were extracted from the rest of the empire, in this case marble, granite, and labor for the "grandiose constructions."[14] An intimate of Stalin's and one of the few members of the Politburo to outlive him, Lazar Kaganovich, who was in charge of Moscow's socialist reconstruction, proclaimed that the subway "far exceeds the bounds of the usual impressions of technological construction. Our metro is a symbol of the new society which is being built." The metro was also a symbol of the victory of Bolshevik organizers over nature.[15]

According to a radical plan adopted in 1935, the center of Moscow would be razed "in an effort to 'rationalize' the new socialist capital." There would be new parks and residential areas, but more attention was paid to the "socialist heart" of Moscow. Huge government buildings would be erected to give workers a feeling of the overpowering authority of the regime. Radial highway arteries would converge at the center of the city, with avenues widened. The 1935 plan was not realized because of disorganization, material shortages, and an undercapitalized construction industry, in part because economic ministries organized around sectors of industry had priority for resources over municipalities. Further, building plans transcended the technological capacities of builders. Only a few radial roads were ever built. The Palace of Soviets had to be abandoned as foundations kept sinking into the mud of the Moscow River floodplain. The Nazi invasion and then Stalin's death saved Moscow's Red Square from "tragic disfigurement." But Stalinist planners succeeded in razing one thousand buildings dating from the fourteenth through the nineteenth centuries.[16]

TECHNOLOGICAL MOMENTUM IN TOTALITARIAN SYSTEMS

Centralization, bureaucratization, an economically determinist philosophy of technology, and Stalinism combined to give great technological momentum[17] to the large-scale systems that were paradigmatic for the Soviet Union. Technological momentum refers to the tendency of large-scale projects to acquire significant social, political, and economic support and of the organizations involved in their construction to become intolerant of obstacles to their diffusion: for example, public opposition. In the absence of market forces, once Soviet construction organizations fulfilled their initial purposes they seemed to take on a life of their own. In a market system, workers and firms might temporarily be displaced but would move to new areas as construction demanded. In the USSR, in order to avoid unemployment and investment in transportation infrastructure, housing, or equipment, Soviet planners sought to provide funding for projects that made use of workers already employed and organization and equipment already in place. Such an approach was needed in the Soviet system to distribute workers because there was no other efficient means. The state and its organizations provided housing, schools, day care, and stores. Indeed, workers preferred working for large organizations with these resources, even though this discouraged their mobility. When projects were finished, institutions and people were transformed into solutions looking for problems.

Technological momentum contributed to premature standardization. The construction of large-scale technologies—subways, hydropower stations, and apartment complexes—required huge capital outlays. Early standardization was a simple way to reduce capital costs. It was no easy matter to establish the specifications for all of the subsystems involved: steel piping, conduit, wiring, prefabricated concrete forms, motors, turbogenerators. Initially, questions concerning construction of large-scale systems required ad hoc decisions by the engineers and managers. Fear of "wrecking" charges made them risk-averse. The engineers naturally gravitated toward accepted practices and norms, rather than innovation, and came to believe one model was enough for the entire empire. Scientific organizations subordinate to a specific branch of industry also focused efforts on increasing short-term production, turning to standardization of component parts. Proletarian aesthetics also contributed to the utilization of fewer redundancies in construction, for example, thicker-gauge pipe or containment vessels for nuclear reactors. The Chernobyl reactor explosion (1986) and the Usinsk oil pipeline disaster (1994–95) serve as reminders of the Soviet technological legacy.[18]

SOVIET TECHNOLOGY: PROLETARIAN AESTHETICS

Proletarian aesthetics grew out of the effort to find economies everywhere in huge engineering projects. In construction, it led to the adoption of simple, prefabricated concrete forms for apartments, offices, and highways. Soviet factories had a universal style, which employed corrugated steel roofs and standard piping, conduit, generators, and machine tools. Engineers in Gosstroi SSSR, the central state construction commission located in Moscow, established codes and specifications for all building materials for the entire empire, irrespective of local geological, meteorological, and other considerations. Industry received appropriations for operations but little for repair and monitoring, making safe operation of these systems a nearly impossible task. One of the most significant manifestations of proletarian aesthetics was shared rights-of-way, where pipelines, highways, railways, and electrical power lines occupied the same thoroughfare. Shared rights-of-way contributed to such disasters as the gigantic Ufa pipeline explosion, which obliterated more than a square kilometer near Tobolsk, Bashkiria, in 1989 and killed more than six hundred people.

Similar kinds of standard techniques were applied universally in the construction industry.[19] Apartment buildings, subways, even the street names (Lenin, October, Revolutionary, Red Banner) were so similar throughout the USSR that you could fall asleep in one city, wake up in another, and not know the difference. As big concrete slabs replaced bricks, problems of aesthetics began to rival those of quality. The large panels produced not apartment buildings but gray houses of cards stacked on top of each other at right angles. It was not surprising that such buildings as these collapsed instantly during earthquakes in Armenia in 1988. Planners and engineers had hoped that the application of these mass production techniques and materials might enable them efficiently to overcome the poor materials and workmanship endemic to the Soviet experience. They overestimated the ability of mass production construction techniques using prefabricated concrete forms and mass-produced slabs to overcome poor workmanship.

In a word, gigantomania, display value (industrial symbolism in competition for prestige with the West), and state control and centralization of R and D characterized technologies in the Soviet Union. So did fascination with economies of scale and mass production. Did Nazi Germany share any of these characteristics?

WEIMAR CULTURE AND TECHNOLOGY

National Socialist ideology was ambivalent about modern technology. On the one hand, technology was central to efforts to rearm Germany and secure the new empire's glorious future. The superweapons its army leaders sought during the war required the input of technological experts. Its Four-Year Plan, adopted on the eve of World War II to prepare the economy for the *Blitzkreig*, was an agglomeration of macroeconomic techniques and state-supported projects geared to produce a great industrial power. Many engineers welcomed the strong central government of the National Socialists for its ability to support modern technology more efficiently than the Weimar regime, which had been plagued by an inexperienced liberal parliamentary government and the chaos of the free market. In 1914, imperial Germany had been the leading scientific and industrial nation in the world. Its engineers saw the new chemical weapons and airplanes of World

War I as signifying the glories that might be achieved by wedding technical knowledge to a strong state power.[20]

On the other hand, German engineers on the whole were conservative individuals who rejected the rationality of Enlightenment social progress. They believed in the ability to understand physical processes empirically, but not in the extension of empirical methods to human problems. Nazi ideologists, for their part, detested the modern symbols of Weimar technology. They perceived in its spare, utilitarian architecture "Bolshevist" designs, which abandoned the natural antiurbanist aesthetic that should have characterized *völkisch* technology. Nazi Germany needed modern technology to achieve its imperial ends but rejected this technology on anachronistic ideological grounds and found great fault with Soviet technological style. Yet the Nazi technological style—characterized by a preference for centralized project management with attendant social and political control and by gigantomania—was paradigmatic for totalitarian regimes.

What was it about Weimar that raised the specter of Bolshevist technology? To many German engineers such vibrant cultural phenomena as the Bauhaus, an artistic and technological movement born in the Weimar Republic, symbolized everything that was non-German about modern technology. Like other modernist architects such as Le Corbusier and Frank Lloyd Wright, Bauhaus architects transformed the house or other building into a tool. They sought to integrate craft, art, and industry in one modern aesthetic. They appealed to such ideals as democracy, optimality, and efficiency in their designs. For Le Corbusier, the park and the skyscraper were united to exalt rational power. Wright, in his utopian vision Broadacre City, saw the telephone and automobile as contributing to the disappearance of the city because these technologies were inherently democratizing in their decentralizing force and their ability to diffuse population, wealth, and power.[21] The democratic ends of such architecture disturbed the conservative Germans.

Bauhaus artists such as Walter Gropius and Ludwig Mies van der Rohe seized upon the aesthetic of the machine, its embodiment of speed, efficiency, and clean lines, in every thing they produced: office buildings and apartments, chairs and other furnishings, even utensils and vases. They took inspiration from factory design. They sought to bring together artists, craftsmen, sculptors, and architects in the common endeavor of tying crafts to industry and mass producing craftwork. Architects secured funding from the Weimar and municipal governments for many of their projects for mass housing and concentrated on the technical aspects of construction technology and planning. Some of them claimed "that construction methods determined style itself," with standardization of building parts—prefabricated concrete forms and modular construction—essential for a uniform aesthetic. They believed that mass housing was not different from mass transportation or any other problem of urban planning.[22] Bauhaus supporters believed that technology could be employed to achieve a diversity of modernist social ends—inexpensive housing, rapid mass transit, etc., and their implicitly democratic ends—through standardized means. No matter what Bauhaus architects believed, socialist, modernist, or totalitarian ends can be achieved through the application of standardized construction techniques.

Most German engineers, however, believed that this style of technology could not be reconciled with German culture in a nationalist ideology. In spite of the fact that the Bauhaus style was recognized internationally as an achievement of German culture, for conservative elements it was "un-German," based on a uniform industrial aesthetic, a proletarian social policy, and a "Bolshevist" political program of helping the masses. They attacked its prefabricated housing and standardized building methods. Conservative architects appealed to national building traditions. They considered that flat roofs, for example, provided inadequate draining of rain and melting snow and were inappropriate to the German climate; according to one, the flat roof was an "oriental form" and equated with flat heads. Other architects declared that standardized building techniques produced "'nomadic architecture,' leading to 'uprootedness, spiritual impoverishment and proletarianization.'" Still others adopted an antiurban theme, attacking skyscrapers and calling for a return to the German soil. Eventually, this kind of criticism of the Bauhaus was incorporated into racial arguments, where the origin of the Bauhaus style was attributed to cultural decadence that had its roots in biological causes. These kinds of explanations, of course, won support in Nazi circles.[23]

NAZI TECHNOLOGY

Were National Socialism a consistent ideology, we would expect efforts to create an agrarian society in which the *Volk* could best prosper. But the Nazi rise to power did not give way to rural nostalgia or to an antimodernist technological ethos that supported the peasant's organic tie to the soil. Rather, Nazism combined anti-Semitism and the embrace of modern technology in a myth, according to which technological advance grew out of a racial battle between Aryan and Jew, blood and gold. The engineer would assist the regime in destroying an unhealthy urban atmosphere, liberating the nation from the "fetters of Jewish materialism. The Nordic race was ideally suited to use technology; the Jew misused it." Like Soviet, Nazi technology embodied service to the nation, not pursuit of profit. Service to the nation meant joining with the state to achieve economic independence. State trade, tariff, tax, price, and wage policies would help to underwrite technological development to achieve autarky and enable the nation to engage in war when cut off from the import of raw materials.[24]

For Hitler himself there was no *völkisch* rejection of technology. If in life and politics the strongest won, so among nations the technologically weak would be defeated. Hitler advocated rearmament and, like Stalin, national autarky. He used new media such as radio and film to praise *völkisch* technology for propaganda ends and sponsored modern highways—the autobahn—and other modern artifacts for economic and military ends.[25] And when the war effort bogged down, Hitler hoped for a technological savior in the form of a new superweapon like the V-2 rocket.

Hitler's writings and speeches criticized Weimar culture for its weaknesses, its decadence, its materialism, its "lack of an heroic ideal," its "Bolshevist" art. He singled out its architecture as the epitome of these problems. Hitler supported gigantomania in Nazi architecture. The Nazi leaders built massive monuments to their rule whose neoclassical style and scale were neither *völkisch* nor humanistic. Hitler believed that a "great" architecture was needed in the Third Reich, since architecture was a vital index of national power and strength. There had to be monuments in cities, not symbols of cultural decay.

When the Nazis began vigorously to oppose the Bauhaus in the 1930s, the foremost party philosopher, Alfred Rosenberg, led the charge. Rosenberg had joined the Nazi Party in 1920 and became editor of *Völkischer Beobachter* in 1921, through which he attacked the Bauhaus. His career had ups and downs, but in 1933 he was put in charge of ideological training of Nazi Party members. From this position, Rosenberg hoped to see the Nazi Party university create a place for natural science, especially to study the biological laws of races to reveal the poisonous influence of the Jews. In 1941, Rosenberg became the Reich's minister for the eastern occupied territories, a position from which he could see Nazi racist policies of *Lebensraum* and the "final solution" put into action. He was hanged for war crimes in 1946.[26]

In 1929, Rosenberg founded the Kampfbund to spread a Nazi gospel of virulent Christian anti-Semitism and racial doctrines. Many of these doctrines were based on the writings of Count Gobineau, who had argued that the rise and fall of civilizations is connected to their racial composition. Those of Aryan stock flourish, while those diluted through miscegenation decline. Rosenberg embraced conspiracy theories and feared the "international Jew," freemasonry, and Jewish control of banking and the media. The Kampfbund was central in spreading Rosenberg's message of the *völkisch* aesthetic. Initially, the Kampfbund had the strong political backing of the Nazi Party. In the same way that Bolshevik organizations had subjugated engineers' professional associations, so the Kampfbund inexorably absorbed smaller rivals. The Kampfbund set forth the party line on cultural values. It attacked the chaos, Russian "Bolshevism," and American "mechanism" allegedly rampant in modern art, the Jewish roots of these problems, and the "Nigger-Culture" that thrived in the Weimar clubs whose excitement and decadence was so well captured in the Broadway musical, and later film, *Cabaret*.[27] Rosenberg despised modern art. He saw in Picasso "Mongrelism," whose "bastardized progeny, nurtured by spiritual syphilis and artistic infantalism was able to represent expressions of the soul"; and he hated the work of such artists as Marc Chagall and Wassily Kandinsky, who was connected to the Bauhaus.[28]

Rosenberg argued that Bauhaus architecture was a symbol of weakened culture, of a mass society whose members had lost their historic identity through urbanization and their economic security through proletarianization and unemployment. The Bauhaus was a "cathedral of Marxism" that resembled a synagogue or a "Bolshevist" building for the nomads of the metropolis.[29] Rosenberg constantly referred to the interconnection of race, art, learning, and moral values in his attacks on the cultural decadence of Weimar, attacks that, to the German people who were suffering through the depression, were appealing.[30] On April 11, 1933, the Berlin police shut down the Bauhaus school by order of the new Nazi government.

Next the Nazis orchestrated the *Gleichschaltung* of municipal building administrations and building societies by purging them of adherents to the modernist Bauhaus movement. The societies were then joined in a central organization under government control, just as all professional societies were subjugated to Communist Party organs under Stalin. In spite of their criticism of "Bolshevist" urban development, the Nazis supported programs for large-scale, low-cost public housing in their appeal for working-class support. This support of public housing resulted in part from Rosenberg's loss of influence to Joseph Göbbels. The Kampfbund was placed under Göbbels's authority, and he established the Reichskulturkammer as a branch of his Propaganda Ministry.[31] The Kulturkammer had sections for film, literature, theater, music, the media, and visual arts, with national and regional offices. Göbbels himself hesitated to purge the new style entirely from the Third Reich. Barbara Lane writes: "If the establishment of the Reichskulturkammer cut short the purges of 1933 and prevented the original leaders of the Kampfbund from gaining control of architectural style, Göbbels's organization never explicitly repudiated the Kampfbund's attacks on the new architecture; and these attacks had a profound effect upon the careers of the radical architects."[32] Many modernist architects were deprived of their livelihood and had to emigrate, like physicists, biologists, and doctors. While depriving Bauhaus architects of influence and dictating issues of style, this did not prevent them from getting new commissions, and many Bauhaus assistants and students received positions in the Nazi government.[33]

GIGANTOMANIA IN NAZI GERMANY

Nazi architectural style, like Soviet, was gigantomanic. Hitler desired immense monuments to his rule and the glory of the Third Reich for millennia to come, buildings of a scale never before seen. The party and its strong central state were the driving force of Nazi architecture. Nazi buildings were intended to express the will of the Nordic people, awaken national consciousness, and contribute to the political and moral unification of the *Volk*.[34] In October 1935, when the major structural frame of the new Luftwaffe building was finished, Hermann Göring addressed gathered workers and functionaries to praise the structure as "a symbol of the new Reich," a building that "shakes our deepest emotions," shows German "will and strength," and would "stand forever like the union of the Volk."[35]

Nazi architecture was not one, historicist style, nor an out and out rejection of the new style, but a variety of styles that reflected the diversity of views of the leadership: public works such as highways and bridges, government buildings, and some apartment buildings; neo-Romanesque; rustic housing projects intended to tie the urban workers to the soil; modern neoclassicism based on the Doric aesthetics of Albert Speer; and even modern.[36]

Albert Speer was the chief architect behind many of the gigantic projects. Hitler desired Speer to make a huge field for military exercises and party rallies, with a large stadium and a hall for Hitler's addresses. While never completed, the planned Nuremburg tract embraced an area of 16.5 square kilometers (roughly 6.5 square miles). All of the structures would have been two to three times the size of the grandiose Greek and Egyptian constructions of antiquity. For example, Speer designed the Nuremburg stadium based on the ancient stadium of Athens, but far larger: six hundred yards by five hundred yards. Speer selected pink granite for the exterior, white for the stands. To the north of the stadium a processional avenue crossed a huge expanse of water in which the buildings would be reflected. When Hitler first saw the designs he was so excited, his adjutant reported, that he "didn't close an eye last night."[37]

In 1939, in a speech to construction workers, Hitler explained his grandiose style: "Why always the biggest? I do this to restore to each individual German his self-respect. In a hundred areas I want to say to the individual: we are not inferior; on the contrary we are the complete equals of every other nation." Hitler's "love of vast proportions," Speer commented, was connected not only with totalitarianism but with a show of wealth and strength and a desire for "stone witnesses to history."[38] Yet these structures could scarcely have instilled in the individual any personal feeling other than insignificance. For anyone but Hitler himself, any sense of glory could come only as an anonymous contributer to the all-powerful state.

Hitler wanted a new chancellery to celebrate his rise in rank to "one of the greatest men in history," with great halls and salons to make an impression on visiting dignitaries. He insisted that it be built within a year. Speer was required to raze an entire neighborhood of Berlin. Forty-five hundred workers labored in two shifts, with several thousand more scattered throughout the Reich producing building materials and furnishings. To meet Hitler's designs, Speer created a great gate, outside staircase, reception rooms, mosaic-clad halls, rooms with domed ceilings, and a gallery twice as long as the Hall of Mirrors at Versailles. The chancellery included an underground air-raid shelter. When it was finished, Hitler "especially liked the long tramp that state guests and diplomats would now have to take before they reached the reception hall."[39]

The future headquarters of the Reich would have been the largest structure of all, with a volume fifty times greater than the proposed Reichstag building. It could have held 185,000 persons standing and was "essentially a place of worship." Its dome opened to admit light. At 152 feet in diameter, it was bigger than the entire dome of the Pantheon (142 feet). A three-tiered gallery was 462 feet in diameter and 100 feet tall. In order to ensure that the structure lasted into the next millennium, engineers calculated, its steel skeleton, from which solid rock walls were suspended, would have to be placed on a foundation of 3.9 million cubic yards of concrete, dozens of feet thick; the engineers did tests to determine how far the monstrous cube would sink into the sandy building site. Hitler was partly motivated by Stalin's projects. "Now this will be the end of their building for good and all," Hitler boasted.[40]

Hitler desired to rebuild Berlin as the capital of "Germania," a new empire that would span the entire Eurasian continent and far outdistance Rome, London, and Paris in grandiosity and history. Hitler had studied the Ringstrasse in Vienna with its prominent public buildings.[41] Speer had to order the heart of the city razed to accommodate the two new axes through the center lined with tall office buildings. Four airports were situated at the terminal points of the axes. A ring Autobahn encircled the new Berlin, incorporating enough space to double the city's population. A four-story copper and glass railway station with steel ribbing and great blocks of stone, elevators, and escalators would surpass Grand Central Station in size. The plans themselves experienced gigantomania, eventually including seventeen radial thoroughfares, each two hundred feet wide, and five rings; the land beyond the last ring would be for recreation, a woodland of artificially planted deciduous trees instead of indigenous pine. The projects required immense effort; SS head Heinrich Himmler offered to supply prisoners to increase production of brick and granite, which were in short supply. Himmler's SS concentration camp operations showed tremendous ignorance of construction techniques and often produced blocks of granite with cracks. They could supply only a small amount of the stone needed; highway construction used the wasted material as cobblestones.[42] Only the demands of war prevented the Nazis from carrying out these radical transformation plans.

Hitler's favorite toy, it seemed, was the model city, a 1:50 scale model that was set up in the former exhibition rooms of the Berlin Academy of Arts. When Speer's father saw the mock-ups he commented, "You've all gone completely crazy." Only later, when in prison, did Speer realize the inhumanity of his designs, the "lifeless and regimented" nature of the avenues, the "complete lack of proportion" of the plans.[43] But we should not think of Speer's designs as unique in the Western world at the time. There was a resurgence of interest in massive neoclassical forms in other Western countries: for example, Rockefeller Center in New York City, the forty-four-story gothic Cathedral of Learning in Pittsburgh, and Stalinist architecture generally.

The heroic Nazi projects pushed to the limits of technology. Very few large projects were carried out, in part because of their astronomical costs and the costs of war. Smaller, more feasible projects became showpieces of Nazi propaganda, with Hitler a prominent figure at groundbreakings. There was constant coverage of the projects, some of which took years, and this propaganda all but obscured the failings of the building program: for example, projects for the masses such as public housing never met demand. Nazi public housing retained the Weimar (and universal) practice of constructing row houses and apartment buildings on the periphery of urban population centers. Only a few projects conformed to the Nazi ideal of tiny houses with sloping roofs, sited on enormous plots of land. The surfacing of these attempted to evoke the countryside: thatched roofs, half-timbering, or vertical wood siding.[44]

There was a contradiction between the designs of the Reich's commissar for public housing to fill the world with lovely peasant houses in the postwar period and the plans of Speer as general building inspector for Berlin to undertake its rebuilding as a "world capital city" of "insane monumentality" whose buildings would be an "imperishable confirmation" of the power of the Third Reich, yet had a "nonstyle of pseudo-antique form, ponderous excess and solemn emptiness." But Hitler recognized this. Referring to his government's new palace he said, "Amid a holy grove of ancient oaks, people will gaze at this first giant among the buildings of the Third Reich in awesome wonder."[45] But this contradiction displays a central contradiction of totalitarianism itself: the superefficient omnipotent state constructed of people kept in isolation and ignorance.

"BIG TECHNOLOGY" IN NAZI GERMANY

The V-2 rocket, the first large guided rocket, was the greatest technological achievement of the Third Reich. Yet it was a poor weapon, unable to carry a large explosive payload, and diverted manpower and other resources from more sensible armaments projects. The V-2 demonstrated the importance of the leader principle in scientific success. Hitler's support set the project off; when he lost faith in the V-2, the program lost priority for material and manpower. Still, the V-2 program foreshadowed the Manhattan Project and "big science" of the postwar years as a paradigm of state mobilization to force the invention of new military technologies. The V-2 program grew out of a military bureaucracy, which rarely considered human or economic costs. This, together with the needs of secrecy and the inadequate technical basis of industry, necessitated the creation of a large government-funded central laboratory.[46]

During the Weimar years, a popular fad for rocketry and space flights produced both stunts and serious liquid fuel experiments that generated national pride for the outcast nation.[47] With the Depression the fad ended, but experimentalists continued their trial-and-error efforts. A group that included Wernher von Braun, who would direct technical aspects of the U.S. manned space program, was established in Berlin in 1930. (Von Braun energetically promoted the U.S. program in the 1950s by touting the possibilities of the moon's colonization by millions of people; his colonization claims resembled Nazi *Lebensraum* ideology.) The Berlin group sought corporate financial backing by stressing commercial applications such as intercontinental transport. The turning point in the V-2 program was the interest of the Army Ordnance Office in the use of rockets to deliver chemical weapons. The office built a large secret facility to maintain the assumed lead in rocket development that Germany had over other nations. It used the Gestapo to impose secrecy and to drive other rocket efforts out of business. Secrecy required outside subcontracting to be abandoned and necessitated the fabrication of one-of-a-kind hardware. The program gained momentum when the Luftwaffe (air force) joined in, securing the political support and resources of Hermann Göring. The Luftwaffe underwrote a new weapons facility, Peenemünde.[48]

At Peenemünde, von Braun and other scientists sought to build an in-house production line of rocket components for the finished weapons. Peenemünde had the advantage over universities in research and development in terms of concentrated scientific interest, commitment to Nazi or national ideology, stable funding, and draft exemptions for key personnel. Von Braun cultivated contacts with university scientists and engineers for manpower. He successfully created an open academic atmosphere in an environment of secrecy. In this environment "big science" was fostered. When the Nazis invaded Poland, appropriations for research rapidly expanded. Leading military figures associated with the V-2 had laid the groundwork for these increases by promising the deployment of missile weapons in short order and by fostering contacts with such high officials as Albert Speer. But at times Hitler withdrew his support, including cutting deliveries of steel in favor of other priority projects. Even relations with the Luftwaffe deteriorated as the war progressed and Germany's prospects worsened.[49]

Speer secured Himmler's intervention in 1943 to push production, utilizing concentration camp labor at newly built underground facilities. In a last-ditch effort to hold off defeat, the Peenemünde facility was pushed to produce as many missiles as possible to use on England. No thought was given to the human cost, in the slaves who toiled without expression to the death in damp, disease-ridden conditions, living, working, and sleeping in the Peenemünde caves.[50]

Hitler's arbitary and autocractic behavior had a negative effect on Nazi "big science." Speer reports that as the military situation deteriorated Hitler made a series of technological blunders: for example, ordering a fighter jet capable of shooting down American bombers to be built instead as a fast but tiny bomber, incapable of holding many bombs. Hitler then insisted that the V-2 be mass produced at a level of nine hundred per month for use as an offensive weapon. Five thousand rockets—five months' production—would deliver perhaps an effective 3,750 tons of explosives; a single combined U.S. and British bomber attack delivered 8,000 tons. But Hitler was determined that some future new weapons would decide the war. So fascinated were the Nazis with a technical fix to their military quandary that they allowed the untried, young von Braun great leeway to pursue the expensive V-2 with only long-term prospects. Yet Speer was also attracted to the romantic possibilities of a superweapon. The Nazis, he later admitted, suffered from an "excess of projects in development," not one of which could ever meet full wartime production, and many of which were rushed "from factory directly into battle" without "customary full testing time."[51] The result, of course, was loss of young life. What is characteristically Nazi about the V-2 technology is not this loss of life but the massive build-up of state support behind the project.[52]

TECHNOLOGY IN AUTHORITARIAN REGIMES

The Soviet Union had great respect for American technology, even as it despised American capitalism. It was utopian in its embrace of technology, assuming that any technology would function smoothly given socialist economic relations. Nazi Germany rejected Western technology in ideological pronouncements, even as it relied on technology for its military rebirth. It had an irrational, dichotomous attitude toward the role technology might play in the national future. Stalin appears to have been more in tune with the limits and prospects of technology than Hitler.

While these are important differences, several common features of authoritarian technology stand out. In authoritarian regimes, technologies are intended to organize workers into malleable individuals devoted to national goals that divert attention from their economic self-interest: for example, from higher wages. The technologies themselves are large-scale, inefficient, and extremely costly in terms of human lives and the natural environment. Technology has great display value, as the efforts to rebuild Moscow and Berlin demonstrate. The scale of such reconstruction dwarfs people. In its radical design and massive thoroughfares, centralized political power is the message. The huge projects garner more than their share of resources, impoverishing other sectors of the economy. Only technological limitations and war prevented socialist Moscow and Nazi Berlin from being realized.

The careers of engineers in authoritarian regimes, too, suggest parallels. In Nazi Germany the engineers, like the physicists of Planck's ilk, opted for political accommodation rather than resistance, assuming that they could control Hitler or, at least, that Hitler would use "legal" means to achieve his ends. Their accommodation involved assistance in economic planning, industrial management, and armaments, all of which they viewed as service to the nation. In the USSR, unrelenting political pressure forced engineers into service to the state. But Soviet technologists also wanted to serve their nation and believed in the power of science to transform nature and society into a better world.

National Socialism lacked a consistent economic or social theory. Instead it put forward a series of disjointed policies, subsidies, retrenchments, plans, etc., an agglomeration of ideas in which race, the concept of blood and soil, *Lebensraum*, the leader principle, and so on were prominent in justification. Soviet Marxism developed its economic policies around the proletarian state, service to the nation, and developing industrial might. But both revered technology: for example, the autobahn and the metro, the V2 rocket and the atomic bomb. And there was little the public or scientists could do to divert the state from its activities as the prime mover behind large-scale technological systems.

NOTES

1. Michael L. Smith, "Selling the Atom: The U.S. Manned Space Program and the Triumph of Commodity Scientism," in *The Culture of Consumption: Critical Essays in American History, 1880–1980*, eds. Richard Fox and T. J. Jackson Lears (New York: Pantheon, 1983); Robert Frost, *Alternating Currents: Nationalized Power in France, 1946–1971* (Ithaca: Cornell University Press, 1991); Thomas P. Hughes, *American Genesis* (New York: Viking, 1989).

2. Jacques Ellul, *The Technological Society* (New York: Vantage, 1964).

3. Thomas P. Hughes, "The Evolution of Large Technological Systems," in *The Social Construction of Technological Systems,* eds. Wiebe E. Bijker, Thomas P. Hughes, and Trevor Pinch (Cambridge: MIT Press, 1989).

4. Jeffrey Herf, *Reactionary Modernism: Technology, Culture, and Politics in Weimar and the Third Reich* (Cambridge and New York: Cambridge University Press, 1984).

5. Kendall Bailes, *Technology and Society under Lenin and Stalin* (Princeton: Princeton University Press, 1978), pp. 381–406.

6. Many of the ideas in this section are developed in Paul Thompson, "'Projects of the Century' in Soviet History: Large-Scale Technologies from Lenin to Gorbachev," *Technology and Culture* 36, no. 3 (July 1995): 519–59.

7. Francis Bacon, *Essays and New Atlantis,* ed. Gordon S. Haight (Toronto, New York, and London: D. Van Nostrand, 1942).

8. Bailes, *Technology and Society under Lenin and Stalin*, pp. 415–17.

9. Rene Folop-Mueller, *The Mind and Face of Bolshevism* (Ann Arbor: University Microfilms, 1965); Richard Stites, *Revolutionary Dreams* (New York and Oxford: Oxford University Press, 1989).

10. Walter Gropius, *The New Architecture and the Bauhaus* (Cambridge: MIT Press, 1965).

11. El. Lissitzky, *Russia: An Architecture for World Revolution,* trans. E. Dluhosch (Cambridge: MIT Press, 1984); Anatole Kopp, *Town and Revolution: Soviet Architecture and City Planning, 1917–1935,* trans. Thomas Burton (New York: G. Braziller, 1970); Blair Ruble, "Failures of Centralized Metropolitanism: Inter-war Moscow and New York," *Planning Perspectives* 9 (1994): 353–76.

12. Lazar Kaganovich, *Za sotsialisticheskuiu rekonstruktsiiu Moskvy i gorodov SSSR.* (Moscow and Leningrad: OGIZ 'Moskovskii rabochii,' 1931)

13. I. Kattsen, *Metro Moskvy* (Moscow: Moskovskii rabochii, 1947); V. L. Makovskii, "Moskovskii metropoliten," *Nauka i zhizn'* 7 (1945): 40–45.

14. Makovskii, "Moskovskii metropoliten."

15. L. M. Kaganovich, *Pobeda metropolitena-pobeda sotsializma* (Moscow: Transzheldorizdat, 1935).

16. Rubble, "Failures of Centralized Metropolitanism," pp. 362–68.

17. Thomas P. Hughes, "Technological Momentum in History: Hydrogenation in Germany, 1898–1933," *Past and Present* 44 (August 1969): 106–32; Albert Teich and W. Henry Lambright, "The Redirection of a Large National Laboratory," *Minerva* 14, no. 4 (winter 1976–77): 447–74.

18. Paul Josephson, "The Historical Roots of the Chernobyl Crisis," *Soviet Union* 13, no. 3 (1986): 275–99.

19. V. P. Moiseenko et al. eds., *Razvitie stroitel'noi nauki i tekhniki v ukrainskoi*

SSSR, vols. 2 and 3 (Kiev: Naukova dumka, 1990).

20. Herf, *Reactionary Modernism,* pp. 152–62.

21. Robert Fishman, *Urban Utopias in the Twentieth Century* (Cambridge: MIT Press, 1982); Frank Whitford, *Bauhaus* (London: Thames and Hudson, 1984); Reyner Banham, *Theory and Design in the First Machine Age,* 2d ed. (Cambridge: MIT Press, 1986).

22. Barbara Miller Lane, *Architecture and Politics in Germany, 1918–1945* (Cambridge and London: Harvard University Press, 1985), pp. 128–30.

23. Ibid., pp. 132–40.

24. Herf, *Reactionary Modernism,* pp. 189–93.

25. Ibid., pp. 194–96.

26. Alfred Rosenberg, *Selected Writings,* ed. and introduced by Robert Pois (London: Jonathan Cape, 1970); Robert Pois, introduction to Rosenberg, *Selected Writings.*

27. Lane, *Architecture and Politics in Germany, 1918–1945,* pp. 149–51.

28. Rosenberg, *Selected Writings,* pp. 128–51.

29. Lane, *Architecture and Politics in Germany, 1918–1945,* pp. 162–63.

30. Ibid., pp. 147–49.

31. Ibid., pp. 169–71.

32. Ibid., pp. 176–84.

33. Ibid., pp. 171–73.

34. Ibid., pp. 185–89.

35. Hermann Göring, *Political Testament of Hermann Göring,* trans. H. W. Blood-Ryan (London: John Long, 1939), pp. 148–50.

36. Albert Speer, *Inside the Third Reich,* trans. Richard and Clara Winston (New York: Collier Books, 1970), pp. 62–63; Lane, *Architecture and Politics in Germany, 1918–1945,* pp. 185–89.

37. Speer, *Inside the Third Reich,* pp. 64–67.

38. Ibid., p. 69.

39. Ibid., pp. 102–103, 113–14.

40. Ibid., pp. 151–55.

41. Carl Schorske, *Fin-de-siècle Vienna: Politics and Culture* (New York: Knopf, 1979).

42. Speer, *Inside the Third Reich,* pp. 73–79, 134–35, 144.

43. Ibid., pp. 132–39.

44. Lane, *Architecture and Politics in Germany, 1918–1945,* pp. 205–13.

45. Karl Dietrich Bracher, *The German Dictatorship,* trans. Jean Steinberg (New York and Washington: Praeger Publishers, 1970), p. 347.

46. Michael J. Neufeld, "The Guided Missile and the Third Reich: Peenemünde and the Forging of a Technological Revolution," in *Science, Technology, and National Socialism,* eds. Monika Renneberg and Mark Walker (Cambridge: Cambridge University Press, 1994), pp. 51–53.

47. Michael J. Neufeld, "Weimar Culture and Futuristic Technology: The Rocketry and Spaceflight Fad in Germany, 1923–1933," *Technology and Culture* 31 (1990): 725–52.

48. Neufeld, "The Guided Missile and the Third Reich," pp. 56–58.

49. Ibid., pp. 59–62.

50. Joseph Goebbels, *The Goebbels Diaries,* trans. and ed. Louis Lochner (London: Hamish Hamilton, 1948), p. 286.

51. Speer, *Inside the Third Reich,* pp. 363–70, 409–10; Neufeld, "The Guided Missile and the Third Reich," pp. 65–66; Goebbels, *The Goebbels Diaries*, p. 219.

52. Neufeld, "The Guided Missile and the Third Reich," pp. 70–71.

ᴢ ᴣ

The Political Impact of Technical Expertise

Dorothy Nelkin

Technologies of speed and power—airports, power generating facilities, highways, dams—are often a focus of bitter opposition. As these technologies become increasingly controversial, scientists, whose expertise forms the basis of technical decisions, find themselves involved in public disputes. This "public" role of science has generated concern both within the profession and beyond; for a scientist's involvement in controversial issues may violate the norms of scientific research, but have considerable impact on the political process. As scientists are called upon to address a wider range of controversial policy questions,[1] "problems of political choice [may] become buried in debate among experts over high technical alternatives."[2]

This [chapter] will discuss some of the implications of the increasing involvement of scientists in controversial areas. What is the role of experts in public disputes? How are they used by various parties to a controversy, and how do scientists behave once involved? Finally, what is their impact on the political dynamics of such disputes?

THE ROLE OF EXPERTS

Scientists play an ambivalent role in controversial policy areas. They are both indispensible and suspect. Their technical knowledge is widely regarded as a source of power. "The capacity of science to authorize and certify facts and pictures of reality [is] a potent source of political influence."[3] Yet experts are resented and feared. While the reliance on experts is growing, we see a revival of Jacksonian hostility toward expertise, and of the belief that common sense is an adequate substitute for technical knowledge.[4]

Reprinted by permission of Sage Publications Ltd., from Dorothy Nelkin, "The Political Impact of Technical Expertise," *Social Studies of Science* 5 (1975): 189–205. Copyright © 1975 Sage Publications Ltd.

The authority of expertise rests on assumptions about scientific rationality; interpretations and predictions made by scientists are judged to be rational because they are based on "objective" data gathered through rational procedures, and evaluated by the scientific community through a rigorous control process. Science, therefore, is widely regarded as a means by which to depoliticize public issues. The increasing use of expertise is often associated with the "end of ideology"; politics, it is claimed, will become less important as scientists are able to define constraints and provide rational policy choices.[5]

Policymakers find that it is efficient and comfortable to define decisions as technical rather than political. Technical decisions are made by defining objectives, considering available knowledge, and analyzing the most effective ways of reaching these objectives. Debate over technical alternatives need not weigh conflicting interests, but only the relative effectiveness of various approaches for resolving an immediate problem. Thus, scientific knowledge is used as a "rational" basis for substantive planning, and as a means of defending the legitimacy of specific decisions. Indeed, the viability of bureaucracies depends so much on the control and monopoly of knowledge in a specific area, that this may become a dominant objective.[6] Recent technological disputes, however, suggest that access to knowledge and expertise has itself become a source of conflict, as various groups realize its growing implications for political choice.

The past decade has been remarkable for the development of "advocacy polittics";[7] consumer advocates, planning advocates, health care advocates, and environmental advocates have mobilized around diverse issues. Key slogans are "accountability," "participation," and "demystification." These groups share common concerns with the "misuse of expertise," the "political use" of scientists and professionals, and the implications of expert decision making for public action. Table 1 presents some statements of these concerns by various groups: radical scientists who have organized to develop "science for the people"; consumer advocates concerned with corporate accountability; advocacy planners who assist communities in expressing their local needs; and environmentalists and health professionals who demand "demystification of medicine."

Their criticism reflects a dilemma. The complexity of public decisions seems to require highly specialized and esoteric knowledge, and those who control this knowledge have considerable power. Yet democratic ideology suggests that people must be able to influence policy decisions that affect their lives. This dilemma has provoked a number of proposals for better distribution of technical information; expertise, it is argued, is a political resource and must be available to communities as well as to corporations, utilities, or developers.[8] The increasing importance of technical information has also prompted analyses of the behavior of scientists as they are diverted to applied and controversial work.

For example, Allan Mazur suggests that the political (i.e., nonscientific) context of controversies crucially affects the activities of scientists, the way they present their findings, and thus their ultimate influence on decisions. Despite norms of political neutrality, claims Mazur, scientists behave just like anyone else when they engage in disputes; their views polarize and as a result the value of scientific advice becomes questionable. Thus, disputes among experts may become a major source of confusion for policymakers and for the public.[9] Guy Benveniste, focusing on the use of scientists by policymakers, suggests that "technical" decisions are basically made on political or economic grounds. Expertise is sought as a means of supporting particular policy programs; the selection of data and their interpretation are thus related to policy goals.[10] Similarly, [Lauriston] King and [Philip] Melanson argue that when knowledge is employed in the resolution of public problems, it is shaped, manipulated, and frequently distorted by the dynamics of the policy arena.[11]

These analyses emphasize the politicization of expertise. Details of two recent disputes in which "experts" were used by both project developers and critics provide an opportunity to develop these arguments, and then to explore the impact of experts on the political process. One of the disputes concerns the siting of an 830 megawatt nuclear power plant on Cayuga Lake in upstate New York; the other is the proposed construction of a new runway at Logan International Airport in East Boston, Massachusetts.

The power plant siting controversy began in June 1967, when the New York State Electric and Gas Company (NYSE&G) first announced its intention to build Bell Station.[12] Groups of scientists and citizens, concerned with the thermal pollution of Cayuga Lake, organized themselves to oppose the plant, and demanded that NYSE&G consider design alternatives that would minimize the damage to the lake caused by waste heat. They forced the utility to postpone its application for a construction permit, and to contract for additional research on the environmental impact of the plant. In March 1973, following consultants' recommendations, NYSE&G announced a power station plan that was essentially the same as its earlier controversial design. The company, however, was now armed with data from one and a half million dollars' worth of environmental research supporting its claim that the heat from Bell Station would not damage the lake. Yet once more there was concerted and well-informed public opposition, this time focused on radiation hazards. Four months later the company was forced to abandon its plan.

The proposed new 9,200-foot runway at Logan Airport was part of a major expansion plan that had been a source of bitter conflict in East Boston for many years.[13] Located only two miles from the center of downtown Boston in an Italian working-class community, this modern convenient airport is a source of extreme irritation, fear, and community disruption. The expansion policies of the Massachusetts Port Authority (Massport) have been opposed, not only by airport neighbors but also by Boston's city government and by state officials concerned with the development of a balanced transportation system. Here, as in the Cayuga Lake power plant siting debate, knowledge was used as a resource both by Massport, seeking justification for its expansion plans, and by those opposed to such plans. Massport's staff was backed by consultants who claimed that without expansion the airport would reach saturation by 1974, and that the new runway would cause no environmental damage. The opponents, primarily from the adjacent working-class neighborhood of East Boston, used technical advice provided by the city of Boston. Following pressure from the governor as well as from the mayor, Massport eventually deleted the proposed runway from the master plan for future airport development.

While this [chapter] will focus on similarities in the dynamics of these two disputes, it is necessary first to point out important differences. The community opposed to the power plant was a college town; the dispute was a middle-class environmental conflict, sustained by expertise from scientists in a nearby university who also lived in the area. In contrast, the opposition to the airport came primarily from a working-class neighborhood dependent on expertise provided by government officials who, for political and economic reasons, chose to oppose the airport development plans.

The technical aspects of the two disputes were also quite different. The power plant issue was embedded in a set of vague uncertainties and intangible fears about radiation; airpot expansion posed the concrete and direct threat of increased noise and land purchase. The main area of technical conflict in the former case was the potential environmental impact of the new power plant, and the experts involved were mostly scientists and engineers. In the latter case the controversial issue was the validity of projections—whether the runway was really necessary at all—and the dispute involved economists and lawyers as well as engineers.

Despite such differences, the two cases have a great deal in common: the use of expertise, the style of technical debate, and the impact of experts on the political dynamics of the dispute are remarkably similar.

THE USE OF EXPERTISE

Opposition to both the power plant and the airport developed in several stages. The developers (utility manager, airport manager) contracted for detailed plans on the construction of their proposed facility. As they applied for the necessary permits, affected groups tried to influence the decision. The developer in each case argued that plans, based on their consultants' predictions of future demands and technical imperatives concerning the location and design of the facility, were definitive, except perhaps for minor adjustments necessary to meet federal standards.

Table 1. Public concern with expertise

Radical scientists	Consumer advocates	Environ-mentalists	Advocacy planners	Medical critics
On the misuse of technology				
[We] feel a deep sense of frustration and exasperation about the use of [our] work. We teach, we do experiments, we design new things—and for what? To enable those who direct this society to better exploit and oppress the great majority of us? To place the technological reins of power in the hands of those who plunder.	What is needed is a sustained public demand for a liberation of law and technology to cleanse the air by disarming the corporate power that turns nature against man.	Many believe that the advantages of our technology compensate for environmental degradation. . . . Some have faith that the laboratories that have delivered miracles can also provide the tools to remedy any problems man may face. But with technology's gifts to improve man's environment has come an awesome potential for destruction.	Advocacy planning may be one of the channels of action through which people may try to humanize their technical apparatus; to prevent the exercise of bureaucratic power from leading to a new diffuse despotism, in which power appears in the image of technical necessity.	Psychiatry and psychology are used as direct instruments of coercion against individuals. Under the guise of "medical methods," people are pacified, punished, or incarcerated.
On the use of expertise				
Skills and talents of potentially enormous usefulness have been bent to destructive ends to guarantee expansion and protect imperialism. For the sake of the ruling class, scientists and engineers have been turned into creators of destruction by the mechanisms of an economic and social system over which they can exert no control.	Too many of our citizens have little or no understanding of the relative ease with which industry has or can obtain the technical solutions.	It doesn't require special training to keep a broad perspective and to apply common sense. Thus, for every technically knowledgeable [person] there is a layman activist. . . . In fact, the technologist's training can stand in his way. There is a growing awareness that civilized man has blindly followed the technologists into a mess.	Even without administrative power, the advocate planner is a manipulator. . . . The planner may not be the first to identify "problems" of an urban area, but he puts them on the agenda and plays a large part in defining the terms in which the problem will be thought about— and those terms in effect play a large part in determining the solution.	Professionals often regard themselves as more capable of making decisions than other people even when their technical knowledge does not contribute to a particular decision. . . . Professionalism is not a guarantor of humane, quality services. Rather it is a code-word for a distinct political posture.

Table 1. (Contd.)

On expertise and public action

What we can hope is not that scientists can provide the people with an objective approach to build a better world, but that in the better world built by the people, scientists will be able to work in a more objective fashion, unfettered by elitism and the worst competitive aspects of present-day science.	An action strategy must embrace the most meticulous understanding of the corporate structure—its points of access, its points of maximum responsiveness, its specific motivational sources, and its constituencies.	The importance of the environmental movement's potential rests not only in what tangible results it can accomplish, but in its acting as a catalyst to start people working together. Alliances possible by organizing around environmental concerns stagger the mind of the seasoned community organizer.	Any plan is the embodiment of particular group interests . . . any group which has interests at stake in the planning process should have those interests articulated. . . . Planning in this view becomes pluralistic and partisan—in a word, overtly political.	Medicine should be demystified. . . . When possible, patients should be permitted to choose among alternative methods of treatment based upon their needs. Health care should be deprofessionalized. Health care skills should be transferred to worker and patient alike.

Sources: These statements are quotations from editorials and the popular literature circulated by such groups as SESPA [Scientists and Engineers for Social and Political Action], Nader's Raiders, Earth Day groups, and the Health Policy Advisory Center.

In the power plant controversy, scientists from Cornell University who lived in the community were the first to raise questions about the NYSE&G plan when it was announced in 1967. By mid-1968, their activity had built up sufficient political support to persuade NYSE&G to postpone its plans, and to undertake further environmental research.

A new sequence of events began in March 1973, when NYSE&G again announced its intention to build the plant and claimed that it was imperative to begin construction promptly. The company's consultants, Nuclear Utilities Services Corporation, had prepared a five-volume technical report. NYSE&G placed copies in local libraries, circulated a summary to its customers, and invited comments. The report supported NYSE&G's earlier plan for a plant involving a General Electric boiling water reactor with a once-through cooling system. The study concluded that cooling towers (which had been recommended by power plant critics in 1968) were economically unfeasible in the size range required for the plant, unsuited to the topography of the area, and would have a tendency to create fog. To develop an optimum design for a once-through cooling system, consultants designed a jet diffuser to provide rapid mixing of the heated discharge with the lake water. With this system, they argued that the plant would have an insignificant effect on the aquatic environment of Cayuga Lake. The consultants only briefly concerned themselves with the issue of radioactive wastes on the grounds that this was not a problem unique to Cayuga Lake; the report only stated that the effect would be substantially below current radiation protection standards.

NYSE&G organized an information meeting attended by 1,000 citizens, and for two hours summarized the highly technical material supporting its plans. This, however, was followed by two and one-half hours of angry discussion, and the utility's president announced that if public protest was likely to cause delay, they would build the plant at another site. He hoped, however, that the decision would be "based on fact and not on emotion."

The first organized response came from twenty-four scientists who volunteered to provide the public with a review and assessment of the utility's massive technical report.[14] Their review was highly critical and NYSE&G's consultants responded in kind (see below). Meanwhile, citizens' groups formed and the community polarized, as the company posed the issue in terms of "nuclear power *or* blackouts."

The airport case also involved experts on both sides of the controversy. Opposing forces mobilized in February 1971, at a public hearing required by the Corps of Engineers in order to approve Massport's request to fill in part of Boston Harbor. One thousand people attended and for ten hours scientists, politicians, priests, schoolteachers, and others debated the priorities which they felt should govern airport decisions. Massport's staff was backed by consultants, who claimed that without the runway the airport would reach saturation by 1974. Consultants provided a brief environmental statement arguing that the new runway would have no direct detrimental effects of ecological significance. The only environmental costs would be the elimination of ninety-three acres of polluted clam flats and two hundred and fifty acres of wildlife preserve—which constituted a hazard in any case because birds interfere with jets. Furthermore, because of the added flexibility, the runway would relieve noise and congestion caused by an expected increase in aircraft operations. Massport's claims were later buttressed by an environmental impact statement commissioned from Landrum and Brown, Airport Consultants, Inc. at a cost of $166,000. The study documented Massport's contention that the new runway was essential for safety and would be environmentally advantageous; it emphasized the positive contributions of Logan Airport—its economic importance to the City of Boston, and the reduction of noise that would result from increased runway flexibility.

The opposition was organized by a coalition of citizens' groups called the Massachusetts Air Pollution and Noise Abatement Committee. The issues raised were diverse. Neighborhood people spoke of the discomfort caused by aircraft operations, and of Massport's piecemeal and closed decision-making procedures. Environmentalists feared the destruction of Boston Harbor, and planners related airport decisions to general urban problems. Legal, economic, and technical experts became involved as the Mayor's office and the Governor evaluated Massport's claims. As in the power plant case, the conflict polarized as Massport posed the issue in terms of "airport expansion *or* economic disaster."

THE STYLE OF TECHNICAL DEBATE[15]

In both cases the technical debate involved considerable rhetorical licence, with many insinuations concerning the competence and the biases of the involved scientists.[16] NYSE&G emphasized that the need for a nuclear power plant on Cayuga Lake was "imperative," that there would be a serious energy shortage if they did not proceed immediately with the plan, and that the impact of the plant on the local environment would be "insignificant." NYSE&G insisted on their unique technical competence to make this decision. "Our study is the most comprehensive study ever made on the lake. Opponents can create delays but are not required to assume responsibility. . . ."

However, the Cornell critics called NYSE&G's data "inadequate," "misleading," "non-comprehensive," and "limited in scope and inadequate in concept." Some of the critics provided data from other research that contradicted NYSE&G's findings. They emphasized that there was simply not enough known about deep-water lakes to assess the risks.

NYSE&G consultants countered by claiming that Cornell critics were unfamiliar with the scope and requirements of an environmental feasibility report; in particular, that the critics' review failed to distinguish between the goals of pure and applied research. "From an academic position a complete ecological model that predicted all possible relationships would be desirable, but this was neither feasible nor necessary for assessing the minor perturbations caused by one plant."

In fact, each group used different criteria to collect and interpret technical data. The two studies were based on diverse premises which required different sampling intervals and techniques. NYSE&G consultants, for instance, claimed that their water quality studies focused on establishing base-line conditions to predict the changes caused by the power plant; Cornell studies focused on limiting factors, such as the impact of nutrients on lake growth.

Scientists attacked each other with little constraint. Cornell reviewers accused NYSE&G consultants of value judgments that led to "glaring omissions," "gross inadequacies," and "misleading interpretations." Consultants referred to the Cornell report's "confusion resulting from reviewers reading only certain sections of the report," and "imaginative, but hardly practicable suggestions." The NYSE&G president accused the Cornell reviewers of bias: "It is of some interest that many of the individuals who participated in the Cornell review have taken a public position in opposition to nuclear plants. Philosophical commitment in opposition to nuclear generation may have made it difficult for these reviewers to keep their comments completely objective."[17]

A similar style of debate characterized the technical dispute over the airport runway. Expansion of Logan was recommended by consultants as "the best opportunity to realize a reduction of current social impact." Failure to expand the airport as proposed would cause delays, increase air pollution, reduce safety margins, and have a "drastic" and "immeasurable" impact on the local economy—"an impact which the Boston area could not afford." Massport's environmental report described and rejected, one by one, alternatives proposed by airport opponents. Banning specific types of aircraft "interferes with interstate commerce." Limiting maximum permissible noise levels is "legally questionable," since the airport functions as part of a coordinated national system. A surcharge for noisy aircraft would be "useless" as economic leverage, since landings fees represent a negligible percentage of total airline expenses. Setting night curfews is "precluded" by the interdependence of flight schedules and aircraft utilization requirements: it would relegate Boston to a "second-class" airport and have "disastrous effects" on service to 65 percent of the 267 cities served by Boston. Moreover, 70 percent of the cargo business would be "negatively affected." Soundproofing neighboring houses and buildings would be "economically prohibitive" and have little effect. The *only* feasible solution to noise and environmental problems, according to the consultants' report, was an expanded runway system that would permit increased flexibility. Massport insisted on the validity of its expertise. "We are closer and more knowledgeable than any other group no matter what their intention may be, on what Logan Airport . . . what Metropolitan Boston, what the entire state of Massachusetts and New England needs."[18] And Massport consultants suggested their agreement with their client when, in a technical analysis of the airport's economic impact, they stated: "It is inconceivable that an enterprise of this magnitude can be treated other than with the most profound respect."[19]

Airport opponents called the Massport technical reports "the logical outcome of efforts directed toward narrow objectives." City consultants contended that authority to restrict aircraft noise was in fact limited neither by the FAA nor by the Massport enabling act, and that the FAA actually encouraged airport operators to restrict airport noise independently. They argued that Massport's assumptions concerning anticipated demand for increased airport capacity were questionable and in any case were subject to modification by consolidating schedules and dispersing general aviation flights. Massport's own raw data suggested that with a reasonable adjustment Logan Airport could accommodate a considerable increase in actual business, for aircraft were operating at an average of just under half capacity. Moreover, projections were based on the growth pattern of the 1960s. The decrease in air travel demand in 1970 could have been regarded either as a new data point or as an anomaly. Massport chose the latter interpretation, ignoring the 1970 slump. Their projections also ignored the possibility of competitive alternatives to air travel.[20]

Massport's figures concerning the economic impact of expansion and the consequences of a moratorium on expansion were debunked by critics as "blatant puffery." As for Massport's contention that the new runway would be environmentally advantageous, city representatives concluded that an expanded airfield would only expose new populations to intolerable noise. Instead, they recommended measures to increase capacity at Logan through scheduling adjustments and efforts to distribute the hours of peak demand by economic controls such as landing fees.

Differences were to be aired at a second round of public hearings scheduled for July 10, 1971. However, on July 8, following a task-force study that recommended alternatives to expansion, Governor Sargent publicly opposed the construction of the new runway. Under these circumstances, the Corps of Engineers was unlikely to approve the project, so Massport withdrew its application for a permit and temporarily put aside its plans for the runway. A year and a half later, in February 1973, Massport deleted the proposed runway from the master plan for future airport development. Citing projections that were close to those used by airport opponents two years earlier, the Port Authority claimed that reevaluation of future needs indicated that the new runway was no longer necessary.

Both disputes necessarily dealt with a great number of genuine uncertainties that allowed divergent predictions from available data. The opposing experts emphasized these uncertainties; but in any case, the substance of the technical arguments had little to do with the subsequent political activity.

THE IMPACT OF EXPERTISE ON POLITICAL ACTION

In both the airport and power plant controversy, it was the *existence* of technical debate more than its *substance* that stimulated political activity.[21] In each case the fact that there was disagreement among experts confirmed the fears of the community and directed attention to what they felt was an arbitrary decision-making procedure in which expertise was used to mask questions of political priorities.

The relationship between technical disputes and political conflict was most striking in the power plant case. Cornell scientists assessed the NYSE&G report with the intention of providing technical information to the public. They focused almost entirely on the issue of thermal pollution—the effect of the plant's heated effluent on Cayuga Lake. The citizens' groups, however, were most concerned with the issue of radiation. They had followed the considerable discussion in the press and in popular journals about the risks associated with the operation of nuclear reactors—risks that had not been as widely publicized at the time of the first controversy in 1968. Thus, the thermal pollution issue (which had dominated earlier controversy) became, in 1973, a relatively minor concern. Citizens, in contrast to the scientists who were advising them, focused on problems of transporting and disposing of nuclear wastes, on the reliability of reactor safety mechanisms, on reactor core defects that would allow the release of radioactive gases, and on the danger of human error or sabotage.

When the citizens' committee first met to establish a position on the issue, its newsletter concentrated entirely on the reactor safety issue.[22] This set the tone of subsequent discussion, in which three possible courses of action were considered: that the committee oppose construction of any nuclear plant on Cayuga Lake until problems of reactor safety and disposal of radioactive wastes were resolved; that it take up its 1968 position and oppose only the *current design* of Bell Station; or that it support NYSE&G plans. The first proposal, one of total opposition, won overwhelming support. The emphasis of citizens' groups thereafter was on the risks associated with nuclear power, despite the fact that the technical debate dealt mainly with the problem of thermal pollution.

The disputes between scientists, however, served as a stimulus to political activity. In the first place, the criticism by Cornell scientists neutralized the expertise of the power company. Simply suggesting that there were opposing points of view on one dimension of the technical problem increased public mistrust of the company's experts, and encouraged citizens to oppose the plant. Second, the involvement of scientists gave moral support to community activists, suggesting that their work would be effective. The citizens' groups called attention to NYSE&G's statement that if there were concerted opposition, the company would not go ahead with its plans. The ready support of local scientists led to substantial expectation in the community that the effort involved in writing letters and going to meetings would not be wasted.

As for the details of the technical dispute, they had little direct bearing on the dynamics of the case. Citizens trusted those experts who supported their position. People who supported NYSE&G voiced their trust in the consultants employed by the power company: "Let us allow the professionals to make the decisions that they get paid to make." And power plant critics used expertise only as a means to bring the issue back to its appropriate political context. The case was one of local priorities, they claimed; it was not a technical decision: "To say that our future is out of our hands and entrusted to scientists and technicians is an arrogant assumption. . . . We suggest that the opinions of area residents who care deeply about their environment and its future are of equal if not greater importance."[23]

In the airport case, the technical arguments served primarily to reinforce the existing mistrust of Massport among those opposed to airport expansion, and they were virtually ignored by those who supported Massport. Opinions about the necessity of the runway were well established prior to the actual dispute. In East Boston, Massport employees and local sports clubs which were supported by an airport community relations program defended the Port Authority's plans for a new runway and maintained their trust in Massport's competence. "In terms of efficient and competent operation, Massport is head and shoulders above other agencies."

Airport opponents, while benefiting from the advice provided by experts from the city of Boston, claimed the issue was a matter of common sense and justice. They defined the problem in terms of values (such as neighborhood solidarity) which are not amenable to expert analysis. "We need no experts. These people will verify themselves the effect of noise. . . . Massport is extremely arrogant. They do not have the slightest conception of the human suffering they cause and could not care less."[24]

Airport critics pointed out various technical errors and problems of interpretation in Massport's predictions and environmental impact statements; but this simply reconfirmed the community's suspicion of Massport, and further polarized the dispute. Later, these same experts who were sympathetic to East

Boston's noise problem failed to convince the community to accept a Massport plan for a sound barrier. Despite advice that this would help to relieve their noise problem, the community chose to oppose construction of the barrier. Local activists feared that this was a diversion, and that if they accepted this project the community would somehow lose out in the long run. Thus, they disregarded expert opinion that this was a favorable decision, and the old mistrust prevailed.

SUMMARY AND CONCLUSIONS

The two conflicts described above, over the siting of a power plant and the expansion of an airport, have several aspects in common. One can trace parallels, for instance, in the way the developers used expertise as a basis and justification of their planning decisions; how experts on both sides of the controversy entered the dispute and presented their technical arguments; and how citizens affected by the plan perceived the dispute. Similarities are evident in public statements, as developers, experts, and citizens expressed their concerns about various aspects of the decision-making process. These are compared in table 2. These similarities, especially with respect to the use of scientific knowledge, suggest several related propositions which may be generalizable to other controversies involving conflicting technical expertise:

First, *developers seek expertise to legitimize their plans and they use their command of technical knowledge to justify their autonomy.* They assume that special technical competence is a reason to preclude outside public (or "democratic") control.

Second, *while expert advice can help to clarify technical constraints, it also is likely to increase conflict,* especially when expertise is available to those communities affected by a plan. Citizens' groups are increasingly seeking their own expertise to neutralize the impact of data provided by project developers.[25] Most issues that have become politically controversial (environmental problems, fluoridation, DDT) contain basic technical as well as political uncertainties, and evidence can easily be mustered to support or oppose a given proposal.

Third, *the extent to which technical advice is accepted depends less on its validity and the competence of the expert, than on the extent to which it reinforces existing positions.* Our two cases suggest that factors such as trust in authority, the economic or employment context in which a controversy takes place, and the intensity of local concern will matter more than the quality or character of technical advice.[26]

Table 2. Perspectives on decision making and expertise

	Power plant dispute	Runway dispute
Developers		
On responsibility and competence for planning	Our study is the most comprehensive study ever made on the lake. Opponents can create delays but are not required to assume responsibility.	We are closer and more knowledgeable than any other group no matter what their intention may be, on what Logan Airport . . . what Metropolitan Boston . . . what New England needs.
On public debate	We have adopted a posture of no public debate.	We have competent staffs. . . . I can't see any sense in having a public hearing . . . if it is to be by consensus that the authority operates.
Experts (consultants)		
On impact of project	Alternate approaches would have undesirable effects on the human environment . . . the proposed design should produce no significant impact. Actual individuals would be exposed to much lower doses than that due to normal habits.	Adverse environmental impact will result from failure to undertake this project as contrasted with the impact if the authority proceeds. Noise measurements of typical urban noise conditions . . . show that street level background noise overshadows taxiway noise.
On planning	Although an ecological model might be desirable from an academic viewpoint, it is not felt to be necessary to provide an adequate assessment of the impact of the minor perturbation introduced by the proposed plant.	A master plan would be nothing more than an academic exercise. . . a study of this magnitude could never be justified for a small project of this nature.
Experts (critics)		
On developers' data	Statements and conclusions were not justified and must therefore be regarded as nothing more than guesses. . . . The data base is not only inadequate, but misleading.	Analysis of the economic impact of Logan Airport shows demonstrated "blatant puffery" in the figures appearing in the report.
Citizens (project supporters)		
On decision-making responsibility	Let us allow the professionals to make the decisions that they get paid to make.	In terms of efficient and competent operation, Massport is head and shoulders above other agencies.

Table 2 (contd.)

Citizens (project opponents)

On decision-making responsibility	To say that our future is out of our hands and entrusted to scientists and technicians is an arrogant assumption. . . . We suggest that the opinions of area residents who care deeply about their environment and its future is of equal if not greater importance.	We need no experts. These people will verify themselves. . . . Massport is extremely arrogant. They do not have the slightest conception of the human suffering they cause and could not care less.
On decision-making process	Were they using the power the people gave them to support their own feelings or those of private concerns? There is representative government in our country, but it sure isn't in our county.	What is really on trial here is not just the Port Authority, it is really the American system. Will it listen to spokesmen for the people and the people who speak for themselves?

Fourth, *those opposing a decision need not muster equal evidence.* It is sufficient to raise questions that will undermine the expertise of a developer whose power and legitimacy rests on his monopoly of knowledge or claims of special competence.

Fifth, *conflict among experts reduces their political impact.* The influence of experts is based on public trust in the infallibility of expertise. Ironically, the increasing participation of scientists in political life may reduce their effectiveness, for the conflict among scientists that invariably follows from their participation in controversial policies highlights their fallibility, demystifies their special expertise, and calls attention to nontechnical and political assumptions that influence technical advice.[27]

Finally, *the role of experts appears to be similar regardless of whether they are "hard" or "soft" scientists.* The two conflicts described here involved scientists, engineers, economists, and lawyers as experts. The similarities suggest that the technical complexity of the controversial issues does not greatly influence the political nature of a dispute.

In sum, the way in which clients (either developers or citizens' groups) direct and use the work of experts embodies their subjective construction of reality—their judgments, for example, about public priorities or about the level of acceptable risk or discomfort. When there is conflict in such judgments, it is bound to be reflected in a biased use of technical knowledge, in which the value of scientific work depends less on its merits than on its utility.

NOTES

1. See discussion of the increased demands for expert decision-making in Garry Brewer, *Politicians, Bureaucrats, and the Consultant* (New York: Basic Books, 1973). Also, Dean Schooler Jr., in *Science, Scientists, and Public Policy* (Glencoe, Ill.: The Free Press, 1971), suggests that in the past, scientific influence has concentrated in government entrepreneurial areas such as space exploration, or in policy areas defined in terms of national security. The participation and influence of scientists has traditionally been rather minimal in policy areas with redistributive implications, e.g., social policy, transportation, and other issues subject to social conflict and competing political interests. As the public seeks technical solutions to social problems, and as scientists themselves become engaged in controversial public issues, this pattern is changing.

2. Harvey Brooks, "Scientific Concepts and Cultural Change," *Daedalus* 94 (winter 1965): 68.

3. Yaron Ezrahi, "The Political Resources of American Science," *Science Studies* 1 (1971): 121. See also Don K. Price, *Government and Science* (New York: New York University Press, 1954).

4. For a discussion of the historical tradition of resentment of experts in the United States see Richard Hofstadter, *Anti-Intellectualism in American Life* (New York: Knopf, 1962).

5. See Robert Lane, "The Decline of Politics and Ideology in a Knowledgeable Society," *American Sociological Review* 31 (October 1966): 649–62, and Daniel Bell, *The End of Ideology* (New York: The Free Press, 1960).

6. See discussion in Michel Crozier, *The Stalled Society* (New York: Viking Press, 1973), chap. 3. A vivid example of the importance of this tendency to monopolize knowledge occurred during the "energy crisis" with the realization that the large oil companies had nearly exclusive knowledge on the state of oil reserves.

7. I am using this term to describe a phenomenon that Orion White and Gideon Sioberg call "mobilization politics," in "The Emerging New Politics in America," in *Politics in the Post-Welfare State*, eds. M. D. Hancock and Gideon Sjoberg (New York: Columbia University Press, 1972), p. 23.

8. Note for example the system of "scientific advocacy" proposed by John W. Gofman and Arthur R. Tamplin, *Poisoned Power* (Emmaus, Penn.: Rodale Press, 1971). A similar system is suggested by Donald Geesaman and Dean Abrahamson in "Forensic Science—A Proposal," *Science and Public Affairs* (*Bulletin of the Atomic Scientists*) 29 (March 1973): 17. Thomas Reiner has proposed a system of community technical services in "The Planner as a Value Technician: Two Classes of Utopian Constructs and Their Impact on Planning," in *Taming Megalopolis*, ed. H. Wentworth Eldridge, vol. 1 (New York: Anchor Books, 1967). Based on systems similar to legal advocacy and expert witness in the courts, such proposals are intended to make technical advice more widely available to citizens' groups—usually through provision of public funds to underwrite the cost of expertise.

9. Allan Mazur, "Disputes between Experts," *Minerva* 11 (April 1973): 243–62.

10. Guy Benveniste, *The Politics of Expertise* (Boston: Glendessary Press, 1912). See also Leonard Rubio, "Politics and Information in the Anti-Poverty Programs," *Policy Studies Journal* 2 (spring 1974): 190–95.

11. Lauriston R. King and Philip Melanson, "Knowledge and Politics," *Public Policy* 20 (winter 1972): 82–101.

12. For a history and analysis of this controversy see Dorothy Nelkin, *Nuclear Power and Its Critics* (Ithaca, N.Y.: Cornell University Press, 1971); "Scientists in an Environmental Controversy," *Science Studies* 1 (1971): 245–61; and "The Role of Experts in a Nuclear Siting Controversy," *Science and Public Affairs* 30 (November 1974): 29–36.

13. Documentation of this conflict can be found in Dorothy Nelkin, *Jetport: The Boston Airport Controversy* (New Brunswick, N.J.: Transaction Books, 1974).

14. Two hundred copies of the critique were sent to libraries, citizens' groups, faculties at universities and colleges in the area, officials in state and federal agencies, political representatives in local, state, and federal government, and newspapers.

15. Unless otherwise noted, the quotations that follow are from local environmental reports, memos, letters, and public hearings. They are statements by the opposing scientists involved in the controversy.

16. Mazur, "Disputes between Experts," also documents the use of rhetoric in technical debates.

17. William A. Lyons, "Recommendations of the Executive Offices of New York State Electric and Gas Corporation to the Board of Directors" (July 13, 1973).

18. Edward King, Massport Executive Director, Testimony at U.S. Corps of Engineers' "Hearings on the Application by the MPA for a Permit to Fill the Areas of Boston Harbor," February 26, 1971, mimeograph, p. 101.

19. Landrum and Brown, Inc., *Boston-Logan International Airport Environmental Impact Analysis*, February 11, 1972, section ix, 3.

20. A systematic critique of Massport's data was made by a commission chaired by Robert Behn (Chairman of Governor's Task Force on Inter-City Transportation), "Report to Governor Sargent," April 1971.

21. For further discussion of this point, see Nelkin, "The Role of Experts in a Nuclear Siting Controversy."

22. CCSL (Citizens Committee to Save Cayuga Lake) *Newsletter* 6 (April 1973). This newsletter reprinted in full a selection of well-informed articles—notably those by Robert Gillette in *Science* 176 (May 5, 1973), 177 (July 28, 1972; September 8, 1972; September 19, 1972; and September 22, 1972), and 179 (January 26, 1973).

23. Statement by Jane Rice cited in the *Ithaca Journal*, May 14, 1973, p. 1.

24. These statements are from testimony at U.S. Corps of Engineers' Hearings (note 18). The ultimate expression of this kind of sentiment was, of course, the remark alleged to have been made by former Vice President Spiro Agnew, responding to the report by the U.S. Presidential Commission on Pornography and Obscenity: "I don't care what the experts say, I *know* pornography corrupts!"

25. For further discussion of the tactics of using expertise within the fluoridation controversy, for example, see Robert Crain et al., *The Politics of Community Conflict* (New York: Bobbs Merrill, 1969); and H. M. Sapolsky, "Science, Voters, and the Fluoridation Controversy," *Science* 162 (October 25, 1968): 427–33.

26. The relation between beliefs and the interpretation of scientific information is analyzed in S. B. Barnes, "On the Reception of Scientific Beliefs," in *Sociology of Science*, ed., Barry Barnes (Hammondsworth, Middlesex, England: Penguin Books, 1972), pp. 269–91.

27. See discussion of how controversy among scientists influences legislators in Barnes, *Sociology of Science*.

From Information
to Communication
The French Experience with Videotex

Andrew Feenberg

INFORMATION OR COMMUNICATION?

otions like "postindustrial society" and the "information age" are forecasts—social science fictions—of a social order based on knowledge.[1]
The old world of coal, steel, and railroads will evaporate in a cloud of industrial
smoke as a new one based on communications and computers is born. The pop-
ularizers of this vision put a cheerful spin on many of the same trends deplored
by dystopian critique, such as higher levels of organization and integration of the
economy and the growing importance of expertise.

Computers play a special role in these forecasts because the management of
social institutions and individual lives depends more and more on swift access to
data. Not only can computers store and process data, they can be networked to
distribute it as well. In the postindustrial future, computer-mediated communication (CMC) will penetrate every aspect of daily life and work to serve the rising
demand for information.

In the past decade, these predictions have been taken up by political and business leaders with the power to change the world. One learns a great deal about a
vision from attempts to realize it. When, as in this case, the results stray far from
expectations, the theories that inspired the original forecast are called into question. This chapter explores the gap between theory and practice in a particularly
important case of mass computerization, the introduction of videotex in France.

Videotex is the CMC technology best adapted to the rapid delivery of data.
Videotex is an on-line library that stores "pages" of information in the memory
of a host computer accessible to users equipped with a terminal and modem.

From *Alternative Modernity*, by Andrew Feenberg, "From Information to Communication: The
French Experience with Videotex (Los Angeles: University of California Press, 1995), pp. 144–66.

Although primarily designed for consultation of material stored on the host, some systems also give users access to each other through electronic mail, "chatting," or classified advertisements. This, then, is one major technological concretization of the notion of a postindustrial society.

The theory of the information age promised an emerging videotex marketplace. Experience with videotex, in turn, tested some of that theory's major assumptions in practice. Early predictions had most of us linked to videotex services long ago.[2] By the end of the 1970s, telecommunications ministries and corporations were prepared to meet this confidently predicted future with new interactive systems. But today most of these experiments are regarded as dismal failures.

This outcome may be due in part to antitrust rulings that prevented giant telephone and computer companies from merging their complementary technologies in large-scale public CMC systems. The Federal Communications Commission's failure to set a standard for terminals aggravated the situation. Lacking the resources and know-how of the big companies, their efforts uncoordinated by government, it is not surprising that smaller entertainment and publishing firms were unable to make a success of commercial videotex.[3]

Disappointing results in the United States have been confirmed by all foreign experiments with videotex, with the exception of the French Teletel system. The British, for example, pioneered videotex with Prestel, introduced three years before the French came on the scene. Ironically, the French plunged into videotex on a grand scale in part out of fear of lagging behind Britain!

Prestel had the advantage of state support, which no American system could boast. But it also had a corresponding disadvantage: overcentralization. At first information suppliers could not connect remote hosts to the system, which severely limited growth in services. What is more, Prestel relied on users to buy a decoder for their television sets, an expensive piece of hardware that placed videotex in competition with television programming. The subscriber base grew with pathetic slowness, rising to only seventy-six thousand in the first five years.[4]

Meanwhile, the successful applications of CMC were all organized by and for private businesses, universities, or computer hobbyists. The general public still has little or no access to the networks aimed at these markets and no need to use such specialized on-line services as bibliographic searches and software banks. Thus after a brief spurt of postindustrial enthusiasm for videotex, CMC is now regarded as suitable primarily for work, not for pleasure; it serves professional needs rather than leisure or consumption.[5]

As I will explain below, the Teletel story is quite different. Between 1981, the date of the first tests of the French system, and the end of the decade, Teletel became by far the largest public videotex system in the world with thousands of services, millions of users, and hundreds of millions of dollars in revenues. Today Teletel is the brightest spot in the otherwise unimpressive commercial videotex picture.

This outcome is puzzling. Could it be that the French are different from everyone else? That rather silly explanation became less plausible as Compuserve and the Sears/IBM Prodigy system grew to a million subscribers. While the final evaluation of these systems is not yet in, their sheer size tends to confirm the existence of a home videotex market. How, then, can we account for the astonishing success of Teletel, and what are its implications for the information-age theory that inspired its creation?

Teletel is particularly interesting because it employs no technology not readily available in all those other countries where videotex was tried and failed. Its success can only be explained by identifying the *social inventions* that aroused widespread public interest in CMC. A close look at those inventions shows the limitations not only of prior experiments with videotex but also of the theory of the information age.[6]

THE EMERGENCE OF A NEW MEDIUM

While Teletel embodies generally valid discoveries about how to organize public videotex systems, it is also peculiarly French. Much that is unique about it stems from the confluence of three factors: (1) a specifically French politics of modernization; (2) the bureaucracy's voluntaristic ideology of national public service; and (3) a strong oppositional political culture. Each of these factors contributed to a result no single group in French society would willingly have served in the beginning. Together they opened the space of social experimentation that Teletel made technically possible.

MODERNIZATION

The concept of modernity is a live issue in France in a way that is difficult to imagine in the United States. Americans experience modernity as a birthright; America does not *strive* for modernity, it *defines* modernity. For that reason, the United States does not treat its own modernization as a political issue but relies on the creative chaos of the market.

France, on the other hand, has a long tradition of theoretical and political concern with modernity as such. In the shadow of England at first and later of Germany and the United States, France has struggled to adapt itself to a modern world it has always experienced at least to some extent as an external challenge. This is the spirit of the famous Nora-Minc Report that President Giscard d'Estaing commissioned from two top civil servants to define the means and goals of a concerted policy of modernization for French society in the last years of the century.[7]

Simon Nora and Alain Minc called for a technological offensive in "telematics," the term they coined to describe the marriage of computers and communications. The telematic revolution, they argued, would change the nature of modern societies as radically as the industrial revolution. But, they added, "'Telematics,' unlike electricity, does not carry an inert current, but rather information, that is to say, power."[8] "Mastering the network is therefore an essential goal. Its framework must therefore be conceived in the spirit of public service."[9] In sum, just as war is too important to be left to the generals, so postindustrial development is too important to be left to businessmen and must become a political affair.

Nora and Minc[10] paid particular attention to the need to win public acceptance of the telematic revolution and to achieve success in the new international division of labor through targeting emerging telematic markets. They argued that a national videotex service could play a central role in achieving these objectives. This service would sensitize the still backward French public to the wonders of the computer age while creating a huge protected market for computer terminals. Leveraging the internal market, France would eventually become a leading exporter of terminals and so benefit from the expected restructuring of the international economy instead of falling further behind.[11] These ideas lay at the origin of the Teletel project, which, as a peculiar mix of propaganda and industrial policy, had a distinctly statist flavor from the very beginning.

OPPOSITION

As originally conceived, Teletel was designed to bring France into the information age by providing a wide variety of services. But is more information what every household needs?[12] And who is qualified to offer information services in a democracy?[13] These questions received a variety of conflicting answers in the early years of French videotex.

Modernization through national service defines the program of a highly centralized and controlling state. To make matters worse, the Teletel project was initiated by a conservative government. This combination at first inspired widespread distrust of videotex and awakened the well-known fractiousness of important sectors of opinion. The familiar pattern of central control and popular "resistance" was repeated once again with Teletel, a program that was "parachuted" on an unsuspecting public and soon transformed by it in ways its makers had never imagined.

The press led the struggle against government control of videotex. When the head of the telephone company announced the advent of the paperless society (in Dallas of all places), publishers reacted negatively out of fear of losing advertising revenues and independence. The dystopian implications of a computer-ruled society did not pass unnoticed. One irate publisher wrote, "He who grasps the wire is powerful. He who grasps the wire and the screen is very powerful. He who will someday grasp the wire, the screen, and the computer will possess the power of God the Father Himself."[14]

The press triumphed with the arrival of the socialist government in 1981. To prevent political interference with on-line "content," the telephone company was allowed to offer only its electronic version of the telephone directory. The doors to Teletel were opened wide by the standards of the day: anyone with a government-issued publishers' license could connect a host to the system. In 1986 even this restriction was abandoned; today anyone with a computer can hook up to the system, list a phone number in the directory, and receive a share of the revenues the service generates for the phone company.

Because small host computers are fairly inexpensive and knowledge of videotex [is] as rare in large as in small companies, these decisions had at first a highly decentralizing effect. Teletel became a vast space of disorganized experimentation, a "free market" in on-line services more nearly approximating the liberal ideal than most communication markets in contemporary capitalist societies.

This example of the success of the market has broad implications, but not quite so broad as the advocates of deregulation imagine. The fact that markets sometimes mediate popular demands for technical change does not make them a universal panacea. The conditions that make such a use of markets possible are quite specific. Frequently, for example, where large corporations sell well-established technologies, they use markets to stifle the demands existing products cannot meet or rechannel them into domains where basic technical change need not occur. Nevertheless, consumers do occasionally reopen the design process through the market. This is certainly a reason to view markets as ambivalent institutions with a potentially dynamic role to play in the development of new technology.

COMMUNICATION

Surprisingly, although phone subscribers were now equipped for the information age, they made relatively little use of the wealth of data available on Teletel. They consulted the electronic directory regularly, but not much else. Then, in 1982, hackers transformed the technical support facility of an information service called Gretel into a messaging system.[15] After putting up a feeble (perhaps feigned) resistance, the operators of this service institutionalized the hackers' invention and made a fortune. Other services quickly followed with names like "Désiropolis," "La Voix du Parano," "SM," "Sextel." "Pink" messaging became famous for spicy pseudononymous conversations in which users sought likeminded acquaintances for conversation or encounters.

Once messaging took off on a national scale, small telematic firms reworked Teletel into a communication medium. They designed programs to manage large numbers of simultaneous users emitting rather than receiving information, and they invented a new type of interface. On entering these systems, users are immediately asked to choose a pseudonym and to fill out a brief "C.V." (curriculum vitae, or *carte de visite*). They are then invited to survey the C.V.'s of those currently on-line to identify like-minded conversational partners. The new programs employ the Minitel's graphic capabilities to split the screen, assigning each of as many as a half-dozen communicators a separate space for their messages. This is where the creative energies awakened by telematics went in France, and not into meeting obscure technical challenges dear to the hearts of government bureaucrats such as ensuring French influence on the shape of the emerging international market in databases.[16]

The original plans for Teletel had not quite excluded human communication, but they certainly underestimated its importance relative to the dissemination of data, on-line transactions, and even video games.[17] Messaging is hardly mentioned in early official documents on telematics.[18] The first experiment with Teletel, at Vélizy, revealed an unexpected enthusiasm for communication. Originally conceived as a feedback mechanism linking users to the Vélizy project team, the messaging system was soon transformed into a general space for free discussion.[19] Even after this experience no one imagined that human communication would play a major role in a mature system. But that is precisely what happened.

In the summer of 1985 the volume of traffic on Transpac, the French packet switching network, exceeded its capacities and the system crashed. The champion of French high tech was brought to its knees as banks and government agencies were bumped off-line by hundreds of thousands of users skipping from one messaging service to another in search of amusement. This was the ultimate demonstration of the new telematic dispensation.[20] Although only a minority of users were involved, by 1987, 40 percent of the hours of domestic traffic were spent on messaging.[21]

Pink messaging may seem a trivial result of a generation of speculation on the information age, but the case can be made for a more positive evaluation. Most important, the success of messaging changed the generally received connotations of telematics, away from information toward communication. This in turn encouraged—and paid for—a wide variety of experiments in domains such as education, health, and news.[22] Television programs, for example, now advertise services on Teletel where viewers can obtain supplementary information or exchange opinions, adding an interactive element to the one-way broadcast. Politicians engage in dialogue with constituents on Teletel, and political movements open messaging services to communicate with their members. Educational experiments have brought students and teachers together for electronic classes and tutoring, for example at a Paris medical school. And a psychological service offers an opportunity to discuss personal problems and seek advice.

Perhaps the most interesting experiment occurred in 1986 when a national student strike was coordinated on the messaging service of the newspaper *Libération*. The service offered information about issues and actions, on-line discussion groups, hourly news updates, and a game mocking the minister of education. It quickly received three thousand messages from all over the country.[23]

These applications reveal the unsuspected potential of the medium for creating surprising new forms of sociability. Rather than imitating the telephone or writing, they play on the unique capacity of telematics to mediate highly personal, anonymous communication. These experiments prefigure a very different organization of public and private life in advanced societies.[24]

SOCIAL CONSTRUCTIVISM

Teletel's evolution confirms the social constructivist approach [to understanding technology]. Unlike determinism, social constructivism does not rely exclusively on the technical characteristics of an artifact to explain its success. According to the "principle of symmetry," there are always alternatives that might have been developed in the place of the successful one. What singles out an artifact is not some intrinsic property such as "efficiency" or "effectiveness" but its relationship to the social environment.

As we have seen in the case of videotex, that relationship is negotiated among inventors, civil servants, businessmen, consumers, and many other social groups in a process that ultimately defines a specific product adapted to a specific mix of social demands. This process is called "closure"; it produces a stable "black box," an artifact that can be treated as a finished whole. Before a new technology achieves closure, its social character is evident, but once it is well established, its development appears purely technical, even inevitable to a naive backward glance. Typically, later observers forget the original ambiguity of the situation in which the black box was first closed.[25]

[Trevor] Pinch and [Wiebe] Bijker illustrate their method with the example of the bicycle. In the late nineteenth century, before the present form of the bicycle was fixed, design was pulled in several different directions. Some customers perceived bicycling as a competitive sport, while others had an essentially utilitarian interest in transportation. Designs corresponding to the first definition had high front wheels that were rejected as unsafe by the second type of rider, who preferred designs with two equal-sized low wheels. Eventually, the low wheelers won out and the entire later history of the bicycle down to the present day stems from that line of technical development. Technology is not determining in this example; on the contrary, the "different interpretations by social groups of the content of artifacts lead via different chains of problems and solutions to different further developments."[26]

This approach has several implications for videotex. First, the design of a system like Teletel is not determined by a universal criterion of efficiency but by a social process that differentiates technical alternatives according to a variety of criteria. Second, that social process is not about the application of a predefined videotex technology, but concerns the very definition of videotex and the nature of the problems to which it is addressed. Third, competing definitions reflect conflicting social visions of modern society concretized in different technical choices. These three points indicate the need for a revolution in the study of technology. The first point widens the range of social conflict to include technical issues which, typically, have been treated as the object of a purely "rational" consensus. The other two points imply that meanings enter history as effective forces not only through cultural production and political action but also in the technical sphere. To understand the social perception or definition of a technology one needs a hermeneutic of technical objects.

Technologies are meaningful objects. From our everyday commonsense standpoint, two types of meanings attach to them. In the first place, they have a function, and for most purposes their meaning is identical with that function. However, we also recognize a penumbra of "connotations" that associate technical objects with other aspects of social life independent of function.[27] Thus automobiles are means of transportation, but they also signify the owner as more or less respectable, wealthy, and sexy.

In the case of well-established technologies, the distinction between function and connotation is usually clear. There is a tendency to project this clarity back into the past and to imagine that the technical function preceded the object and called it into being. The social constructivist program argues, on the contrary, that technical functions are not pregiven but are discovered in the course of the development and use of the object. Gradually certain functions are locked in by the evolution of the social and technical environment, as for example the transportation functions of the automobile have been institutionalized in low-density urban designs that create the demand automobiles satisfy. Closure thus depends in part on building tight connections in a larger technical network.

In the case of new technologies, there is often no clear definition of function at first. As a result, there is no clear distinction between different types of meanings associated with the technology: a bicycle built for speed and a bicycle built for safety are both functionally and connotatively different. In fact, connotations of one design may be functions viewed from the angle of the other. These ambiguities are not merely conceptual, since the device is not yet "closed" and no institutional lock-in ties it decisively to one of its several uses. Thus ambiguities in the definition of a new technology must be resolved through technical development itself. Designers, purchasers, and users all play a role in the process by which the meaning of a new technology is finally fixed.

Technological closure is eventually consolidated in a technical code. Technical codes define the object in strictly technical terms in accordance with the most general social meanings it has acquired. For bicycles, this was achieved in the 1890s. A bicycle safe for transportation could only be produced by conforming to a code which dictated a seat positioned well behind a small front wheel. When consumers encountered a bicycle produced according to this code, they immediately recognized it for what it was: a "safety" in the terminology of the day. That definition in turn connoted women and older riders, trips to the grocery store, and so on, and negated associations with the young sportsman out for a thrill.

Technical codes are interpreted with the same hermeneutic procedures used to interpret texts, works of art, and social actions.[28] But the task gets complicated when codes become the stakes in significant social disputes. Then ideological visions are sedimented in technical design. This is what explains the "isomorphism, the formal congruence between the technical logics of the apparatus and the social logics within which it is diffused."[29] These patterns of congruence explain the impact of the larger sociocultural environment on the mechanisms of closure.[30]

Videotex is a striking case in point. In what follows I will trace the pattern from the highest level of worldviews down to the lowest level of technical design.

THE SOCIAL CONSTRUCTION OF THE MINITEL

As we have seen, the peculiar compromise that made Teletel a success was the resultant of these forces in tension. I have traced the terms of that compromise at the macrolevel of the social definition of videotex in France, but its imprint can also be identified in the technical code of the system interface.

WIRING THE BOURGEOIS INTERIOR

The Minitel is a sensitive index of these tensions. Those charged with designing it feared public rejection of anything resembling a computer, typewriter, or other professional apparatus and worked to fit it into the social context of the domestic environment. They carefully considered the "social factors" as well as the human factors involved in persuading millions of ordinary people to admit a terminal into their home.[31]

This is a design problem with a long and interesting history. Its presupposition is the separation of public and private, work and home, which begins, according to Walter Benjamin,[32] under the July Monarchy: "For the private person, living space becomes, for the first time, antithetical to the place of work. The former is constituted by the interior; the office is its complement. The private person who squares his accounts with reality in his office demands that the interior be maintained in his illusions."

The history of design shows these intimate illusions gradually shaped by images drawn from the public sphere through the steady invasion of private space by public activities and artifacts. Everything from gas lighting to the use of chrome in furniture begins life in the public domain and gradually penetrates the home.[33] The telephone and the electronic media intensify the penetration by decisively shifting the boundaries between the public and the private sphere.

The final disappearance of what Benjamin calls the "bourgeois interior" awaits the generalization of interactivity. The new communications technologies promise to attenuate and perhaps even to dissolve the distinction between the domestic and the public sphere. Telework and telemarketing are expected to collapse the two worlds into one. "The home can no longer pretend to remain the place of private life, privileging noneconomic relations, autonomous with respect to the commercial world."[34]

The Minitel is a tool for accomplishing this ultimate deterritorialization. Its modest design is a compromise on the way toward a radically different type of interior. Earlier videotex systems had employed very elaborate and expensive dedicated terminals, television adapters, or computers equipped with modems. So far, outside France, domestic CMC has only succeeded where it is computer based, but its spread has been largely confined to a hobbyist subculture. No design principles for general distribution can be learned from these hobbyists, who are not bothered by the incongruous appearance of a large piece of electronic equipment on the bedroom dresser or the dining room table. Functionally, the Minitel is not even a computer in any case. It is just a "dumb terminal," that is, a video screen and keyboard with minimal memory and processing capabilities and a built-in modem. Such devices have been around for decades, primarily for use by engineers to operate mainframe computers. Obviously designs suitable for that purpose would not qualify as attractive interior decoration.

The Minitel's designers broke with all these precedents and connoted it as an enhancement of the telephone rather than as a computer or a new kind of television, the two existing models.[35] Disguised as a "cute" telephonic device, the Minitel is a kind of Trojan horse for rationalistic technical codes.

It is small, smaller even than a Macintosh, with a keyboard that can be tilted up and locked to cover the screen. At first it was equipped with an alphabetical keypad to distinguish it from a typewriter. That keypad pleased neither nontypists nor typists and was eventually replaced with a standard one; however, the overall look of the Minitel remained unbusinesslike.[36] Most important, it has no disks and disk drives, the on-off switch on its front is easy to find, and no intimidating and unsightly cables protrude from its back, just an ordinary telephone cord.

The domesticated Minitel terminal adopts a telephonic rather than a computing approach to its users' presumed technical capabilities. Computer programs typically offer an immense array of options, trading off ease of use for power. Furthermore, until the success of Windows, most programs had such different interfaces that each one required a special apprenticeship. Anyone who has ever used early DOS communications software, with its opening screens for setting a dozen obscure parameters, can understand just how inappropriate it would be for general domestic use. The Minitel designers knew their customers well and offered an extremely simple connection procedure: dial up the number on the telephone, listen for the connection, press a single key.

The design of the function keys also contributed to ease of use. These were intended to operate the electronic telephone directory. At first there was some discussion of giving the keys highly specific names suited to that purpose, for example, "City," "Street," and so on. It was wisely decided instead to assign the function keys general names such as "Guide," "Next Screen," "Back," rather than tying them to any one service.[37] As a result, the keyboard imposes a standard and very simple user interface on all service providers, something achieved in the computing world by Windows, but only with much more elaborate equipment.

The Minitel testifies to the designers' original skepticism with regard to communication applications of the system: the function keys are defined for screen-oriented interrogation of data banks, and the keypad, with its unsculptured chiclet keys, is so clumsy it defies attempts at touch typing. Here the French paid the price of relying on a telephonic model: captive telephone company suppliers ignorant of consumer electronics markets delivered a telephone-quality keypad below current international standards for even the cheapest portable typewriter. Needless to say, export of such a terminal has been difficult.

AMBIVALENT NETWORKS

So designed, the Minitel is a paradoxical object. Its telephonic disguise, thought necessary to its success in the home, introduces ambiguities into the definition of telematics and invites communications applications not anticipated by the designers.[38] For them the Minitel would always remain a computer terminal for gathering data, but the domestic telephone, to which the Minitel is attached, is a social, not an informational medium. The official technical definition of the system thus enters into contradiction with the telephonic practices that immediately colonize it once it is installed in the home.[39]

To the extent that the Minitel did not rule out human communication altogether, as have many videotex systems, it could be subverted from its intended purpose despite its limitations. For example, although the original function keys were not really designed for messaging applications, they could be incorporated into messaging programs, and users adapted to the poor keyboard by typing in a kind of on-line shorthand rich in new slang and inventive abbreviations. The Minitel thus became a communication device.

The walls of Paris were soon covered with posters advertising messaging services. A whole new iconography of the reinvented Minitel replaced the sober modernism of official Telecom propaganda. In these posters, the device is no longer a banal computer terminal, but is associated with blatant sexual provocation. In some ads, the Minitel walks, it talks, it beckons; its keyboard, which can flap up and down, becomes a mouth, the screen becomes a face. The silence of utilitarian telematics is broken in a bizarre cacophony.

In weakening the boundaries of private and public, the Minitel opens a two-way street. In one direction, households become the scene of hitherto public activities, such as consulting train schedules or bank accounts. But in the other direction, telematics unleashes a veritable storm of private fantasy on the unsuspecting public world. The individual still demands, in Benjamin's phrase, that the "interior be maintained in his illusions." But now those illusions take on an aggressively erotic form and are broadcast over the network.

The technical change in the Minitel implied by this social change is invisible but essential. It was designed as a client node, linked to host computers, and was not intended for use in a universally switched system which, like the telephone network, allows direct connection of any subscriber with any other. Yet as its image changed, the Telecom responded by creating a universal electronic mail service, called Minicom, which offers an electronic mailbox to everyone with a Minitel. The Minitel was finally to be fully integrated to the telephone network.

Despite the revenues earned from these communications applications, the Telecom grumbles that its system is being misused. Curiously, those who introduced the telephone a century ago fought a similar battle with users over its definition. The parallel is instructive. At first the telephone was compared to the telegraph and advertised primarily as an aid to commerce. There was widespread resistance to social uses of the telephone, and an attempt was made to define it as a serious instrument of business.[40] In opposition to this "masculine" identification of the telephone, women gradually incorporated it into their daily lives as a social instrument.[41] As one telephone company official complained in 1909, "The telephone is going beyond its original design, and it is a positive fact that a large percentage of telephones in use today on a flat rental basis are used more in entertainment, diversion, social intercourse and accommodation to others than in actual cases of business or household necessity."[42]

In France erotic connotations clustered around these early social uses of the telephone. It was worrisome that outsiders could intrude in the home while the husband and father were away at work. "In the imagination of the French of the Belle Epoque, the telephone was an instrument of seduction."[43] So concerned was the phone company for the virtue of its female operators that it replaced them at night with males, presumably proof against temptation.[44]

Despite these difficult beginnings, by the 1930s sociability had become an undeniable referent of the telephone in the United States. (In France the change took longer.) Thus the telephone is a technology which, like videotex, was introduced with an official definition rejected by many users. And like the telephone, the Minitel too acquired new and unexpected connotations as it became a privileged instrument of personal encounter. In both cases, the magic play of presence and absence, of disembodied voice or text, generates unexpected social possibilities inherent in the very nature of mediated communication.

CONCLUSION: THE FUTURE
OF THE COMMUNICATION SOCIETY

In its final configuration, Teletel was largely shaped by the users' preferences.[45] The picture that emerges is quite different from initial expectations. What are the lessons of this outcome? The rationalistic image of the information age did not survive the test of experience unchanged. Teletel today is not just an information marketplace. Alongside the expected applications, users invented a new form of human communication to suit the need for social play and encounter in an impersonal, bureaucratic society. In so doing, ordinary people overrode the intentions of planners and designers and converted a postindustrial informational resource into a postmodern social environment.

The meaning of videotex technology has been irreversibly changed by this experience. But beyond the particulars of this example, a larger picture looms. In every case, the human dimension of communication technology emerges only gradually from behind the cultural assumptions of those who originate it and first signify it publicly through rationalistic codes. This process reveals the limits of the technocratic project of postindustrialism.

NOTES

1. Daniel Bell, *The Coming of Post-industrial Society* (New York: Basic Books, 1973).

2. Herbert S. Dordick, Helen G. Bradley, and Burt Nanus, eds., *The Emerging Network Marketplace* (Norwood, N.J.: Albex, 1979).

3. Anne Branscomb, "Videotext: Global Progress and Comparative Policies," *Journal of Communication* 38, no. 1 (1988).

4. Jean-Marie Charon, "Teletel, de l'interactivité home/machine à la communication médiatisée," in *Les Paradis Informationnels,* ed. Marie Marchand (Paris: Masson, 1987), pp. 103–106; Renate Mayntz and Volker Schneider, "The Dynamics of System Development in a Comparative Perspective: Interactive Videotex in Germany, France, and Britain," in *The Development of Large Technical Systems,* eds. Renate Mayntz and Thomas Hughes (Boulder: Westview Press, 1988), p. 278.

5. James Ettema, "Interactive Electronic Text in the United States: Can Videotex Ever Go Home Again?" in *Media Use in the Information Society,* eds. J. C. Salvaggio and J. Bryant (Hillsdale, N.J.: Lawrence Erlbaum, 1989).

6. Andrew Feenberg, *Critical Theory of Technology* (New York: Oxford University Press, 1991), chap. 5.

7. Simon Nora and Alain Minc, *L'Informatisation de la société* (Paris: Editions du Seuil, 1978).

8. Ibid., p. 11.

9. Ibid., p. 67.

10. Ibid., pp. 41–42.

11. Ibid., pp. 94–95.

12. Raymon-Stone Iwaasa, "Télématique grand public: l'information ou la communication? Les cas de Grétel et de Compuserve," *Le Bulletin de l'DATE*, no. 18 (1985): 49.

13. Marie Marchand, *La Grande Aventure du Minitel* (Paris: Larousse, 1987), pp. 40ff.

14. Ibid., p. 42.

15. Thierry Bruhat, "Messageries electroniques: Grétel a Strasbourh et Teletel a Vélizy," in *Télématique: promenade dans les usages,* eds. Marie Marchand and Claire Ancelin (Paris: Documentation français, 1984), pp. 54–55.

16. Nora and Minc, *L'Informatisation de la société, p. 72.*

17. Marchand, *La Grande Aventure du Minitel, p. 136.*

18. Henri Pigeat et al., *Du téléphone a la télématique: rapport du groupe de travail* (Paris: Documentation français, 1979).

19. Jean-Marie Charon and Eddy Cherky, *Le Vidéotex: un nouveau média local: enquete sur l'experimentation de Vélizy* (Paris: Centre d'Etude des Mouvements Sociaux, 1983), pp. 81–92; Marchand, *La Grande Aventure du Minitel, p. 72.*

20. Marchand, *La Grande Aventure du Minitel,* pp. 132–34.

21. J. L. Chabrol and Pascal Perin, *Usages et usagers du vidéotex: les pratiques domestiques du vidéotex en 1987* (Paris: DGT, 1989), p. 7.

22. Marchand, *La Grande Aventure du Minitel*; Catherine Bidou, Marc Guillaume, and Véronique Prévost, *L'Ordinaire de la télématique: offre et usages des services utilitaires grand-public* (Paris: Editions de l'Iris, 1988).

23. Marchand, *La Grande Aventure du Minitel*, pp. 155–58.

24. Andrew Feenberg, "A User's Guide to the Pragmatics of Computer Meditated Communication," *Semiotica* 75, no. 3/4 (1989): 271–75; J. Jouet and P. Flichy, *European Telematics: The Emerging Economy of Words*, trans. D. Lytel (Amsterdam: Elsevier, 1991).

25. Bruno Latour, *Science in Action* (Cambridge: Harvard University Press, 1987), pp. 2–15.

26. Trevor Pinch and Wiebe Bijker, "The Social Construction of Facts and Artefacts: or How the Sociology of Science and the Sociology of Technology Might Benefit Each Other," *Social Studies of Science* 14, no. 3 (1984): 423.

27. Jean Baudrillard, *Le Système des objects* (Paris: Gallimard, 1968), pp. 16–17.

28. Paul Ricoueur, "The Model of the Text: Meaningful Action Considered as a Text," in *Interpretive Social Science: A Reader*, eds. P. Rabinow and W. Sullivan (Berkeley: University of California Press, 1979).

29. Bidou, Guillaume, and Prévost, *L'Ordinaire de la télématique: offre et usages des services utilitaires grand-public*, p. 18.

30. Pinch and Bijker, "The Social Construction of Facts and Artefacts: or How the Sociology of Science and the Sociology of Technology Might Benefit Each Other," p. 409.

31. Andrew Feenberg, "The Written Word," in *Mindweave: Communication, Computers, and Distance Learning*, eds. A. Kaye and R. Mason (Oxford: Pergamon Press, 1989), p. 29.

32. Walter Benjamin, "Paris, Capital of the Nineteenth Century," in *Reflections*, ed. P. Dementz, trans. E. Jephcott (New York: Harcourt Brace Janovich, 1978), p. 154.

33. Wolfgang Schivelbusch, *Disenchanted Light*, trans. A. Davies (Berkeley: University of California Press, 1988); Adrian Forty, *Objects of Desire* (New York: Pantheon, 1986), chap. 5.

34. Marie Marchand, "Conclusion: vivre avec le Videotex," in *Télématique: promenade dans les usages*, eds. Marie Marchand and Claire Ancelin (Paris: Documentation français, 1984), p. 184.

35. Alain Giraud, "Une Lente Emergence," in *Télématique: promenade dans les usages*, eds. Marie Marchand and Claire Ancelin (Paris: Documentation français, 1984), p. 9.

36. Marchand, *La Grande Aventure du Minitel*, p. 64; Donald Norman, *The Psychology of Everyday Things* (New York: Basic Books, 1988), p. 147.

37. Marchand, *La Grande Aventure du Minitel*, p. 65.

38. Christian Weckerlé, *Du téléphone au Minitel: acteurs et facteurs locaux dans la constitution des images er usages sociaux de la télématique*, 2 vols. (Paris: Groupe de Recherche et d'Analyse du Social et de la Sociabilité, 1987), pp. 1, 14–15.

39. Ibid., pp. 1, 26.

40. Claude Fischer, "'Touch Someone': The Telephone Industry Discovers Sociability," *Technology and Culture* 29, no. 1 (1988); Jean Attali and Yves Stourdze, "The Birth of the Telephone and Economic Crisis: The Slow Death of the Monologue in French Society," in *The Social Impact of the Telephone*, ed. Ithiel de Sola Pool (Cambridge: MIT Press, 1977).

41. Claude Fischer, "Gender and the Residential Telephone, 1890–1940: Technologies of Sociability," *Sociological Forum* 3, no. 2 (1988).

42. Quoted in Fischer, "'Touch Someone': The Telephone Industry Discovers Sociability," p. 48.

43. Catherine Bertho, *Télégraphes et téléphones: de Valmy au microprocesseur* (Paris: Livre de Poche, 1981), p. 243.

44. Ibid., pp. 242–43.

45. Jean-Marie Charon, "Teletel, de l'interactivité home/machine à la communication médiatisée," p. 100.

Citizen Virtues in a Technological Order

Langdon Winner

As it ponders important social choices that involve the application of new technology, contemporary moral philosophy works within a vacuum. The vacuum is created, in large part, by an absence of widely shared understandings, reasons, and perspectives that might guide societies as they confront the powers offered by new machines, techniques, and large-scale technological systems. Which computer applications are desirable and which ought to be avoided? How can one weigh the risks of introducing a new chemical into the environment as compared to benefits of its use? Should there be limits placed upon the ability of biotechnology to alter the genetic structure of plant and animal life? As we ponder issues of this kind, it is not always clear which principles, policies, or forms of moral reasoning are suited to the choices at hand.

The vacuum is a social as well as intellectual one. Often there are no persons or organizations with clear authority to make the decisions that matter. In fact, there may be no clearly defined social channels in which important moral issues can be addressed at all. Typically, what happens in such cases is that, as time passes, a mixture of corporate plans, market choices, interest group activities, lawsuits, and government legislation takes shape to produce jerrybuilt policies. But given the number of points at which technologies generate significant social stress and conflict, this familiar pattern is increasingly unsatisfactory.

Philosophers sometimes rush in to fill the void, offering advice that matches their training and competence. They examine cases in which some feature of a present or emerging technology raises questions about right and wrong in individual choices and social policies. They take note of properties of the new technology that have important consequences for social life, properties that raise interesting philosophical issues; for example, issues about the rights and responsibilities of those who

Reprinted from "Citizen Virtues in a Technological Order," by Langdon Winner, from *Inquiry* (www.tandf.no/inquiry), 1992, Volume 35, pp. 341–61, by permission of Taylor & Francis AS.

develop or use the technology in question. From there they can develop a variety of theories, principles, and arguments that may help people decide what to do.

Proceeding in this way, philosophers may find themselves involved in an exercise that is essentially technocratic. The complicated business of research, development, and application in modern life includes a moment where the "value issues" need to be studied and where the contributions of knowledgeable, degree-carrying experts can be enlisted. In the United States, for example, the National Science Foundation has for many years included a program on "ethical and value studies" that supports university scholars who do research of this kind. The underlying assumption seems to be that this is an important area that the nation needs to cultivate. The sponsors may hope that officially designated "values experts" can eventually provide "solutions" to the kinds of "problems" whose features are ethical rather than solely technical. This can serve as a final tune-up for working technological models about to be rolled out the showroom door. "Everything else looks good. What are the results from the ethics lab?"

Philosophers sometimes find it tempting to play along with these expecta-tions, gratifying to find that anyone cares about what they think, exhilarating to notice that their ideas might actually have some effect. But is it wise to don the mantle of values expert? Although philosophers may be well equipped to help fill the intellectual emptiness caused by the lack of moral understandings, ethical reasoning, and community guidelines, there remains the social and political vacuum that so often surrounds discussions about the moral dimensions of tech-nological choice. After one has addressed the range of social theories, empirical analyses, philosophical arguments, and ethical principles about the possibilities of Technology X, there remains the embarrassing question: Who in the world are we talking to? Where is the community in which our wisdom will be welcome?

Consider the following passages from two prominent writers addressing urgent ethical questions for our time. The first is from a well-known biologist reflecting about the ethical dimensions of developments in his own field.

> Given the nature of our society, which embraces and applies any new tech-nology, it appears that there is no means, short of unwanted catastrophe, to pre-vent the development of [human] genetic engineering. It will proceed. But this time, perhaps we can seek to anticipate and guide its consequences.[1]

The second passage was written by a professional philosopher, exploring avenues for the new field of computer ethics.

> We are open to invisible abuse or invisible programming of inappropriate values or invisible miscalculation. The challenge for computer ethics is to formulate policies which will help us deal with this dilemma. We must decide when to trust computers and when not to trust them.[2]

Both of these passages are notable for the way they employ the term *we* in contexts where moral issues about technology are open for discussion. But who is the "we" to whom the writers refer? Both writers seem to mean something like "people in general" or "society as a whole." Or perhaps they mean something like "those who work in a particular field of technical development and have privileged access to the decisions that matter."

I raise this point not to call attention to the way writers, including this one, loosely deploy first-person plural pronouns. What matters here is that this lovely "we" suggests the presence of a moral community that may not, in fact, exist at all, at least not in any coherent, self-conscious form. If "we" scholars find ourselves talking about a collectivity of others who are not in fact engaged in decisions, then it is time for "us" to look around and find out where "they" have gone. That is the important first task for the contemporary ethics of technology. It is time to ask: What is the identity and character of the moral communities that will make the crucial, world-altering judgments and take appropriate action as a result?

This question is, in my view, one about politics and political philosophy rather than a question for ethics considered solely as a matter of right and wrong in individual conduct. For the central issues here concern how the members of society manage their common affairs and seek the common good. Because technological things so often become central features in widely shared arrangements and conditions of life in contemporary society, there is an urgent need to think about them in a political light. Rather than continue the technocratic pattern in which philosophers advise a narrowly defined set of decision makers about ethical subtleties, today's thinkers would do better to reexamine the role of the public in matters of this kind. How can and should democratic citizenry participate in decision making about technology?

Unfortunately, the Western tradition of moral and political philosophy has little to recommend on this score, almost nothing to say about the ways in which persons in their roles as citizens might be involved in making choices about the development, deployment, and use of new technology. Most thinkers in our tradition have placed technology and politics in separate categories, defining citizen roles as completely isolated from the realities of technical practice and technical change. There have been two distinctive paths to this conclusion, one characteristic of thinkers in antiquity, another strongly advanced in modern times. But whether we are pondering ancient *techne* or today's megatechnics, any attempt to discuss technology as a topic in political and moral philosophy needs to pause long enough to appreciate how this crucial separation occurred and how it impairs our sense of possibilities.

TECHNOLOGY AND CITIZEN: THE ANCIENT VIEW

At the beginning of Western moral and political philosophy, speculation about *techne*, the realm of the practical arts, plays a prominent but largely negative role.

As Socrates, Plato, and Aristotle seek to define the nature of knowledge, the good, political society, justice, rulers and citizens, and the form of the best state, they frequently draw comparisons to *techne*, the realm of the arts and crafts, viewing it with a mixture of awe and suspicion. Foremost among their concerns is the belief that technical affairs constitute an inferior realm of objects, knowledge, and practice, one that threatens to infect all who aspire to higher things.[3] Plato goes even further, specifying why the realm of *techne* is both inferior and potentially dangerous. True knowledge, he argues, is not that of worldly, mutable, material things, but knowledge of the realm of unchanging ideas, *eidos*.[4]

Arguing a position that was to become commonplace in antiquity and throughout much of the Middle Ages, Plato also criticizes the practical arts for their tendency to produce innovations, a source of harmful, potentially boundless change in human affairs. Political philosophy seeks to establish good order and to maintain it against the world's tendency toward chaos and decay. "Change, we shall find, is much the most dangerous thing in everything except what is bad—in all the seasons, in bodily habits, and in the characters of souls."[5] In the first century B.C.E. Lucretius echoes these sentiments, lamenting the destructive role of new techniques in warfare. "Tragic discord gave birth to one invention after another for the intimidation of the nations' fighting men and added daily increments to the horrors of war."[6]

Of all classical arguments calling for the separation of technology from political affairs, the most significant is Aristotle's. For unlike Plato, Aristotle explores the possibilities of a broadly based citizenship in political societies of many different kinds, perhaps even ones that resemble our own. As he defines the roles and virtues of a citizen, however, the crucial differences between technical and political life stand out.

Aristotle's view that "man is by nature a political animal" means that humans are creatures naturally suited to live in a *polis* or city-state.[7] Drawing upon studies of some one hundred and fifty city-states of his time, the *Politics* argues that the *polis* is the highest form of human organization, one that completes the development of other forms of association, the household and the village. Political life is a gathering of freemen and equals. Each person is free in the sense that there is no master to dictate one's activities. Each one is equal as well, equal in legal standing, access to public office, and right to speak in political matters. Political life concerns matters that all citizens have in common. In the public sphere one's attention moves beyond personal or family interests to seek the good of the whole community. "One citizen differs from another, but the salvation of the community is the common business of them all."[8] Citizenship, active participation in public life, fulfills man's highest potential. The *bios politicos* realizes a greater good than more primitive forms of human existence ever attain.

Having defined politics in this manner, Aristotle goes on to explore the specific roles and virtues of the citizen. He notes the traditional distinction between the rulers and the ruled and concludes that the citizen must be different from both. Citizenship in his view must include both roles within each person.[9]

The excellence of the two is not the same, but the good citizen ought to be capable of both; he should know how to govern like a freeman, and how to obey like a freeman—these are the excellences of a citizen. And although the temperance and justice of a ruler are distinct from those of a subject, the excellence of a good man will include both.

Looking at a range of existing constitutions, Aristotle concludes that a good constitution will allow the rotation of citizens in office so the "excellences" or "virtues" he recommends will become common in actual practice.

In the same passages that offer his definition of citizenship, Aristotle takes care to specify which persons are not capable of holding this role. He points to the menial duties and craft work that were handled by slaves and foreign workers in Greek city-states of the time. Physical toil and use of the practical arts bind one to the realm of material necessity, a condition incompatible with the unencumbered freedom needed for citizenship. While slaves and craftsmen are necessary for the existence of the state and while some city-states recognize them as citizens, a good society will not extend citizenship in this way, "for no man can practice excellence who is living the life of a mechanic or laborer."[10]

Aristotle goes even further, arguing that citizens should avoid learning the practical arts because that would be degrading. "Certainly the good man and the statesman and the good citizen ought not to learn the crafts of inferiors except for their own occasional use; if they habitually practice them, there will cease to be a distinction between master and slave."[11] Thus, the making of useful things and the activities of public life must forever remain separate.

While the ideas of Socrates, Plato, and Aristotle did not by themselves define the understanding of the Greeks and Romans on such matters, entirely similar notions about technology and economics were common in antiquity. The sphere of technical affairs was closely associated with slavery and menial labor and was, therefore, something that persons of the ruling classes sought to avoid. In fact, wealthy Romans normally left the day-to-day handling of private economic affairs to their slaves, the origins of what we today call "management."[12] While Romans sought material wealth, it was usually gained through landed property and commercial trade, economic sources that did not require recurring technical change. Indeed, technological innovation was widely regarded with suspicion. Suetonius tells of a time when a creative soul came to the emperor Vespasian with a device for carrying heavy columns into Rome at a low cost. Although Vespasian rewarded the man for his invention, he refused to use it, exclaiming, "How will it be possible for me to feed the populace?"[13] As the historian M. I. Finley concludes in *The Ancient Economy*, "Economic growth, technical progress, increasing efficiency are not 'natural' virtues; they have not always been possibilities or even desiderata, at least not for those who controlled the means by which to try to achieve them."[14]

TECHNOLOGY AND CITIZEN: THE MODERN VIEW

With the renewal of political theory in the sixteenth century and since, the prospects for social and political life are gradually redefined. Concepts of power, authority, order, liberty, equality, and the state are deployed in ways that we now consider distinctly modern. The attempts of [Niccolò] Machiavelli, [Thomas] More, [Thomas] Hobbes, [John] Locke, [Baron de] Montesquieu, [Jeremy] Bentham, and [Karl] Marx to create a new understanding of politics corresponded to pathbreaking work in natural sciences that produced new ways of thinking about the physical world. Strongly associated with these intellectual movements is a thoroughgoing reevaluation of the sphere of technical practice and its economic settings, a reevaluation in which the pessimism of ancient and medieval views eventually yields to an unbridled optimism. In this ferment of ideas, the traditional view of the relationship between politics and technology was overthrown and a new one imagined.

A leader in promoting respect for technical activity was Francis Bacon. In *The New Organon*, Bacon surveys the state of knowledge in his time, criticizing the hold of the ancient philosophers over the minds of moderns. He argues that the supposed wisdom of the Greeks is suspect precisely because it lacks any practical, material value: "[I]t can talk, but it cannot generate, for it is fruitful of controversies but barren of works."[15] As an alternative Bacon sets forth a new program of knowledge and practice, one based upon careful study of particular phenomena, adherence to method, inductive logic, controlled experiment, naturalistic explanation, and a specialized division of labor among scientists. The ultimate purpose of such activity, he makes clear, ought to be the conquest of nature and expansion of human powers. Natural philosophy must go beyond the quest for knowledge as an end in itself and seek fulfillment in the practical arts.

As a former politician who had fallen from power in disgrace, Bacon enthusiastically praises the superiority of the new scientific and technical pursuits in contrast to affairs of state. Comparing the contributions of history's political heroes to those who have made wonderful discoveries and inventions, Bacon concludes that the highest honors go to scientific and technical innovators, "For the benefits of discoveries may extend to the whole race of man, civil benefits only to particular places; the latter last not beyond a few ages, the former through all time."[16]

Although Bacon's expectations about the directions the arts and sciences ought to pursue were not always prescient, his promotional views won numerous followers in later generations. Explicitly taking his advice, many French philosophers of the eighteenth century took great care to stress not only the practical value of technical pursuits but their intellectual strengths as well. In his *Preliminary Discourse to the Encyclopedia of Diderot*, Jean Le Rond D'Alembert notes the widespread contempt that surrounds the mechanical arts, an outlook that even the artisans themselves seem to share. He argues that, in fact, "it is perhaps in the artisan that one must seek the most admirable evidences of the sagacity, the patience, and the resources of the mind."[17]

Closely linked to a more favorable view of the practical arts and technical innovation is a change in attitude toward commerce and material self-interest. During the Middle Ages, avarice was often identified as both a sin and a source of civil unrest. While medieval societies were often quite open in their quest for wealth, the dominant view among church, political, and intellectual elites was that such motives should be carefully contained. A significant development in modern social and political thought was to annul this distrust and to recast ideas about wealth and commerce in an entirely favorable light. The pursuit of economic gain, some philosophers began to argue, is actually a force for moderation, helping to nurture more rational, peace-loving attitudes among both rulers and subjects. Persons with an economic stake in such trade and manufacturing were now thought to be healthy contributors to stability and justice in political society.[18] As Baron de Montesquieu argues in *The Spirit of the Laws*, "[T]he spirit of commerce is naturally attended with that of frugality, economy, moderation, labor, prudence, tranquillity, order, and rule. So as long as this spirit subsists, the riches it produces have no bad effect."[19] Commerce, he argues, has another beneficial effect, binding nations together in a pattern of mutual need that discourages conflict.

Ideas of this sort, increasingly common in seventeenth- and eighteenth-century political theories, helped justify the modern optimism about economic self-interest and faith in the beneficence of economic growth which lie at the foundation of modern liberal thought. In the new understanding, wealth is good not only for its material benefits but also because its pursuit produces better rulers and better citizens.

The idea that self-interested economic activity is fundamental to politics is strongly expressed in the writings of John Locke. In *The Second Treatise of Government*, Locke's conception of man is that of an acquisitive creature who subdues nature and makes it his property. Men leave the "state of nature" when they come to realize that their possessions are insecure. They form a society and, as a second step, submit to the rule of a government which recognizes their rights, particularly the right of property. From this point of view, the function of political society and government is that of defending the holdings of what are in essence private individuals. If it turns out that government is not useful in achieving these purposes, it can be rightfully overturned in revolution.

At the center of Locke's theory of political society and of modern liberal theory in general is a conception of human life that C. B. MacPherson has called "possessive individualism."[20] In this vision, acquisitiveness emerges as a positive, civilizing force. For as people pursue material gain, they become more rational, industrious, peaceful, and law-abiding. Hence the purely private virtues appropriate to a market society and capitalism are the virtues that build a stable political order. Of the activities that help produce a good society, none are superior to technical pursuits. As David Hume explains in his essay "Of Refinement in the Arts," "In times when industry and the arts flourish, men are kept in perpetual occupation, and enjoy, as their reward, the occupation itself, as well as those pleasures which are the fruit of their labor. The mind acquires new vigor; enlarges its powers and faculties."[21] For that reason Hume advises rulers to encourage the development of manufacturing even in preference to agriculture. Dynamic new enterprises are more civilizing than the bucolic traditions of farming.

An important feature of this persuasion in contrast to classical notions is that politics is assigned a relatively low position in the broader scheme of human affairs. For Locke, government is an instrument with no intrinsic value. Its role is to protect the rights of "life, liberty, and property" by serving as an umpire when disputes arise. Attending to governmental matters is certainly not a sphere in which a person can realize one's highest potential. Locke finds no higher meaning in the realm of citizen action. One enters the public realm merely to express one's private interests. In contrast to Aristotle's view, Lockean liberalism recognizes neither goods nor virtues that stem from one's being as a public person.

In *The Wealth of Nations* Adam Smith develops the belief in the primacy of private affairs to its logical conclusion, viewing all public interference with scorn. Government measures, he argues, have "retarded the natural progress of England towards wealth and improvement."[22] Government is the source of extravagance, misconduct, and countless ill-conceived projects while the "uniform, constant and uninterrupted effort of every man to better his condition"[23] he identifies as the wellspring of most private and public good.

> It is the highest impertinence and presumption, therefore, in kings and ministers to pretend to watch over the economy of private people, and to restrain their expense, either by sumptuary laws, or by prohibiting the importation of foreign luxuries. They are themselves always, and without any exception, the greatest spendthrifts in the society. Let them look well after their own expense, and they may safely trust private people with theirs.[24]

Ideas of this kind underlie basic institutions of politics and economics in modern liberal democracies, posing strong barriers to attempts to think about the public dimensions of technological choice. Technological change, defined as "progress," is seen as an ineluctable process in modern history, one that develops as the result of the activities of men and women seeking private good, activities which include the development of inventions and innovations that benefit all of society. To encourage progress is to encourage private inventors and entrepreneurs to work unimpeded by state interference. As later theorists in the liberal tradition modify this understanding, they notice "market externalities" that cause stress in the social system or environment. This does not alter the fundamental attitude toward economic and technical choices. The burden of proof rests on those who would interfere with beneficent workings of the market and processes of technological development.

If one compares liberal ideology about politics and technology with its classical precursors, an interesting irony emerges. In modern thought the ancient pessimism about *techne* is eventually replaced by all-out enthusiasm for technological advance. At the same time basic conceptions of politics and political membership are reformulated in ways that help create new contexts for the exercise of power and authority. Despite the radical thrust of these intellectual developments, however, the classical separation between the political and the technical spheres is strongly preserved, but for entirely new reasons. Technology is still isolated from public life in both principle and practice. Citizens are strongly encouraged to become involved in improving modern material culture, but only in the market or other highly privatized settings. There is no moral community or public space in which technological issues are topics for deliberation, debate, and shared action.

TECHNOLOGY AND THE QUALITY OF CONTEMPORARY CITIZENSHIP

The hollowness of modern citizenship, the paucity of citizen roles and lack of opportunities for direct participation in politics, is now a general condition, not limited to technology policy-making alone. Many writers have lamented structures of representative democracy that effectively exclude ordinary people from significant involvement in public affairs. Thus, Hannah Arendt notes with approval Thomas Jefferson's proposals that American government include "elementary republics" that might have brought small-scale political assemblies into the realm of everyday life. "What he perceived to be the mortal danger to the republic was that the Constitution had given all power to the citizens, without giving them the opportunity of *being* republicans and of *acting* as citizens."[25]

In contemporary political science, low voter turnout, citizen apathy, the triviality of political campaigns are often cited as consequences of the failure of modern democracies to include citizens in meaningful activities. Much of the recent discussion among social scientists about "participatory democracy" and "strong democracy" speculates about ways to remedy these shortcomings.[26] But other than noticing the pungent effects of television upon election campaigns and the pervasive effects of modern consumerism, social scientists seldom take note of the connection between the hollowness of modern citizenship and the social relations of technology.

In fact, the political vacuum evident in the lack of citizen roles, citizen awareness, and citizen speech within liberal democratic society is greatly magnified within today's technology-centered workplace. Devices and systems commonly used in factories, fields, shops, and offices seek productivity and profit by controlling human behavior. In such settings the spontaneity and variability of workers' activities are regarded as a cause of uncertainty and a risk for business. For that reason the physical movements and decision-making abilities of employees are subject to rational planning and centralized guidance. Rather than encourage personal autonomy, creativity, and moral responsibility, many jobs and machines are designed to eliminate these qualities altogether.[27]

One might suppose that the technical professions offer greater latitude in dealing with the moral and political dimensions of technological choice. Indeed, the codes of engineering societies mention the higher purposes of serving humanity and the public good, while universities often offer special ethics courses for students majoring in science and engineering.[28] As a practical matter, however, the moral autonomy of engineering and other technical professionals is highly circumscribed. The historical evolution of modern engineering has placed most practitioners within business firms and government agencies where loyalty to the ends of the organization is paramount. During the 1920s and 1930s there were serious attempts to change this pattern, to organize the various fields of engineering as truly independent professions similar to medicine and law, attempts sometimes justified as ways to achieve more responsible control of emerging technologies. These efforts, however, were undermined by the opposition of business interests that worked to establish company loyalty as the engineer's central moral concern.[29] Calls for a higher degree of "ethical responsibility" among engineers are still heard in courses in technical universities and in obligatory after-dinner speeches at engineering societies. But pleas of this sort remain largely disingenuous, for there are few legitimate roles or organized settings in which such responsibility can be strongly expressed.

One could expand the inventory of social vocations in which moral issues in technological choice might be deliberated and decided, to include business managers, public officials, and the citizenry at large. Alas, there is little evidence that anything about these roles adds qualities of ethical reflection or action missing in ordinary workers or technical professionals. The responsibility of business managers is to maintain the profitability of the firm, a posture that usually excludes attention to the ethics of technological choice. Where questions of responsibility arise, businessmen usually listen to hired lawyers who explain their legal liabilities. Elected officials, similarly, find little occasion to consider the moral dimensions of technological choices. Their standard approach is to consult the opinions of scientific and technical experts, judging this information in ways that reflect a variety of economic and political interests. The general public may have a vague awareness of policy choices in energy, transportation, biomedical technology, and the like. But its response is increasingly apathetic, reactive, and video-centered.

Under such circumstances it is not surprising to find that people who call for moral deliberation about specific technological choices find themselves isolated and beleaguered, working outside or even in defiance of established channels of power and authority. At the level of individual action one finds the hero of much contemporary writing about technology and ethics—the "whistle-blower," an employee who notices something troubling in the day-to-day workings of a sociotechnical system and tries to call it to the attention of a reluctant employer or the news media. By all accounts, such behavior is often severely punished by the organizations whose actions and policies the whistle-blowers criticize. When they cannot be simply ignored, whistle-blowers are isolated, fired from their jobs, and then blackballed within their professions. Their lives become embroiled in exhausting efforts to show the truth of their claims and reestablish their value as employees.[30] For career-minded students who study the stories of whistle-blowers in university ethics courses, the underlying message is (regardless of what their teachers may intend): this is what happens if you speak out.

At the level of collective social action the method commonly used for expressing moral concerns about technological matters is that of "public interest" or "citizens" groups. Organized around key issues of the day, such groups take it upon themselves to express the interests and concerns of an otherwise silent populace about such matters as the arms race, nuclear power, environmental degradation, abortion, and many other issues. Ralph Nader, Helen Caldicott, and Jeremy Rifkin are among the contemporary figures who have become skillful in using this persuasive approach. It is characteristic of interest groups of this kind to be external to established, authoritative channel of decision-making power. The explicit purpose of groups identifying themselves with the "public interest" and "social responsibility" is to apply pressure, external pressure, upon political processes that otherwise move in what group members see as undesirable directions.

While the activities of public interest groups are clearly an exercise of the right of free speech, and while they are obviously important to the effective operation of modern democracy, the very existence of these groups points to the lack of any clear, substantive meaning for the term *public*. In this conception, the "public" arises ad hoc around certain points of social stress. One can claim to speak for "the public" simply by staging a demonstration or appearing on morning television news programs. The ease with which activists appropriate the word *public* leads to charges that particular groups are, in fact, unrepresentative, that "they don't represent my idea of the public interest." Nevertheless, public interest organizations offer the most direct means liberal democracies now have for focusing and mobilizing the concerns of ordinary people about controversial technologies.

The lack of any coherent identity for the "public" or of well-organized, legitimate channels for public participation contributes to two distinctive features of contemporary policy debates about technology, (1) futile rituals of expert advice and (2) interminable disagreements about which choices are morally justified.

Disputes about technology policy often arise in topic areas that seem to require years of training in fields of highly esoteric, science-based knowledge. A widely accepted notion about science is that it offers a precise, objective understanding of the world. Because technology is regarded as "applied science," and because the consequences of these applications involve such matters as complicated scientific measurements and the interpretation of arcane data, a common response is to turn to experts and expert research findings in hope of settling key policy questions.

This faith in scientific and technical advice involves much frustration in actual practice. Often it turns out that deep-seated uncertainties cannot be dispelled by consulting the experts. For the search for an objective answer brings a plurality of responses rather than a simple consensus. Studying the probable effects of background radiation, for example, different fields of scientific research give very different estimates of possible hazards. Problems of this kind are compounded by the fact that expertise is often indelibly linked to and biased by particular social interests. For example, looking at the problem of toxic waste disposal at Love Canal near Niagara Falls, New York, in the late 1970s, different social interests proposed different scientific models of the boundaries of the question and produced drastically different estimates of the hazards to citizens living in the area.[31] If, as contemporary sociologists claim, scientific knowledge is socially constructed, then scientific findings used in policy deliberations are doubly so. To an increasing extent, lawmakers and bureaucrats see scientific studies merely as resources to be deployed in ongoing power struggles.

What this suggests is that political disputes about technology are seldom if ever settled by calling upon the advice of experts. At public hearings held before legislative bodies, different social interests parade carefully chosen scientists and technical professionals. All of them speak with a confident air of "objectivity," but the experts often do not agree. Even where there is agreement about the "facts," there are still bound to be disagreements about how the "facts" are to be interpreted or what action is appropriate as a consequence.

Another characteristic of contemporary discussions about technology policy is that, as Alasdair MacIntyre might have predicted, they involve what seem to be interminable moral controversies. In a typical dispute, one side offers policy proposals based upon what seem to be ethically sound moral arguments. Then the opposing side urges entirely different policies using arguments that appear equally well grounded. The likelihood that the two (or more) sides can locate common ground is virtually nil. Consider the following arguments, ones fairly typical of today's technology policy debates.

1a. Conditions of international competitiveness require measures to reduce production costs. Automation realized through the computerization of office and factory work is clearly the best way to do this at present. Even though it involves eliminating jobs, rapid automation is the way to achieve the greatest good for the greatest number in advanced industrial society.

  b. The strength of any economy depends upon the skills of people who actually do the work. Skills of this kind arise from traditions of practice handed down from one generation to the next. Automation that de-skills the work process ought to be rejected because it undermines the well-being of workers and harms their ability to contribute to society.

2a. A great many technologies involve risks of one kind or another.

Judging the risks of chemical pesticides, one must balance the social benefits they bring against the risks they pose to human health and the environment. Considering the whole spectrum of benefits and risks involved, the good in using pesticides far outweighs their possible dangers.

b. Persons have a right to be protected from harm, including possible harm that may stem from useful technological applications. The use of pesticides subjects consumers to health hazards over which they have little or no control. Regardless of the larger good that the use of pesticides might bring, their use should be curtailed to prevent the risk of harm to individual consumers.

Positions of this kind involve a mixture of what may be highly uncertain empirical claims combined with philosophical arguments about which there is little consensus. Parties who square off in disputes of this kind usually believe that their side draws upon the very best data available and strong moral principles as well. But as the combatants circle each other in the ring, there is often a gnawing feeling that the various lines of moral reasoning have been concocted on the spot, used to justify positions that could be better described as emotional judgments or matters of sheer self-interest. In this way debates about technology policy confirm MacIntyre's argument that modern societies lack the kinds of coherent social practice that might provide firm foundations for moral judgments and public policies.[32]

What usually happens in such cases is a process of "muddling through." Interest groups apply pressure on politicians, gaining influence in proportion to the amount of money a group has to spend on the effort. Lawsuits are filed on one side or the other or both. Lawyers and judges sort through the flagrantly one-sided legal briefs, seeking precedents that might be patched together to provide a framework for deciding the case at hand. Television ads bombard viewers with flashy images and ten-second "sound bytes." Public opinion polls monitor the level of support for various proposals. Candidates for election sometimes take stands on issues that can then be included among the influences that sway voters in one direction or another. Eventually a policy outcome of some kind evolves, but it is seldom one that contains any experience of social learning that might be applied to similar episodes in the future.

REDEFINING CITIZENSHIP

In summary, I have argued that as moral philosophy confronts contemporary technology-related issues, it does so in an intellectual and social vacuum, one located in a deep gap between the technical and political spheres established by both ancient and modern philosophers. I have pointed to some of the consequences of this situation for thinking about technological choices and technology policies in our time. From this point of view, the technocratic approach I mentioned earlier—rushing forward with philosophical expertise to clarify moral categories, theories, and arguments in the hope that policymakers or the public will find them decisive—is a forlorn strategy. For the trouble is not that we lack good arguments and theories, but rather that modern politics simply does not provide appropriate roles and institutions in which the goal of defining the common good in technology policy is a legitimate project.

Under these circumstances a more fruitful path for philosophy is to begin exploring ways in which publics suited to renewed discussion about technological choices and policies might be constituted. Rather than echo the judgments of Aristotle and Adam Smith that political and technical affairs are essentially different, contemporary philosophers need to examine that question anew.

Some interesting possibilities arise in the fact that at long last the conceptual and practical boundaries between technology and politics upheld in both ancient and modern theory have begun to collapse. In the world of the late twentieth century, the spheres of technical and political life have merged in a variety of ways, woven together in situations in which common forms of human living have become dependent upon and shaped by technological devices and systems in telecommunications, computing, medicine, mass production, transportation, agriculture, and the like. To an increasing extent the qualities of technical artifacts reflect the possibilities of human living, what human beings are and aspire to be. At the same time, people mirror the technologies which surround them. Each day we see a widening of the kinds of human activities and consciousness that are technically embedded and technically mediated.

Although this rapidly growing, planetary technopolis strongly influences what our lives contain, few have tried to imagine forms of citizenship appropriate to this way of being. Some observers are content to point out the obvious, namely that technology is already highly politicized, that the development, introduction, and use of technologies of various kinds are always shaped by conflicts, negotiations, and machinations among powerful social interests. But to notice this fact is by no means to acknowledge the technopolitical sphere as a public space where citizen deliberation and action ought to be encouraged. To take that step, one must move beyond supposedly neutral sociological descriptions and explanations of how technologies arise and begin raising questions about the proper relationship between democratic citizenship and the shaping of technological order.[33]

Attempts of this kind have been launched recently in several modest experiments within the Scandinavian social democracies. These experiments are interesting in their own right, but also show the promise of creating citizen roles in places where private calculations of efficiency and effectiveness, costs, risks, benefits, and profits usually rule the day. A prototype of this variety of technological citizenship took shape at a research institute in Stockholm, the Center for Working Life. The basic goal of the center's work was to expand the scope of Scandinavian ideals of worker democracy in which technological innovation was likely to occur. They were encouraged by Swedish laws passed in the middle 1970s that recognized the right of all parties in the workplace, managers and workers alike, to negotiate about matters that affect the quality of working life. The "co-determination laws" cover such areas as job allocation, training, and work environment. Beginning in the 1970s, legal rights of this kind were carried in a novel direction by a group of labor unions working with university-educated computer scientists and systems designers. Realizing that computerization was likely to transform Swedish factories, shops, and offices, fearing the loss of jobs and workers' skills, the teams set out to investigate the new technologies and to explore possible alternatives.[34]

In one such case, the UTOPIA project of the early 1980s, workers in the Swedish newspaper industry—typesetters, lithographers, graphic artists, and the like—joined with representatives from management and with university computer scientists to design a new system of computerized graphics used in newspaper layout and typesetting. The first phase of the project was to survey existing work practices, techniques, and training in the graphic industries. The group then formed a design workshop to consider possibilities for a new system, using a paper-and-plywood mock-up as the model of a newspaper workstation. From there they produced a forty-eight-page technical document giving precise design specifications to the computer suppliers.

The pilot system, installed at the Stockholm daily newspaper *Aftonbladet*, offers a pattern of hardware, software, and human relationship very different from what would have been produced by managers and engineers alone. It allows graphics workers considerable latitude in arranging texts and images, retaining many of their traditional skills, but realizing them in a computerized form. In their deliberations, project members considered but rejected the prepacked graphics programs promoted by vendors from the United States because they reflected an "anti-democratic and de-skilling approach."[35] As project member and computer scientist Pelle Ehn observes, "What was new was that these technical requirements were derived from the principle that the equipment should serve *as tools for skilled work* and for production of *good use quality products*."[36]

The "Scandinavian approach" to participation in design is interesting not only for its tangible results but also for what it suggests about a positive politics of technology seen in broader perspective. In a small and tentative manner, the UTOPIA project created a public space for the political deliberation about the qualities of an emerging technical artifact. A diverse set of needs, viewpoints, and priorities came together to determine which material and social patterns would be designed, built, and put into operation. As Pelle Ehn points out, the important step in this process was to find a "project language game" in which all the participants from very different vocations, professions, and social backgrounds could speak to each other.[37] True, it was a fairly limited public that was constituted here. But it was far more inclusive than is normally the case in the printing industry or elsewhere.[38]

The creation of public spaces of this kind is, of course, predicated on modifying the right of owners of private property to have exclusive or even primary control of the shape of new technologies that affect how others live. That condition is, to a great extent, an accomplishment peculiar to Scandinavian social democracy, a product of political conflicts and agreements over the past several decades. It is now a condition sustained by the fact that more than 80 percent of Swedish workers are union members.[39]

Another achievement of the "Scandinavian approach" is to eliminate what I noted earlier as one of the most troubling features in contemporary technology policy: the ritual of expertise. In the UTOPIA project and others similar to it, a person's initial lack of knowledge of a domain of complex technical knowledge does not create a barrier to participation. The information and ideas needed to participate are mastered as part of a process in which the equality of team members is the established norm. Working from the opposite direction, those who came to the process with university degrees and professional qualifications explicitly rejected the idea that they were the designated, authoritative problem-solvers. Instead they offered themselves as persons whose knowledge of computers and systems design could contribute to discussions conducted in democratic ways.

This approach may also help dispel the second disturbing feature of contemporary technology policy debates, the interminable moral controversies they tend to generate. Here the guiding assumption is that if people with diverse viewpoints and conflicting social interests come together as equals in a situation that presents a common problem to be solved, an agreement will eventually evolve. As Ehn describes a typical predicament, "Management introduces new technology to save manpower. Journalists, graphics workers, and administrative staff confront each other in the struggle over a decreasing number of jobs. Is there a basis for solving these demarcation disputes across professional and union-based frontiers? Can a new way of organizing work create peaceful coexistence in the borderland?"[40] The answer seems to be yes. However, the answer is never as simple as one set of philosophically well grounded prescriptions winning out over another. Instead what happens is a negotiated political agreement among those whose interests will be affected by the change.

What the Scandinavian projects have done in an experimental way is to institute technopolitical practices from which new citizen virtues can emerge. Within small communities constituted for the purpose, choices about technologies that will influence the quality of social life are carefully studied and debated. This involves no expectations of political heroism, only the sense that ordinary people, regardless of background or prior expertise, are capable of taking a turn making decisions of this kind.[41] The vision of knowledge and social policy that underlies these efforts strongly resembles Paul Feyerabend's anarchistic proposals for "committees of laymen" involved in science.[42] In this instance, however, there was an opportunity to test the ideas in actual practice.

As revealed by Ehn's engaging treatise *Work-Oriented Design of Computer Artifacts*, the role of philosophy in this process is a limited but useful one. It attempts to clarify the basic conditions that undergird practices of work and discourse within the design projects. By seeking to understand these practices at a deeper, more general level, philosophical inquiry may shed light on ongoing negotiations as they occur. Thus, Ehn draws upon the writings of [Martin] Heidegger, [Ludwig] Wittgenstein, [Jürgen] Habermas, and other philosophers to illuminate his central concerns.[43] In the ideal case, philosophical reflection becomes one element in the process, although not one given privileged status. For it is understood that the key insights, lessons, and prescriptions must arise from a process in which project members, regarded as equals, join to explore the properties of both technical artifacts and social arrangements in a variety of configurations.

A criticism that might be raised about approaches like that pursued by Ehn and his Scandinavian colleagues is that they work at a superficial level within the technologies they confront. As the historian of technology Ulrich Wengenroth has noted, there is today a widening gap between "professionalization" and "trivialization" in many fields of technological development. Deeper, more complex levels of technical design and operation—the making of computer chips, for example— are accessible to and acted upon by only a handful of technical professionals. The same technologies are, however, restructured at the level of the user interface and present themselves in a deceptively friendly form. As Wegenroth observes, "If a new technology is met by suspicion and resistance in society, its acceptance is not won by reducing its complexity to make it intelligible and thus controllable by the general public, but by reengineering its interface to trivialize it."[44]

Do the Scandinavian projects merely retailor interfaces to make them more agreeable to workers while leaving the deeper structures of the technology as something given? The question cannot be answered in this brief overview. It is worth noting, however, that within the domain of computer programming the innovations of the Scandinavian researchers appear to be fairly deep-seeking. As noted, members of the UTOPIA project rejected an American firm's software package because it contained entrenched forms of hierarchical work organization, features that the group found "anti-democratic and de-skilling." Rather than try to weed out the deep-seated authoritarianism of American computer programs, the UTOPIA project elected to start from scratch.[45]

It is perhaps too early to characterize the virtues of citizen participation that might emerge from practices of this kind, too soon to specify whether this experience might be successfully applied to realms of technological choice usually governed by the merciless logic of economic and technical rationalization.[46] Members of the UTOPIA project appear to have developed a sense of cooperation, caution, and concern for the justice of their decisions. They were especially conscientious in trying to find effective designs that could take advantage of computer power while preserving the qualities of traditional workmanship. The members realized that conditions expressed in the design of a new system were conditions they would eventually have to live with. In that way their work echoes Aristotle's definition of the virtue of the good citizen, namely an understanding of both how to rule and be ruled. At a time in which politics and technology are thoroughly inter-woven, perhaps a similar definition of the virtue of citizens is that they know both how to participate in the shaping of technologies of various kinds and how to accept the shaping force that these technologies will eventually impose.

From this viewpoint the creation of arenas for the politics of technological choice is much more than a way of solving unsettling problems that arise in the course of technological change, although steps of this kind certainly might do that. It is also more than finding alternatives to the increasingly absurd logic of efficiency, productivity, and control that now drives technological choices in the global economy, although there is certainly a need for such alternatives. Even more important, the creation of new spaces and roles for technological choice might lead us to affirm a missing feature in modern citizenship: the freedom experienced in communities where making things and taking action are one and the same.

NOTES

1. Robert Sinsheimer, "Genetic Engineering: Life as a Plaything," in *Contemporary Moral Controversies in Technology*, ed. A. Pablo Iannone (New York: Oxford University Press, 1987), p. 131.

2. James H. Moor, "What Is Computer Ethics?" *Metaphilosophy* 16, no. 4 (1985): 275.

3. My treatment of classic and modern attitudes toward technology draws upon Carl Mitcham's excellent survey, "Three Ways of Being-With Technology," in *From Artifact to Habitat: Studies in the Critical Engagement of Technology*, ed. Gayle Ormiston (Bethlehem, Penn.: Lehigh University Press, 1990).

4. As Plato explains in *The Republic*, the real table is not that made by a craftsman, but the table that exists as an ideal form in the transcendent realm. Attempts to define the good society must understand this, seeking true rather than debased foundations for political practice. For that reason, Plato places the arts and crafts in the lowest of three social classes, and removes from them any chance of holding power. While he recognizes that agriculture, medicine, architecture, and the other practical arts are necessary to the life of the state, they offer nothing of value in ruling a good society. In both *The Republic* and *The Laws*, Plato advises those who would rule to stay as far away from mundane technical activities as possible. See my discussion of Plato's views in *The Whale and the Reactor: A Search for Limits in an Age of High Technology* (Chicago: University of Chicago Press, 1986), chap. 3.

5. Plato, *The Laws of Plato*, trans. Thomas L. Pangle (Chicago: University of Chicago Press, 1980), 797d.

6. Lucretius, *The Nature of Things*, trans. Ronald Latham (Baltimore: Penguin Books, 1951), p. 211.

7. Aristotle, *Politics*, trans. Benjamin Jowett, in *The Complete Works of Aristotle*, vol. 2, ed. Jonathan Barnes (Princeton: Princeton University Press, 1984), p. 1987.

8. Ibid., p. 2026.

9. Ibid., p. 2027.

10. Ibid., pp. 2028f.

11. Ibid., p. 2027.

12. M. I. Finley, *The Ancient Economy* (Berkeley: University of California Press, 1973), pp. 75–76.

13. Ibid., p. 75.

14. Ibid., p. 84.

15. Francis Bacon, *The Great Instauration*, in *The New Organon and Related Writings*, ed. Fulton Anderson (Indianapolis: Bobbs-Merrill Co., 1960), p. 8.

16. Ibid., p. 117.

17. Jean Le Rond D'Alembert, *Preliminary Discourse to the Encyclopedia of Diderot*, trans. Richard N. Schwab (Indianapolis: Bobbs-Merrill, 1963), p. 42.

18. See Albert O. Hirschman, *The Passions and the Interests: Political Arguments for Capitalism before Its Triumph* (Princeton: Princeton University Press, 1977).

19. Baron de Montesquieu, *The Spirit of Laws*, trans. Thomas Nugent, rev. ed., vol. 1 (New York: P. F. Collier & Son, 1900), p. 46.

20. C. B. MacPherson, *The Theory of Possessive Individualism: Hobbes to Locke* (Oxford: Clarendon Press, 1962).

21. David Hume, "Of Refinements in the Arts," in *The Philosophical Works*, vol. 3, eds. T. H. Green and T. H. Grouse (1882; reprint, Aalen: Scientific Verlag, 1964), p. 301.

22. Adam Smith, *The Wealth of Nations*, bks I–III with introduction by Andrew Skinner (Harmondsworth: Penguin, 1970), p. 446.

23. Ibid., p. 443.

24. Ibid., p. 446.

25. Hannah Arendt, *On Revolution* (Harmondsworth: Penguin Books, 1977), p. 253.

26. See, for example, Benjamin Barber, *Strong Democracy: Participatory Politics for a New Age* (Berkeley: University of California Press, 1984).

27. For poignant descriptions of circumstances that often face workers, see Barbara Garson, *Electronic Sweatshop: How Computers Are Transforming the Office of the Future into the Factory of the Past* (New York: Simon & Schuster, 1988).

28. See, for example, Peter Windt et al., eds., *Ethical Issues in the Professions* (Englewood Cliffs, N.J.: Prentice-Hall, 1989), and Deborah G. Johnson, ed., *Ethical Issues in Engineering* (Englewood Cliffs, N.J.: Prentice-Hall, 1991). My essay "Engineering Ethics and Political Imagination," in *Broad and Narrow Interpretations of Philosophy of Technology*, ed. Paul T. Durbin (Dordrecht: Kiuwer Academic Publishers, 1990), pp. 53–64, criticizes the approaches often used to teach ethics for technical professionals.

29. Edwin Layton, *Revolt of the Engineers: Social Responsibility and the American Engineering Profession* (Cleveland: Case Western Reserve University, 1971), chaps. 1–2.

30. Myron Glazer and Penina Glazer, *The Whistleblowers: Exposing Corruption in Government and Industry* (New York: Basic Books, 1989).

31. Beth Savan, *Science under Siege: The Myth of Objectivity in Scientific Research* (Montreal: CBC Enterprises, 1988).

32. Alasdair MacIntyre, *After Virtue: A Study in Moral Theory*, 2d ed. (Notre Dame: University of Notre Dame Press, 1984), chaps. 14 and 15.

33. For a critique of the new sociology of technology, see my "Social Constructivisim: Opening the Black Box and Finding It Empty," *Science as Culture*, no. 16 (autumn 1992).

34. For a description of Scandinavian experiments in democratic participation in

design, see Pelle Ehn, *Work-Oriented Design of Computer Artifacts* (Stockholm: Arbet-slivcentrum, 1988).

35. Ibid., p. 345.

36. Ibid., p. 339 (italics in the original text).

37. Ibid., p. 17.

38. In fact, problems arose within the UTOPIA project because it was not inclusive enough, excluding the participation of journalists. As Ehn notes, the future of the project "depends upon whether the graphic workers and journalists succeed in overcoming their professional clash of interests, and together develop a common strategy." Ehn, *Work-Oriented Design of Computer Artifacts*, p. 357.

39. Peter Lawrence and Tony Spybey, *Management and Society in Sweden* (London: Routledge & Kegan Paul, 1986), p. 85. For an overview of the relationship between technology and work in Sweden, see Ake Sandberg, *Technological Change and Co-Determination in Sweden: Background and Analysis of Trade Union and Managerial Strategies* (Philadelphia: Temple University Press, 1992). An excellent discussion of the moral issues confronting Scandinavian social democracy can be found in Alan Wolfe, *Whose Keeper?: Social Science and Moral Obligation* (Berkeley: University of California Press, 1989).

40. Ehn, *Work-Oriented Design of Computer Artifacts*, p. 342.

41. For a general exploration of tensions between technical expertise and direct democracy, see Langdon Winner, ed., *Democracy in a Technological Society* (Dordrecht: Kluwer Academic Publishers, 1992), and Frank Fischer, *Technocracy and the Politics of Expertise* (Newbury Park, Calif.: Sage Publications, 1990).

42. See Paul K. Feyerabend, *Science in a Free Society* (London: NLB, 1978), and his suggestions in "Democracy, Elitism, and Scientific Method," *Inquiry* 23, no. 1 (1980): 3–18.

43. Arguments and conclusions similar to Pelle Ehn's can be found in Terry Winograd and Fernando Flores, *Understanding Computers and Cognition* (Reading, Mass.: Addison-Wesley, 1987).

44. Ulrich Wengenroth, "The Cultural Bearings of Modern Technological Development," in *Humanistic Perspectives on Technology, Development, and Environment*, eds. Francis Sejersted and Ingunn Moser (Oslo: Centre for Technology and Culture, Report Series No. 3, 1992).

45. Ehn, *Work-Oriented Design of Computer Artifacts*, pp. 344–45.

46. Methods of organizing people and machinery in the mode of "just-in-time" and "lean production," now gaining momentum in the global market economy, point in directions much different from those pursued by Scandinavian workplace reformers. The workplace regimes created within this mode of production could well achieve levels of rationalization and centralization that would make Frederick W. Taylor and Jacques Ellul blush. See J. P. Womack, D. T. Jones, and D. Roos, *The Machine that Changed the World* (New York: Rawson Associates, 1990).

2 6

Technological Ethics
in a Different Voice

Diane P. Michelfelder

The rapid growth of modern forms of technology has brought both a threat and a promise for liberal democratic society. As we grapple to understand the implications of new techniques for extending a woman's reproductive life or the spreading underground landscape of fiber-optic communication networks or any of the other developments of contemporary technology, we see how these changes conceivably threaten the existence of a number of primary goods traditionally associated with democratic society, including social freedom, individual autonomy, and personal privacy. At the same time, we recognize that similar hopes and promises have traditionally been associated with both technology and democracy. Like democratic society itself, technology holds forth the promise of creating expanded opportunities and a greater realm of individual freedom and fulfillment. This situation poses a key question for the contemporary philosophy of technology. How can technology be reformed to pose more promise than threat for democratic life? How can technological society be compatible with democratic values?

One approach to this question is to suggest that the public needs to be more involved with technology not merely as thoughtful consumers but as active participants in its design. We can find an example of this approach in the work of Andrew Feenberg. As he argues, most notably in his recent book *Alternative Modernity: The Technical Turn in Philosophy and Social Theory*, the advantage of technical politics, of greater public participation in the design of technological objects and technologically mediated services such as health care, is to open up this process to the consideration of a wider sphere of values than if the design process were to be left up to bureaucrats and professionals, whose main concern is with preserving efficiency. Democratic values such as personal autonomy and individual agency are part of this wider sphere. For Feenberg, the route to technolog-

Diane P. Michelfelder, "Technological Ethics in a Different Voice," in *Technology and the Good Life?* eds., Eric Higgs, Andrew Light, and David Strong, pp. 219–33. Copyright © 2000 University of Chicago Press. Reprinted by permission.

ical reform and the preservation of democracy thus runs directly through the intervention of nonprofessionals in the early stages of the development of technology.[1]

By contrast, the route taken by Albert Borgmann starts at a much later point. His insightful explorations into the nature of the technological device—that "conjunction of machinery and commodity"[2]—do not take us into a discussion of how public participation in the design process might result in a device more reflective of democratic virtues. Borgmann's interest in technology starts at the point where it has already been designed, developed, and ready for our consumption. Any reform of technology, from his viewpoint, must first pass through a serious examination of the moral status of material culture. But why must it start here, rather than earlier, as Feenberg suggests? In particular, why must it start here for the sake of preserving democratic values?

In taking up these questions in the first part of this [chapter], I will form a basis for turning in the following section to look at Borgmann's work within the larger context of contemporary moral theory. With this context in mind, in the third part of this [chapter] I will take a critical look from the perspective of feminist ethics at Borgmann's distinction between the thing and the device, a distinction on which his understanding of the moral status of material culture rests. Even if from this perspective this distinction turns out to be questionable, it does not undermine, as I will suggest in the final part of this [chapter], the wisdom of Borgmann's starting point in his evaluation of technological culture.

PUBLIC PARTICIPATION AND TECHNOLOGICAL REFORM

One of the developments that Andrew Feenberg singles out in *Alternative Modernity* to back up his claim that public involvement in technological change can further democratic culture is the rise of the French videotext system known as Teletel.[3] As originally proposed, the Teletel project had all the characteristics of a technocracy-enhancing device. It was developed within the bureaucratic structure of the French government–controlled telephone company to advance that government's desire to increase France's reputation as a leader in emerging technology. It imposed on the public something in which it was not interested: convenient access from home terminals (Minitels) to government-controlled information services. However, as Feenberg points out, the government plan for Teletel was foiled when the public (thanks to the initial assistance of computer hackers) discovered the potential of the Minitels as a means of communication. As a result of these interventions, Feenberg reports, general public use of the Minitels for sending messages eventually escalated to the point where it brought government use of the system to a halt by causing it to crash. For Feenberg, this story offers evidence that the truth of social constructivism is best seen in the history of the computer.

Let us imagine it does offer this evidence. What support, though, does this story offer regarding the claim that public participation in technical design can further democratic culture? In Feenberg's mind, there is no doubt that the Teletel story reflects the growth of liberal democratic values. The effect generated by the possibility of sending anonymous messages to others over computers is, according to Feenberg, a positive one, one that "enhances the sense of personal freedom and individualism by reducing the 'existential' engagement of the self in its communications."[4] He also finds that in the ease of contact and connection building fostered by computer-mediated communication, any individual or group of individuals who is a part of building these connections becomes more empowered.[5]

But as society is strengthened in this way, in other words, as more and more opportunities open up for electronic interaction among individuals, do these opportunities lead to a more meaningful social engagement and exercise of individual freedom? As Borgmann writes in *Technology and the Character of Contemporary Society* (or *TCCL*): "The capacity for significance is where human freedom should be located and grounded."[6] Human interaction without significance leads to disengagement; human freedom without significance leads to banality of agency. If computer-mediated communications take one where Feenberg believes they do (and there is little about the more recent development of Internet-based communication to raise doubts about this), toward a point where personal life increasingly becomes a matter of "staging . . . personal performances,"[7] then one wonders what effect this has on other values important for democratic culture: values such as self-respect, dignity, community, and personal responsibility.

The Teletel system, of course, is just one example of technological development, but it provides an illustration through which Borgmann's concern with the limits of public participation in the design process as a means of furthering the democratic development of technological society can be understood. Despite the philosophical foundations of liberal democracy in the idea that the state should promote equality by refraining from supporting any particular idea of the human good, in practice, he writes, "liberal democracy is enacted as technology. It does not leave the question of the good life open but answers it along technological lines."[8] The example we have been talking about illustrates this claim. Value-neutral on its surface with respect to the good life, Feenberg depicts the Teletel system as encouraging a play of self-representation and identity that develops at an ever-intensifying pace while simultaneously blurring the distinction between private and public life. The value of this displacement, though, in making life more meaningful, is questionable.

To put it in another way, for technology to be designed so that it offers greater opportunities for more and more people, what it offers has to be put in the form of a commodity. But the more these opportunities are put in the form of commodities, the more banal they threaten to become. This is why, in Borgmann's view, technical politics cannot lead to technical reform.

For there truly to be a reform of technological society, Borgmann maintains, it is not enough only to think about preserving democratic values. One also needs to consider how to make these values meaningful contributors to the good life without overly determining what the good life is. "The good life," he writes, "is one of engagement, and engagement is variously realized by various people."[9] While a technical politics can influence the design of objects so that they reflect democratic values, it cannot guarantee that these values will be more meaningfully experienced. While a technical politics can lead to more individual freedom, it does not necessarily lead to an enriched sense of freedom. For an object to lead to an enriched sense of freedom, it needs, according to Borgmann, to promote unity over dispersement, and tradition over instantaneity. Values such as these naturally belong to objects, or can be acquired by them, but cannot be designed into them.

To take some of Borgmann's favorite examples, a musical instrument such as a violin can reflect the history of its use in the texture of its wood;[10] with its seasonal variations, a wilderness area speaks of the natural belonging together of time and space.[11] We need to bring more things like these into our lives, and use technology to enhance our direct experience of them (as in wearing the right kinds of boots for a hike in the woods), for technology to deliver on its promise of bringing about a better life. As Borgmann writes toward the end of TCCL, "So counterbalanced, technology can fulfill the promise of a new kind of freedom and richness."[12]

Thus for Borgmann the most critical moral choices that one faces regarding material culture are "material decisions":[13] decisions regarding whether to purchase or adopt a technical device or to become more engaged with things. These decisions, like the decisions to participate in the process of design of an artifact, tend to be inconspicuous. The second type of decision, as Wiebe E. Bijker, Thomas P. Hughes, and Trevor Pinch have shown,[14] fades from public memory over time. The end result of design turns into a "black box" and takes on the appearance of having been created solely by technical experts. The moral decisions Borgmann describes are just as inconspicuous because of the nature of the context in which they are discussed and made. This context is called domestic life. "Technology," he observes, "has step by step stripped the household of substance and dignity."[15] Just as Borgmann recalls our attention to the things of everyday life, he also makes us remember the importance of the household as a locus for everyday moral decision making. Thus Borgmann's reflections on how technology might be reformed can also be seen as an attempt to restore the philosophical significance of ordinary life.

BORGMANN AND THE RENEWAL OF PHILOSOPHICAL INTEREST IN ORDINARY LIFE

In this attempt, Borgmann does not stand alone. Over the course of the past two decades or so in North America, everyday life has been making a philosophical comeback. Five years after the publication of Borgmann's *TCCL* appeared Charles Taylor's *Sources of the Self*, a fascinating and ambitious account of the history of the making of modern identity. Heard throughout this book is the phrase "the affirmation of everyday life," a life characterized in Taylor's understanding by our nonpolitical relations with others in the context of the material world. As he sees it, affirming this life is one of the key features in the formation of our perception of who we are.[16] Against the horizons of our lives of work and play, friendship and family, we raise moral concerns that go beyond the questions of duties and obligations familiar to philosophers. What sorts of lives have the character of good lives, lives that are meaningful and worth living? What does one need to do to live a life that would be good in this sense? What can give my life a sense of purpose? In raising these questions, we affirm ordinary life. This affirmation is so deeply woven into the fabric of our culture that its very pervasiveness, Taylor maintains, serves to shield it from philosophical sight.[17]

Other signs point as well to a resurgence of philosophical interest in the moral dimensions of ordinary life. Take, for example, two fairly recent approaches to moral philosophy. In one of these approaches, philosophers such as Lawrence Blum, Christina Hoff Summers, John Hartwig, and John Deigh have been giving consideration to the particular ethical problems triggered by interpersonal relationships, those relationships among persons who know each other as friends or as family members or who are otherwise intimately connected. As George Graham and Hugh LaFollette note in their book *Person to Person*, these relationships are ones that almost all of us spend a tremendous amount of time and energy trying to create and sustain.[18] Such activity engenders a significant amount of ethical confusion. Creating new relationships often means making difficult decisions about breaking off relationships in which one is already engaged. Maintaining interpersonal relationships often means making difficult decisions about what the demands of love and friendship entail. In accepting the challenge to sort through some of this confusion in a philosophically meaningful way, those involved with the ethics of interpersonal relationships willingly pay attention to ordinary life. In the process, they worry about the appropriateness of importing the standard moral point of view and standard moral psychology used for our dealings with others in larger social contexts—the Kantian viewpoint of impartiality and the distrust of emotions as factors in moral decision making—into the smaller and more intimate settings of families and friendships.

Another, related conversation about ethics includes thinkers such as Virginia Held, Nel Noddings, Joan Tronto, Rita Manning, Marilyn Friedman, and others whose work has been influenced by Carol Gilligan's research into the development of moral reasoning among women. I will call the enterprise in which these theorists are engaged feminist ethics, since I believe that description would be agreeable to those whom I have just mentioned, all of whom take the analysis of women's moral experiences and perspectives to be the starting point from which to rethink ethical theory.[19] Like interpersonal ethics, feminist ethics (particularly the ethics of care) places particular value on our relationships with those with whom we come into face-to-face contact in the context of familial and friendly relations. Its key insight lies in the idea that the experience of looking out for those immediately around one, an experience traditionally associated with women, is morally significant, and needs to be taken into account by anyone interested in developing a moral theory that would be a satisfactory and useful guide to the moral dilemmas facing us in all areas of life. Thus this approach to ethics also willingly accepts the challenge of paying philosophical attention to ordinary life. This challenge is summed up nicely by Virginia Held: "Instead of importing into the household principles derived from the marketplace, perhaps we should export to the wider society the relations suitable for mothering persons and children."[20]

On the surface, these three paths of ethical inquiry—Borgmann's ethics of modern technology, the ethics of interpersonal relationships, and feminist ethics— are occupied with different ethical questions. But they are united, it seems to me, in at least two ways. First, they are joined by their mutual contesting of the values upon which Kantian moral theory in particular and the Enlightenment in general are based. Wherever the modernist project of submitting public institutions and affairs to one's personal scrutiny went forward, certain privileges were enforced: that of reason over emotion, the "naked self" over the self in relation to others, impartiality over partiality, the public realm over the private sphere, culture over nature, procedural over substantive reasoning, and mind over body. In addition to the critique of Kantian ethics already mentioned by philosophers writing within a framework of an ethics of interpersonal relationships, feminist ethics has argued that these privileges led to the construction of moral theories insensitive to the ways in which women represent their own moral experience. Joining his voice to these critiques, Borgmann has written (while simultaneously praising the work of Carol Gilligan), "Universalism neglects . . . ways of empathy and care and is harsh toward the human subtleties and frailties that do not convert into the universal currency. . . . The major liability of moral universalism is its dominance; the consequence of dominance is an oppressive impoverishment of moral life."[21]

A second feature uniting these relatively new forms of moral inquiry is a more positive one. Each attempts to limit further increases in the "impoverishment of moral life" by calling attention to the *moral* aspects of typical features of ordinary life that have traditionally been overlooked or even denied. The act of mothering (for Virginia Held), the maintenance of friendships (for Lawrence Blum), and the loving preparation of a home-cooked meal (for Borgmann) have all been defended, against the dominant belief to the contrary, as morally significant events.[22]

Despite the similarities and common concerns of these three approaches to moral philosophy, however, little engagement exists among them. Between feminist ethics and the ethics of interpersonal relationships, some engagement can be found: for instance, the "other-centered" model of friendship discussed in the latter is of interest to care ethicists as part of an alternative to Kantian ethics. However, both of these modes of ethical inquiry have shown little interest in the ethical dimensions of material culture. Nel Noddings, for example, believes that while caring can be a moral phenomenon when it is directed toward one's own self and that of others, it loses its moral dimension when it is directed toward things. In her book *Caring*, she defends the absence of discussion of our relations to things in her work: "As we pass into the realm of things and ideas, we move entirely beyond the ethical. . . . My main reason for setting things aside is that we behave ethically only through them and not toward them."[23]

And yet in ordinary life ethical issues of technology, gender, and interpersonal relationships overlap in numerous ways. One wonders as a responsible parent whether it is an act of caring to buy one's son a Mighty Morphin Power Ranger. If I wish to watch a television program that my spouse cannot tolerate, should I go into another room to watch it or should I see what else is on television so that we could watch a program together? Is a married person committing adultery if he or she has an affair with a stranger in cyberspace? Seeing these interconnections, one wonders what might be the result were the probing, insightful questioning initiated by Borgmann into the moral significance of our material culture widened to include the other voices mentioned here. What would we learn, for instance, if Borgmann's technological ethics were explored from the perspective of feminist ethics?

In the context of this [chapter] I can do no more than start to answer this question. With this in mind, I would like to look at one of the central claims of *TCCL*: the claim that the objects of material culture fall either into the category of things or devices.

FEMINISM AND THE DEVICE PARADIGM

As Borgmann describes them, things are machines that, in a manner of speaking, announce their own narratives and as a result are generous in the effects they can produce. For example, we can see the heat of the wood burning in the fireplace being produced in front of our eyes—the heat announces its own story, its own history, in which its relation to the world is revealed. In turn, fireplaces give us a place to focus our attention, to regroup and reconnect with one another as we watch the logs burn. In this regard, Borgmann speaks compellingly not only of the fireplace but also of wine: "Technological wine no longer bespeaks the particular weather of the year in which it grew since technology is at pains to provide assured, i.e., uniform, quality. It no longer speaks of a particular place since it is a blend of raw materials from different places."[24]

Devices, on the other hand, hide their narratives by means of their machinery and as a result produce only the commodity they were intended to produce. When I key the characters of the words I want to write into my portable computer, they appear virtually simultaneously on the screen in front of me. I cannot see the connection between the one event and the other, and the computer does not demand that I know how it works in order for it to function. The commodity we call "processed words" is the result. While things lead to "multi-sided experiences," devices produce "one-sided experiences."[25] Fireplaces provide warmth, the possibility of conviviality, and a closer tie to the natural world; a central heating system simply provides warmth.

What thoughts might a philosopher working within the framework of feminist ethics have about this distinction? To begin with, I think she would be somewhat uneasy with the process of thinking used to make decisions about whether a particular object would be classified as a thing or a device. In this process, Borgmann abstracts from the particular context of the object's actual use and focuses his attention directly on the object itself. The view that some wine is "technological," as the example described above shows, is based on the derivation of the wine, the implication being that putting such degraded wine on the table would lead to a "one-sided experience" and further thwart, albeit in a small way, technology's capability to contribute meaningfully to the good life. In a feminist analysis of the moral significance of material culture, a different methodology would prevail. The analysis of material objects would develop under the assumption that understanding people's actual experiences of these objects, and in particular understanding the actual experiences of women who use them, would be an important source of information in deciding what direction a technological reform of society should take.

The attempt to make sense of women's experience of one specific technological innovation is the subject of communication professor Lana Rakow's book *Gender on the Line: Women, the Telephone, and Community Life*.[26] As its title suggests, this is a study of the telephone practices of the women residents of a particular community, a small midwestern town she called, to protect its identity, Prospect.

Two features of Rakow's study are of interest with regard to our topic. One relates to the discrepancy between popular perceptions of women's use of the telephone, and the use revealed in her investigation. She was well aware at the beginning of her study of the popular perception, not just in Prospect but widespread throughout American culture, of women's use of the telephone. In the popular perception, characterized by expressions such as "Women just like to talk on the phone" and "Women are on the phone all the time"; telephone conversations among women appear as "productivity sinks," as ways of wasting time. Understandably from this perception the telephone could appear as a device used for the sake of idle chatter that creates distraction from the demands of work and everyday life. This is how Borgmann sees it:

> The telephone network, of course, is an early version of hyperintelligent communication, and we know in what ways the telephone has led to disconnectedness. It has extinguished the seemingly austere communication via letters. Yet

this austerity was wealth in disguise. To write a letter one needed to sit down, collect one's thoughts and world, and commit them laboriously to paper. Such labor was a guide to concentration and responsibility.[27]

Rakow's study, however, did not support the popular perception. She found that the "womentalk" engaged in by her subjects was neither chatter nor gossip. Rather, it was a means to the end of producing, affirming, and reinforcing the familial and community connections that played a very large role in defining these women's lives. Such "phone work," very often consisting of exchanges of stories, was the stuff of which relations were made: "Women's talk holds together the fabric of the community, building and maintaining relationships and accomplishing important community relations."[28]

Let me suggest some further support for this view from my own experience. While I was growing up, I frequently witnessed this type of phone work on Sunday afternoons as my mother would make and receive calls from other women to discuss "what had gone on at church." Although these women had just seen each other at church several hours before, their phone calls played exactly the role that Rakow discovered they played in Prospect. At the time, they were not allowed to hold any positions of authority within the organizational structure of this particular church. The meaning of these phone calls would be missed by calling them idle talk; at least in part, these phone visits served to strengthen and reinforce their identity within the gendered community to which these women belonged.

Another interesting feature of Rakow's study was its discovery of how women used the telephone to convey care:

> Telephoning functions as a form of care-giving. Frequency and duration of calls . . . demonstrate a need for caring or to express care (or a lack of it). Caring here has the dual implication of caring *about* and caring *for*—that is, involving both affection and service. . . . While this [care-giving role] has been little recognized or valued, the caring work of women over the telephone has been even less noted.[29]

As one of the places where the moral status of the care-giving role of women has been most clearly recognized and valued, feminist ethics is, of course, an exception to this last point. Rakow's recognition of the telephone as a means to demonstrate one's caring for speaks directly to Nel Noddings's understanding of why giving care can be considered a moral activity.[30] In caring one not only puts another's needs ahead of one's own, but, in reflecting on how to take care of those needs, one sees oneself as being related to, rather than detached from, the self of the other. In commenting that not only checking on the welfare of another woman or phoning her on her birthday but "listening to others who need to talk is also a form of care,"[31] Rakow singles out a kind of caring that well reflects Nodding's description. More often one needs to listen to others who call one than one needs to call others; and taking care of the needs of those who call often involves simply staying on the phone while the other talks. As Rakow correctly points out, this makes this particular practice of telephone caring a form of work. Those who criticize the ethics of care for taking up too much of one's time with meeting the needs of individual others might also be critical of Rakow's subjects who reported that

> they spend time listening on the phone when they do not have the time or interest for it. . . . One elderly woman . . . put a bird feeder outside the window by her telephone so she can watch the birds when she has to spend time with these phone calls. "I don't visit; I just listen to others," she said.[32]

As these features of telephone conversations came to light in the interviews she conducted with the women of Prospect, Rakow began to see the telephone as "a gendered, not a neutral, technology."[33] As a piece of gendered technology, the telephone arguably appears more like a thing than like a device, allowing for, in Borgmann's phrase, the "focal practice" of caring to take place. Looking at the telephone from this perspective raises doubts about Borgmann's assessment of the telephone. Has the telephone in fact become a substitute for the thing of the letter, contributing to our widespread feelings of disconnectedness and to our distraction? Rakow's fieldwork provides support for the idea that phone work, much like letter writing, can be "a guide to concentration and responsibility." By giving care over the phone, the development of both these virtues is supported. Thus, on Borgmann's own terms—"The focal significance of a mental activity should be judged, I believe, by the force and extent with which it gathers and illuminates the tangible world and our appropriation of it."[34]—it is difficult to see how using the telephone as a means of conveying care could not count as a focal concern.

Along with the question of whether a particular item of our material culture is or is not a device, looking at the device paradigm from a feminist point of view gives rise to at least two other issues. One is connected to an assumption on which this paradigm rests: that the moral significance of an object is directly related to whether or not that object is a substitute for the real thing. This issue is also connected to the idea that because technological objects are always substitutes for the real thing, the introduction of new technology tends to be a step forward in the impoverishment of ordinary life.

Certainly technological objects are always substitutes for *something or another*. A washing machine is a substitute for a washing board, dryers are substitutes for the line out back, krab [*sic*] is often found these days on salad bars, and so forth. In some cases, the older object gradually fades from view, as happened with the typewriter, which (but only as of fairly recently) is no longer being produced. In other cases, however, the thing substituted for is not entirely replaced, but continues to coexist alongside the substitute. In these cases, it is harder to see how the technological object is a substitute *for the real thing*, and thus harder to see how the introduction of the new object threatens our sense of engagement with the world. While it is true that telephones substitute for letter writing, as Borgmann observes, the practice of letter writing goes on, even to the point of becoming intertwined with the use of the telephone. Again, from Rakow:

> The calls these women make and the letters they send literally call families into existence and maintain them as a connected group. A woman who talks daily to her two nearby sisters demonstrated the role women play in keeping track of the well-being of family members and changes in their lives. She said, "If we get a letter from any of them (the rest of the family) we always call and read each other the letters."[35]

Perhaps, though, the largest question prompted by Rakow's study has to do with whether Borgmann's distinction itself between things and devices can hold up under close consideration of the experiences and practices of different individuals. There are many devices that can be and are used as the women in this study used the telephone. Stereos, for example, can be a means for someone to share with someone else particular cuts on a record or songs from a CD to which she or he attaches a great deal of personal significance. In this way, stereos can serve as equipment that aid the development of mutual understanding and relatedness, rather than only being mechanisms for disengagement. The same goes, as Douglas Kellner points out for the use of the computer as a communicative device. Empirical investigations into the gendered use of computer-mediated communications suggest that while women do not necessarily use this environment like the telephone, as a means of promoting care, they do not "flame" (send electronic messages critical of another individual) nearly as much as do men, and they are critical of men who do engage in such activity.[36]

In particular, from a feminist perspective one might well wonder whether, in Borgmann's language, the use of those "conjunctions of machinery and commodity" inevitably hamper one's efforts at relating more to others and to the world. Borgmann argues that because devices hide their origins and their connections to the world, they cannot foster our own bodily and social engagement with the world. But as I have tried to show here, this is arguably not the case. Whether or not a material object hides or reveals "its own story" does not seem to have a direct bearing on that object's capacity to bind others together in a narrative web. For instance, older women participating in Rakow's study generally agreed that telephones improved in their ability to serve as a means of social support and caregiving once their machinery became more hidden: when private lines took the place of party lines and the use of an operator was not necessary to place a local call. To generalize, the machinery that clouds the story of a device does not appear to prevent that device from playing a role in relationship building.

DEVICES AND THE PROMISE OF TECHNOLOGY

While a child growing up in New Jersey, I looked forward on Friday evenings in the summer to eating supper with my aunt and uncle. I would run across the yard separating my parents' house from theirs to take my place at a chair placed at the corner of the kitchen table. The best part of the meal, I knew, would always be the same, and that was why I looked forward to these evenings. While drinking lemonade from the multicolored aluminum glasses so popular during the 1950s, we would eat Mrs. Paul's fish sticks topped with tartar sauce. With their dubious nutritional as well as aesthetic value, fish sticks are to fresh fish as, in a contrast described eloquently by Borgmann, Kool Whip is to fresh cream.[37] One doesn't know the seas in which the fish that make up fish sticks swim. Nearly anyone can prepare them in a matter of minutes. Still, despite these considerations, these meals were marked by family sociability and kindness, and were not hurried affairs.

I recall these meals now with the following point in mind. One might be tempted by the course of the discussion here to say that the objects of material culture should not be divided along the lines proposed in *TCCL* but divided in another manner. From the perspective of feminist ethics, one might suggest that one needs to divide up contemporary material culture between relational things, things that open up the possibility of caring relations to others, and nonrelational things: things that open up the possibility of experience but not the possibility of relation. Telephones, on this way of looking at things, would count as relational things. Virtual reality machines, such as the running simulator Borgmann imagines in *Crossing the Postmodern Divide*, or golf simulators that allow one to move from the green of the seventeenth hole at Saint Andrews to the tee of the eighteenth hole at Pebble Beach, would be nonrelational things. One can enjoy the experiences a virtual golf course makes possible, but one cannot in turn, for example, act in a caring manner toward the natural environment it so vividly represents. But the drawback of this distinction seem similar to the drawback of the distinction between things and devices: the possibility of using a thing in a relational and thus potentially caring manner seems to depend more on the individual using that thing and less on the thing itself. Depending on who is playing it, a match of virtual golf has the potential of strengthening, rather than undoing, narrative connections between oneself, others, and the world.

But if our discussion does not lead in this direction, where does it lead? Let me suggest that although it does not lead one to reject the device paradigm outright, it does lead one to recognize that while any device does use machinery to produce a commodity, the meaning of one's experience associated with this device does not necessarily have to be diminished. And if one can use technology (such as the telephone) to carry out focal practices (such as caregiving), then we might have cause to believe that there are other ways to recoup the promise of technology than Borgmann sees. As mentioned earlier, his hope is that we will give technology more of a supporting role in our lives than it has at present,[38] a role he interprets as meaning that it should support the focal practices centered around focal things. But if devices can themselves support focal practices, then the ways in which technology can assume a supporting role in our lives are enhanced.

But if the idea that devices can support focal practices is in one way a challenge to the device paradigm, in another way it gives additional weight to the notion that there are limits to reforming technology through the process of democratic design. When they are used in a context involving narrative and tradition, devices can help build engagement and further reinforce the cohesiveness of civil society. Robert Putnam has pointed out the importance of trust and other forms of "social capital" necessary for citizens to interact with each other in a cooperative manner. As social capital erodes, democracy itself, he argues, is threatened.[39] While this [chapter] has suggested that devices can under some conditions further the development of social capital, it is difficult to see how they can be deliberately designed to do so. In thinking about how to reform technology from a democratic perspective, we need to remember the role of features of ordinary life such as narrative and tradition in making our experience of democratic values more meaningful. Borgmann's reminder to us of this role is, it seems to me, one of the reasons why *TCCL* will continue to have a significant impact in shaping the field of the philosophy of technology.

NOTES

1. Andrew Feenberg, *Alternative Modernity: The Technical Turn in Philosophy and Social Theory* (Berkeley and Los Angeles: University of California Press, 1995).

2. Albert Borgmann, "The Moral Significance of the Material Culture," *Inquiry* 35 (1992): 296.

3. Feenberg, *Alternative Modernity,* pp. 144–66. [See chapter 24 in this volume.]

4. Ibid., p. 159.

5. Ibid., p. 160.

6. Albert Borgmann, *Technology and the Character of Contemporary Life: A Philosophical Inquiry* (Chicago: University of Chicago Press, 1984), p. 102; hereafter referred to as *TCCL.*

7. Feenberg, *Alternative Modernity,* p. 160.

8. Borgmann, *TCCL,* p. 92.

9. Ibid., p. 214.

10. Borgmann, "Moral Significance of the Material Culture," p. 294.

11. Borgmann, *TCCL,* p. 191.

12. Ibid., p. 248.

13. Albert Borgmann, *Crossing the Postmodern Divide* (Chicago: University of Chicago Press, 1992), p. 112.

14. Wiebe E. Bijker, Thomas P. Hughes, and Trevor Pinch, eds., *The Social Construction of Technological Systems* (Cambridge: MIT Press, 1987).

15. Borgmann, *TCCL,* p. 125.

16. Charles Taylor, *Sources of the Self* (Cambridge: Harvard University Press, 1989), p. 13.

17. Ibid., p. 498.

18. George Graham and Hugh LaFollette, eds., *Person to Person* (Philadelphia: Temple University Press, 1989), p. 1.

19. I am not using "feminist ethics" in a technical sense, but as a way of referring to the philosophical approach to ethics that starts from a serious examination of the moral experience of women. For philosophers such as Alison Jaggar, the term feminist ethics primarily means an ethics that recognizes the patriarchal domination of women and the need for women to overcome this system of male domination. Thus she and others might disagree that the ethics of care, as I take it here, is an enterprise of feminist ethics.

20. Virginia Held, "Non-contractual Society: A Feminist View," in *Science, Morality, and Feminist Theory,* eds. Marsha Hanen and Kai Nielsen (Calgary: University of Calgary Press, 1987), p. 122.

21. Borgmann, *Crossing the Postmodern Divide,* pp. 54–55.

22. For example, Virginia Held has written: "[Feminist moral inquiry] pays attention to the neglected experience of women and to such a woefully neglected though enormous area of human moral experience as that of mothering. . . . That this whole vast region of human experience can have been dismissed as 'natural' and thus as irrelevant to morality is extraordinary," ("Feminist Moral Inquiry and the Feminist Future," in *Justice and Care: Essential Readings in Feminist Ethics,* ed. Virginia Held [Boulder: Westview Press, 1995], p. 160).

23. Nel Noddings, *Caring* (Berkeley and Los Angeles: University of California Press, 1984), pp. 161–62.

24. Borgmann, *TCCL,* p. 49.

25. The term "multi-sided experiences" is used by Mihaly Csikzentmihalyi and Eugen Rochbert-Halton in their work, *The Meaning of Things,* discussed in Borgmann, "The Moral Significance of the Material Culture."

26. Lana F. Rakow, *Gender on the Line: Women, the Telephone, and Community Life* (Urbana: University of Illinois Press, 1992).

27. Borgmann, *Crossing the Postmodern Divide,* p. 105.

28. Rakow, *Gender on the Line,* p. 34.

29. Ibid., p. 57.

30. Noddings, *Caring.*

31. Rakow, *Gender on the Line,* p. 57.

32. Ibid., p. 57.

33. Ibid., p. 33.

34. Borgmann, *TCCL,* p. 217.

35. Rakow, *Gender on the Line,* p. 64.

36. See, for example, Susan Herring, "Gender Differences in Computer-Mediated Communication: Bringing Familiar Baggage to the New Frontier" (unpublished paper).

37. Albert Borgmann, "The Invisibility of Contemporary Culture," *Revue internationale de philosophie* 41 (1987): 239–42.

38. Borgmann, *TCCL,* p. 247.

39. Robert D. Putnam, "Bowling Alone: America's Declining Social Capital," *Journal of Democracy* 6 (1995): 67.

PART 7
APPROPRIATE TECHNOLOGY

INTRODUCTION

The concept of appropriate technology integrates many of the strands of thought we have been considering. Contrary to some critics, advocates of appropriate technology are not antitechnology but opposed to those forms of technology that they consider in some sense to be inappropriate in the modern world. Thus, in contrast to large-scale, complex, centralized systems of technology, they propose an alternative.

Although there is considerable controversy among the proponents of appropriate technology (AT), implicit in some of these perspectives is a vision of the good life, the good society, and the good earth—a vision strikingly different from the traditional values of industrial capitalism or totalitarianism. While not denigrating technology as a means of satisfying our basic needs, it rejects the view that the meaning of life is to be found in the consumption of ever-increasing "goods" pouring forth in mindless profusion from modern industry. The good life is characterized instead by meaningful work, harmony with one's fellowman, and a nonexploitative attitude toward nature—a life in which we are actively engaged in shaping our own life and justly concerned with the welfare of distant peoples and future generations. Human fulfillment on this vision requires a renewed sense of community.

The orthodox faith that technological growth would continue to satisfy increasing consumer demands in an endless cycle received a serious blow in the early seventies. Using state-of-the-art computer technology, a study called *Limits to Growth* predicted catastrophic disaster if present trends continued. Given the exponential growth in population, there was no way future demands on resources could be met, even assuming the development of new technologies to provide, for example, unlimited sources of energy. There are finite, and not too distant, limits to growth. Technological fixes merely postpone, and ultimately exacerbate, the eco-

417

logical crisis. While the methodology behind this study was severely criticized, it accomplished its major purpose, which was to stimulate public discussion.

The energy crisis (actual or contrived) during this period gave real bite to the idea that there might be limits to "progress." The *Global 2000 Report*, completed at the behest of U.S. President Jimmy Carter, reaffirmed the central conclusions of *Limits to Growth*, moving Carter to urge Americans to conserve energy by setting back their thermostats. Urging such belt-tightening measures was a major contributory factor in Carter's loss of the 1980 election to Ronald Reagan. Some asked: Did Carter not believe in the American dream of unlimited growth, of continuously increasing prosperity?

Apparent shortages of natural resources, growing awareness of the ecological consequences of industrial production, the failure of advanced military technology to win the war in Vietnam—all combined to transform the idea of appropriate technology into a grass-roots movement. Organizations devoted to developing and using small-scale, decentralized technology sprang up in many sections of the country. Although public concern about finite resources fluctuates wildly in response to events in the Persian Gulf and the apparent supply of oil, increasing awareness about environmental problems has kept the appropriate technology movement alive despite the continued reliance in some sectors on oil production as the best means for achieving "energy independence."

We begin this section with two reminders, aimed at strikingly different targets, of how technologies need a new design paradigm as much as they need a new ethics and politics. The idea that there is an authoritarian aspect to technology can be traced to Lewis Mumford and the publication in 1934 of his *Technics and Civilization*. According to Mumford, the roots of modern technology are to be found prior to the Industrial Revolution in the harnessing of human power. The model for large-scale technological systems can be seen in the system of monarchy and in the efficient organization of manpower in military armies with their division of labor and hierarchical chains of command. Employed in "civil engineering," it was this hierarchically organized "megamachine" that built the Pyramids. The "secret" to its efficient running were whips and chains, not new mechanical inventions.

Visible whips and chains have disappeared in modern technology, only to be replaced by the system itself. Having rejected the divine right of kings, we now come to worship at the shrine of technological progress. Even the high priests of science, the technological elite, and the captains of industry are trapped in their own system. When Mumford says, "The center of authority . . . now lies in the system itself," he is subscribing to a version of the autonomous technology thesis. According to Mumford, authoritarian technics represent a clear and present danger to our democratic institutions, but the future is not determined. What may give the appearance of determinism is that we have bought into the system, the consumer society, sacrificing our democratic ideals for material prosperity. But we could choose otherwise. Alternatives exist.

Democratic technics existed prior to authoritarian technics and continue to this day. These humanly scaled technologies are relatively weak, but are most resourceful and durable. They offer viable alternatives to the large-scale systems that threaten them. While not wanting to deny the material benefits of large-scale systems, Mumford thinks it is time to put people at the center of our concerns. Heralding the later work of the authors in the previous section such as Langdon Winner, Andrew Feenberg, and Diane Michelfelder, Mumford writes, "There are large areas of technology that can be redeemed by the democratic process, once we have overcome the infantile compulsions and aufomations that now threaten to cancel out our real gains."

Ruth Schwartz Cowan turns from the broader picture of the construction of empires to the more intimate picture of the technological construction of the household. In a brief and elegant essay, she summarizes the case against the common misperception that all technological changes are good. Her particular target is the "laborsaving devices" of the home, such as the vacuum cleaner and the washing machine. While these technologies were sold on the basis of their ability to make housework easier and more expedient, the result has not been one of freeing more time for women to engage in other pursuits outside the household. American housewives in the 1960s, 1970s, and 1980s continue to log the same number of hours maintaining a household as their progenitors did in the 1910s, 1920s, and 1930s. The reason is that these devices did little more than replace household servants and cleaning help with electronic gadgets. As Cowan puts it, "What Maggie had once done with a broom, Mrs. Smith was now doing with a vacuum cleaner." And because more women are now working outside the home, they are essentially holding down two jobs. Any move toward an alternative technology must keep changes like this in mind and consider, as Diane Michelfelder argued in the previous section, what sorts of freedoms are really being produced by new technologies.

But what, exactly, is an appropriate or alternative technology? Many critics have argued that there is no coherent notion of what constitutes an "appropriate" technology. Reviewing the literature, they simply find various authors pushing various technological fixes for various reasons. The movement seems to have degenerated into a fixation with do-it-yourself hardware and a romantic back-to-the-earth idealism. Contrary to these critics, Thomas Simon thinks that an adequate definition can be constructed for "appropriate" technology and lists what he takes to be the four fundamental conditions: (1) environmental soundness, (2) labor intensity, (3) small scale, and (4) decentralization. Each condition, he claims, is necessary, and together they are sufficient for a technology to be deemed appropriate. These are relatively clear criteria, but they are "interdependent conditions that must be explained together when evaluating a particular technique."

The crucial problem with appropriate technology is not a matter of specifying the criteria for what "appropriate" means, says Simon, but political. He agrees with Langdon Winner (and many of the people whose work we have examined) that "technical choices are, in fact, political." Simon thinks Winner errs, however, when he argues that some form of appropriate technology offers an alternative political philosophy. If appropriate technology's defenders discuss politics at all, Simon thinks they embrace the status quo, and thus are not much different from the advocates of high technology (such as Samuel Florman).

In defending his position, Simon analyzes the conditions he has stated for appropriate technology in order to show that the debate is not about the nature of various technologies, but about values—about what political philosophy we ought to adopt. All too often proponents of appropriate technology assume a kind of technological determinism, as though technology were an alternative to politics. From Simon's political perspective, conditions for appropriate technology ought to be: (1) ecological, (2) liberatory, (3) indigenous, and (4) politically radical.

The importance of Simon's discussion may lie not so much in the value of his claim that appropriate technology does not give us an alternative political philosophy—for a wide diversity of views are found in this movement—but in his emphasis on the fact that our choices are fundamentally normative. "Philosophically, the first and most important goal is an adequate politics of technological choice."

Authoritarian and Democratic Technics

Lewis Mumford

"Democracy" is a term now confused and sophisticated by indiscriminate use, and often treated with patronizing contempt. Can we agree, no matter how far we might diverge at a later point, that the spinal principle of democracy is to place what is common to all men above that which any organization, institution, or group may claim for itself? This is not to deny the claims of superior natural endowment, specialized knowledge, technical skill, or institutional organization: all these may, by democratic permission, play a useful role in the human economy. But democracy consists in giving final authority to the whole, rather than the part; and only living human beings, as such, are an authentic expression of the whole, whether acting alone or with the help of others.

Around this central principle clusters a group of related ideas and practices with a long foreground in history, though they are not always present, or present in equal amounts, in all societies. Among these items are communal self-government, free communication as between equals, unimpeded access to the common store of knowledge, protection against arbitrary external controls, and a sense of individual moral responsibility for behavior that affects the whole community. All living organisms are in some degree autonomous, in that they follow a life-pattern of their own; but in man this autonomy is an essential condition for his further development. We surrender some of our autonomy when ill or crippled: but to surrender it every day on every occasion would be to turn life itself into a chronic illness. The best life possible—and here I am consciously treading on contested ground—is one that calls for an ever greater degree of self-direction, self-expression, and self-realization. In this sense, personality, once the exclusive attribute of kings, belongs on democratic theory to every man. Life itself in its fullness and wholeness cannot be delegated.

Lewis Mumford, "Authoritarian and Democratic Technics," *Technology and Culture* 5, no. 1 (1964): 1–8. © Society for the History of Technology. Reprinted with permission of the Johns Hopkins University Press.

In framing this provisional definition I trust that I have not, for the sake of agreement, left out anything important. Democracy, in the primal sense I shall use the term, is necessarily most visible in relatively small communities and groups, whose members meet frequently face to face, interact freely, and are known to each other as persons. As soon as large numbers are involved, democratic association must be supplemented by a more abstract, depersonalized form. Historical experience shows that it is much easier to wipe out democracy by an institutional arrangement that gives authority only to those at the apex of the social hierarchy than it is to incorporate democratic practices into a well-organized system under centralized direction, which achieves the highest degree of mechanical efficiency when those who work it have no mind or purpose of their own.

The tension between small-scale association and large-scale organization, between personal autonomy and institutional regulation, between remote control and diffused local intervention, has now created a critical situation. If our eyes had been open, we might long ago have discovered this conflict deeply embedded in technology itself.

I wish it were possible to characterize technics with as much hope of getting assent, with whatever quizzical reserves you may still have, as in this description of democracy. But the very title of this [chapter] is, I confess, a controversial one; and I cannot go far in my analysis without drawing on interpretations that have not yet been adequately published, still less widely discussed or rigorously criticized and evaluated. My thesis, to put it bluntly, is that from late neolithic times in the Near East, right down to our own day, two technologies have recurrently existed side by side: one authoritarian, the other democratic, the first system-centered, immensely powerful, but inherently unstable, the other man-centered, relatively weak, but resourceful and durable. If I am right, we are now rapidly approaching a point at which, unless we radically alter our present course, our surviving democratic technics will be completely suppressed or supplanted, so that every residual autonomy will be wiped out, or will be permitted only as a playful device of government, like national balloting for already chosen leaders in totalitarian countries.

The data on which this thesis is based are familiar; but their significance has, I believe, been overlooked. What I would call democratic technics is the small-scale method of production, resting mainly on human skill and animal energy but always, even when employing machines, remaining under the active direction of the craftsman or the farmer, each group developing its own gifts, through appropriate arts and social ceremonies, as well as making discreet use of the gifts of nature. This technology had limited horizons of achievement, but, just because of its wide diffusion and its modest demands, it had great powers of adaptation and recuperation. This democratic technics has underpinned and firmly supported every historical culture until our own day, and redeemed the constant tendency of authoritarian technics to misapply its powers. Even when paying tribute to the most oppressive authoritarian regimes, there yet remained within the workshop or the farmyard some degree of autonomy, selectivity, creativity. No royal mace, no slave-driver's whip, no bureaucratic directive left its imprint on the textiles of Damascus or the pottery of fifth-century Athens.

If this democratic technics goes back to the earliest use of tools, authoritarian technics is a much more recent achievement: it begins around the fourth millennium B.C.E. in a new configuration of technical invention, scientific observation, and centralized political control that gave rise to the peculiar mode of life we may now identify, without eulogy, as civilization. Under the new institution of kingship, activities that had been scattered, diversified, cut to the human measure, were united on a monumental scale into an entirely new kind of theological-technological mass organization. In the person of an absolute ruler, whose word was law, cosmic powers came down to earth, mobilizing and unifying the efforts of thousands of men, hitherto all-too-autonomous and too decentralized to act voluntarily in unison for purposes that lay beyond the village horizon.

The new authoritarian technology was not limited by village custom or human sentiment: its herculean feats of mechanical organization rested on ruthless physical coercion, forced labor, and slavery, which brought into existence machines that were capable of exerting thousands of horsepower centuries before horses were harnessed or wheels invented. This centralized technics drew on inventions and scientific discoveries of a high order: the written record, mathematics and astronomy, irrigation and canalization: above all, it created complex human machines composed of specialized, standardized, replaceable, interdependent parts—the work army, the military army, the bureaucracy. These work armies and military armies raised the ceiling of human achievement: the first in mass construction, the second in mass destruction, both on a scale hitherto inconceivable. Despite its constant drive to destruction, this totalitarian technics was tolerated, perhaps even welcomed, in home territory, for it created the first economy of controlled abundance: notably, immense food crops that not merely supported a big urban population but released a large trained minority for purely religious, scientific, bureaucratic, or military activity. But the efficiency of the system was impaired by weaknesses that were never overcome until our own day.

To begin with, the democratic economy of the agricultural village resisted incorporation into the new authoritarian system. So even the Roman Empire found it expedient, once resistance was broken and taxes were collected, to consent to a large degree of local autonomy in religion and government. Moreover, as long as agriculture absorbed the labor of some 90 percent of the population, mass technics were confined largely to the populous urban centers. Since authoritarian technics first took form in an age when metals were scarce and human raw material, captured in war, was easily convertible into machines, its directors never bothered to invent inorganic mechanical substitutes. But there were even greater weaknesses: the system had no inner coherence: a break in communication, a missing link in the chain of command, and the great human machines fell apart. Finally, the myths upon which the whole system was based—particularly the essential myth of kingship—were irrational, with their paranoid suspicions and animosities, and their paranoid claims to unconditional obedience and absolute power. For all its redoubtable constructive achievements, authoritarian technics expressed a deep hostility to life.

By now you doubtless see the point of this brief historical excursus. That authoritarian technics has come back today in an immensely magnified and adroitly perfected form. Up to now, following the optimistic premises of nineteenth-century thinkers like Auguste Comte and Herbert Spencer, we have regarded the spread of experimental science and mechanical invention as the soundest guarantee of a peaceful, productive, above all democratic, industrial society. Many have even comfortably supposed that the revolt against arbitrary political power in the seventeenth century was causally connected with the industrial revolution that accompanied it. But what we have interpreted as the new freedom now turns out to be a much more sophisticated version of the old slavery: for the rise of political democracy during the last few centuries has been increasingly nullified by the successful resurrection of a centralized authoritarian technics—a technics that had in fact for long lapsed in many parts of the world.

Let us fool ourselves no longer. At the very moment Western nations threw off the ancient regime of absolute government, operating under a once-divine king, they were restoring this same system in a far more effective form in their technology, reintroducing coercions of a military character no less strict in the organization of a factory than in that of the new drilled, uniformed, and regimented army. During the transitional stages of the last two centuries, the ultimate tendency of this system might be in doubt, for in many areas there were strong democratic reactions; but with the knitting together of a scientific ideology, itself liberated from theological restrictions or humanistic purposes, authoritarian technics found an instrument at hand that has now given it absolute command of physical energies of cosmic dimensions. The inventors of nuclear bombs, space rockets, and computers are the pyramid builders of our own age: psychologically inflated by a similar myth of unqualified power, boasting through their science of their increasing omnipotence, if not omniscience, moved by obsessions and compulsions no less irrational than those of earlier absolute systems: particularly the notion that the system itself must be expanded, at whatever eventual cost to life.

Through mechanization, automation, cybernetic direction, this authoritarian technics has at last successfully overcome its most serious weakness: its original dependence upon resistant, sometimes actively disobedient servomechanisms, still human enough to harbor purposes that do not always coincide with those of the system.

Like the earliest form of authoritarian technics, this new technology is marvelously dynamic and productive: its power in every form tends to increase without limits, in quantities that defy assimilation and defeat control, whether we are thinking of the output of scientific knowledge or of industrial assembly lines. To maximize energy, speed, or automation, without reference to the complex conditions that sustain organic life, have become ends in themselves. As with the earliest forms of authoritarian technics, the weight of effort, if one is to judge by national budgets, is toward absolute instruments of destruction, designed for absolutely irrational purposes whose chief by-product would be the mutilation or extermination of the human race. Even Ashurbanipal and Genghis Khan performed their gory operations under normal human limits.

The center of authority in this new system is no longer a visible personality, an all-powerful king; even in totalitarian dictatorships the center now lies in the system itself, invisible but omnipresent: all its human components, even the technical and managerial elite, even the sacred priesthood of science, who alone have access to the secret knowledge by means of which total control is now swiftly being effected, are themselves trapped by the very perfection of the organization they have invented. Like the pharaohs of the Pyramid Age, these servants of the system identify its good with their own kind of well-being: as with the divine kind, their praise of the system is an act of self-worship; and again like the king, they are in the grip of an irrational compulsion to extend their means of control and expand the scope of their authority. In this new systems-centered collective, this Pentagon of power, there is no visible presence who issues commands: unlike Job's God, the new deities cannot be confronted, still less defied. Under the pretext of saving labor, the ultimate end of this technics is to displace life, or rather, to transfer the attributes of life to the machine and the mechanical collective, allowing only so much of the organism to remain as may be controlled and manipulated.

Do not misunderstand this analysis. The danger to democracy does not spring from any specific scientific discoveries or electronic inventions. The human compulsions that dominate the authoritarian technics of our own day date back to a period before even the wheel had been invented. The danger springs from the fact that, since Francis Bacon and Galileo defined the new methods and objectives of technics, our great physical transformations have been effected by a system that deliberately eliminates the whole human personality, ignores the historical process, overplays the role of the abstract intelligence, and makes control over physical nature, ultimately control over man himself, the chief purpose of existence. This system has made its way so insidiously into Western society that my analysis of its derivation and its intentions may well seem more questionable—indeed more shocking—than the facts themselves.

Why has our age surrendered so easily to the controllers, the manipulators, the conditioners of an authoritarian technics? The answer to this question is both paradoxical and ironic. Present-day technics differs from that of the overtly brutal, half-baked authoritarian systems of the past in one highly favorable particular: it has accepted the basic principle of democracy, that every member of society should have a share in its goods. By progressively fulfilling this part of the democratic promise, our system has achieved a hold over the whole community that threatens to wipe out every other vestige of democracy.

The bargain we are being asked to ratify takes the form of a magnificent bribe. Under the democratic-authoritarian social contract, each member of the community may claim every material advantage, every intellectual and emotional stimulus he may desire, in quantities hardly available hitherto even for a restricted minority: food, housing, swift transportation, instantaneous communication, medical care, entertainment, education. But on one condition: that one must not merely ask for nothing that the system does not provide, but likewise agree to take everything offered, duly processed and fabricated, homogenized and equalized, in the precise quantities that the system, rather than the person, requires. Once one opts for the system no further choice remains. In a word, if one surrenders one's life at source, authoritarian technics will give back as much of it as can be mechanically graded, quantitatively multiplied, collectively manipulated and magnified.

"Is this not a fair bargain?" those who speak for the system will ask. "Are not the goods authoritarian technics promises real goods? Is this not the horn of plenty that mankind has long dreamed of, and that every ruling class has tried to secure, at whatever cost of brutality and injustice, for itself?" I would not belittle, still less deny, the many admirable products this technology has brought forth, products that a self-regulating economy would make good use of. I would only suggest that it is time to reckon up the human disadvantages and costs, to say nothing of the dangers, of our unqualified acceptance of the system itself. Even the immediate price is heavy; for the system is so far from being under effective human direction that it may poison us wholesale to provide us with food or exterminate us to provide national security, before we can enjoy its promised goods. Is it really humanly profitable to give up the possibility of living a few years at Walden Pond, so to say, for the privilege of spending a lifetime in *Walden Two?* Once our authoritarian technics consolidates its powers, with the aid of its new forms of mass control, its panoply of tranquilizers and sedatives and aphrodisiacs, could democracy in any form survive? That question is absurd: life itself will not survive, except what is funneled through the mechanical collective. The spread of a sterilized scientific intelligence over the planet would not, as Teilhard de Chardin so innocently imagined, be the happy consummation of divine purpose: it would rather ensure the final arrest of any further human development.

Again: do not mistake my meaning. This is not a prediction of what *will* happen, but a warning against what *may* happen.

What means must be taken to escape this fate? In characterizing the authoritarian technics that has begun to dominate us, I have not forgotten the great lesson of history: prepare for the unexpected! Nor do I overlook the immense reserves of vitality and creativity that a more humane democratic tradition still offers us. What I wish to do is to persuade those who are concerned with maintaining democratic institutions to see that their constructive efforts must include technology itself. There, too, we must return to the human center. We must challenge this authoritarian system that has given to an underdimensioned ideology and technology the authority that belongs to the human personality. I repeat: Life cannot be delegated.

Curiously, the first words in support of this thesis came forth, with exquisite symbolic aptness, from a willing agent—but very nearly a classic victim!—of the new authoritarian technics. They came from the astronaut, John Glenn, whose life was endangered by the malfunctioning of his automatic controls, operated from a remote center. After he barely saved his life by personal intervention, he emerged from his space capsule with these ringing words: "Now let man take over!"

That command is easier to utter than obey. But if we are not to be driven to even more drastic measures than Samuel Butler suggested in *Erewhon*, we had better map out a more positive course: namely, the reconstitution of both our science and our technics in such a fashion as to insert the rejected parts of the human personality at every stage in the process. This means gladly sacrificing mere quantity in order to restore qualitative choice; shifting the seat of authority from the mechanical collective to the human personality and the autonomous group; favoring variety and ecological complexity, instead of stressing undue uniformity and standarization; above all, reducing the insensate drive to extend the system itself, instead of containing it within definite human limits and thus releasing man himself for other purposes. We must ask, not what is good for science or technology, still less what is good for General Motors or Union Carbide or IBM or the Pentagon, but what is good for man: not machine-conditioned, system-regulated, mass-man, but man in person, moving freely over every area of life.

There are large areas of technology that can be redeemed by the democratic process, once we have overcome the infantile compulsions and automatisms that now threaten to cancel out our real gains. The very leisure that the machine now gives in advanced countries can be profitably used, not for further commitment to still other kinds of machine, furnishing automatic recreation, but by doing significant forms of work, unprofitable or technically impossible under mass production: work dependent upon special skill, knowledge, aesthetic sense. The do-it-yourself movement prematurely got bogged down in an attempt to sell still more machines; but its slogan pointed in the right direction, provided we still have a self to do it with. The glut of motor cars that is now destroying our cities can be coped with only if we redesign our cities to make fuller use of a more efficient human agent: the walker. Even in childbirth, the emphasis is already happily shifting from an officious, often lethal, authoritarian procedure, centered in hospital routine, to a more humane mode, which restores initiative to the mother and to the body's natural rhythms.

The replenishment of democratic technics is plainly too big a subject to be handled in a final sentence or two: but I trust I have made it clear that the genuine advantage our scientifically based technics has brought can be preserved only if we cut the whole system back to a point at which it will permit human alternatives, human interventions, and human destinations for entirely different purposes from those of the system itself. At the present juncture, if democracy did not exist, we would have to invent it, in order to save and recultivate the spirit of man.

28

Less Work for Mother?

Ruth Schwartz Cowan

Things are never what they seem. Skimmed milk masquerades as cream. And laborsaving household appliances often do not save labor. This is the surprising conclusion reached by a small army of historians, sociologists, and home economists who have undertaken, in recent years, to study the one form of work that has turned out to be most resistant to inquiry and analysis—namely, housework.

During the first half of the twentieth century, the average American household was transformed by the introduction of a group of machines that profoundly altered the daily lives of housewives; the forty years between 1920 and 1960 witnessed what might be aptly called the "industrial revolution in the home." Where once there had been a wood- or coal-burning stove there now was a gas or electric range. Clothes that had once been scrubbed on a metal washboard were now tossed into a tub and cleansed by an electrically driven agitator. The dryer replaced the clothesline; the vacuum cleaner replaced the broom; the refrigerator replaced the icebox and the root cellar; an automatic pump, some piping, and a tap replaced the hand pump, the bucket, and the well. No one had to chop and haul wood any more. No one had to shovel out ashes or beat rugs or carry water; no one even had to toss egg whites with a fork for an hour to make an angel food cake.

And yet American housewives in 1960, 1970, and even 1980 continued to log about the same number of hours at their work as their grandmothers and mothers had in 1910, 1920, and 1930. The earliest time studies of housewives date from the very same period in which time studies of other workers were becoming popular—the first three decades of the twentieth century. The sample sizes of these studies were usually quite small, and they did not always define housework in precisely the same way (some counted an hour spent taking children to the playground as "work," while others called it "leisure"), but their results were more or

less consistent: whether rural or urban, the average American housewife performed fifty to sixty hours of unpaid work in her home every week, and the only
variable that significantly altered this was the number of small children.

A half century later not much had changed. Survey research had become
much more sophisticated, and sample sizes had grown considerably, but the
results of the time studies remained surprisingly consistent. The average American housewife, now armed with dozens of motors and thousands of electronic
chips, still spends fifty to sixty hours a week doing housework. The only variable
that significantly altered the size of that number was full-time employment in the
labor force; "working" housewives cut down the average number of hours that
they spend cooking and cleaning, shopping and chauffeuring, to a not insignificant thirty-five—virtually the equivalent of another full-time job.

How can this be true? Surely even the most sophisticated advertising copy-
writer of all times could not fool almost the entire American population over the
course of at least three generations. Laborsaving devices must be saving something or Americans would not continue, year after year, to plunk down their hard-
earned dollars for them.

And if laborsaving devices have not saved labor in the home, then what is it
that has suddenly made it possible for more than 70 percent of the wives and
mothers in the American population to enter the workforce and stay there? A brief
glance at the histories of some of the technologies that have transformed housework in the twentieth century will help us answer some of these questions.

The portable vacuum cleaner was one of the earliest electric appliances to
make its appearance in American homes, and reasonably priced models appeared
on the retail market as early as 1910. For decades prior to the turn of the century,
inventors had been trying to create a carpet-cleaning system that would improve
on the carpet sweeper with adjustable rotary brushes (patented by Melville Bis-
sell in 1876), or the semiannual ritual of hauling rugs outside and beating them,
or the practice of regularly sweeping the dirt out of a rug that had been covered
with dampened, torn newspapers. Early efforts to solve the problem had focused
on the use of large steam, gasoline, or electric motors attached to piston-type
pumps and lots of hoses. Many of these "stationary" vacuum-cleaning systems
were installed in apartment houses or hotels, but some were hauled around the
streets in horse-drawn carriages by entrepreneurs hoping to establish themselves
as "professional housecleaners."

In the first decade of the twentieth century, when fractional-horsepower electric motors became widely—and inexpensively—available, the portable vacuum
cleaner intended for use in an individual household was born. One early model—
invented by a woman, Corrine Dufour—consisted of a rotary brush, an electrically driven fan, and a wet sponge for absorbing the dust and dirt. Another,
patented by David E. Kenney in 1907, had a twelve-inch nozzle, attached to a
metal tube, attached to a flexible hose that led to a vacuum pump and separating
devices. The Hoover, which was based on a brush, a fan, and a collecting bag,

was on the market by 1908. The Electrolux, the first of the canister types of cleaner, which could vacuum something above the level of the floor, was brought over from Sweden in 1924 and met with immediate success.

These early vacuum cleaners were hardly a breeze to operate. All were heavy, and most were extremely cumbersome to boot. One early home economist mounted a basal metabolism machine on the back of one of her hapless students and proceeded to determine that more energy was expended in the effort to clean a sample carpet with a vacuum cleaner than when the same carpet was attacked with a hard broom. The difference, of course, was that the vacuum cleaner did a better job, at least on carpets, because a good deal of what the broom stirred up simply resettled a foot or two away from where it had first been lodged. Whatever the liabilities of the early vacuum cleaners may have been, Americans nonetheless appreciated their virtues; according to a market survey done in Zanesville, Ohio, in 1926, slightly more than half the households owned one. Eventually improvements in the design made these devices easier to operate. By 1960 vacuum cleaners could be found in 70 percent of the nation's homes.

When the vacuum cleaner is viewed in a historical context, however, it is easy to see why it did not save housewifely labor. Its introduction coincided almost precisely with the disappearance of the domestic servant. The number of persons engaged in household service dropped from 1,851,000 in 1910 to 1,411,000 in 1920, while the number of households enumerated in the census rose from 20.3 million to 24.4 million. Moreover, between 1900 and 1920 the number of household servants per thousand persons dropped from 98.9 to 58.0, while during the 1920s the decline was even more precipitous as the restrictive immigration acts dried up what had once been the single most abundant source of domestic labor.

For the most economically comfortable segment of the population, this meant just one thing: the adult female head of the household was doing more housework than she had ever done before. What Maggie had once done with a broom, Mrs. Smith was now doing with a vacuum cleaner. Knowing that this was happening, several early copywriters for vacuum cleaner advertisements focused on its implications. The vacuum cleaner, General Electric announced in 1918, is better than a maid: it doesn't quit, get drunk, or demand higher wages. The switch from Maggie to Mrs. Smith shows up, in time-study statistics, as an increase in the time that Mrs. Smith is spending at her work.

For those—and they were the vast majority of the population—who were not economically comfortable, the vacuum cleaner implied something else again: not an increase in the time spent in housework but an increase in the standard of living. In many households across the country, acquisition of a vacuum cleaner was connected to an expansion of living space, the move from a small apartment to a small house, the purchase of wall-to-wall carpeting. If this did not happen during the difficult 1930s, it became more possible during the expansive 1950s. As living quarters grew larger, standards for their upkeep increased; rugs had to be vacuumed every week, in some households every day, rather than semiannually, as had been customary. The net result, of course, was that when armed with a vacuum cleaner, housewives whose parents had been poor could keep more space cleaner than their mothers and grandmothers would have ever believed possible. We might put this everyday phenomenon in language that economists can understand: The introduction of the vacuum cleaner led to improvements in productivity but not to any significant decrease in the amount of time expended by each worker.

The history of the washing machine illustrates a similar phenomenon. "Blue Monday" had traditionally been, as its name implies, the bane of a housewife's existence—especially when Monday turned out to be "Monday . . . and Tuesday to do the ironing." Thousands of patents for "new and improved" washers were issued during the nineteenth century in an effort to cash in on the housewife's despair. Most of these early washing machines were wooden or metal tubs combined with some kind of hand-cranked mechanism that would rub or push or twirl laundry when the tub was filled with water and soap. At the end of the century, the Sears catalog offered four such washing machines, ranging in price from $2.50 to $4.25, all sold in combination with hand-cranked wringers.

These early machines may have saved time in the laundering process (four shirts could be washed at once instead of each having to be rubbed separately against a washboard), but they probably didn't save much energy. Lacking taps and drains, the tubs still had to be filled and emptied by hand, and each piece still had to be run through a wringer and hung up to dry.

Not long after the appearance of fractional-horsepower motors, several enterprising manufacturers had the idea of hooking them up to the crank mechanisms of washers and wringers—and the electric washer was born. By the 1920s, when mass production of such machines began, both the general structure of the machine (a central-shaft agitator rotating within a cylindrical tub, hooked up to the household water supply) and the general structure of the industry (oligopolistic—with a very few firms holding most of the patents and controlling most of the market) had achieved their final form. By 1926 just over a quarter of the families in Zanesville had an electric washer, but by 1941 fully 52 percent of all American households either owned or had interior access (which means that they could use coin-operated models installed in the basements of apartment houses) to such a machine. The automatic washer, which consisted of a vertically rotating washer cylinder that could also act as a centrifugal extractor, was introduced by the Bendix Home Appliance Corporation in 1938, but it remained expensive, and therefore inaccessible, until after World War II. This machine contained timing devices that allowed it to proceed through its various cycles automatically; by spinning the clothes around in the extractor phase of its cycle, it also eliminated the wringer. Although the Bendix subsequently disappeared from the retail market (versions of this sturdy machine may still be found in Laundromats), its design principles are replicated in the agitator washers that currently chug away in millions of American homes.

Both the early wringer washers and their more recent automatic cousins have released American women from the burden of drudgery. No one who has ever tried to launder a sheet by hand, and without the benefits of hot running water, would want to return to the days of the scrubboard and tub. But "labor" is composed of both "energy expenditure" and "time expenditure," and the history of laundry work demonstrates that the one may be conserved while the other is not.

The reason for this is, as with the vacuum cleaner, twofold. In the early decades of the century, many households employed laundresses to do their wash; this was true, surprisingly enough, even for some very poor households when wives and mothers were disabled or employed full-time in field or factory. Other households—rich and poor—used commercial laundry services. Large, mechanized "steam" laundries were first constructed in this country in the 1860s, and by the 1920s they could be found in virtually every urban neighborhood and many rural ones as well.

But the advent of the electric home washer spelled doom both for the laundress and for the commercial laundry; since the housewife's labor was unpaid, and since the washer took so much of the drudgery out of washday, the one-time expenditure for a machine seemed, in many families, a more sensible arrangement than continuous expenditure for domestic services. In the process, of course, the time spent on laundry work by the individual housewife, who had previously employed either a laundress or a service, was bound to increase.

For those who had not previously enjoyed the benefits of relief from washday drudgery, the electric washer meant something quite different but equally significant: an upgrading of household cleanliness. Men stopped wearing removable collars and cuffs, which meant that the whole of their shirts had to be washed and then ironed. Housewives began changing two sheets every week, instead of moving the top sheet to the bottom and adding only one that was fresh. Teenagers began changing their underwear every day instead of every weekend. In the early 1960s, when synthetic no-iron fabrics were introduced, the size of the household laundry load increased again; shirts and skirts, sheets and blouses that had once been sent out to the dry cleaner or the corner laundry were now being tossed into the household wash basket. By the 1980s the average American housewife, armed now with an automatic washing machine and an automatic dryer, was processing roughly ten times (by weight) the amount of laundry that her mother had been accustomed to. Drudgery had disappeared, but the laundry hadn't. The average time spent on this chore in 1925 had been 5.8 hours per week; in 1964 it was 6.2.

And then there is the automobile. We do not usually think of our cars as household appliances, but that is precisely what they are since housework, as currently understood, could not possibly be performed without them. The average American housewife is today more likely to be found behind a steering wheel than in front of a stove. While writing this article, I interrupted myself five times: once to take a child to field hockey practice, then a second time, to bring her back when practice was finished; once to pick up some groceries at the supermarket; once to retrieve my husband, who was stranded at the train station; once for a trip to a doctor's office. Each time I was doing housework, and each time I had to use my car.

Like the washing machine and the vacuum cleaner, the automobile started to transform the nature of housework in the 1920s. Until the introduction of the Model T in 1908, automobiles had been playthings for the idle rich, and although many wealthy women learned to drive early in the century (and several participated in well-publicized auto races), they were hardly the women who were likely to be using their cars to haul groceries.

But by 1920, and certainly by 1930, all this had changed. Helen and Robert Lynd, who conducted an intensive study of Muncie, Indiana, between 1923 and 1925 (reported in their famous book *Middletown*), estimated that in Muncie in the 1890s only 125 families, all members of the "elite," owned a horse and buggy, but by 1923 there were 6,222 passenger cars in the city, "roughly one for every 7.1 persons, or two for every three families." By 1930, according to national statistics, there were roughly 30 million households in the United States—and 26 million registered automobiles.

What did the automobile mean for the housewife? Unlike public transportation systems, it was convenient. Located right at her doorstep, it could deposit her at the doorstep that she wanted or needed to visit. And unlike the bicycle or her own two feet, the automobile could carry bulky packages as well as several additional people. Acquisition of an automobile therefore meant that a housewife, once she had learned how to drive, could become her own door-to-door delivery service. And as more housewives acquired automobiles, more businessmen discovered the joys of dispensing with delivery services—particularly during the Depression.

To make a long story short, the iceman does not cometh anymore. Neither does the milkman, the bakery truck, the butcher, the grocer, the knife sharpener, the seamstress, or the doctor. Like many other businessmen, doctors discovered that their earnings increased when they stayed in their offices and transferred the responsibility for transportation to their ambulatory patients.

Thus a new category was added to the housewife's traditional job description: chauffeur. The suburban station wagon is now "Mom's Taxi." Children who once walked to school now have to be transported by their mothers; husbands who once walked home from work now have to be picked up by their wives; groceries that once were dispensed from pushcarts or horse-drawn wagons now have to be packed into paper bags and hauled home in family cars. "Contemporary women," one time-study expert reported in 1974, "spend about one full working day per week on the road and in stores compared with less than two hours per week for women in the 1920s." If everything we needed to maintain our homes and sustain our families were delivered right to our doorsteps—and every member of the family had independent means for getting where she or he wanted to go—the hours spent in housework by American housewives would decrease dramatically.

The histories of the vacuum cleaner, the washing machine, and the automobile illustrate the varied reasons why the time spent in housework has not markedly decreased in the United States during the last half century despite the introduction of so many ostensibly laborsaving appliances. But these histories do not help us understand what has made it possible for so many American wives and mothers to enter the labor force full-time during those same years. Until recently, one of the explanations most often offered for the startling increase in the participation of married women in the workforce (up from 24.8 percent in 1950 to 50.1 percent in 1980) was household technology. What with microwave ovens and frozen foods, washer and dryer combinations and paper diapers, the reasoning goes, housework can now be done in no time at all, and women have so much time on their hands that they find they must go out and look for a job for fear of going stark, raving mad.

As every "working" housewife knows, this pattern of reasoning is itself stark, raving mad. Most adult women are in the workforce today quite simply because they need the money. Indeed, most "working" housewives today hold down not one but two jobs; they put in what has come to be called a "double day." Secretaries, lab technicians, janitors, sewing machine operators, teachers, nurses, or physicians for eight (or nine or ten) hours, they race home to become chief cook and bottle washer for another five, leaving the cleaning and the marketing for Saturday and Sunday. Housework, as we have seen, still takes a lot of time, modern technology notwithstanding.

Yet household technologies have played a major role in facilitating (as opposed to causing) what some observers believe to be the most significant social revolution of our time. They do it in two ways, the first of which we have already noted. By relieving housework of the drudgery that it once entailed, washing machines, vacuum cleaners, dishwashers, and water pumps have made it feasible for a woman to put in a double day without destroying her health, to work full-time and still sustain herself and her family at a reasonably comfortable level.

The second relationship between household technology and the participation of married women in the workforce is considerably more subtle. It involves the history of some technologies that we rarely think of as technologies at all—and certainly not as household appliances. Instead of being sheathed in stainless steel or porcelain, these devices appear in our kitchens in little brown bottles and bags of flour; instead of using switches and buttons to turn them on, we use hypodermic needles and sugar cubes. They are various forms of medication, the products not only of modern medicine but also of modern industrial chemistry: polio vaccines and vitamin pills; tetanus toxins and ampicillin; enriched breads and tuberculin tests.

Before any of these technologies had made their appearance, nursing may well have been the most time-consuming and most essential aspect of housework. During the eighteenth and nineteenth centuries and even during the first five decades of the twentieth century, it was the woman of the house who was expected (and who had been trained, usually by *her* mother) to sit up all night cooling and calming a feverish child, to change bandages on suppurating wounds, to clean bed linens stained with excrement, to prepare easily digestible broths, to cradle colicky infants on her lap for hours on end, to prepare bodies for burial. An attack of the measles might mean the care of a bedridden child for a month. Pneumonia might require six months of bed rest. A small knife cut could become infected and produce a fever that would rage for days. Every summer brought the fear of polio epidemics, and every polio epidemic left some group of mothers with the perpetual problem of tending to the needs of a handicapped child. Cholera, diphtheria, typhoid fever—if they weren't fatal—could mean weeks of sleepless nights and hard-pressed days. "Just as soon as the person is attacked," one experienced mother wrote to her worried daughter during a cholera epidemic in Oklahoma in 1885, "be it ever so slightly, he or she ought to go to bed immediately and stay there; put a mustard [plaster] over the bowels and if vomiting over the stomach. See that the feet are kept warm, either by warm iron or brick, or bottles of hot water. If the disease progresses the limbs will begin to cramp, which must be prevented by applying cloths wrung out of hot water and wrapping round them. When one is vomiting so terribly, of course, it is next to impossible to keep medicine down, but in cholera it must be done."

These were the routines to which American women were once accustomed, routines regarded as matters of life and death. To gain some sense of the way in which modern medicines have altered not only the routines of housework but also the emotional commitment that often accompanies such work, we need only read out a list of the diseases to which most American children are unlikely to succumb today, remembering how many of them once were fatal or terribly disabling: diphtheria, whooping cough, tetanus, pellagra, rickets, measles, mumps, tuberculosis, smallpox, cholera, malaria, and polio.

And many of today's ordinary childhood complaints, curable within a few days of the ingestion of antibiotics, once might have entailed weeks, or even months, of full-time attention: bronchitis; strep throat; scarlet fever; bacterial pneumonia; infections of the skin, or the eyes, or the ears, or the airways. In the days before the introduction of modern vaccines, antibiotics, and vitamin supplements, a mother who was employed full-time was a serious, sometimes life-endangering threat to the health of her family. This is part of the reason why life expectancy was always low and infant mortality high among the poorest segment of the population—those most likely to be dependent upon a mother's wages.

Thus modern technology, especially modern medical technology, has made it possible for married women to enter the workforce by releasing housewives not just from drudgery but also from the dreaded emotional equation of female employment with poverty and disease. She may be exhausted at the end of her double day, but the modern "working" housewife can at least fall into bed knowing that her efforts have made it possible to sustain her family at a level of health and comfort that not so long ago was reserved only for those who were very rich.

29

Appropriate Technology and Inappropriate Politics

Thomas Simon

INTRODUCTION

Over the past two decades "appropriate technology" has been used as code for new ways of thinking about the social implications of technological choice. Appropriate technology is seen variously: as a means of ushering in a New Age, as an alternative to high technology, as a social movement, and, by some, as utopian delusion. The debate over appropriate technology is raising some of the most difficult questions facing a philosophy of technology, including: "the relationship between technology and development, between ideology and industrialization, and more fundamentally, between man and machine."[1] Yet, the most critical question is seldom explicitly addressed in this debate: What is and what should be the political philosophy underlying the appropriate technology movement?

Before providing my own answer to that question we need to get clear about just what appropriate technology is. Some critics charge that appropriate technology is a grab bag of vague ideas, which resist any attempts to provide definitional coherence. According to this view,

> [Appropriate] technology is not a coherent philosophy. It is a collection of a large number of ideas and concepts, many of them quite incoherent, almost as diverse as the name of the outlook: intermediate technology, humane technology, a new alchemy, peoples' technology, radical hardware, biotechnics, etc. Each of these names emphasize a different aspect of the new technology: workers' control, demystification of expertise, reform of work rules, low specialization, development under the condition of low capital, local or regional self-sufficiency, balanced economic development, resource conservation, low energy use, reduced technological risks, and so on.[2]

From *Technology and Contemporary Life*, ed. Paul T. Durbin, pp. 107–28. Copyright © 1988 Kluwer Academic Publishers, Dordrecht, Holland. Reprinted by permission.

Contrary to this assessment, I will argue that methodological coherence can be given to the appropriate technology (AT) proposals. To do this, I will first look at what proponents of AT indicate as their agenda, and then I will undertake a rational reconstruction of the aims and assumptions of this agenda. In other words, prevailing AT ideas and practices can be used to define a methodologically adequate concept of AT.

Armed with a clear definition of AT we should be in a position to differentiate appropriate technology from its close kin—ecotechnology, liberatory, indigenous, intermediate, and radical technologies. Each of these might be seen to emphasize a different definitional feature than AT emphasizes. This approach, however, will be shown to be misleading.

In the final section of the [chapter] I will demonstrate that the problem with AT (and its kindred conceptions) is not methodological adequacy but rather the problem is the political foundation of the particular philosophy of technology. Appropriate technologists debate high technologists in terms of the features of their differing technologies. Likewise some ecotechnologists, liberatory technologists, etc., debate appropriate technologists in terms of the features of their differing technologies. This, however, is the wrong debate. The debate should not be over the features of the technologies. The debate is covertly and should be overtly a political debate. Failure to recognize the political nature of the debate leads critics to see a hodgepodge among these differing conceptions of technology. Recognizing the political nature of these different conceptions results in AT being judged politically inadequate. Before making the case for its political inadequacy let us first demonstrate the methodological adequacy of AT.

DEFINING APPROPRIATE TECHNOLOGY

When you read the AT literature and come across definitions of appropriate technology such as "technology with a human face"[3] and "whatever is appropriate,"[4] you can immediately see the need for some definitional clarity. Those AT proponents with some methodological sophistication use three definitional tactics: providing ostensive definitions, listing evidentiary features, and specifying necessary and sufficient conditions. While I will argue that the last tactic is the most convincing, the others are worth describing since the discussion sheds further light on the AT movement.

Ostensive Definitions

Some appropriate technologies readily concede the difficulties in providing a clear definition of AT: "Although we do not know how to define it, we do know an appropriate technology when we see one."[5] There is something to this attitude. After all, whatever the definitional complexities, the differences between a bulldozer and an oxplough or between a nuclear power plant and kerosene lanterns are readily apparent.

Nevertheless, there are cases where we cannot simply see the difference between an appropriate technology and other forms of technology. What, for example, is the most appropriate technology for maize-grinding in Kenya— mortar and pestle, hand operated mill, hammer mills, or roller mills?[6] Making this determination by simply looking at the candidate technologies is doomed from the outset.

The ostensive-definitional strategy reflects an attitudinal bias which needs to be borne in mind when assessing AT. The attitude manifests itself in the impatient "Let's get on with it" directive typically issued by technologists of all varieties. Generally, the technologist wants to give priority to doing over knowing. Problems cannot await the plodding intellectualization of philosophers and theoreticians. It is important to note that there are many proponents of appropriate technologies who subscribe to this attitude. This should underscore the point that appropriate technologists are first and foremost technologists, a point which is developed further in this [chapter].

While I have some sympathy with the technologists' attitude, this attitude rules out a certain fundamental line of inquiry. If I say "Let's get on with it" to practically any kind of project, I am thereby cutting off the questioning of the project itself, its underlying assumptions and implications. For appropriate technologists this can involve a refusal to examine the broader social/political context of technology introduction.

The tactic of providing ostensive definitions avoids the task of developing a political philosophy of technology, implying that techniques, at least "appropriate" ones, are politically benign and, perhaps, inherently good. This political avoidance behavior is also evident, although less blatantly so, in the other definitional approaches. Before showing that, I want briefly to describe a "Let's get on with it" attitude among philosophers, comparable to the one found among technologists. Philosophers have little patience with extended treatments of ostensive definitions since they are so quickly dismissed in the philosophical literature. However, what is obvious in one discipline is not so obvious in another. Many practically oriented technologists take ostensive definitions seriously, and philosophers should at least be willing to give the technologists the benefit of the doubt.

Evidentiary Features

Other commentators offer as a definitional strategy a shopping list of features which, supposedly, can be expected to be found in the appropriate technology store. No one feature or set of features is regarded as necessary or sufficient for appropriate technology, but the addition of each feature is argued to increase our inductive confidence that a candidate technology is appropriate. A solar satellite would presumably have relatively few of these features, while active solar heating and, certainly, passive solar design can be comfortably situated in the AT camp. What are these discriminating features?

No single list is found in the AT literature. Huelphinil proposes fifteen features;[7] Robin Clark proposes thirty-five, the longest list. Both lists seek to distinguish an appropriate technology society from a hard technology one. An AT-organized society emphasizes functionality for all time, communal units, local bartering, and integration of young and old. In contrast, hard technology breeds functionality for a limited time only, nuclear families, worldwide trade, and alienation of young from old. [Romesh K.] Diwan and [Dennis] Livingston, trying to order the chaos, group the features from various lists in the following way:

> In terms of material aspects of AT production, "Appropriateness" connotes the use of renewable sources of energy and recyclable materials, minimum destructive impact on the environment, and maximum utilization of local resources. In terms of the modes of production, AT fabrication should take place close to the resource base, using processes which are labor intensive (capital saving), small scale, amenable to user participation and/or worker management, and located close to points of consumption. In terms of application, AT connotes assimilation with local environmental and cultural conditions. It does not overwhelm the community, but is comprehensible, accessible, and easy to maintain.[8]

Despite some progress achieved by this grouping, the list approach in general confronts a number of methodological problems. One is "how incompatible [features] are to be combined."[9] Let us take two features from Diwan and Livingston's list, viz., minimum destructive impact on the environment, and small scale. [Witold] Rybczynski[10] proposes "the flush toilet, the throwaway container, the aerosol can, . . . and deforestation" as counterexamples to the claim made by AT proponents, like [E. F.] Schumacher, that small-scale devices are environmentally benign compared to large-scale ones. Small-scale devices can have a considerable negative environmental impact. (Later we will see how a better definitional strategy can meet this objection.)

Moreover, some items that we might find intuitively unacceptable as appropriate technologies qualify under the list approach. For example, do we find a description of an appropriate technology in the following?

> It will be large enough for the family but small enough for the individual to run and care for. It will be constructed of the best materials, by the best men to be hired, after the simplest designs that engineering can devise. But it will be so low in price that no man will be unable to own one—and enjoy with his family the blessing of hours of pleasure in God's great open spaces.[11]

Appropriate technology features abound in this description: small scale, simplicity, economical, amenable to mastery and maintenance by local people, etc. Yet, this was Henry Ford's description of the universal car, the Model T, hardly a fitting exemplar for AT enthusiasts. Similar descriptions can be found among nuclear power proponents, who are typically archenemies of AT advocates. As an example take the 1946 claim of [Alvin] Weinberg, [Eugene] Wigner, and [Gale] Young, that "nuclear power plants [would] make feasible a greater decentralization of industry."[12]

Again, as in the case of ostensive-definitional tactics, something more than methodology is amiss here. [Langdon] Winner[13] correctly derides the paradise-for-Christmas versus the sordid-perils-of-modernity lists as shallow social criticism. Oddly enough, all the good things in life are found under the appropriate technology tree while all the evils group themselves under the "high technology" label. Just as we found beneath the "Let's get on with it" technologist's impatience with theory a refusal to deal with social and political issues, so we find here a social-political criticism disguised as a debate over types of technologies. When proponents of appropriate technology present these lists, they presuppose both a particular critique of current society and a proposal for a better society. Yet, the critique and proposal are never made explicit. Instead, we are to accept or reject them in terms of technological choices.

One final methodological problem besetting the evidentiary feature strategy is the inability to determine which features are central. For example, is using local resources, which results in destroying a rain forest, more important than environmental preservation? In the next section we will examine a strategy that helps overcome some of the methodological problems apparent in the other two strategies. Yet, this third strategy, while making some progress, also fails to confront the problem of providing a political foundation.

Conditions

The following is a sample of typical definitions of appropriate technology in the literature:

—Technology which is decentralized, small in scale, labor intensive, amenable to mastery and maintenance by local people, and harmonious with local cultural and environmental conditions.[14]

—We emphasize four criteria: smallness; simplicity; capital cheapness; and non-violence.[15]

—Besides its [AT's] ability to offer to every member of society the fullness of the *summum bonum*, it offers to everyone also the dignity of work needed to attain it.[16]

—It is conducive to decentralization, compatible with the laws of ecology, gentle in its use of scarce resouces, and designed to serve the human person instead of making him the servant of machines.[17]

—AT as content has come to be applied to a special class of technology as systems: those incorporating energy-efficiency, labor-intensive, small-scale, decentralized technologies.[18]

From this widely variegated list, it is easy to sympathize with the charge that AT is an incoherent hodgepodge. Nevertheless, an adequate definition, acceptable to most AT proponents, can be salvaged from these proposals. If the above proposals are examined not as an unmanageable list of features but as alternative sets of conditions, it is possible, through selecting the best set of conditions, to devise an adequate definition of AT. However, as I will try to show, this definition, despite its methodological adequacy, falls far short of providing a much-needed political critique. Moreover, when the political foundations are fleshed out, they are found wanting.

My strategy in this section is to focus on [Paul H.] DeForest's proposed set of conditions. Why I have chosen DeForest's set will become more evident when a final assessment of the definition is given in the next section. At this stage, suffice it to say that DeForest provides a manageable list of conditions, which overlap to some extent or another with almost every other definitional proposal of AT that I have seen. Nevertheless, DeForest's list needs fine-tuning. Each condition is therefore examined to determine how the set might contribute to a manageable and defensible framework for AT. The final step is to develop my critique of AT in terms of its being either politically impotent or politically inadequate.

Energy Efficiency

This condition is central to two of AT's kindred spirits, soft technologies and biotechnics. The key feature of these technologies is (unlike their counterparts, hard technologies) that "they are matched in *energy quality* to end-use needs."[19] In accord with this analysis, nuclear power is a highly inefficient means of heating homes since the end-use does not require such a high-grade form of energy. Similarly, home water heating is better done with natural gas than electricity since natural gas heats the water directly whereas a conversion process is required to produce electricity. Mismatching end-use typifies hard energy technologies.

Environmental Soundness

However valuable an analysis in terms of energy efficiency might be, it is not broad enough to capture the concerns of many AT proponents. Energy efficiency is important to appropriate technologists but not as important as environmental soundness. For example, the insulating and ventilating properties of Third World homes is not an issue for AT proponents. In contrast, there is a great deal of discussion within the AT literature about the conceivable environmental impact of introducing tractors into a region.[20] Accordingly, a technology is evaluated in terms of its impact on the immediate environment. Therefore, there are good grounds for rejecting energy efficiency as too narrow a condition for defining AT. Furthermore, where energy efficiency is a consideration, it can be subsumed under environmental soundness.

Labor Intensity

Almost every proponent of AT includes some form of labor—as opposed to capital—intensity as a minimum requirement (i.e., as a necessary condition) for AT. For example:

> Instead of concentrating on labor-saving devices, which has been the whole trend of modern technology, can you turn your attention to capital-saving devices, because it is capital that is lacking in developing countries, not labor.[21]

Nuclear power and large farm tractors are capital-intensive technologies, resulting in few workers employed. AT proponents oppose these options when other technologies can be deployed to carry out work which expands employment.

> Because more people are employed, the benefits of growth will be spread more widely, and this wider distribution of income will contribute greatly to sparking demand for marketable goods in other industries.[22]

(The founder of the AT movement, E. F. Schumacher,[23] makes the case for a Buddhist economics which is not preoccupied with marketable production. A thorough reading of the AT literature, however, reveals that Schumacher is an exception on this point.)

Small Scale

Schumacher outlined, in *Small Is Beautiful*, the parameters of the AT movement, which [George] McRobie sought to implement in *Small Is Possible*.[24] Since the publication of Schumacher's book, homage is repeatedly paid to the virtues of smallness; bigness takes on an almost immoral character. But recognizing the importance of smallness for AT is one thing; getting a clear sense of what smallness means is quite another. Schumacher regards smallness as a relative concept, necessary to offset the currently reigning "idolatry of giantism." Beyond that consensus smallness is used ambiguously throughout the AT literature to apply to production,[25] equipment,[26] or markets.[27] AT proponents see these three senses of smallness in terms of factors enabling people to attain control and as meeting the demands of the market.

Decentralization

Again, it is difficult to find a clear-cut meaning of this term. Does decentralization mean that production units are widely dispersed spatially rather than concentrated in special regions? Does it mean putting factories in the rice paddies?[28] Or does it mean that decisions about work and production are to be made at the lowest levels of production? AT proponents use "decentralization" in two senses, viz., as a decision about where the technology should be placed, and as [a] decision about worker control.

My Definition

Summarizing the analysis so far, a modified version of DeForest's definition can be proposed. Appropriate technology will be taken to incorporate the following characteristics:
  (1) environmental soundness,
  (2) labor intensity,
  (3) small scale, and
  (4) decentralization.

These are conjunctively sufficient conditions for defining AT with no one condition sufficient in itself. Also, these conditions are disjunctively necessary for AT. Thus, this is a coherent set of conditions which capture (for the most part) what AT proponents mean when they talk about AT.

Adequacy of the Definition

The proposed definition more or less captures what most proponents are trying to express about the features and qualities of AT. As a further defense of the modified DeForest proposal, I would argue that any proposed alternative condition would be either vaguer than any of the accepted conditions, subsumable under one of those conditions, or objectionable in its own right. For example, McRobie's claim that AT should be nonviolent is vague. Labor-intensive technologies are easy to recognize: nonviolent ones are difficult to discern.

Many other proposed conditions can be subsumed under one or more of the accepted ones. For example, operating according to the accepted conditions will involve the use of renewable sources of energy and maximum use of local resources. Furthermore, adhering to these conditions generally yields a flexible, low-cost technology with a rural emphasis. In fact, many AT advocates use small scale as a condition because it is thought to result in a rural emphasis.

Finally, other candidate conditions can be readily dismissed. "Simplicity of operation" is just too problematic. The division of labor certainly simplifies operations. Are assembly lines, then, models of appropriate technology? I hardly think so. In summary, the accepted conditions include most of the features AT proponents want and exclude the undesirables.

More importantly, these conditions enable AT theorists to rebut previously telling charges. A number of critics argue against AT by taking each condition in isolation and then showing how that condition is not acceptable. This tactic has been used, for example, against the small-is-beautiful thesis. The smallness of a production process does not prevent environmental and human destruction on a massive scale. However, this violates the environmental soundness condition. Similarly, the criticism that a small slave system is more environmentally sound than a large factory farm is refuted by noting that a slave system violates the decentralization condition, giving workers some control over the process.

Also, some critics have sought to pit the conditions against one another, showing their mutual inconsistency.[29] As noted above, Rybczynski[30] proposes the aerosol can and deforestation as counterexamples to Schumacher's claim that small-scale devices are always more environmentally sound than large-scale ones. [Harvey] Brooks[31] follows a similar line, arguing the incompatibility of environmental soundness and decentralization. Response to these criticisms is straightforward. While the conditions are related (small-scale is generally, but not always, more environmentally sound than large-scale production), they are inter-dependent conditions that must be explained together when evaluating a particular technique. To take one of the alleged counterexamples, the aerosol can does not qualify as AT because it violates the condition of environmental soundness, since it is partially responsible for ozone depletion.

Thus far, I have demonstrated that prevailing AT ideas and practices can be used to define a methodologically adequate concept of AT. At this stage, it should be at least clear what people are talking about when they use the term "appropriate technology." From a definitional vantage point, AT can avoid the charge of representing an incoherent hodgepodge by appeal to a relatively clear set of conditions. Accepting that a defensible definition of AT can be rendered, we are now in a position to examine more fundamental criticisms of AT. Difficulties with AT are found, not in specifying what types of technology qualify as AT, but in constructing adequate philosophical, particularly political, foundations of AT.

THE POLITICAL FOUNDATIONS OF APPROPRIATE TECHNOLOGY AND RELATED TECHNOLOGIES

The real problem with AT is not methodological. The problem is political. On the one hand, Winner is correct in claiming that technological choices are, in fact, political.

> Choices about supposedly neutral technologies—if "choices" they ever merit being
> called—are actually choices about the kind of society in which we shall live.[32]

On the other hand, Winner is incorrect in thinking that AT offers an alternative political philosophy. AT proponents generally explicitly refuse to address political questions. To the extent that they do, AT political philosophy is not altogether that different from the political philosophy espoused by high technology advocates.

There are exceptions to this political failure within the AT movement. However, within AT that debate often takes on the character of a technological, not a political, debate. Ecotechnology, liberatory, indigenous, and intermediary technologies are thought to emphasize different features of the technologies. So, Sardar and Rosser-Owen,[33] quoted at the outset of this [chapter], are partially on target in criticizing these formulations for simply emphasizing different aspects of the new technology. Some of the proponents of these alternative formulations are guilty of that. But what they are really trying to do (some self-consciously) is construct an alternative philosophy of technology, emphasizing particularly the political aspects.

In what follows, I will try to articulate that philosophy in the course of showing that the debate between AT and these related technologies has been miscast as a debate over features of technology. Although I will examine each condition separately, my purpose is not to isolate them unwittingly but rather to reveal the political complex underlying each. I will be contrasting aspects of political philosophies, not features of technologies.

Environmental Soundness versus Ecotechnology

It is not simply a quirk in linguistic habits when AT proponents speak of environmental rather than ecological impact. The choice in phrasing reflects a specific political philosophy. Yet, even AT's commitment to environmental concerns is questionable. Environmental concerns are given lip service, but seldom is there an extended description of actual environmental impact in AT case studies. But let us assume environmental awareness in AT practices.

In coining the term "ecotechnology," [Murray] Bookchin[34] makes an important distinction between the environmental and the ecological, not as features of the technology but rather as philosophies. Environmentally, a technology is regarded as a relatively isolated device whose impact on the environment can be more or less precisely calculated. Ecologically, a technology is regarded as "functionally integrated with human communities as part of a shared biosphere of people and nonhuman life forms."[35] Bookchin reveals the important philosophical difference between these when he characterizes the concern of AT proponents over the impact of technology as

> within the context of environmentalism which tends to reflect instrumentalist sensibility in which nature is viewed merely as a passive habitat, an agglomeration of external objects and forces, that must be made more serviceable for human use irrespective of what these uses may be.[36]

When environmental issues are discussed in the AT literature, they are couched in this instrumental framework and not in a broader ecological context, whose conscious goal is promoting the integrity of the biosphere.

Environmental soundness and ecological integrity are not two comparable features of technologies. Underlying each are different political philosophies. AT proponents, for the most part, do not challenge the existing political power structure except for advocating some degree of popular control over the technology. According to Bookchin:

> To speak of "Appropriate Technology" . . . without *radically* challenging the political "technologies," the media "tools," and the bureaucratic "complexities" that have turned these concepts into elitist "art forms" is to completely betray their revolutionary promise as a challenge to the existing social structure.[37]

In other words, without an explicit alternative political philosophy AT offers little alternative beyond accommodation.

Labor Intensiveness versus Liberatory Technology

While putting people to work is a noble undertaking, the labor intensity condition becomes problematic when AT is compared to liberatory technology, also developed by Bookchin.[38] A primary liberating function of technology is that it can free people from toil and drudgery. Bookchin bemoans the ineffectiveness of earlier revolutionary movements in their inability to alter conditions and structures of scarcity. Failures of the past might be avoided, he argues, as we enter a post-scarcity phase of history where technology is available to relieve people of having to slave for subsistence.

Some of Bookchin's liberatory technology examples are highly debatable; for example, controlled thermonuclear reactions for mining, and genetic improvements of food plants. It is unclear how truly liberatory these are, and I have other reservations about this analysis. Notwithstanding challenges that can be made to his liberatory technology philosophy, it seems clear that AT is not predicated on the same labor-technology relation.

For example, AT exhibits a myopic emphasis on manual labor rather than on the liberation potential of technology. Work is glorified as evidenced by this description of ten beggars employed through the use of AT:

> They are completely rehabilitated now. They are well-dressed; they are happy; and are now completely different people. . . . It just shows that if one has the imagination, almost anything can be done.[39]

Technology is cast in the role of rehabilitation, which translates as "put people to work." Unemployment, from this perspective, is taken as a consistently overriding factor in technological choice. But a technology freeing people of arduous tasks can be just as important as one creating jobs. After reading AT literature it is difficult to avoid the feeling that major laborsaving technologies are reserved for the First and not the Third World.

Further problems arise with the AT view of manual labor. For not only is there an emphasis on production, but even more narrowly there is a focus on the male productive functions of technology. Only the male aspects of technological development—innovation, design, construction, supply—are given attention. The other side of technology—daily operation, maintenance, use, care, responsibility—are ignored. A heroic image is fostered which finds little if any serious value in the nurturing functions. There is clearly a sexual division of labor between those features of technologies most commonly associated with males and those associated with females.

Devastating consequences, including the failure of many AT projects, result from the sexual division of technology in the AT movement. In the 1970s, windmills were sprouting up all over the northeastern part of the United States, including one atop a rehabilitated building in the New York City Loisada project. That was the exciting phase. Now few of these windmills are turning because, among other things, no one thought to maintain them. The dull, nurturing phase was bypassed. Likewise those technologies impacting primarily on women, such as reproduction, cooking, sanitation, and education, take a back seat to the gadgetry and engineering feats needed to build appropriate technologies.

Much of the critique just presented comes from [Arnold] Pacey's *The Culture of Technology*.[40] Yet, Pacey refuses to apply this critique to AT. I find nothing in AT practices to exonerate it from his charges.

It would appear that AT is not liberatory by design. Freeing people from work and addressing nurturing activities are not a central concern in the AT framework. So, labor intensity betrays something deeper about the AT movement than its concern to employ people. The AT movement has certainly been influenced by the environmental movement. Comparable influences, however, from the libertarian and feminist movements are not as evident or prominent.

Small Scale versus Indigenous Technologies

A convenient way to highlight the political bias underlying the small-scale condition is to introduce another variant of AT, intermediate technology. Although there are differences between appropriate and intermediate technologies, these do not need to concern us here. The important thing is that intermediate technology stands politically in marked contrast to indigenous technology.

Commonsensically, intermediate means between low and high. The lower rung of the ladder—cheap, primitive, indigenous technology—is contrasted with the higher rung—expensive, mass-produced, foreign-based technology. Intermediate technology, recognizing "the economic boundaries and limitations of poverty,"[41] presumably lies somewhere between these two extremes. It is allegedly a progressive step above traditional technologies, and a step below high technology in that it is "production by the masses," not mass production.

So, when AT advocates proclaim that small is beautiful, the smallness condition is generally thought not applicable to indigenous technologies, despite the fact that these indigenous technologies generally satisfy all of the conditions. The reason for this is that despite Schumacher and despite some direct AT disclaimers to the contrary, the AT movement, in the final analysis, is capitalist. Generally, appropriate technologists focus on questions of productivity and expanding market structures. Production of useful commodities for the market is positively valued. Accordingly, crafting religious artifacts, which might encourage community cohesion and discourage marketable production, does not constitute an AT activity.

The AT bias against indigenous technologies is nicely illustrated by the following example of grain storage silos in Africa:

> Many different types of traditional granary or silo exist, most of them built with mud walling. However, in some places there has been heavy loss of grain through the depradations of rats, insects, dampness, and mold, and this has contributed to food shortages and malnutrition. Initially, it was assumed that such inefficiency was an inevitable part of the traditional technology, which was dismissed as almost worthless.[42]

Because of the bias against traditional technologies, the immediate action taken was to design concrete and metal alternatives to the indigenous ones. However, in this particular case AT innovations proved a failure, and attention was turned to the more nurturing task of suggesting "detailed improvements that would make maintenance easier"[43] on the traditional silos. Among AT practitioners this was a rare return to indigenous techniques.

Again, what is at stake here is not a choice between two kinds of technology, small-scale appropriate or indigenous technologies. Rather constrasting philosophies about indigenous people are at stake. Paternalism in the form of "knowmg what is best for the natives" is readily apparent among AT practitioners. AT proponents emphasize learning and adopting techniques which replace or notably upgrade a community's technical capacity. But the reverse is hardly ever the case: learning from and adopting indigenous technologies. This is most evident in the area of medicine. Western medicine does not hold a monopoly over healing. We could learn a great deal from indigenous medical theories and practices. Yet, as with almost all appropriate technologies, the important thing is not what we can learn from the indigenous people but what we can teach them.

Small scale is not an innocent condition of AT. Small scale, when conceived principally in terms of the production of marketable goods, not only yields a bias against indigenous technologies; it also hides an antipathy toward the powers and abilities of the very people it claims to help.

Decentralization versus Radical Technology

Of the related versions—ecotechnology, liberatory, and indigenous technology—radical technology is preferable. This is not because it emphasizes a superior feature of technologies. In fact, it is just the opposite. Radical technology is least likely to be construed as only a philosophy about technology. On the surface, it makes little sense to search for the radical features of technological things. Radical technology is preferable because it is first and foremost a political philosophy which encompasses the political aspects of ecotechnology, liberatory, and indigenous technology.

At this stage I need to counter a possible objection. By isolating the AT conditions I seem guilty of the very charge I lodged against the critics of AT. This claim is easily met by noting that, unlike the case with replies to other critics, AT proponents in this case do not have the option of appealing to the virtues of one condition in order to offset the criticized vices of another. For example, the ecotechnologist's criticism of AT, that it only considers environmental problems in isolation, is not countered by appeal to labor intensity or any of the other conditions. Moreover, radical technologists try to overcome the problem of pitting one form of domination against another by searching for the common roots of domination.

Although some attention is paid to the importance of political considerations in making technological choices, politics is regarded by AT proponents as secondary to the technology itself. Indeed, technology seems to be conceived as a substitute for politics.

> The choice of technology is the most important collective decision confronting any country. *It* is a choice that determines who works and who does not; that is, who gets income, new skills, self-reliance. *It* determines where work is done, whether concentrated in cities or more decentralized in smaller units; that is, *it* determines the kind of infrastructure required, and the whole quality of people's lives. *It* determines the ownership of industry; huge technologies are available only to the rich and powerful, whereas small technologies are tools in the hands of the poor.[44]

In ascribing these awesome powers to "it," McRobie's characterization of technological choice certainly has undertones of technological determinism: supply the Third World with the right kind of technology, and their lot will be improved. Form (engineering, technology) can be substituted for content (values, politics). Instead of the murky waters of values and politics people can turn to the relatively clear waters of engineering and technology. Seen in this light, it is not at all clear whether AT is an alternative to the technocracy esteemed within high technology circles; both appropriate technology and high technology advocates seem to be cut from the cloth of "technology pushers."

Radical technologies reject the technocratic approach, proposing that political choices of what type of society to live in, not the choice of a technology, is "the most important collective decision confronting any country."[45] Technological choice is seen as first and foremost a political choice. Technology does not simply appear on the scene as an exogenous circumstance necessitating a particular social choice.

> For the process of technological development is essentially social, and thus there is always a large measure of indeterminacy, of freedom, within it. Beyond the very real constraints of energy and matter exists a realm in which human thoughts and actions remain decisive. Therefore, technology does not necessitate. It merely consists of an evolving range of possibilities from which people choose.[46]

The preoccupation of AT proponents with the qualities of technique ignores the existence of political power, of who decides and for whose benefit.

While AT philosophy seeks to avoid direct engagement of the problem of political power, this does not mean that it is without a specific political stance. AT proponents may shun political debates and feign political neutrality, but as I have tried to show there is a political philosophy and agenda underlying the AT movement. In general, by refusing to challenge and confront the political *status quo*, the politics of AT is largely one of accommodation and reform.

Some AT advocates interpret the decentralization condition in the explicitly political sense of "amenable to mastery and maintenance by local people."[47] Nevertheless, closer examination of this interpretation shows that the idea of *control* by local people is entirely absent, and in AT practice local people seldom, if ever, take control of technological development. Decentralization in AT projects seems to be largely managerial and presupposes a technical hierarchy of AT innovators forming one class and AT implementers another. So, even where the political philosophy of AT is made relatively explicit, it no more actually challenges the political power structure than does high technology.

The aim of AT is to increase the number of technological choices, not necessarily to challenge any existing political structures.

> What the [AT] proponents are trying to do is to open up the spectrum and find solutions which are better suited to local conditions. The aim is generally not to replace the existing industrial system—but to promote technological innovation in the areas where it has been, until now, either weak or ineffective.[48]

AT differs from high technology more in terms of the range of technological choices presented than in the array of political options debated.

Reformist politics coupled with the failure to take into account or challenge the political/social context of technical application has resulted in some AT projects becoming instruments of class domination. For example, in India only wealthy farmers in one region are able to make use of a methane gas plant.[49] The two people able to afford a solar pump now monopolize the sale of previously free water in a Mauritanian village.[50] Small irrigation machines in Gao, Mali, enable farmers to sell enormous quantities of melons to Parisian tables, while food crops for local people are in short supply. (For further examples, see Lappé and Collins.)[51]

An example of the failure to account for political context but with a different result is provided by [David] Elliott.[52] While working for an environmental group, he invented a handheld radiation backscatter device which would measure the water content of a coal pile, resulting in more efficient utilization of coal. The device, however, proved highly unpopular among plant managers since their promotions depended on their coal-burning efficiency rating. They could easily improve this rating on their own by reducing the water level in the coal. An "objective" measuring device would spoil this political maneuvering. In this case, an AT device's introduction was resisted because it would diminish the political power of management.

Without explicit nonreformist politics, the AT movement will always remain indifferent to political outcomes, and this exacerbates the problem of technological hierarchy and dominance. As a condition of AT, "decentralization is reduced to a mere technical stratagem for concealing hierarchy and domination."[53]

Advantages of Radical Technology

I do not want to leave the impression that the radical technology proposal only gets its strength from its critique of AT. Radical technology is a rich and powerful political philosophy in its own right. Although obviously I do not have the space to provide a full rendering of its theory and practices, an outline of its basic tenets can be given.

How does radical technology combine ecotechnology, liberatory, and indigenous technologies as political philosophies? There are two main foundation stones supporting the radical technology framework. The first consists of a theoretical critique of and concerted opposition to domination. Domination comes in many different forms. Economic, political, gender, and race forms are a few of the more commonly recognized varieties of domination. Radical technology develops an analysis of each of these and their interrelationships. Moreover, radical technologists seek to reveal the connection between these forms of human domination and domination of nature: "The attempt to dominate nature stems from the domination of human by human."[54] Any political philosophy choosing to ignore the domination of nature is doomed to foster the growth of domination structures rather than combatting them. So, radical technology combines ecotechnology and liberatory technology by providing a general critique of domination.

Empowerment, the second foundation stone of radical technology, means providing people with control over their institutions and practices. As we have seen AT practices can result in wresting technological control from people. The AT movement's major concession to popular control is a very weak interpretation of decentralization. In contrast empowerment is central to radical technology.

However, by promoting empowerment radical technologists do not thereby romanticize indigenous technologies and their accompanying political/social structures. Traditional practices can not only engender the domination of nature, they can also be forms of domination in themselves. Yet, my own hypothesis is that the traditional examples of domination count for little in the overall structure of domination. Nevertheless, the radical technologists should be aware of all forms of domination.

Radical technology (RT) is superior to appropriate technology because unlike the AT program the RT agenda is first and foremost an explicit political philosophy. Anything else evades the basic issues confronting any philosophy of technology. The issues are political and not technological in nature, where politics is defined as "any persistent pattern of human relationships that involve to a significant extent power, rule, or authority."[55] To see the issue of technological choice as not involving these relationships is to avoid seeing reality.

Moreover, RT is superior to AT in its political stance. Although AT proponents do not make a habit of making their political beliefs explicit, they do have them. Overall the AT political philosophy is to advocate very moderate piecemeal reform of the existing power structures. Yet, it is just these power structures which are impediments to a just, egalitarian, and democratic development of technology.

CONCLUSION

Contrary to Sardar and Rosser-Owen, AT is a coherent philosophy. It is not "a collection of a large number of ideas and concepts, many of them quite incoherent."[56] A coherent program can be found in the prevailing ideas and practices of the AT movement.

Yet, the problem, by this analysis, is not the development of a coherent AT program. Attention to each definitional strategy showed that the AT movement has failed to develop its political side. Under the guise of insisting on ostensive definitions, a technocratic impatience with political questions was uncovered. Listing evidentiary features turned out to be an inappropriate way of giving political criticism, and qualities of techniques were elevated as political ends in themselves.

Finally, political problems were found even with the methodologically adequate set of conditions of AT that I proposed. With its environmental soundness condition AT, unlike ecotechnology, treats environmental problems as subsidiary, as isolated ones for which technical fixes can be found. AT proposals refuse to confront the broader, political problem of dominating nature. By emphasizing the labor-intensive quality of technology, AT is not liberatory in the sense of freeing people from arduous tasks nor in the sense of developing nurturing aspects of technology. The small-scale condition is prone to an emphasis on marketable production and consequently devalues indigenous technologies and alternative economic forms and goals. The AT program as a whole exhibits a very muted and reformist sense of politics which neglects the political/social implications of its use.

Appropriate technology is often called alternative technology. To an extent, that claim is valid. High technologists and appropriate technologists deal with different technologies. However, it is delusional to think that AT offers very much of an alternative philosophy of technology—particularly when one factors in the political components of that philosophy. A truly alternative philosophy of technology requires the development of an ecological, liberatory, and thereby radical political foundation for technological choice. Philosophically, the first and most important goal is an adequate politics of technological choice.

NOTES

1. Witold Rybczynski, *Paper Heroes* (Garden City, N.Y.: Doubleday, 1980), p. v.

2. Ziauddin Sardar and Dawud G. Rosser-Owen, "Science Policy and Developing Countries," in *Science, Technology, and Society*, eds. I. Spiegel-Rösing and D. Prive (Beverly Hills, Calif.: Sage, 1977), p. 564.

3. E. F. Schumacher, *Small Is Beautiful* (New York: Harper & Row, 1973).

4. W. W. Ndongho and S. O. Anyang, "The Concept of 'Appropriate Technology,'" *Monthly Review* 37, no. 9 (1981).

5. Nicholas Jequier, *Appropriate Technology: Problems and Promises* (Paris: Organization for Economic Cooperation and Development, 1976).

6. Frances Stewart, *Technology and Underdevelopment* (Boulder, Colo.: Westview Press, 1977), chap. 9.

7. In Tom De Wilde, "Some Social Criteria for Appropriate Technology," in *Introduction to Appropriate Technology*, ed. R. Congdon (Emmaus, Penn.: Rodale Press, 1977), pp. 161–63.

8. Romesh K. Diwan and Dennis Livingston, *Alternative Development Strategies and Appropriate Technology* (New York: Pergamon Press, 1979).

9. J. van Brakel, "Appropriate Technology: Facts and Values," in *Research in Philosophy & Technology*, ed. P. Durbin (Cambridge: MIT Press, 1977), p. 392.

10. Rybczynski, *Paper Heroes*, p. 20.

11. Quoted in ibid., p. 19.

12. As quoted in Zolmay Khalilzad and Cheryl Bernard, "Energy: No Quick Fix for a Permanent Crisis," *The Bulletin of the Atomic Scientists* 36, no. 10 (1980): 15–20.

13. Langdon Winner, "Building a Better Mousetrap: Appropriate Technology as a Social Movement," in *Appropriate Technology and Social Values*, eds. F. Long and A. Oleson (Cambridge, Mass.: Bollinger, 1980), p. 40.

14. Rashid Shaikh, "Commentary: Reflections of AT in India," *Science for the People* 13, no. 2 (1981): 26.

15. George McRobie, *Small Is Possible* (New York: Harper & Row, 1981), p. 36.

16. Leopold Kohn, "Appropriate Technology," in *The Schumacher Lectures*, ed. S. Kumar (New York: Harper & Row, 1980), p. 190.

17. Schumacher, *Small Is Beautiful*, p. 145.

18. Paul H. DeForest, "Technology Choice in the Context of Social Values—A Problem of Definition," in Long and Oleson, *Appropriate Technology and Social Values*, p. 13.

19. Amory B. Lovins, *Soft Energy Paths* (New York: Harper & Row, 1974), p. 39.

20. Anil Date, "Understanding Appropriate Technology," in *Appropriate Technology in Third World Development*, ed. P. Ghosh (Westport, Conn.: Greenwood Press, 1984).

21. McRobie, *Small Is Possible*, p. 4.

22. Sarah Jackson, "Economically Appropriate Technologies for Developing Countries: A Survey," in *Appropriate Technology in Third World Development*, ed. P. Ghosh, p. 80.

23. Schumacher. *Small Is Beautiful*.

24. See McRobie, *Small Is Possible*, p. 36.

25. David Dickson, *The Politics of Alternative Technology* (New York: Universe Books, 1974), p. 106.

26. McRobie, *Small Is Possible*, p. 36.

27. Steward, *Technology and Underdevelopment*, pp. 102–103.

28. Frances Moore Lappé and Joseph Collins, *Food First* (New York: Ballantine, 1978), p. 170.

29. Langdon Winner, "Building a Better Mousetrap: Appropriate Technology as a Social Movement."

30. Rybczynski, *Paper Heroes*, p. 21.

31. Harvey Brooks, "A Critique of the Concept of Appropriate Technology," in *Appropriate Technology and Social Values*, eds. Long and Oleson.

32. Langdon Winner, "The Political Philosophy of Alternative Technology," in *Technology and the Future*, ed. A. Teich (New York: St. Martin's Press, 1986), p. 315.

33. Sardar and Rosser-Owen, "Science Policy and Developing Countries."

34. Murray Bookchin, *Toward an Ecological Society* (Montreal: Black Rose Books, 1980).

35. Ibid., p. 109.

36. Ibid., p. 58.

37. Murray Bookchin, *The Ecology of Freedom* (Palo Alto, Calif.: Cheshire Books, 1982), p. 243.

38. Murray Bookchin, *Post-Scarcity Anarchism* (Palo Alto, Calif.: Ramparts Press, 1971).

39. Paul R. Lofthouse, "Industrial Liaison," in *Introduction to Appropriate Technology*, ed. Congdon, p. 156.

40. Arnold Pacey, *The Culture of Technology* (Oxford: Basil Blackwell, 1983).

41. Schumacher, *Small Is Beautiful*, p. 179.

42. Pacey, *The Culture of Technology*, p. 151.

43. Ibid.

44. McRobie, *Small Is Possible* (emphasis mine—Ed.).

45. Peter Harper and Godfrey Boyle, eds., *Radical Technology* (New York: Pantheon, 1976), p.8.

46. David Noble, *Forces of Production: A Social History of Industrial Automation* (New York: Knopf, 1984), p. xiii.

47. Shaikh, "Commentary: Reflections on AT in India."

48. Jequier, *Appropriate Technology: Problems and Promises.*

49. Shaikh, "Commentary: Reflections on AT in India."

50. Jequier, *Appropriate Technology: Problems and Promises.*

51. Lappé and Collins, *Food First.*

52. David Elliott, "Energy Policy: Gut Reactions and Rationalists," unpublished manuscript, 1985.

53. Bookchin, *Toward an Ecological Society.*

54. Ibid., pp. 66–67; William Leiss, *Domination of Nature* (Boston: Beacon, 1977).

55. Robert Dahl, *After the Revolution* (New Haven, Conn.: Yale University Press, 1970), p. 16.

56. Sardar and Rosser-Owen, "Science Policy and Developing Countries," p. 564.

PART 8
COMPUTERS, INFORMATION, AND VIRTUAL REALITY

INTRODUCTION

In our final section, we offer a type of case study that reflects the broad understanding of the relationship between technology and humanity discussed so far in this book. Many of these essays represent examples of the theoretical and historical problems developed in the previous parts, but the focus is on computer information technology (inclusive of virtual reality).

There can be no doubt that the revolution in information technology that has swept most of the Western world since the mid-1980s is one of the most important historical changes in our technological culture. The meaning of these changes is, however, in much dispute. Do they represent a radically new development, or are they merely novel and flashy extensions of previous forms of information technology—such as the printed word, the radio, or the telephone? The hype of a publication like *Wired* magazine is that technologies, such as computer-mediated communication, the World Wide Web, and the development of virtual reality, have changed our lives forever and taken us into a brave new world only dimly related to our past. Others, such as philosopher Gordon Graham, see more smoke and less fire in the pronouncements of new forms of community and communication on the Web.

Whatever the case, at least one thing is certain: Our daily lives are now so wrapped up in these technologies that we could not begin to pretend that we are not significantly affected by new developments in this area. Take, for example, the massive coverage and public interest in the potential Y2K problem, shortly before the turn of this century. As the new millennium dawned, computer experts and policymakers feared the threat of worldwide computer breakdown. The worry was that because most computer systems had been built to recognize years in two-digit formats, the rollover on January 1, 2000, to "00" would confuse these systems, causing them to shut down or worse. Predictions were made that the new year would see planes falling out of the sky, bank accounts being wiped out, and huge blackouts in many major cities leading to possible riots. Such worries were not voiced simply as idle speculation. Government agencies and private companies spent millions of dollars rewriting code to meet the New Year's Eve deadline and avoid these possible catastrophes.

But in the end, the Y2K problem went off with a whimper instead of a bang. No major consequences were reported. Nonetheless, the coverage of Y2K arguably sobered a broader audience of people to the breadth of the saturation of technology in both our forms of social organization and in our everyday lives. Through Y2K, many learned the lesson that the theorists of technology represented in this collection have known for quite some time: We live in a technologically mediated world. Still, this pass on the issue does not quite get it. We live in a world increasingly indistinguishable from the human-made artifacts and systems that we have created and that, in turn, have begun to create us. For if the worst expectations of the results of Y2K had turned out to be true, the entire culture would have been altered by a fairly simple programming glitch, a glitch whose importance is measured by the sheer pervasiveness of computing systems in our world today.

In a sense, Y2K marked the end of our technological innocence. Many will demur that the end of our innocence came at the end of World War II, with the first use of the atom bomb. But the difference between the bomb and Y2K is that Y2K revealed the *everyday* nature of our technological world, rather than the terrible power of technology at particular and unique times, such as during a war. Whatever we want to say then about the consequences of computer-mediated technologies in our lives, there is no doubt that through them we have become more consciously aware now of the technological world that we have come to inhabit.

Once we admit the pervasiveness of technology in our lives through our acknowledgement of the role that computer technologies play in this picture, what is to be done? How do we respond critically and responsibly to the new booms in computer technologies? In the first selection, David Shenk, noted media critic and commentator in a variety of public forums on science and technology issues, offers a guidepost he calls "technorealism." Seeking a middle ground between "technoutopianism" (it's all good) and "neo-Luddism" (it's all dreadful), Shenk offers eight basic principles of technorealism from which we can build a more responsible form of technology criticism. Similar to many of the theorists in the previous sections, Shenk's hope is to provide a platform from which we can connect the public discussion of technology to issues of "basic human values." Shenk provides, therefore, a helpful and publicly accessible path to investigate the ground between determinists and antideterminists.

The next four authors could all be described as technorealists about computer technologies to one degree or another. All are critical of different aspects of these technologies without failing to appreciate the importance of these technologies to human life. Oftentimes their target is not so much the technologies themselves, but the exaggerated claims that some producers and users of these technologies make concerning the potentials of these systems and tools to transform everything from ourselves to our social systems.

Edward Tenner begins by offering a sustained and lengthy critique of the benefits of computer technology in the workplace and for the individual consumer. While no one would doubt the possible consequences of computing technologies and computer-mediated communication in more personal contexts, the development of many of these technologies is driven by their possible commercial application. If those commercial applications are not successful, then the prospects that other impacts could result from their development is dim. Tenner begins with an overview of the now familiar cycle of planned obsolescence in the computer industry. The machines we purchased only a few years ago now cannot run the software upgrades that are constantly produced. And because so many of these systems are linked, any addition of any bit of software usually produces a need for upgrades in programs or equipment across the board. We are left at the mercy of the push for further "improvements."

But are each of these improvements really better? Is Word 8 really better than Word 5.1? It is not always clear that it is. Further, and more important, the increases in efficiency heralded by changes in computing power do not always translate into a smoother-running workplace. Banks may be able to get by with fewer tellers as they install more automated teller machines, but now they must hire more programmers. The scientific typesetting program TEX made it possible to more easily set into print the complex equations of physicists and mathematicians, but the time it takes to learn the program is considerable. With every increase in the power of complexity of these programs and machines comes more and more problems in maintaining the programs, working out the bugs, and preparing for the next upgrade. Can we plausibly say that it is all good?

Gordon Graham leaps from the now mundane world of the PC and word processor to the exotic leaps of the imagination envisioned by the culmination of these different technologies in the creation of "virtual reality" (VR). But like Tenner and Shenk, Graham urges caution in believing both that what is being produced here is really new, and even if it is, whether it is worth surrendering pursuit of real experiences for virtual ones.

The potential of VR, Graham argues, is not really so different from the creation of other worlds of "make-believe" represented in films, plays, novels, or other forms of artistic imagination. In all of these venues we place ourselves, at least mentally, in a place where we can imagine what it is like to have a certain kind of experience, such as meeting a tiger in the depths of a Southeast Asian jungle, without the risks. (Quoting Kendall Walton, Graham says that we get the experience "for free.") But even with the advantages of VR technology embodied in the vision of a full bodynet that allows us to more accurately feel the experience, we are not experiencing the real thing, since, at the least, the experience is for free. What we have then here, claims Graham, is a potential difference in degree of make-believe experience rather than a difference in kind. In the case of virtual communities on the Web, Graham sees more potential for something more akin to "real" experiences. The question then becomes one of whether the pursuit of those experiences outside of the contrast with reality really makes our lives better. Anticipating Albert Borgmann's general argument on much of these same issues, Graham concludes that "to live one's life primarily on the Internet would be a poor way to be."

Suzanne K. Damarin takes up the feminist consequences of these new technologies, looking even further down the road than Graham's perusal of VR to those that will eventually press the boundaries of what we understand to be human individuality. Her focus is "technologies of the individual," or those that supplant functions normally undertaken by individuals or change the relationship of the individual to society. A more mundane example of such technologies, in terms of our everyday experience but not in terms of its effects, are eyeglasses, which deny the effects of aging on eyesight. But we can imagine an upper limit of such technologies in the much heralded potential of eventually combining computer and human in the figure of the cyborg: the person who has fully merged his identity with the machine.

What interests Damarin however is how these technologies enter the lives of women and change both their self-understanding and relation to society. Echoing Diane Michelfelder's analysis of the use that women have made of the telephone in their homes, Damarin finds powerful ways in which these technologies are poised to change the lives of women and so are susceptible to a specifically gendered critique of their development and propagation. When it comes to the often hyperbolic rhetoric of roboticists (such as Hans Morawec) who envision a world where humans have completely merged with machines and shrugged off the confines of their bodies, Damarin sees, in the end, the reproduction of gender hierarchies contributing to the domination of women. Imagining a world without bodily differences (by the elimination of bodies as we know them) does not necessarily entail elimination of these traditional concerns.

Finally, Albert Borgmann ends this section with a sometime poetic and always powerful plea for more attention to reality—to hold onto our sense of the world despite the lure of appearances provided by computer-mediated technologies. In some sense though, Borgmann's synopsis of the relationship between information technology and our understanding of reality is not so much a normative argument that we should be resisting the VR tide and its associated technologies, but a hopeful reporting that, perhaps, people are not as convinced of the lure of the technologies of "hyperinformation" (akin to the "devices" identified by Borgmann in section 2) as its proponents would have us believe.

Borgmann links the concerns over computer and information technologies to our understanding of our environment. Moving eloquently from examples involving the protection of wilderness in his home state of Montana to the rich cultural experiences available in a place like New York City, Borgmann finds in both locales natural and cultural environments worthy of preservation. Here, he argues, we can still make choices to form relationships with the persons and places around us that resist the tide to replace them with technological reproductions. We can and should seek relationships of fidelity to those places. In doing so, we seek to maintain not only our own character in cyberspace, but that of the places which we inhabit. It is not simply a struggle against VR and the like, but also the spin-offs of our car culture, replaceable strip malls, and such, which denude all places of their specificity.

∃ ０

Technorealism
An Overview

David Shenk

In this heady age of rapid technological change, we all struggle to maintain our bearings. The developments that unfold each day in communications and computing can be thrilling and disorienting. One understandable reaction is to wonder: Are these changes good or bad? Should we welcome or fear them?

The answer is both. Technology is making life more convenient and enjoyable, and many of us healthier, wealthier, and wiser. But it is also affecting work, family, and the economy in unpredictable ways, introducing new forms of tension and distraction, and posing new threats to the cohesion of our physical communities.

Despite the complicated and often contradictory implications of technology, the conventional wisdom is woefully simplistic. Pundits, politicians, and self-appointed visionaries do us a disservice when they try to reduce these complexities to breathless tales of either high-tech doom or cyber-elation. Such polarized thinking leads to dashed hopes and unnecessary anxiety, and prevents us from understanding our own culture.

Over the past few years, even as the debate over technology has been dominated by the louder voices at the extremes, a new, more balanced consensus has quietly taken shape. This document seeks to articulate some of the shared beliefs behind that consensus, which we have come to call *technorealism*.

Technorealism demands that we think critically about the role that tools and interfaces play in human evolution and everyday life. Integral to this perspective is our understanding that the current tide of technological transformation, while important and powerful, is actually a continuation of waves of change that have taken place throughout history. Looking, for example, at the history of the automobile, television, or the telephone—not just the devices but the institutions they became—we see profound benefits as well as substantial costs. Similarly, we anticipate mixed blessings from today's emerging technologies, and expect to

From *The End of Patience: Cautionary Notes on the Information Revolution*, by David Shenk. Copyright © 1999 Indiana University Press. Reprinted with permission.

forever be on guard for unexpected consequences—which must be addressed by thoughtful design and appropriate use.

As technorealists, we seek to expand the fertile middle ground between techno-utopianism and neo-Luddism. We are technology "critics" in the same way, and for the same reasons, that others are food critics, art critics, or literary critics. We can be passionately optimistic about some technologies, skeptical and disdainful of others. Still, our goal is neither to champion nor to dismiss technology, but rather to understand it and apply it in a manner more consistent with basic human values.

Below are some evolving basic principles that help explain technorealism.

PRINCIPLES OF TECHNOREALISM

Technologies are not neutral.

A great misconception of our time is the idea that technologies are completely free of bias—that because they are inanimate artifacts, they don't promote certain kinds of behaviors over others. In truth, technologies come loaded with both intended and unintended social, political, and economic leanings. Every tool provides its users with a particular manner of seeing the world and specific ways of interacting with others. It is important for each of us to consider the biases of various technologies and to seek out those that reflect our values and aspirations.

The Internet is revolutionary, but not utopian.

The Net is an extraordinary communications tool that provides a range of new opportunities for people, communities, businesses, and government. Yet as cyberspace becomes more populated, it increasingly resembles society at large, in all its complexity. For every empowering or enlightening aspect of the wired life, there will also be dimensions that are malicious, perverse, or rather ordinary.

Government has an important role to play on the electronic frontier.

Contrary to some claims, cyberspace is not formally a place or jurisdiction separate from Earth. While governments should respect the rules and customs that have arisen in cyberspace, and should not stifle this new world with inefficient regulation or censorship, it is foolish to say that the public has no sovereignty over what an errant citizen or fraudulent corporation does online. As the representative of the people and the guardian of democratic values, the state has the right and responsibility to help integrate cyberspace and conventional society.

Technology standards and privacy issues, for example, are too important to be entrusted to the marketplace alone. Competing software firms have little interest in preserving the open standards that are essential to a fully functioning interactive network. Markets encourage innovation, but they do not necessarily ensure the public interest.

Information is not knowledge.

All around us, information is moving faster and becoming cheaper to acquire, and the benefits are manifest. That said, the proliferation of data is also a serious challenge, requiring new measures of human discipline and skepticism. We must not confuse the thrill of acquiring or distributing information quickly with the more daunting task of converting it into knowledge and wisdom. Regardless of how advanced our computers become, we should never use them as a substitute for our own basic cognitive skills of awareness, perception, reasoning, and judgment.

Wiring the schools will not save them.

The problems with America's public schools—uneven funding, social promotion, bloated class size, crumbling infrastructure, lack of standards—have almost nothing to do with technology. Consequently, no amount of technology will lead to the educational revolution prophesied by President Clinton and others. The art of teaching cannot be replicated by computers, the Net, or "distance learning." These tools can, of course, augment an already high quality educational experience. But to rely on them as any sort of panacea would be a costly mistake.

Information wants to be protected.

It's true that cyberspace and other recent developments are challenging our copyright laws and frameworks for protecting intellectual property. The answer, though, is not to scrap existing statutes and principles. Instead, we must update old laws and interpretations so that information receives roughly the same protection it did in the context of old media. The goal is the same: to give authors sufficient control over their work so that they have an incentive to create, while maintaining the right of the public to make fair use of that information. In neither context does information want "to be free." Rather, it needs to be protected.

The public owns the airwaves; the public should benefit from their use.

The recent digital spectrum giveaway to broadcasters underscores the corrupt and inefficient misuse of public resources in the arena of technology. The citizenry should benefit and profit from the use of public frequencies, and should retain a portion of the spectrum for educational, cultural, and public access uses. We should demand more for private use of public property.

***Understanding technology should be an essential component of global
citizenship.***

In a world driven by the flow of information, the interfaces—and the underlying
code—that make information visible are becoming enormously powerful social
forces. Understanding their strengths and limitations, and even participating in
the creation of better tools, should be an important part of being an involved citizen. These tools affect our lives as much as laws do, and we should subject them
to a similar democratic scrutiny.

The Computerized Office
Productivity Puzzles

Edward Tenner

T he subtle risks of computers have not been their only unpleasant surprise. Their greatest paradox may be that it is still hard to evaluate their benefits where these promised to be greatest: in the most rapidly growing parts of the economy. There is little debate about the value of electronics in manufacturing and distribution. Robots still have limited applications, but nobody doubts that they sometimes can do things faster, better, and more cheaply than human workers. It is unlikely that any old-fashioned purchasing agent or warehouse manager could match the performance of good software. For the widget-masters and metal-bashers of the world, computers usually improve speed and quality and reduce waste.

But in manufacturing, too, consequences can be unexpected, especially the results of improved communication. Consider the U.S. garment industry. It has always been cheaper to make textiles and clothing in Asia than in rural America, let alone in New York City. What protected American manufacturing as much as tariffs and quotas was the flow of information. Even with airmail, it might take too long to send designs and patterns overseas. Fax and electronic mail have changed that, taking weeks off turnaround times. With air freight an option for higher-priced merchandise, a major advantage of North American and European manufacturers—closeness to the end customer—has been seriously eroded. By buying time cheaply, new technology works against those who used to have time on their side.

The introduction of the IBM PC as the standard for personal computers in 1981 (older rivals soon faded, though the Apple Macintosh made its debut as an alternative standard in 1984) launched a new era for office work and the service sector. Mainframes had long controlled the operation of large corporations and government bodies, from manufacturing control to the processing of tax forms.

From *Why Things Bite Back*, by Edward Tenner. Copyright © 1996 by Edward Tenner. Used by permission of Alfred A. Knopf, a division of Random House.

Microcomputers promised to transform medium-sized and small businesses that could never have considered buying a minicomputer costing tens of thousands of dollars. Book-length documents could be edited with ease. Complex financial calculations, previously demanding hours of work, could be completed, and when necessary revised, in seconds. Records that only a few years earlier would have needed a variety of awkward colored tabs or punch cards plus long needles for sorting could now be stored on inexpensive disks or tapes. Programming no longer required months or years of apprenticeship in the old FORTRAN and COBOL languages; schoolchildren were mastering BASIC. And for those who worried that BASIC as an introduction led to bad programming habits, there were PASCAL, Turbo-PASCAL, and a host of more exotic entries from ADA to XENIX.

Throughout most of the 1980s, it was hard to imagine services and white-collar work in general *not* becoming more productive. The cost of everything from random-access memory (RAM) to disk storage space seemed to be shrinking by half every eighteen months. One of the first things that computer users have discovered is that not trading up to the latest processor means forgoing many new products and software versions, and even losing support for older software. Word-Perfect 5.1, introduced in 1989, runs fast on 286-class computers from the mid-1980s; WordPerfect 6.0 almost demands a 486-class machine, generally four or more times as fast, as many Windows 3.1 applications did in the early 1990s. Microsoft's Windows 95 needs at least a *fast* 486 machine to run efficiently.

If the automobile metaphor is apt at all, it would probably be best expressed as a salesperson having to take on a larger and larger territory, paying the same amount every three to five years for a vehicle (ultimately an airplane) with greater and greater range, but getting an ever-smaller trade-in price for the last model. In 1992 I spent about the same for a Hewlett-Packard LaserJet 4 printer capable of eight pages per minute that I had paid seven years earlier for a "fast" C. Itoh daisy-wheel printer with less than one-eighth the speed. But unlike other consumer hardware (automobiles, cameras), my printer had lost nearly all of its resale value in seven years, even though it was still in excellent working order and could produce Courier 10 text that was actually sharper and clearer than copy in the same font from my LaserJet. Even the retail trade-in value of the most unpopular automobile models is a bastion of stability by comparison—all the more surprising because aging cars need costly overhauls that most mechanical printers don't. (Heavy-duty daisy-wheel printers are the massive, almost immortal elephants of office electronics. A school is now using mine.)

My experience was hardly isolated. The revenge effect of the explosion in computer performance is, as every excited new system buyer soon discovers, an *im*plosion in value. In February 1994, a Compaq 486/66M machine with a list price of $4,654 was projected to keep only 17 percent of its retail value over two years and only 6 percent after three years, 3 percent at wholesale. The wholesale and retail value in three years was listed as "salvage." Meanwhile, as of 1995 the Internal Revenue Service still required computers to be depreciated over five years.[1]

Not only does resale value decline precipitously; the price of replacement systems actually does not go down. There was a saying in the 1980s that "the system you want always costs $5,000," and for all the breakthroughs of the 1990s this still seems to be the case for state-of-the-art products, especially portables. A new entry-level system still costs around $1,500, as it did in the days of the beloved Apple II, except that the product has many times the processor speed, memory, and storage capacity. Of course, cheap computers can still do a lot. But they can be the most expensive of all because they can become socially obsolete—unable to run new releases of important software efficiently—within a year if not months. Some popular online services are already unavailable for DOS-only computers. Upgrades may be disappointing. The price of additional memory chips, for example, has not declined as much as the prices of other components. It fluctuates with demand and is most likely to increase when users need it most, to support more powerful operating systems.[2]

By the mid-1990s a 340- or 528-megabyte hard drive is a near necessity for handling bulky program and graphics files. These numbers in turn demand a high-capacity tape backup, and many new software packages are now written to be delivered on CD-ROM, requiring yet another hardware addition. The wish to run several programs concurrently using graphic interfaces like Microsoft Windows or OS/2 prods users to upgrade from 14- and 15-inch to 17-inch and larger monitors. Meanwhile, business travelers who at first demanded smaller and lighter notebook computers can now add portable printers (with batteries good for only about thirty pages), wireless transmitters, CD-ROM drives, and speakers, bringing total weight and price back up to the days of the original Compaq luggables and beyond.[3]

The huge investment in computing in the 1980s and early 1990s, then reflected one of the great cultural inversions of our time: North American and European corporations, and millions of professionals, small businesspeople, academics, and students, learned to stop worrying and love planned obsolescence. If the watchword of the 1970s was "survival," that of the 1980s was "empowerment." People felt autonomous, in control, more powerful, and absolutely more productive. But toward the end of the 1980s, the sentiment grew that something was not right. Throughout the decade, scholars like the sociologist Rob Kling and the political scientist Langdon Winner were pointing out that computerization was a movement that reflected social conflict and organizational infighting as much as technical change. By the 1990s, a new set of critics had joined them. Unlike the social scientists, philosophers, and technophobic humanists who had long challenged the cult of productivity at all costs, the new skeptics approved of capitalism, higher output, technology, and all the rest. They had emerged from within technocratic culture: from economics departments, business schools, and consulting firms. And their message was that things weren't happening according to plan. The service sector, where the potential for gains was greatest, was strangely lagging.[4]

Intuitively it is hard to believe that huge investments in improving the quality of anything will not pay off in the long run. There is a brilliant, well-supported argument that we are actually in the early stages of a more fundamental revolution. In a famous article, the economist Paul A. David has compared the early results of microcomputer use with the introduction of small electric motors to an industrial world formerly dependent on mills requiring massive power sources driven by water and later steam. Decades after the introduction of electricity, most plants still relied on networks of shafts, pulleys, and leather belts that were not so far from the world of the late-eighteenth-century millwright. With large investments in functional older systems, manufacturers often added electric motors controlling a group of machines while keeping the centralized power plants and drive trains in place. This may have stretched the useful life of equipment, but it prevented the most efficient use of either old or new technology. Even managers optimistic about the ultimate benefits of powering each machine with its own electric motor may have rationally preferred to introduce electric power in a series of smaller-scale, exploratory applications.[5]

By the 1880s, the advantages of powering each machine with its own electric motor were apparent. As a result, factories could run efficiently on a single, skylit level. The (literal) overhead of noisy, oil-splattering drive-train equipment could be eliminated. Materials and semifinished work could circulate with less delay, as heavier machinery no longer had to be placed near the main shafts and drive belts. Workers could control machine speeds more precisely. Factories could make new goods that would have been difficult to produce with shaft-driven power. All these advantages were known. Yet it took forty years or more for all the elements to fall into place and for electrified factories to transform industrial processes and raise workers' standards of living and conditions in the 1920s. Replacing a big steam engine in the basement with a large electric motor was only a small first step toward more productivity. Not only the investment in older plants but utility rates and the price of both electric motors and machinery helped determine the curve of investment in, and benefits from, electricity. And it took time to bring new, single-story plants on line. All this leads David to conclude that as different as information and electricity may seem, we may be looking too closely at the early problems of shifting from one set of structures to another.

Statistics have nevertheless been discouraging. According to one study by the economist Stephen Roach, investment in advanced technology in the service sector grew by over 116 percent per worker between 1980 and 1989, while output increased by only 0.3 percent to 1985 and 2.2 percent to 1989. Two other economists, Daniel E. Sichel of the Brookings Institution and Stephen D. Oliner of the Federal Reserve, have calculated the contribution of computers and peripherals as no more than 0.2 percent of real growth in business output between 1987 and 1993. Meanwhile, manufacturing productivity was growing rapidly. One possible explanation, so far not systematically studied, is that the mediocre performance of the service sector is due to the average of results from companies that had done well along with others that had lost ground. Precisely because service companies now depend so much on computer systems, there is more to be lost as well as gained if technical bugs get out of hand. This is consistent with the shaky job security of chief information officers (CIOs), natural scapegoats for computer-related debacles even when other managers may be more at fault. Some banks and brokerage houses seem to fire their CIOs almost every year, and more than a third of CIOs surveyed in the early 1990s reported that their predecessors had been forced out (or down). But even if the problem has been high gains offset by high risks, rather than overall mediocrity of performance, something has not gone according to plan.[6]

Could the problem be system failures? Could hardware defects and software bugs be responsible for much of the loss in production? Computer catastrophes are definitely worrisome, especially for computer professionals. The details of known electronic disasters have filled whole books.

Software code, like laws and sausages, should never be examined in production. Indeed, it seldom is—if only because today's developers write code in teams, each member producing a small part of an increasingly bulky and complex whole. There is no software equivalent of Tracy Kidder's hardware odyssey, *The Soul of a New Machine*. There is, however, an excellent contrast in Lauren Ruth Wiener's *Digital Woes*, the best general-interest explanation to date of why and how computer software is inherently unreliable. Some of the stories she and others tell are chilling. In the mid-1980s a computer at the Bank of New York began overwriting data on government securities transactions. Before the bug was discovered, the bank owed the Federal Reserve $32 billion without knowing who had bought what securities, and it ultimately lost $5 million in interest on a $23.6 billion loan it had to take from the Fed, pledging all of the bank's assets as collateral. At about the same time, cancer patients in Texas, Georgia, Washington State, and Ontario, Canada, received lethal X-ray doses from a software-controlled radiation therapy machine that failed after a specific sequence of commands to change its mode, as instructed, from a high-intensity X-ray beam to a low-intensity electron beam—while it nonetheless removed the tungsten shield that protects patients during X–ray sessions. Death and serious injury were the consequences. The *Risks Digest*, an Internet bulletin board expertly moderated by Peter G. Neumann of SRI International, leaves little doubt about the hazards of software as readers throughout the world contribute cautionary tales.[7]

The ghastly consequences of some of these failures, and the existence of computer codes that might automate nuclear strikes, make complacency about electronic risks nearly criminal. But talking to computer support and maintenance professionals changes the picture. They believe that at least microcomputer hardware has if anything become steadily more reliable over time. Floppy disk drives and even hard drives fail much less often, even if they always seem to crash at the worst possible time. Some manufacturers in the early 1990s have announced MBTF (mean time between failure) rates equivalent to thirty years of continuous use—not bad for equipment that may be socially obsolescent in three years because of new standards and capacities. As computer parts get cheaper, it is easier to protect systems by building in two or more of all crucial hardware items, especially hard drives. A computer with a RAID (Random Array of Inexpensive Disks) stores two or more copies of data. At a price, companies can regularly back up data to secure off-site storage services. Even if an office has to be evacuated for days or weeks, as happened after the World Trade Center bombing in 1993, employees can resume work in temporary quarters—a resilience impossible in paper-based offices.[8]

For both corporate and individual users, software failure is still not only a likelihood but a certainty. Yet here, too, technology has taken the edge off the sudden disaster. The cost and ease of backing up large data sets continues to improve. Few computer buyers pay the $180 to $300 that a tape drive costs, but this is a trifling price for the security of being able to store hundreds of megabytes of data on a $25 tape. It's easy to program a computer to back itself up unattended, during a lunch break or in the evening. There is still, of course, a burden of vigilance: shuffling floppy disks for the majority of users who still don't have tape backups, maintaining the discipline of backing up every single day, comparing backup data regularly with original files, and performing all the little rituals that are part of the technological strategy against catastrophe. Most network administrators now routinely back up files that reside on individual computers.

Even viruses are less menacing now than they at first appeared to be. Not every virus is a serious threat to every computer user; most users exchange programs and data intensively with a relatively small group of other users. According to researchers at IBM, few viruses reach the level—very similar to the infection threshold in medical epidemiology—at which they become stubbornly established in a population. To this social barrier is added the protection of antivirus software. This may be far from perfect, it's true. Once more, the price of protection is chronic vigilance. Here it means regularly downloading definitions for new viruses—dozens are identified every month, and there are even underground toolkits, like legitimate software development packages, for creating new ones. A few viruses have been designed to attach themselves to popular antivirus software. It is also true that locating and eradicating a known virus on every single node in a corporate computer network might cost tens of thousands of dollars, because each workstation must be disinfected individually. Still, deleting and overwriting files accidentally, spilling beverages, forgetting one of those proliferating passwords (some managers must memorize dozens), and other mundane mistakes do far more damage (as yet) than all the world's diabolical code. One of the biggest computer outages so far, the crash of the AT&T telephone network in January 1990, resulted from a single ambiguous statement in the C programming language that controlled the system's switching center—not the criminal conspirators sought by police investigators at the time. The world's angry hackers will continue to be a chronic nuisance and from time to time the source of local disasters, but not the great enemies of productivity some have portrayed them to be.[9]

As always, no technology (including manual and mechanical systems) is entirely safe. Nor do people invariably use the safest systems, or use them with sufficient vigilance. But considering our dependence on electronic hardware, the important thing is not that it fails but that off-the-shelf hardware has become more and more stable. We can rule out system failures as a big component of the productivity paradox. The problem is really in the software—and its use and misuse.

Computers tend to replace one category of worker with another. There are two ways to get something done. You can find one group trained to accomplish things the old-fashioned way. Or you can pay another group to set up and maintain machines and systems that will do the same work with fewer employees—of the older category of worker. You are not really replacing people with machines; you are replacing one kind of person-plus-machine with another kind of machine-plus-person. When IBM persuaded corporations to modernize their bookkeeping in the 1950s, businesses were able to get along with far fewer accountants, as they expected, but they had to hire more programmers than they had anticipated. Automatic teller systems also require programmers and technicians paid four times as much as bank tellers. If things go well, banks need less than a quarter of the staff, and they come out ahead. But it is notoriously difficult to predict all problems, or their levels of difficulty, in advance. And one mark of newer technology is that while it is cheap in routine operation, it is expensive to correct and modify.[10]

We have all seen the sign "The Difficult We Do Immediately; the Impossible Takes Time." Computerization turns this manifesto on its ancient head. Software can devour highly complex tasks with ease if they fit well into its existing categories. But even a simple change illustrates the revenge effect of recomplicating. The scientific typesetting program TEX, developed by the computer scientist Donald S. Knuth and now the standard in many branches of physics and mathematics, makes short work of the most fearsomely complex equations that once cost publishers up to $60 per page to typeset. An author proficient in TEX—and I have had the good fortune to work with several of them—can prepare camera-ready copy that stands up to most commercially available systems. But making small changes, alterations that might require dropping in a metal slug or pasting in a new line in traditional systems, can sometimes take costly programmers' time. A hairline rule can take more time and money than pages of author-for-matted proofs brimming with integration signs, sigmas, deltas, and epsilons.

Minor incompatibilities between authors' TEX programs and publishers' typesetting equipment can delay book-length manuscripts for weeks and run up costs well beyond those of conventional typesetting. Worse still from the publisher's point of view, some inexperienced and unskilled TEX-using authors—including distinguished scientists—blame the publisher and typesetter when their work is held up.

As editors of conventional manuscripts, my colleagues and I could identify problems early and request changes before texts went into production. Even experienced electronic manuscript specialists cannot evaluate a TEX manuscript

reliably just by looking at the author's laser-printed version. Messy or nonstan-
dard coding may fail to reproduce the same beautiful output when fed into professional typesetting equipment. Consequently, there are real hidden productivity
costs associated with an "inexpensive" TEX manuscript; it may require open-
heart surgery rather than a haircut. Publishers and typesetters discovering such
insurmountable glitches have been known quietly to set the author's electronic
manuscript aside and dispatch the hard copy to Asian compositors for conventional keyboarding. This may be speedier than waiting for the author to learn the
fine points of TEX, but it inevitably delays production, embarrasses author and
publisher alike, and introduces new errors. What computerization offers—or simplifies—with its right hand it can withdraw—or recomplicate—with the left.

TEX demonstrates the additional burdens of vigilance that advanced technology imposes. It may slash production times and costs for a scientific or engineering publisher, but only if either (1) the whole burden of typesetting is shifted
to the author, who then has to be knowledgeable and vigilant about levels of
detail that copy editors and typesetters otherwise would supervise, or (2) the
author's editor is prepared to spend hours learning the fine points of TEX, adding
technical support to his or her job description.

What about less esoteric programs—the word processors, spreadsheets, and
databases that ordinary office workers use? Surely these have become easier? Text
appears onscreen more or less as it will on paper (What You See Is What You Get).
Graphic user interfaces (GUIs) with colorful icons have replaced most of the mys-
teries of DOS command lines, which usually required constant reference to a
manual—a good thing, because new documentation is notoriously skimpy. If
everything is now so easy, why have there not been more gains in productivity?[11]

In fact, another set of revenge effects is at work. Graphic interfaces like the
Macintosh System 7 and above, Microsoft Windows, OS/2, NextStep, and others
actually do not simplify underlying programs. What they do is add all kinds of
programs and files behind an organized facade. If things go right, as on a luxu-
rious cruise ship, nobody worries about the engine room. When a problem does
arise, though, the sheer number of elements and the unforeseeable ways in which
they interact means that a high price may have to be paid for the seeming ease of
everyday use—just as the levees, locks, and channels of the Army Corps of Engineers may tame the fair-weather river while shifting and even magnifying the
destruction of floods when they come.

In the world of computers, there are two solutions to this problem. One,
adopted by Apple Computer for the Macintosh, is to police the system closely
and enforce high compatibility and integration. Consistency has made Macintosh
computers the easiest to learn and use, even when they have been somewhat
slower in performing tasks than PCs running DOS with comparable processors.
Apple's refusal to license the Macintosh operating system until the mid-1990s
kept prices high but also helped it maintain a more consistent user interface. On
the other hand, it has always been more expensive to develop applications that

meet these standards than to program for the original DOS, so there has been less choice as well as less grief for Macintosh users. (Graphic artists and scientists are the exceptions; their Macintosh applications are more abundant and powerful than DOS or Windows counterparts.)

In the mid-1990s, processor speed and storage capacity continue to grow exponentially, but operating systems and programs are multiplying their demands, too. IBM's OS/2 2.1 took no less than 40 megabytes for full installation—more than many users' whole hard disks. Its successor OS/2 Warp 3 needs 65 megabytes. Microsoft Windows 3.1 plus MS-DOS needed 20 megabytes; Windows 95 requires 60. Rarer, more powerful systems like elegant NextStep 3.2 demand 120. The inflation of code, also reflected in 25-megabyte application programs, illustrates a chronic problem of computing that contributes to the productivity paradox: as swiftly as resources grow, the demands of software tend to expand even faster. Programmers and developers are understandably working for the future, preparing for the next generation of machines. Most users are living in the past because either their budgets or their bosses won't let them upgrade hardware to the current standard. Meanwhile the planned obsolescence of software means that earlier versions of major programs may no longer be supported. Computer managers are moving users to Windows without always giving them the processor power, memory, or disk space to work effectively. For many users, computer software seems related to hardware as Thomas Malthus believed population responded to food resources: software does not just match but eventually outstrips hardware in their respective rates of growth. Users who do not have enough computer power spend more and more time waiting for their systems to finish work or struggling with messages that protest, accusingly, "out of memory."[12]

Even when one's computer has power to spare, the techniques that make computers more "user-friendly" have revenge effects of their own. Take the icons that are replacing typed commands in controlling computers. Computer icons have the advantage of being more readily recognizable and distinguishable than strings of text. They also require less space. Pointing to an icon is also defended as more "natural" and "intuitive" than spelling out a request. In the normal course of education one is taught to spell and not to point, but much of the appeal of icons (and even more elementary graphic interfaces like Microsoft's Bob) is their playful innocence. The trouble is that simplification can have its own recomplicating effects. At first, with only a small number of icons, and a single organization (Apple) largely controlling their use, the idea proved a smashing success.

The real problem surfaced when open standards were adopted to encourage independent software producers, while manufacturers (especially Apple) lost part of their control over the interface. Meanwhile, applications of software swelled in size with new features as developers sought to please users in many occupations with each program. Today there are hundreds and perhaps thousands of icons in use, not to mention popular packages for creating and modifying icons. There are even animated and sound-enhanced icons, and no doubt some will soon be able to sing in 24-bit sound and four-part harmony. The recomplicating effect is that while some commands and programs are much clearer as symbols than as words, others are resolutely and sometimes inexplicably nongraphic. (The world has at least two serviceable Stop designs, but still no decent Push and Pull symbols for doors.)

"I love standards; there are so many to choose from" is a computer industry joke that applies to many icon designs. It probably does not matter too much that Apple Computer has copyrighted the Macintosh icon of the garbage can as a symbol to which unwanted files can be dragged for deletion. Windows software uses a wastebasket instead. But what does a shredder mean, then? Does it discard files, and/or does it do what paper shredders are supposed to: make the original text impossible to recognize or reassemble? And some symbols mean different things in programs written for the same operating system. A magnifying glass can call for enlarging type as it does in some Apple software, but it can also begin searching for something or looking up a file in Macintosh as well as in Windows applications. A turning arrow can mean Rotate Image, but a similar arrow can denote Undo; Microsoft tried a hundred icons for Undo and finally gave up. People might think of Stop or Greetings when they see an extended hand, and in some countries this might be an obscene gesture, but it is supposed to mean Move Through Document. No wonder software producers are starting to add text to icons, sometimes in the form of pop-up balloons that appear when the cursor is moved over the icon—a recomplicating effect if ever there was one. And while the computer industry is rightly proud of the special features it has provided for people with disabilities, current reliance on graphic interfaces is at least a temporary setback for vision-impaired users, who will probably have to rely on voice synthesis. A survey by London's Royal National Institute for the Blind revealed that nearly three-quarters of software firms producing text-based software were planning to shift to graphic interfaces.[13]

Adding the bugs and glitches of software to the limits and quirks of hardware is a recipe for problems that can easily wipe out the productivity gains in any system. This need not happen, of course, but the triumphs of computing are understandably reported more frequently than disasters that most managers would rather keep from customers as well as from the competition. The problem is usually not catastrophic but—once again—gradual. What occurs is a slow leakage of staff time that does not show up in most statistics, but is nevertheless a drag on output.

Sociologists of science and technology have long realized that getting both experiments and machinery to work takes skills that don't appear in textbooks or manuals. So-called high-technology professionals learn these as artisans always have, by working with masters of the craft, and of course by trial and error. Some companies are lucky enough to have staff specialists who know how to coax programs, machines, and people into working well together, just as manufacturers have millwrights who, with nothing more than an ordinary toolbox, can get moving a stalled production line that would baffle managers and even most engineers. Despite twenty years or more of electronic diagnostics, automobile repairs still depend primarily on basic mechanical know-how and real-world expertise. Yet many organizations cannot find—or afford—the technical support they require. As the cognitive psychologist Thomas K. Landauer has observed, software is written by programmers whose logical, spatial, and mathematical skills exceed those of most users, and whose experience often blinds them to user needs. Some features, like the Windows games Minesweeper and Solitaire, have become major corporate timekillers, and conscientious employees sometimes struggle for hours to get columns to align on their printers—behavior that a tongue-in-cheek survey by one software producer, SBT Accounting Systems, dubbed the "PC futz factor." And according to the 3M Company, 30 percent of business users lose data every year, costing an estimated 24 million business days to replace.[14]

As both software features and the numbers of users have multiplied, support by manufacturers has not kept up. Long waits and high charges are becoming more common, even when a bug in the product is at fault. Nearly 80 percent of companies increased their support costs between 1990 and 1995, but even in-house support cannot be ubiquitous. Filling the gap is peer support, the person down the hall who becomes a resource without any amendments to the job description. Many of these people can beat outside specialists. After all, they are likely to work with similar hardware configurations, using the same software for the same purposes. But there are also hidden costs involved. To help other users, relatively high paid executives and professionals are spending time as peer consultants, time they have to take from their real jobs. Although the time thus lost never turns up as a line item on any financial statement, it means that these Good Samaritans may never get around to refining a strategy further, to making an extra client presentation, or to doing any of a number of other income-producing things.[15]

The Boston consulting firm Nolan, Norton & Co. has documented what happens when organizations—among them AT&T, Bell Laboratories, Ford Motor Company, Harvard University, and Xerox Corporation—add workstations and networks without increasing support staff. Peer support expands to fill the gap. While journalists continue to celebrate the ever-lower prices of computers, the total costs of *computing* suggest that there is no such thing as a free menu. Nolan, Norton found that a single PC workstation can cost up to $20,000 a year. Of this, $2,000 to $6,500 appears on the organization's information services budget as the cost of equipment, supplies, and in-house technical assistance. Peer support turns out to cost two to three times as much, or $6,000 to $15,000. The hours needed to help end users learn things and resolve problems seem to reach a natural equilibrium. Networking multiplies the need for support. And even among professional end-user computing staff, these needs change priorities. They spend most of their time reacting to problems and have less time for planning and implementing more effective enterprise-wide computing. Most peer-support time is reactive, too.[16]

Peer support often turns out to be a rearranging effect. We save time by having computers accomplish things for us, but before they can do them we take time from our colleagues. A local area network (LAN) may help a department work more effectively together, but it can also take hours of time as employees become administrators and assume the care and feeding of cables, interface cards, and networking software. The problem is more serious for those who do not have a formal role. Most amateur gurus do not mind interrupting what they are supposed to be doing. They may like computers more than their real jobs. Even if they love their work, they probably find peer help psychologically rewarding and politically valuable in accumulating allies and favors owed. Professional computer managers appreciate the volunteers for relieving their budgets and of course for knowing their colleagues' needs better than a full-time technical person could. Since some peer helpers are world-class scientists, engineers, and other professionals, the information services department is getting a bargain— and the organization is losing a part of the high-priced time it is paying for. Yet without them, as William M. Bulkeley wrote in the *Wall Street Journal*, "millions of U.S. workers might turn their PCs into expensive hatracks."[17]

One obvious solution is to bring support costs into the open by expanding budgets for information services to reflect the hidden demand. In the mid-1990s, however, corporations appear to be moving in the opposite direction: buying more and more "bargain" hardware and expecting the same staff to support it. Managers think that because local area networks are cheaper than mainframes they should cost less to maintain. Actually, one consultant for network strategy, Janet Hyland, points out, "For client/server, ongoing support is much more costly than a mainframe." And she adds that "no one will ever know how expensive it is" to support networks throughout a corporation.[18]

Despite industry call for usability, then, the computing world of the 1990s turns out to be a patchwork of stand-alone machines and networks, professionals and amateurs, always in a state of tension between the productivity benefits greater power brings and the learning and support costs that it requires. Whether by planning or by chance, some organizations have the people needed to make computers work brilliantly. These become the legends of the movement. Other organizations groan under the weight of their new machinery, because they don't have people with the right skills in the right places. It comes as no surprise that computing, like most other forms of contemporary technology, is neither a miracle weapon nor a dud, but a set of tools that need constant attention and maintenance. The software crisis is yet another example of how the mastery of great tasks increases rather than reduces the chronic burdens of vigilance.

What about analysis? Of course computers take time to learn, and some people may miss the point. But who can deny that people think better, work better, analyze better, and write better with computers? Actually there is growing evidence that computers may not be consistently better for real work after all. This suggestion infuriates many computer professionals. Industry columnists arguing vehemently with one another about almost everything else concur in denouncing anyone so perversely goofy as to doubt that computers really do make everyone more productive. (For computer magazine editors and writers, there is no doubt that a rapidly changing industry promotes productivity; they would have very little to produce without a monthly shot of new hardware, programs, environments, and even operating systems.)

From 1981 to the present, American businesses have spent hundreds of millions of dollars on software to help improve decision making. Software producers are now disbursing millions of dollars on research to make their programs even better and easier to use. Spreadsheets and other decision-making aids are now ubiquitous in academia and the professions as well as in business. Thus it is all the more remarkable how little research has been devoted to the effects of computers on the quality of decision making. Given more data and more powerful analytic tools, is it reasonable to assume that the quality of decisions will be correspondingly higher? This may seem plausible, and possible. It may even be true. Yet only a handful of studies exist. After all, why try to prove what everybody knows intuitively and what is built into the curriculum of business schools? The problem is that there is growing evidence that software doesn't necessarily improve decision making.

Jeffrey E. Kottemann, Fred D. Davis, and William E. Remus, specialists in business decisions, tested the ability of a group of students to control a simulated production line under conditions of demand with a strong random element. They could add more workers with the risk of idle time on one hand, or maintain a smaller workforce with the risk of higher overtime payments on the other. They could restrict output relative to demand, possibly losing sales if orders jumped, or they could maintain higher output at the cost of maintaining possibly excessive inventory.[19]

The subjects, M.B.A. student volunteers and experienced spreadsheet users, were divided into two groups. One group was shown a screen that prompted only for the number of units to produce and workforce level. The other group could enter anticipated sales for each proposed set of choices and immediately run a simulation program to show the varying results for projected inventory. They could immediately see the costs of changing workforce levels, along with overtime, idle time, and nonoptimal inventory costs, all neatly displayed within seconds after entering their data. They had no more real information than the first group, but a much more concrete idea of the consequences of every choice they made.

If all we have read about the power of spreadsheets is right, the "what-if" group should have outperformed the one with more limited information, unable to experiment with a wide variety of data. Actually the what-if subjects incurred somewhat higher costs than those without the same analytical tools, although the difference between the two groups was, as expected, not statistically significant. The most interesting results had less to do with actual performance than with the subjects' confidence. The "non-what-ifs" rated their own predictive ability fairly accurately. The ratings that subjects in the what-if group assigned to their own performance had no significant relationship to actual results. The correlation was little better than if they had tossed coins.

Even more striking was how the subjects in the what-if group thought of the effects of their decision-making tools. What-if analysis improved cost performance for only 58 percent of the subjects who used it, yet 87 percent of them thought it had helped them, 5 percent believed there was no difference, and only 1 percent thought it had hurt. This last-mentioned subject was actually one of those it had helped. Another experiment had an even more disconcerting result: "decision makers were indifferent between what-if analysis and a quantitative decision rule which, if used, would have led to tremendous cost savings." In other words, the subjects preferred what-if exploration to a proven technique.[20]

Why the attraction of what-if analysis? Perhaps the subjects, being M.B.A. student-volunteers, had a strong belief in their ability to manipulate numbers. But that does not explain why they also rejected a carefully formulated equation in the second experiment in favor of free-form interaction. The real reason, Davis and Kottemann surmise, must be what psychologists call the illusion of control: the way we can easily convince ourselves, given the proper setting, that we are making things happen when in reality they are chance events. Both the beauty and the risk of computerized analysis is the concreteness it can give our plans—even when our underlying data are doubtful and our models untested or even wrong. We have seen the theatrical power of computers; in the illusion of control, we turn it on ourselves to reassure ourselves that the powers we possess are indeed real.

Computer advertising appeals to this sense of control, and in fact it often *is* real control, at least in the right hands. But consider all the hardware and software producers who have failed or lost tens or hundreds of millions of dollars. Surely their employees and at least some of their executives were among the world's most proficient users of computers of all kinds—with IBM at their head. And surely they used the best models and techniques available. But the losers faced situations in which what-if questions are of limited value, in which politics, distribution, the evolution of standards, and sheer bluff matter as much as technical excellence. Of course, we can't go back to graph paper, slide rules, pencil marks, and eraser shavings, but we do not seem to be moving forward as rapidly beyond the present categories of software as we might, despite the extravagant early promises of artificial intelligence research. In fact, some old-style financial executives still astound younger colleagues by using slide rules and mental techniques at meetings to get faster results than the others obtain from their calculators and notebook computers.[21]

Spreadsheets are still valuable tools, but they are more like lathes than can openers. They need careful setup, adjustment, and oversight. The decision scientists Raymond R. Panko and Richard P. Halverson Jr., found that while spreadsheet users have a low error rate at the cell level (0.9 to 2.4 percent), these errors cascade into disturbingly high error rates at the bottom line (53 to 80 percent). As the logical intricacy of spreadsheets increases, the error rate also grows—a recomplicating effect. When a bug in mathematical operations of the Pentium chip in 1994 alarmed users, few understood that their own mistakes in entering data and failure to debug their programs were a far greater risk to their results than the problems of Intel's processor could ever be. Spreadsheet users, like other beneficiaries of technology, have not realized that the way errors propagate from step to step in a program is yet another chronic problem of technology. The answer is that once more, a labor-saving technology needs a surprising amount of time-consuming vigilance to work properly.[22]

To return to Paul David's comparison of electric motors and electronic workstations: Could it be that only now are we starting to see the beginnings of new structures that will make possible the kind of productivity increases that manufacturing enjoyed in the 1920s? Networks seem to be rebuilding organizations in new ways. If only there were at least basic agreement on what is happening. On the one hand, networks are said to inject grassroots democracy and equality, ushering in a kind of corporate New Age where rank and title matter less than the quality of a contribution, bare as electronic mail is of the usual trappings of executive office. Documents can be transmitted and processed with new speed, sent between employees with different computer types and operating systems. Electronic meetings can open the floor to ideas of those who might otherwise have been pushed to the sidelines by superiors or more aggressive colleagues in a conventional face-to-face session. Regrettably, an in-house computerized meeting room costs $50,000 to $200,000, and outside providers charge thousands of dollars a day for their use.[23]

The flip side of the innately democratic network is the malignant authoritarian network. Where some find inventiveness percolating up and correspondingly rewarded, others find discipline and punishment raining down and privacy trampled underfoot. If networks appear to open channels previously barred—and it's not clear how having to put ink on paper ever prevented sending a message to top management—they also make it possible to read files surreptitiously, monitor activities, and even trace message traffic to discover clusters of malcontents. Aside from such ethical lapses, a collegial style doesn't necessarily mean a flattening of power. The InfoWorld magazine columnist "Robert X. Cringely" suggested not so long ago that the casual culture of Microsoft masked a management style that was at heart not so different from the hard-driving ways of the robber barons. ("You can work any eighty hours a week you want.") Of course, an authoritarian organization can be extremely productive, whether or not the iron fist is in a faded denim glove, but there is no evidence that networks as such make managers different. Rather, people seem to build networks in their own image.[24]

Apart from issues of democracy and authority, computers have encouraged a trend that may have unintended results for corporate efficiency: reducing support staff. The American Manufacturing Association, in a much-noticed study, found that staff reductions in general did not reliably increase profits. At downsized companies, profits increased for only 43 percent of firms, and actually dropped in 24 percent of those studied. Almost as many reported a drop in worker productivity as reported an increase.[25]

If computers really made it possible for a smaller number of people to accomplish the same amount of work, there would be little outcry about longer hours for middle managers and professionals. In fact, after examining twenty case studies in five major U.S. corporations, Peter G. Sassone, an economist and management consultant, concluded that computerization has helped reduce rather than promote the amount of time that these employees spend performing their highest and best work. Sassone found that many highly paid people were spending a significant amount of their time performing what amounted to secretarial and clerical functions, usually working with computers but often not doing what they spent years at college and graduate school learning to do. Why not? One reason is that companies tend to expand by adding professionals and to cut back by laying off support staff, giving them a top-heavy structure. Another is that computer systems make it possible for managers to do more things for themselves. Their jobs become more diverse in a negative way, including things like printing out letters that their secretaries once did. Sassone's analysis suggests that the increased productivity of office technology has an "indirect and unintended effect on staffing" that "may cause overall organizational productivity to decline." This he calls the "law of diminishing specialization."[26]

This is just fine for many people. Either you like the division of labor or you don't. Some social thinkers, and some people in business, believe that doing your own word processing, filing, and so forth builds character. But economic theory suggests that it doesn't do much for profits. The most rational way to deploy staff is to have the most skilled and specialized employees working as much of the time as possible at the highest level of specialization, while less skilled and lower-paid staff should take over the rest of the work. A famous precomputer textbook example is the story of the best lawyer in town who was also the best typist in town, but who would look foolish not employing a typist. Computers can create an illusion that the machine is doing all the work, but a surprising amount actually remains: formatting letters, replenishing and unjamming paper, replacing toner, and of course addressing and stuffing envelopes. Professionals who are doing these things aren't doing other, more productive things. Of course, top executives understand this pitfall—when it comes to themselves. They rarely cut back on their immediate support staff. Sassone claims that by reversing the formula and employing fewer managers and professionals and more support staff, corporations could achieve savings of $7,400 per employee. What keeps them from doing this is another form of the illusion of control, in this case the misconception that now support functions can take care of themselves.[27]

The relentless speed and efficiency promised by microcomputers and networks, their computation capacity doubling every eighteen months, has a catch. The more powerful systems have become, the more human time it takes to maintain them, to develop the software, to resolve bugs and conflicts, to learn new versions, to fiddle with options. Once again, the intensity of a given technology may bring repeating and rearranging effects in its wake. Early computer-minded historians were overjoyed at their new powers of quantitative analysis—until they realized that they would probably have to spend months or years entering data by hand before a machine could spend its few minutes processing and analyzing the data.

One approach to computers and productivity is to insist that many benefits of computing defy conventional measurement. If anything should change, it is the methods of economists, not the claims of the computer industry. A more radical form of this argument is that the point of the computer is the pleasure of mastering and using it, its responsiveness when things go well. In this view, that game of Minesweeper on an employee's monitor should not make the boss scowl; there probably is something to learn from it. Computerization is as much an end as a means. Nearly every computer user must feel such joy when things go right— seeing the first gorgeous page of text from a new program or printer, for example—that it seems crass to talk profit and loss. But this approach ignores the times of correspondingly great frustration. Some programmers are repeat keyboard smashers. And if personal experience is paramount, what of the countless people who still enjoy manual ways of doing things?[28]

For both technophiles and technophobes, the best, and perhaps the only, way to avoid the revenge effects of computing is to maintain skills and resources that are independent of the computer. We can learn back-of-the-envelope calculation to beware of misplaced decimal points. We can work on face-to-face and telephone relationships with colleagues and outsiders to avoid the misunderstandings of excessive reliance on electronic mail. In a way, the problems of computers, like so many other revenge effects, have reminded people of the value of other things. In the first decade of computing, handwriting and handwritten letters began to rebound from years of neglect. Bugs, glitches, and crashes have a positive side: they are the machine's way of telling us to diversify our attention, not to put all of our virtual eggs in one electronic basket.

NOTES

1. Julie Hart, "Buy Smart: Will Your Latest Purchase Hold Its Value?" *Computerworld*, 21 February 1994, 103.

2. Brooke Crothers and Rob Guth, "Strong Demand, Weak Dollar Squeeze DRAM," *InfoWorld*, 17 April 1995, 1.

3. See James R. Chiles, "Now That Everything's Portable, Getting Around Can Really Be a Drag," *Smithsonian* (January 1994): 110.

4. Rob Kling and Suzanne Iacono, "The Mobilization of Support for Computerization: The Role of Computerization Movements," *Social Problems* 35, no. 3 (June 1988): 226–43; Langdon Winner, "Mythinformation," *Whole Earth Review* (January 1985): 22–28, and reprinted in Langdon Winner, *The Whale and the Reactor: A Search for Limits in the New Age of High Technology* (Chicago: University of Chicago Press, 1986), pp. 98–117. Rob Kling, ed., *Computerization and Controversy: Value Conflicts and Social Choices*, 2d ed. (San Diego: Academic Press, 1995), is the most comprehensive anthology of reprinted and original papers on social aspects of computing.

5. The most recent statement is Paul A. David, "Computer and Dynamo: The Modern Productivity Paradox in a Not-Too-Distant Mirror," reprinted from *Technology and Productivity: The Challenge for Economic Policy* (Paris: OECD, 1991), pp. 315–47.

6. Don L. Boroughs, "Desktop Dilemma," *U.S. News & World Report*, 24 December 1990, 46–48; Dean Foust, "Is the Computer Boost That Big?" *Business Week*, 16 January 1995, 24; Glenn Rifkin, "Heads That Roll If Computers Fail," *New York Times*, 14 May 1991, D1; Garry Ray, "The Productivity Chase," *Computerworld*, 27 December 1993–3 January 1994, 56. For the strongest defense of the computer industry's record, see U.S. National Research Council, *Information Technology in the Service Society* (Washington, D.C.: National Academy Press, 1994); for a powerful claim that poor usability is indeed undermining productivity, see Thomas K. Landauer, *The Trouble with Computers* (Cambridge: MIT Press, 1995).

7. Lauren Ruth Wiener, *Digital Woes: Why We Should Not Depend on Software* (Reading, Mass.: Addison-Wesley, 1993), pp. 10–13. See also Steven Casey, *Set Phasers on Stun and Other True Tales of Design, Technology, and Human Error* (Santa Barbara, Calif.: Aegean, 1993), pp. 13–22, on how the medical technician's inability to see or hear the patient compounded the injury.

8. John Holusha, "The Painful Lessons of Disruption," *New York Times*, 17 March 1993, D1.

9. William M. Bulkeley, "To Read This, Give Us the Password . . . Oops! Try It Again," *Wall Street Journal*, 19 April 1995, A1; John Markoff, "Computer Viruses: Just Uncommon Colds After All?" *New York Times*, 1 November 1992; Darrel Ince, "Nasty Virus, but Not Fatal," *Independent* (London), 6 September 1993, 14; Christopher O'Malley, "Stalking Stealth Viruses," *Popular Science*, January 1992, 54ff; Mike Holderness, "On the Trail of the Cyberspace Pirates," *New Scientist* 137, no. 1860 (13 February 1993): 46–47.

10. I am indebted to Michael S. Mahoney for identifying the substitution of programmers for accountants.

11. Kelly Conatser, "Where Has All the Documentation Gone?" *InfoWorld*, 17 April 1995, 1ff.

12. Randall Kennedy, "OS/2 Warp Goes Light Years Ahead of 2.1," *InfoWorld*, 14 November 1994, 167–70; Esther Schindler, "Third Time the Charm? OS/2 3 Goes to Warp Speed," *Computer Shopper*, January 1995, 586ff.

13. Dave Kansas, "The Icon Crisis: Tiny Pictures Cause Confusion," *Wall Street Journal*, 17 November 1993, B1; Joe Lazzaro, "Adapting GUI Software for the Blind Is No Easy Task," *Byte*, May 1994, 33.

14. Landauer, *Trouble with Computers*, pp. 210–20; Joel Garreau, "Labor-Saving Devices," *Washington Post*, 9 March 1994, C1; William M. Bulkeley, "Data Trap: How Using Your PC Can Be a Waste of Time, Money," *Wall Street Journal*, 4 January 1993, B5; Lamont Wood, "Office Futz Factor Is a Threat to PC Productivity," *Chicago Tribune*, 3 October 1993, Technology and the Workplace, 3.

15. Gary H. Anthes and William Braundel, "Quality Questioned," *Computerworld*, 24 April 1995, 1.

16. Nolan, Norton & Co., "Managing End-User Computing," research report (Boston: Nolan, Norton, 1992).

17. William M. Bulkeley, "Study Finds Hidden Costs of Computing," *Wall Street Journal*, 2 November 1992, B4.

18. Deborah Asbrand, "Lean Budgets Put the Squeeze on IS Departments," *InfoWorld*, 7 February 1994, 55.

19. Jeffrey E. Kottemann, Fred D. Davis, and William E. Remus, "Computer-Assisted Decision Making: Performance, Beliefs, and the Illusion of Control," *Organizational Behavior and Human Decision Processes* 57, no. 1 (January 1994): 26ff.

20. Fred D. Davis and Jeffrey E. Kottemann, "User Perceptions of Decision Support Effectiveness: Two Production Planning Experiments," *Decision Sciences* 25, no. 1 (January 1994): 57ff.

21. Arthur Howe, "No Calculator Required: Executives Flex Their Math Minds," *Philadelphia Inquirer*, 10 July 1986, 1A.

22. Raymond R. Panko and Richard P. Halverson Jr., "Patterns of Errors in Spreadsheet Development," unpublished paper, December 1994.

23. William M. Bulkeley, "'Computerizing' Dull Meetings Is Touted as an Antidote to the Mouth That Bored," *Wall Street Journal*, 28 January 1992, B1.

24. Robert X. Cringely, *Accidental Empires* (New York: HarperBusiness, 1993), p. 115.

25. Susan Cohen, "White Collar Blues," *Washington Post Magazine*, 17 January 1993, 10–13, 24–27; Andrea Knox, "By Downsizing, Do Firms Ax Themselves in the Foot?" *Philadelphia Inquirer*, 17 February 1992.

26. Peter G. Sassone, "Survey Finds Low Office Productivity Linked to Staffing Imbalances," *National Productivity Review* 2, no. 2 (spring 1992): 147–58.

27. Ibid., p. 154.

28. See, for example, Herb Brody, "The Pleasure Machine," *Technology Review* 95, no. 3 (April 1992): 31–36; also a thoughtful and exuberant letter by Andrew Paul Grell, "Life Is Sweeter on the Electronic Superhighway," *New York Times*, 13 July 1993, A18.

Virtual Reality
The Future of Cyberspace

Gordon Graham

With the concept of "virtual reality" (VR) we arrive at the world of science fiction and of fantasy. Or so many suppose, and it may indeed seem that if in the discussion of virtual reality every limit on the imagination is removed, then all the factual and conceptual constraints which control our speculations are lifted also. Such a lack of constraint may suit the novelist or storyteller, but it is a freedom we do not want in the sort of inquiry we are engaged in here, because it makes any speculation as good as every other and hence marks the end of serious critical investigation. Actually, although in discussing virtual reality we are indeed entering the realms of the highly speculative, the position should not be construed as one quite without constraint. What is true is that we need to be careful about the constraints we observe. The sheer speed at which the technology of the Internet is developing is probably unprecedented in human history, and for this reason there is a constant danger of declaring something to be impossible a priori, only to find that in a very short time it is a reality. The same thing can be said of VR technology, which has largely been developed independently of the Internet.

THE "BODYNET" AND THE "SMARTROOM"

Nevertheless, though the future undoubtedly holds many hitherto undreamt of technical marvels, what is *imaginable* is not necessarily *conceivable*. The distinction is illustrated with great regularity in traditional fairy stories. We can imagine, easily enough, that the prince is turned into a frog, but there are well-known philosophical obstacles in the way of making such an imaginary event properly

From *The Internet: A Philosophical Inquiry*, by Gordon Graham, "Virtual Reality: The Future of Cyberspace" (London: Taylor & Francis/Routledge, 1999), pp. 151–66.

intelligible. The same sort of thing is observable in science fiction. There are many tales of time travelers, and very entertaining and diverting they can be. But philosophers know well that whether time travel is conceptually possible is a vexed and difficult question. A large part of its difficulty lies in the fact that the idea of time travel seems to admit the flatly contradictory—the traveler existing before he was born, for instance, or encountering on his travels a younger person who both is and is not himself. Contradictory states of affairs are as good a mark of the impossible as one can hope to find, but even so we can *imagine* the contradictory without any great difficulty, as when we picture (either mentally or with pen and ink) the older and the younger time traveler meeting each other.

From this it follows that not all that is imaginable is conceivable. We can imagine things which logically could not happen, and if they are indeed logically impossible, necessarily they are empirically impossible also. To make any headway with the idea of virtual reality and hence with the future of cyberspace, then, we need to restrict ourselves to the realms of the conceptually possible and ignore the impossible imaginings of science-fiction writers. At the same time, if we are to think about the future of cyberspace to any real purpose, then we should allow full rein to speculation about the *technically* possible. At present, life on the Internet does not amount to much more than interactive television. People have speculated, however, that whenever the devices of VR technology that might be developed in the future can be plugged into the Internet, the result will be something much more like a "total" experience. One of the steps along the way to this is the "bodynet," described by one of the most enthusiastic futurists in the world of information technology as follows:

> The bodynet is the brainchild of Olin Shivers at the MIT Laboratory for Computer Science. . . . The Shivers bodynet builds on a pair of "magic glasses" that you wear. They have clear lenses that let you see where you are going but also present miniature inset displays that show color images to each eye. The images are generated by a computer the size of a cigarette pack on your belt or in your purse. The glasses also have photodiode sensors that monitor the whites of your eyes in order to detect where you're looking. Miniature microphones and earphones attached to the glasses let you speak to and hear from your equipment. . . . The gadgets communicate with one another in a language called "bodytalk," which is transmitted via low-power radio waves that are confined to an invisible envelope around your body—the body network, or bodynet.[1]

The bodynet, or the "smartroom" (another very hi-tech device described by Dertouzos), is not necessarily linked to the pursuit of virtual reality, but clearly has or is expected to have implications in this connection. For convenience devices such as these and speculations about their possible uses can be combined into what is sometimes known as a "VR Body Zone." The aspiration behind the VR Body Zone is that it would make Internet encounters far less like watching television and more like experiencing the real thing. And it is to the world of the

Internet *with* such a device, rather than that now currently prevailing, that the term "virtual reality" is normally applied.

Let us grant the Internet enthusiasts and futurists all the advances and advantages that these technological speculations permit and let us for the moment deploy the phrase "virtual reality" to indicate the kind of experience that electronic communication up to and including such devices would make possible. We can then ask the following questions. Is virtual reality in any sense a new mode of being, and insofar as it is, is it better or worse than the everyday world with which we are all familiar?

THE "VIRTUAL" AND THE "REAL THING"

Consider other examples of the use of "virtual." In the expression "virtual certainty," the word "virtual" implies "as good as." When something is a virtual certainty, then, even though it is *not* certain, it can be taken to be so, at any rate for the purposes in hand. Insofar as it has any currency, the expression "virtual reality" seems to function in a similar way. It signals something not the same as, but as good as, the real thing, at least for certain purposes. Of course, in making sense of this a lot turns on what we mean by the "real thing," but I do not think we need to spell this out in the abstract. In any individual case it will be clear enough. To have met a virtually real Marilyn Monroe is as good as having met her in the flesh; to have climbed the Eiger in virtual reality is as good as having climbed the Eiger; to have "virtually" encountered a man-eating tiger in the depths of the Southeast Asian jungle is as good as having encountered a real one. To each of these examples we must add, of course, the qualification "for certain purposes." What might those purposes be? The last example gives us a clue. For the purposes of knowing "what it is like" to meet a man-eating tiger, a virtual one is as good as a real one and it has the further advantage of none of the normal risks attaching. One can extend the example certainly and imagine being "virtually eaten" by the virtual tiger, and hence getting to know what it is like to be eaten by a tiger. Still, no death results and this has to be an advantage virtual reality enjoys over the real thing.

The extension of the man-eating tiger example might be thought to bring us not so much to the realms of the fantastic as the ludicrous. And so perhaps it does. But it allows us to raise two important doubts about virtual reality. First, do we really find out "what it is like" to face a man-eating tiger if we know that there is no danger of actually being eaten? Second, do we need all the apparatus of the bodynet and the smartroom to accomplish this? It will be most convenient to begin with the second of these questions.

Are people in the networked VR Body Zone so very different from the person stimulated by and caught up in far simpler forms of make-believe? At a minimum, it is not obvious that they are, because the ordinary power of make-believe is very considerable. This is a fact about the arts in general that has been taken up by Kendall Walton, author of a major study in aesthetics entitled *Mimesis as Make-believe.* An important part of his thesis is that the world of make-believe as embodied in films, plays, novels, and other forms of artistic imagination allows us to "enjoy" the experiences depicted without the normal costs of doing so.

> Make-believe . . . is a truly remarkable invention. We can . . . make sure the good guys win, or see what it is like for the bad guys to win. . . . There is a price to pay in real life when the bad guys win, even if we learn from experience. Make-believe provides the experience—something like it anyway—for free. The divergence between fictionality and truth spares us pain and suffering we would have to experience in the real world. We realize some of the benefits of hard experience without having to undergo it.[2]

Whether this is a good explanation of the value of works of art is open to question, but that is not the issue here. Rather, we want to know whether the vaunted advantages of an Internet capable of full-blooded virtual reality effects constitutes something radically new. And the answer seems to be that it is not: certainly it may be true that virtual reality "provides the experience—something like it anyway—for free," but this is no more than Walton claims for the "remarkable invention" of make-believe as exhibited in less technically advanced media such as plays, novels, and films.

To see that there really is no very great difference here despite the futuristic nature of cyberspace, something more needs to be said about the relation between, on the one hand, having the experience "what it is like to . . ." and, on the other, any given medium that induces this. Suppose that I am reading a narrative of encountering a tiger in a jungle. It might be fiction, or it might be the reporting of fact, but in the hands of a good author either might induce in me a sense of "knowing what it is like." Of course there is certainly a difference between knowing what it is like in virtue of reading an account and knowing what it is like because I have myself been in this circumstance. An important part of the difference lies in this simple fact: in the first case, though I "know what it is like," it is not true of me that I have encountered a tiger; in the second case, it *is* true that I have encountered the real thing, and it is precisely because this is true that I know what it is like. However, this difference applies just as well to my encounter with a virtual tiger as to my merely reading about it. In this case too, however lifelike the virtual experience of a tiger, it is not *actually* an experience of a tiger and no increase in technical sophistication will make it so.

There are prospective differences between storytelling and VR experiences, certainly, just as there are differences between a film and a novel. But since in both cases the impression, however vivid, necessarily falls short of the reality, any such differences, whether between storytelling and virtual reality or between film and literature, must lie in the nature of the alternative media. "Knowing what it is like" is the same in both cases; all that differs is the means by which this has been induced.

It seems plausible to characterize this difference in the following way. The storybook case relies upon my imagination, aided no doubt by the imaginative and descriptive power of the author. The virtual reality case, by contrast, leaves much less to my imagination, perhaps hardly anything at all, because the gap between experience and reality which imagination normally bridges is filled by the technical devices of the electronic medium.

It is tempting to try to strengthen this point by saying: reading about tigers is *not at all like* encountering tigers, whereas virtual experience of a tiger is *very like* actual experience of a tiger. Let us agree that this is true. Nevertheless, if the point or value of the virtual experience is coming to know "what it is like," this is also the case in the imaginative experience prompted by my reading. What is called virtual reality may have the advantage of inducing this sense of "what it is like" more easily, and perhaps more vividly, especially among the less imaginative, but it still remains the case that if the merit of virtual reality is said to lie in its being "as good as" the real thing for the purpose of coming to know "what it is like," this is true of other much less technically complex media also.

Despite the difficulty of discerning any very great distinction between media such as novels and films with which we are thoroughly familiar and Virtual Body Zones yet to be invented, the idea is likely to persist, I think, that a virtual reality experience of X is in some important and interesting way *closer* to the actual experience of X than is merely reading or seeing a film about X. But on what might this greater closeness be based? One common answer is that virtual reality experiences are (or would be at any rate) "just like" the real thing. Is this true? Here there is reason to return to the first of the two questions posed earlier. Do we really find out "what it is like" to face a man-eating tiger if we know that there is no danger of actually being eaten? Suppose we answer "No" to this question with respect to novels and plays. Then we must also answer "No" in the case of virtual reality. This is because however vivid the experience inside our wraparound goggles and data gloves, we know we are not in danger of being eaten just as much as we know it when we sit and read the book. Many people, I imagine, would feel inclined to answer "Yes" with respect to virtual reality machines, to say that in the case of virtual reality we really *do* get to know what it is like to face a man-eating tiger even though we know there is no chance of our being eaten. The temptation to say this is because we can predict that under such circumstances people are likely to experience the same emotions and exhibit the same reactions that they would in the presence of an actual tiger—terror, anxiety, and so on. But on exactly the same grounds we would have to answer "Yes" in the case of other media also because it is a fact (much discussed by philosophers of art) that novels, films, and plays can stir strong emotional reactions on the part of audiences—horror at horror films, sadness at tragedies, and so on—even in the familiar surroundings of their own homes. It follows, it seems to me, that either way we have no reason to attribute any difference *in kind* to virtual reality experiences over other make-believe experiences. At most, we have reason to attribute a difference of degree of vividness or intensity.

The reason for this is that on the analysis so far, even the most technically sophisticated virtual reality setups must be interpreted as experiences without "the reality." What is missing, it appears, is any ingredient that would incline us to say that virtual reality is more than a simulacrum—that it is a different *kind* of reality, different but just as good as (or better than) any other for certain purposes.

"VIRTUAL" AS A KIND OF REALITY

Internet communities are frequently referred to as "virtual communities." It seems, though, that the use of "virtual" in "virtual community" signals something a little different from its use in the more general expression "virtual reality." A virtual community is not a community that is an experientially indistinguishable copy of the real thing, but rather a community of a different kind. In this sense "virtual" signals not a simulacrum of reality, but a different kind of reality, and this is just the idea we want to explore.

It was Howard Rheingold, author of *The Virtual Community: Homesteading on the Electronic Frontier*, who first brought this use of the term to prominence and who provides a definition that has gained a certain currency:

> Virtual communities are social aggregations that emerge from the Net when enough people carry on ... public discussions long enough, with sufficient human feeling, to form webs of personal relationships in cyberspace.[3]

The first thing to be observed about this definition is that it contains no reference to (or even hint of) substitute experience. Nothing said here suggests that the emergence of a virtual community depends on or would even be specially advanced by the development of hi-tech Virtual Body Zones. The contrast, it appears, is not between that of a "real" community of flesh-and-blood people and an experiential simulacrum which conveys what the real thing is like. The difference lies elsewhere. Where could this be? The answer will be found, I think, by remembering the origins of the virtual community. MUDS and MOOS, out of which those groups called virtual communities have emerged, were first devised as many-player games. So the question is: When does interacting on the Internet in a structured way cease to be merely a game? When does it take on the seriousness (rather than the feel) of community life? The answer in outline is: when a *virtual* community is formed, and this is precisely the point which Rheingold's definition aims to capture or isolate (whether or not it succeeds).

In this sense of "virtual," the virtual is not a semblance of something else, but an alternative to it—an alternative type of entity with properties both similar and dissimilar to that with which it is contrasted. If this is correct, the question is whether this alternative form of community is a kind of entity in its own right, and of a sort that allows us to attribute to it a distinctive form of reality.

Why should anyone deny this, or deny that virtual communities, though different are real enough in their own way? The crucial difference between "encountering a tiger" or "climbing the Eiger" by means of a VR Body Zone and doing these things in real life can, as we saw, be expressed in terms of make-believe. It is possible (in theory) by means of VR to have the experience of encountering a tiger (or something like it) but this is really no different to having "made-believe" with the help of a novel or a film that I have met a tiger. In the absence of the real thing it remains true of me that I have never *as a matter of fact* encountered a tiger. Imagine, now, someone whose sole experience of interacting with others lay, or at least had come to lie, in relationships formed within a GeoCity or virtual community. Would it be true of such a person that they had never "really" interacted with anyone, that they merely knew "what it was like" to do so? The answer is much less obvious than in the VR case and this is why there is some reason to attribute to virtual communities a reality of their own.

Why is the answer only *much less* clear? Why is it not plain that they have indeed formed real relationships, though of a different sort? The residual doubt arises from this important fact: It seems possible that indefinitely many of the relationships they enter into by this means are fictional.

A virtual community could in theory comprise the one real person we are hypothesizing (for convenience, our real person can be a woman called Technos) and indefinitely many invented personalities. If Technos does not know this, and believes that she has friends and acquaintances with whom she exchanges gossip, from whom she learns and to whom she turns for advice, is she not deluded? There *are* no friends out there, only invented personalities, so how can she have real relationships with them? Alter the case so that Technos has herself invented the larger part of the character by which she represents herself on the Internet. The whole thing now seems a delusion, no better than a game and possibly worse insofar as it has generated a degree of deception.

But is the resultant community life truly a delusion? The case in which all the characters in a virtual community are significantly different to the real-life "players" who "operate" them, is certainly the most interesting case to explore. If under these circumstances we could find a way of attributing a certain sort of reality to the community they comprise, we should indeed have found a virtual reality of an interestingly different kind. The principal difficulty in the way of spelling this possibility out convincingly lies, of course, in the thought that such a community would in fact have returned to that from which it had its origin— namely a mere game. What we need to ask, therefore, is whether the wholly fictional community just imagined could have the sort of seriousness that Rheingold builds into his definition, or whether it would be nothing more than a game of very great sophistication.

"VIRTUAL" ACHIEVEMENTS

To simplify the task of exploring this question I shall introduce some stipulative terminology. Let us use the term "community" to refer to the normal case—the villages, associations, cities, and countries with which we are familiar. And, ignoring the fact that many of the Internet groups currently referred to as virtual communities are "inhabited" by real people, let us reserve the term "virtual community" for an Internet group in which all the on-screen personalities are the *alter egos* of those who, so to speak, sit at their terminals. Is a virtual community in this limited sense anything more than a game?

Computer games have something in common with VR. In a computer game I "kill," let us say, a number of evil invaders. Of course, in an obvious sense, it is true that even so I have never actually killed anyone. That is to say, just as in the earlier example where my "VR encountering" a tiger falls short of actually encountering a tiger, however lifelike the experience, in the game there is a similar falling short with respect to "killing" invaders. Nevertheless, there is a difference and one which, we might say, makes all the difference, for killing enough invaders in the game counts not as an imaginary win, but a *real* one. In other words "killing game invaders," though different from "killing real invaders," is still an achievement. It is not the *same* achievement, but an achievement nonetheless—an achievement within the game. Similarly, a relationship established by Technos, though not a relationship with the person at the other terminal, is a relationship "within the virtual community." Two questions arise. First, does this remain true if Technos is herself an invented persona? Second, if it does, is this enough to make the relationship thereby achieved something other than a move in a game?

We can expand the example in ways which allow us to explore these questions further. Imagine that Technos gets elected mayor of the virtual community (let us call it Cyberville) and initiates a building program. A "law" is established by which every time someone comes online, their persona has to contribute to the emerging communal construction by the addition of a suitable building icon. Failure to do so results in exclusion. Over time, this cooperative activity results in Cyberville's having the most sophisticated architectural construction of any virtual community to be found on the Internet. Then Technos starts to make "political" misjudgements and the policies she pursues lead to dissension. So great is this dissension that more and more personae leave Cyberville until, finally, it is abandoned and the virtual community known as Cyberville ceases to exist. It becomes, in other words, a "ghost town on the electronic frontier" just like so many on the original Frontier. Could we not say that the architectural construction which lasted for a time was a significant achievement and that Technos's mismanagement of the virtual community was, correspondingly, a significant failure? If we can say this then, as it seems to me, we have given sense to the idea that Cyberville is (or was) a community in a different order of reality and thus something more substantial than a mere game.

To decide whether this is the right conclusion to draw, we have to describe the history of Cyberville with some care. Despite the way I have just posed the question, the crucial point is not so much whether an achievement or a failure "within the virtual community" is a move in a game, but whether either the achievement or the failure can properly be attributed to the personae who inhabit this community. Did the inhabitants build an impressive structure? Did Technos fail the community as mayor? It seems there is reason to doubt this, because it seems odd to say that the public policy which proved disastrous was decided by Technos. Isn't it more plausible to hold that the policy was decided by the *author* of Technos—the person who sat at the terminal? Suppose that Technos has an arrogant and autocratic temperament, while the author of Technos is not like this at all, and that it is because of this arrogance and autocracy that the policy failed. Even so, we have not eliminated the ultimate responsibility of the author, because it is the author who has decided that Technos shall have this character, and who, step by step, has authored the reactions and responses which best express this temperament. In short, it appears that at some point or other we must make reference to real persons and cannot explain the whole history of Cyberville entirely in terms of the invented personae who constitute the virtual community.

This does not show, however, that it is all a game. For a time the author of Technos has successfully presided over a functioning community which has to its credit an electronic construction of impressive proportions. The truth of this is not altered by the fact that the people of Cyberville are fictions or that the basis of the construction is digital information held on servers rather than bricks and mortar existing in physical space. Asked the question: "What have you ever achieved?" the author of Technos can truthfully point to a community and an architectural construction which really existed—*really* existed in cyberspace. This reveals a difference with the earlier virtual reality examples. There the person who, thanks to the VR Body Zone, had the experience of encountering a tiger can say "I know what it is like to encounter a tiger" but cannot say "I have encountered a tiger." By contrast, the author of Technos can say *both* "I know what it is like to preside over a community" *and* "I have presided over a community." Moreover, she can say the former precisely because she can truly say the latter.

I have chosen this particular example with care because it is not clear that the same point could be made about all relationships established within a virtual community. Imagine another invented persona: Webman. Let us suppose that, on some level of description or other, Webman and Technos fall in love, marry, and set up a new home in Cyberville. On the strength of this it seems plausible to think that the author of Technos could claim that she knew what it was like to be in love, but not that she had been in love. From this it follows that only *some* Internet relationships and activities can count as real achievements as opposed to simulated achievements like those we experience in the VR Body Zone, and in turn this implies that the distinctive reality which is to be found in virtual communities may be importantly limited.

Still, the general point is this. Virtual reality—interpreted not as simulation but as the sort of world realized in the virtual community just described—can properly be conceived of as a distinctive mode of existence, a mode that is not just a game, but a world of its own in which a significant if limited range of things can be accomplished and lost. If we now add to this minimal claim the reminder that the Internet is at a very early stage of development, there is reason to think the future of cyberspace will bring metaphysical novelties—that virtual reality interpreted via the virtual community is to some extent a new world and one that we are on the edge of.

To make this notion of a metaphysical novelty clearer it may be useful to consider what has been called the "cyberstore"—an Internet site where a wide variety of goods can be inspected and ordered. According to Neil Barrett, the cyberstore

> can be represented as a virtual reality implementation of the largest, flagship megastore. Consumers can . . . move through the cyberstore using a mouse or keyboard to direct their search. Shelves full of goods can be displayed and the shopper can select and sample goods—perhaps simply by "clicking" on the image.[4]

Internet shopping is now fairly commonplace and growing rapidly. The Internet bookstore Amazon.com has reported annual increases in sales of over 480 percent. Much of it, however, is simply a matter of doing by other means what we have done for centuries. We order a book, say, but the book is dispatched from an ordinary warehouse by ordinary means. More interesting are the cyberstores which sell only electronic goods—texts, films, music—that are downloaded direct to the purchaser's PC and paid for by a credit-card system that is wholly electronic, with earnings attributed to the earner through BCTS (the computerized Bank Credit Transfer System). Such a cyberstore operates wholly within the world of the Internet and the point at which it "touches" the ordinary world is only at the point of consumption and consumer satisfaction. What it sells, however, is substantial enough. Though neither the goods, nor their transfer, nor the financial transaction has any ordinary physical embodiment, these are *real* goods and sales and transfers—virtually real perhaps but real nonetheless. The point is, they are of a kind that could not have existed before the advent of the Internet and the World Wide Web.

THE POVERTY OF CYBERSPACE

Virtual reality interpreted in this way, then, is a kind of reality and not merely a copy or simulation of something else. There remains this all-important issue. Even if there is indeed a new world coming into view, not just in a metaphorical but in a metaphysical sense, have we any special reason to welcome it? The answer to this question turns, as it seems to me, on what we can say about the value of this new medium relative to other media that it might replace. Once more, in order to simplify the discussion, I shall make a stipulation and from here on use the expression "virtual reality" to mean the medium which the examples of Technos, Cyberville, and the virtual megastore have isolated, ignoring any special connection it normally has with VR Body Zones and the like.

Why would we favor community life in virtual reality rather than community life on the streets, so to speak? One possible explanation lies with the account Walton gives of make-believe. His claim, it will be recalled, is that "there is a price to pay in real life when the bad guys win, even if we learn from experience. Make-believe provides the experience . . . for free." It is not hard to see how this line of thought applies to Cyberville. "For free" is not quite right, however. While there is clearly a price to pay when real communities break up in dissension and strife, there is also a price to pay for the breakup of a virtual community. But it is not the same and certainly not as high. No actual bones are broken, no blood spilt, no buildings or businesses destroyed. What is destroyed is the virtual community itself and the virtually real accomplishments which have resulted from its existence. If we take Rheingold's definition seriously, the building of communities in virtual reality includes "sufficient human feeling" and this feeling will be dissipated and frustrated. It is not itself "virtual" feeling, but the emotion invested by the authors of Technos, Webman, and all the other personae. This is a real cost. Still, it is a cost that falls short of the losses sustained by those whose ordinary homes and communities are the casualties of civil strife.

One thought which this prompts is that, given a choice, human beings would risk less by engaging in virtual relationships than in ordinary ones. Might this be the attraction of the world of virtual reality? The answer depends on a trade-off between the diminished risk and the more limited character of such engagements. In advance of knowing the future character of cyberspace, this is a calculation that is hard to make. Nevertheless, there seems reason to suppose that the limitations of cyberspace will always be greater than those which operate on corresponding relations in ordinary life and consequently that the reduced risks will not outweigh the loss of possible benefits. Anyone who has followed the exploration of this chapter, in fact, could hardly conclude anything but that virtual communities are relatively poor substitutes for real ones. Their poverty does not lie in their being mere simulacra, as is the case for VR Body Zone experiences, which as we saw lack the crucial element of reality. But on present reckoning, and as far into the future as one can reasonably see, they are impoverished nonetheless. The possibilities of virtual reality may well bring added benefits, but without the context of ordinary life, to live one's life primarily on the Internet would be a poor way to be.

NOTES

1. Michael L. Dertouzos, *What Will Be: How the New World of Information Will Change Our Lives* (London: Piatkus, 1997), pp. 64–65.

2. Kendall Walton, *Mimesis as Make-Believe* (Cambridge: Harvard University Press, 1990), p. 68.

3. Howard Rheingold, *The Virtual Community: Homesteading on the Electronic Frontier* (Reading, Mass.: Addison-Wesley, 1993), p. 5.

4. Neil Barrett, *The State of the Cybernation* (London: Kogan Page, 1996), p. 112.

Technologies of the Individual
Women and Subjectivity in the Age of Information

Suzanne K. Damarin

> When we can no longer declare what is True in the registers of morality, cos-
> mology, or politics, the spaces evacuated by such Truths do not remain empty
> but, to the contrary, grow crowded with technical truths—instrumentalist dis-
> course dangerously cut loose from regulating values and substantive, account-
> able aims.
> —Wendy Brown, "Feminist Hesitations, Postmodern Exposures," 1991

By technologies of the individual, I mean those technologies which disrupt beliefs about the nature of the human individual, supplant functions or activities ordinarily or historically performed by the individual, extend (and/or delimit) individual capability, and/or change the relationship of the individual to the larger society. Today the computer can be seen as typifying these effects and, in many senses, as the archetypal technology of the individual.

To some extent, virtually all technologies can be seen to have one or more of the effects listed above; however, the technologies of the individual not only have all of these but have them as direct and intentional effects. In Don Ihde's phe-nomenology, these technologies are among those identified as "technics embodied" and are characterized by "the symbiosis of artefact and user within a human action."[1] As Ihde notes, the perfection of such technologies is not related to the machine alone but to the integration of machine and person; a deep con-tradiction is entailed in the perfection of these technologies.

> The desire is, at best, contradictory. I want the transformation that the tech-
> nology allows, but I want it in a way that I am basically unaware of its presence.
> I want it in such a way that it becomes me. Such a desire both secretly rejects
> what technologies are and overlooks the transformational effects which are nec-

Suzanne K. Damarin, "Technologies of the Individual: Women and Subjectivity in the Age of Infor-mation," *Research in Philosophy and Technology* 13 (1993): 183–98. Copyright © 1997 Suzanne K. Damarin. Reprinted with permission of the author.

essarily tied to human technology relations. This illusory desire belongs equally
to pro- and anti-technology interpretations of technology.[2]

Ihde's first illustrative example of "technics embodied" is eyeglasses. A little
reflection reveals that eyeglasses participate in this contradiction; they work best
when the wearer is not conscious of their presence. Moreover, in some measure
at least, glasses have the characteristics listed above as definitive of "technologies of the individual." For example, they deny the effects of aging on eyesight
by supplanting the focal lengths of human eyes and thus increasing the focal
capacities of the wearer; these changes support reading, fine detail resolution,
and other visual activity which affect the relationship of the person to the society.

The primary argument of this [chapter] is that technologies of the individual
have gender-specific effects; that is, these technologies are created, implemented,
used and abused, valued and devalued in ways that are not gender neutral but,
rather, define and affect the sexes differentially. In particular, it will be argued
that technologies of the individual enter the lives of women in ways that challenge their subjectivity and change their relations to the world in gender-specific
ways. A second argument follows on the first and addresses Ihde's "contradiction" as stated above by moving it from the private domain to the public and
political domain. In this larger domain, patriarchy, the desired "transformation(s)
that the technology allows" are achieved by the mainstream/male-stream which
"rejects . . . and overlooks the transformational effects which are necessarily tied
to human technology relations" by displacing the latter effects from itself and
assigning them to women and other marginalized groups.

For women, technologies of the individual cannot be transparent because
their creation, forms, interpretations, and uses encode and reproduce sexist biases
that go well beyond the explicit or stated intent of the technics, and that affect all
dimensions of women's lives. The nontransparency of Ihde's eyeglasses, for
example, is revealed in the well-known words of Dorothy Parker, "Men seldom
make passes at girls who wear glasses." Clearly, this nontransparency affects the
lives of the sexes differentially; for Parker's men, the glasses *of others* simply
add to the criteria for selecting women for sexual (in)attention; but for women
their own glasses mark their membership among the unchosen and unchoosable.

While eyeglasses, in their technological and intentional simplicity, can provide an instructive example, they can also mislead us as to the multiplicity and
complexity of intentions, interrelations, and effects of current state-of-the-art
technologies of the individual. These technologies include the "cyborg technologies" of artificial intelligence, knowledge engineering, reproductive technology,
genetic engineering, and various other technologies of the body such as the multitude of organ transplant technologies, hormone replacement therapies, and cos-
metic engineering among others. As discussed and theorized by Donna Haraway
in *A Manifesto for Cyborgs*, these technologies come together in our postmodern
era to define new networks of relation among conditions, persons, artifacts,

assumptions, theories, and politics; together they determine what Haraway calls an informatics of domination.[3] In the following discussion, I shall take as a central focus the technological discourses of computers and artificial intelligence. Examining the directions of their development, often in conjunction with other technologies, I explore their effects both on the place(s) of women within the patriarchy and on women's subjectivity.

COMPUTER TECHNOLOGIES AND TECHNOVISIONS

The relations between computing machines and human beings have been addressed by philosophers, social scientists, historians, and writers of fiction. While some of these writers posit the computer as a condition, context, or environment of human activity,[4] others see the computer as an actor situated in the same plane of activity as the person. For the latter group, questions of comparison and contrast, distance, and interaction between computers and persons are of philosophical and theoretical importance. The literatures of relation between human and machine actors are too vast for review here, but several points are of general importance to the subsequent discussion.

The conception of the relation between persons and machines is asymmetrical in that each comparison discusses one of these in terms of the other, with the latter considered to be primary. In recent years, there has been apparent a shift in the direction of this asymmetry. In Alan Turing's 1950 discussion, for example, a machine is argued to be intelligent if, when given the Turing test, it proves to be indistinguishable from a person.[5] Thus, for Turing, human intelligence is axiomatic to the definition of machine intelligence. Similarly, the positioning of persons and human attributes as primary in the discussion is apparent in the texts, and even in the titles of many philosophers' writings on the subject: Justin Lieber's *Can Animals and Machines Be Persons?* Hubert L. Dreyfus's *What Computers Can't Do*, Pamela McCorduck's *Machines Who Think*, and Stuart E. and H. L. Dreyfus's *Mind over Machine*.[6] The central position of the person in framing discussions of computer capability is not universal, however. In the work of some social scientists, there is evident a shift in direction. Following Allen Newell and Herbert S. Simon's 1972 analysis of *Human Problem Solving*[7] using the terms and framework of (then current) machine information processing, numerous cognitive scientists describe human behavior in machine terms, and philosophers use computer programming concepts to describe human thought: Jerry Fodor's *Modularity of Mind* and John Haugeland's *Mind Design*.[8] Sociologist Sherry Turkle argues that the computer provides a "sustaining myth" through which we can understand ourselves and view ourselves as machines.[9] In the visions discussed below, the ambivalence, at every level, as to the primacy of machine versus person in the conceptualization and actualization of relations between person and machine is apparent.

A second general trend in the literature is a decrease in distance between person and machine. With the proliferation of silicon chips, computer hardware is evident in all parts of the physical and social environment, and with the proliferation of computer programs, computer software intrudes into all aspects of our lives as social and intellective beings. Supercomputers, parallel processing, advanced computer graphics techniques, hypertext, and cyberspace decrease the visible and/or superficially detectable differences between computer simulations and the phenomena (such as scientific experiments or conversations) they simulate. The psychological and social distance of persons from computers is, thus, decreasing, and with these the philosophical distance closes as well (at least for some philosophers). In the following paragraphs, four visions of the relationship between computers and persons are discussed briefly in relation to their effects on women. These visions are considered in descending order of the assumed distance between computer and person.

THE MASTER/SLAVE VISION:
MAN AND MACHINE IN SEPARATE SPHERES

Many of the social myths surrounding computers have as their function the preservation of the idea of separateness of persons and machines. Schoolchildren are taught that computers can do only what they are programmed to do and that the programmer (or user) is not only distinct from the machine but in control of it. The myth of the household robot, assigned primarily to tasks that are currently assigned to women, conveys the idea of a robot that is not only separate from its owner but totally obedient to him (her?). Moreover, this robot is not necessarily thought of as very intelligent in order to do the mundane tasks of cleaning, ironing, and feeding the dog; the mythical household robot does not preempt the rights and responsibilities of male family members. Created in the image of R2D2, robots envisioned in households, schools, and other social settings are anthropomorphized as cute, useful, and tractable.

The vision of the machine as "slave" to a human "master" is evident in many robot narratives, both scientific and fictional. The original robotic theater piece, Karel Čapek's *R. U. R.*, depicted slave robots as gendered but without sex; that is, whereas Rossum's Universal Robots were designed with neither sexual reproduction nor sexual pleasure, the robots assigned to secretarial and other "feminine" work were created with superficial similarity to their gendered human counterparts. In Čapek's rewriting of the Adam and Eve myth, the introduction (at the insistence of a human woman) of female sexuality among the robots spells the beginning of chaos and the overthrow of the male masters by the robot slaves.[10]

In current narratives, robots are without explicit sexual or gender identity and without anthropomorphism at the level of appearance; they are simply "other" to their male creators. The descriptors assigned to robotic activities and "natures" are derived primarily from the domain traditionally preserved for the feminine. In particular, machines are argued to be valuable when they are (user) friendly, patient, attentive to detail, able to work long hours, uncomplaining, and dependent. Robots are never described as macho, aggressive, or virile; indeed Isaac Asimov's three rules of robot behavior proscribe their aggressive potential.[11] In these senses the construction of robot as other to man mimics and thus displaces the construction of woman as other to man.

This ascription to robots and other machines of characteristics traditionally assigned to the female displaces women at many levels. In many instances the feminization of work is accomplished through the disassembly of tasks into components requiring rational decision making, routine operation, and physical strength. While retaining the decision making for the male, the remaining components are assigned to "others," that is, to the woman and the robot, respectively.

With regard to cognition, different otherings apply in the patriarchal move to claim powers not available to computers. With the development of artificially intelligent computers, the world is increasingly posited as a series of problems for solution.[12] Many mathematical procedures have been mechanized, and their performance no longer serves to "validate" superior male accomplishment. Correspondingly, these procedures have given way to "higher order thinking" and "problem solving ability" as the domain for measuring sex differences in achievement. Reports on mathematics tests generally indicate that whereas females are superior on "lower-level computational skills," males excel at "higher-order problem solving skills." These changes in psychometric literature of sex differences are consistent with shifts in the discourse of gender and brain lateralization; here the characterization of the right brain (male) has changed from "spatial and mathematical" to "spatial and intuitive" while descriptions of the left brain (female) have changed from "verbal" to "verbal and symbolic." Thus, as the artificially intelligent computer is developed as ever more capable at symbol manipulation, this manipulation becomes other to "humans" and is shifted from male to female descriptors; at the same time, "intuition," previously maligned as feminine, becomes valuable (as a problem-solving heuristic) and is adopted as a male descriptor. In these and related ways, "gendered human characteristics" are resorted so as to assure that both machines and women are other to men.

As a final example, in the evolutionary mythology of educational computing, the computer emerges as nurturing and mothering, displacing the female teacher as a guide and support for students.[13] As I discuss at some length in another article, the computer system delivering instruction is posited as providing direction, praise, motivation, and rewards for the student, while the teacher (and, for that matter, the mother) is reduced to functioning as a monitor and a "scold."[14] Here, the computer is slave to the system of educational objectives, to behaviorist theories of learning, and to the child who operates it; the woman is irrelevant to these operations.

In sum, the master/slave vision of the evolution of computers and their uses constructs the computer as other to male by assigning it to tasks which were previously the domain of the socially constructed woman. In the myths of this vision, the woman is simply displaced; neither man nor machine, she becomes irrelevant.

THE DYADIC VISION:
HUMAN AND MACHINE IN SYMBIOSIS

In this vision of the high-tech future, increasingly powerful computers are used by already powerful individuals and institutions; in theory, the computer and the computer-user each does "what it does best," as they work together in a powerful symbiotic dyad. Of the arrangements discussed here, this is the most commonly implemented in business today. It is also highly romanticized by those who argue that the computer will liberate workers from dull and mundane tasks and by those educators who argue that the classroom computer can be used to provide disempowered children with a more liberating education. These romantic visions are precluded from becoming reality by the unequal access of individuals both to the power *of* computer use and to power *over* computer use.

The effects of computer implementation in offices, in manufacture, in businesses ranging from McDonald's to brokerage houses, and in social service agencies have been documented by a number of social scientists.[15] In all cases, the functions of the majority of workers are changed to accommodate the efficiency of computer operations; work becomes increasingly fragmented, meaningless, and repetitive, and a large number of jobs, especially women's jobs, become nothing more than the preparation and entry of data for computer manipulation. In the hierarchy of United States jobs, black women are found most frequently in the lowest paid and most tedious.

Increasingly, the most routine, boring, and health-threatening jobs that support the growth and use of computer technology are assigned to women of color. Whether these jobs are exported to Third World countries or performed by Chicana women lured to the Silicon Valley with promises of high-tech employment, they exert control over the lives and subjectivities of the women workers. The gender and racial logics with which management controls and exploits these women are based on the fragmentation of their identities in order to deal with them separately as ethnics, as women, or as workers, whichever allows greater control in the particular situation.[16] As Aihwa Ong writes:

> Management seeks to control the women's self-perception by talking about the "natural" ability of "oriental" women's fingers, eyes, and passivity to withstand this low-skilled, mind-deadening work. The reduction of the social person to an organism subordinated to technological instrumentality is no mere mystifica-

tion. It constitutes everyday reality for the perception and treatment of workers.
. . . Women's fingers and eyes coded as extensions of electronic instrumentalities and women's capabilities and subjectivities reduced to pure sexuality.[17]

In some conceptualizations of computers, the fully operational machine consists of hardware, software, wetware, and orgware, the last being the (technology-based) organizational system which schedules work, monitors progress, and oversees the operation. In their relation to the machine, women in the positions described above are defined as "wetware," biological material essential to the operation of the hardware and software. In this framework, the women workers are seen, not as complementary to, but as an integral part of the machine.

An analogous phenomenon occurs with "higher-level" employees in fields such as sales, consultation of various sorts, and social service who are increasingly "assisted" in their work by computer-based artificially intelligent expert systems. Knowledge which was once a personal asset of the worker is encoded in a data base in the machine. The worker, no longer permitted to speak or act from her or his own expertise, becomes an "input/output device" comparable to a digitizing camera, a printer, or other "computer peripheral."[18] The design of the human-machine interface deliberately limits the scope of interaction between the computer and its human user.[19] Menu-driven software puts the computer in control of communication, forcing the articulate and imaginative human user to limit communication and restructure inquiry to accommodate the multiple choice directives of the computer. As many jobs that previously required human expertise are limited to serving as a computer interface, these jobs are redefined and made less lucrative. New classifications of employment as "data workers" are identified, and these jobs become feminized, that is, low paid "women's work."

Thus, as the computer-human dyad approach to the workplace integration of humans and computers matures, increasingly women are absorbed into computer systems; race and class distinctions are preserved as women of color and lower-class women become "wetware" while middle-class white women become more highly paid "peripherals." In both cases, electronic supervision of workers, increasing pressure for productivity, and poor working conditions prevail. In both cases, the women who fill these jobs are reduced to machine components whose claim upon an essential personhood is reduced to their sexuality; that is, their otherness to the electronic machine lies in their sexual functioning. Theirs is a post-gender world in that they are constructed not as gendered social beings but as machine parts which function mechanically.

Reproductive technology extends the construction of the female-as-machine well beyond the workplace. The construction of woman as part of the reproductive machine denies the wisdom and intuition of mothers (and pregnant women), replacing it with computer printouts, sonograms, and other technowise directives that monitor, that is to say control, the activities of the woman and the development of the child. Here, as in the workplace, the woman becomes part of the artificially intelligent system. Absorbed into the machine for the purposes of both productive and reproductive labor, the only specifically human identity allowed to the woman becomes her sexuality; her otherness to the machine lies in her eroticism, her sexual appeal, and vulnerability. The catch phrase *high tech–high touch*, which was common among the microcomputer avant garde of the late 1970s, takes on new meaning and becomes synonymous with the term *pornotechnics*, which identifies "the prostitution of technology, paralleling the term pornography for the prostitution of art."[20]

In summary, the implementation of computers in the mode of the human-computer dyad leads to the absorption of the machine user into the architecture and the control of the machine. As machines become more powerful, the level of intelligence demanded *and allowed* from the human user is decreased. As a result, employment in machine-related jobs becomes restricted to women, people of color, and other workers who have been marginalized in the job markets of the past.[21] Although this may create employability problems for those currently more privileged, for the women (as well as persons of color and all but the elite) who are dependent upon these jobs, the situation requires a reduction of individual subjectivity. As a machine part, the person is less than the machine; if she is to be more than the machine, it must be through her erotic potential.

THE CYBORG VISION: HUMANS INVADED BY THE MACHINE

The cyborg woman is not absorbed into the machine; rather, in the cyborg vision the machine intrudes into the woman. Her biological organs are replaced or enhanced with silicon implants, permanent contact lenses, artificial kidneys and lungs, electronic pacemakers, and a virtual hardware store of plug-in electronic parts. Not only are her reproductive functions increasingly monitored and enhanced through technology, but her every aspect is replaced by its technological equivalent. Her voice becomes her digitized voiceprint and her essential uniqueness becomes her human genome. Donna Haraway has rendered vividly a vision of the cyborg and the postmodern world that she inhabits.[22]

Cyborg activity is coded by command-control-communication-intelligence, the "C^3I," which is the raison d'être of the invisible microelectronic computer that invades the body and captures the mind. The continuing evolution of the biologically perfect, but imperfectly biological cyborg body demands increasing computer power and resources. The cyborg needs the $3 billion Human Genome Project (HGP), which will map every gene in the biological body, primarily for the purposes of conquering disease; but more broadly to control the individual body, ridding it of "genetic abnormalities"; in order to conduct HGP, the director needs new methods of automation, that is, new computers. In this project, as in others, computer technology and biotechnology are intimately and circularly related, each both a cause and an effect of the other. In the cyborg world the computer and the body become indistinguishable and their discourses merge. Today's computer talk refers to mother(boards), mice, apples, clones, and environments, the stuff of old-fashioned biology and newfangled biosystems; and the discourse of computer viruses is a direct analogue to the discourse on AIDS.[23] On the other hand, biology is becoming an information science, with talk of systems, feedback, loops, codes, programs, and the other stuff of computer science. In the cyborg world, the distinction between biology and information loses its meaning and our biological selves become the images of the information coded in our DNA.

In Haraway's vision, the cyborg world is a "world without gender."[24] But, at least in its birth, the world of the cyborg looks (to me) like one in which the traditional patriarchal social construction of gender gains a new lease on the future as computers (and related technologies) are used to help women become more "feminine." The CLINIQUE computer is "more personal" as it provides direction for personal skin care; it "fine tunes your treatment and makeup to your skin's needs."[25] Revlon and other cosmetics concerns offer computer analyses of our ideal colors so we might purchase the most attractive lipstick, mascara, and blush. Computer imaging in the beauty salon is advertised as taking the risk out of haircuts and much more. Larry Blumsack's New Image Center in Natick, Massachusetts, for example, "is proof positive that you can try it before you buy it and that includes hairstyles, colors, streaks, tips, as well as all manner of makeup to create greater beauty. . . . Blumsack insists the service is not limited to women."[26] Computer imaging is not limited in its application to the superficial and transient matters of cosmetics and hairdos, but is also used in the sale of new noses, facelifts, and other plastic surgery available to the imperfectly beautiful woman. Nor has the weight loss industry ignored the power of computers to project new and improved body images, to measure specific densities of fatty tissues, and to plan a dietary regimen that will ensure for women an ideal figure by the desired date. Clearly the computer has found many uses in the business of helping women to reconstruct themselves as desirable sex objects, and as excellent specimens for this "gender role."

Sex itself decreases in importance as the cyborg world approaches. As Haraway observes, sex (in the sense of sexual reproduction) gives way to genetic engineering;[27] numerous feminists are currently studying the phenomena and effects of the reproductive technologies which presage the separation of sexual identification from reproductive role. At the same time, the path to the cyborg world leads through "the transsexual empire,"[28] a world in which (biological) male or female sex is optional, dependent upon the preferred gender role of the cyborg-person. Essential to transsexual discourse, the very idea of "a woman trapped in a man's body" is dependent upon the notion of a constructed gendered woman who is somehow independent of the body "she" inhabits. The development of the technology to liberate this gendered woman from the sexed male furthers the reification (and not the demise) of gender as a basic social component or category. Reports from male-to-constructed-female transsexuals indicate that their postoperative pleasures frequently reflect their enjoyment of precisely those attitudes toward women which feminists of the 1970s found most symbolic of the social construction of woman as both incompetent and glorified.[29] Thus, at least insofar as current research can inform us, in the postsex world of the transsexual, valorization of traditional patriarchal gendered female subjectivities will increase with the number of male-to-constructed-female transsexuals.

In short, in the postmodern cyborg vision, sex is neither permanent nor consequential with respect to reproduction. If cyberpunk science fiction is any guide, sexual pleasure will be replaced by drug-induced merging of individual bodies and will be enjoyed, not only by couples of the same or opposite sex, but also by larger groups.[30] The cyborg world is clearly postheterosexual and, indeed, postsex. However, the cyborg does not seem to be constructing itself as postgender; instead, with the support of computers traditional gender identities and practices would appear to be defining the subjectivities of individual cyborgs. One might say that the cyborg world is postfeminist; but it is perhaps more appropriate to find the cyborg a return to the prefeminist era.

THE VISION OF SUPERIOR MAN:
HUMANS REPLACED BY MACHINES

Some scientists not only believe that we are on the brink of a postbiological era but also work actively toward the attainment of that condition. Hans Morawec, director of the Mobile Robot Laboratory at Carnegie Mellon University and a leading researcher in the field of robotics, writes:

> Our biological genes and the flesh and blood bodies they build, will play a rapidly diminishing role. . . . In the present conditions we are uncomfortable halfbreeds, part biology, part culture, with many of our biological traits out of step with the inventions of our minds. . . . It is easy to imagine thought freed from bondage to a mortal body.[31]

Tom Stonier, of the University of Bradford (United Kingdom), goes further in his statement, which not only summarizes his own view but is consonant with the views of many roboticists and evolutionists.

> Machine intelligence will outpace human intelligence and very likely will do so in the lifetime of our children. The mix of advanced machine intelligence with human individual and communal intelligence will create an evolutionary discontinuity as profound as the origin of life. It will presage the end of the human species as we know it. The question, in the author's view, is not whether this will happen, but when, and what should be our response.[32]

The justifications for this stance are numerous and are related to aspects of patriarchal thought that have been addressed by feminists in the last decade. Pure (masculine) scientific rationality is seen as the essence of "mankind." Therefore, the development of entities which are pure mind, unhindered by biology, is seen not only as desirable but also as necessary to scientific progress, which entails (for example) interstellar and intergalactic travel, an endeavor that would clearly be easier without messy, demanding biological bodies to worry about. Thus, the postmodern world of these roboticists is one in which mind-body dualism is eliminated by the elimination of the body. The electronic mind housed in microelectronic chips is argued to be superior to the biological thinking person in two important regards: (1) as opposed to human brains, which are seen as presently operating at or close to capacity, the robot has theoretically unlimited potential for (rational deductive) thought; and (2) the robot is superior to humans with respect to *reproduction*.[33]

From a feminist perspective, the latter argument positions the work of these roboticists in the continuing series of practices, institutions, and scientific developments, that Mary O'Brien has identified as male efforts to control reproduction.[34] In O'Brien's analysis the biological connection of males to the next generation is not material but abstract; rules governing the behavior of women were designed not so much to control women as to concretize and certify the link of the male (father) to its progeny. Likewise, research on biological reproduction was designed with the intent of providing "concrete" evidence (generally, statistical evidence) of the male fathering of other humans. Reproductive technologies, as currently described and analyzed by many feminists, extend and implement this technological desire, as does the technology of cloning.

The creation and manufacture of robots complete with programs for their own replication can be seen as complementing, continuing, and completing this effort. O'Brien's research is supportive of the argument made by Sally Hacker that the purpose of patriarchy is not to control women but to control other men;[35] women are regulated and controlled, not as an end of patriarchy, but as a means of maintaining patriarchal systems, including systems of reproduction. With the ascendancy of robot-man in control of nonsexual robot reproduction, there is no longer any use for women in the patriarchal system.

Given these analyses, the world as envisioned by roboticists must be seen as postbiological, postgender, and (most assuredly) postfeminist, but *not* as postpatriarchal. The exploitation and control of all that is other to the masculine rationality encoded on silicon chips constitute the primary goal of robotic development.

READING ACROSS THE VISIONS

As each of the visions sketched above struggles to become realized through technological rationality, development, and implementation, certain commonalities among them become apparent. Each vision is posited on an implicit or explicit view of the nature of male man as generic Man; appropriately to the postmodern era, the nature of woman is not initially theorized or essentialized. In the absence of an idea or regulating myth concerning the nature of women, however, there is no boundary condition to shape or to stem the intrusion of the actualization of these robotic visions into the day-to-day lives of women. As described above, the rendering of these visions in the "here and now" positions women as displaced by, and even irrelevant to, the technological development. Each of these visionary myths participates in a social reconstruction of both the individual woman and the collective of women; each, in its own way, proscribes the productive and repro-ductive labor of women. In different ways, the master/slave and superior man visions deny women; they proceed without reference to women, ignoring the his-torical contribution of women to the performance of functions soon to be the domain of the machine. Both the dyadic and cyborg visions, on the other hand, confound woman with the machine by essentializing her nature: her fingers and "passivity" in the dyadic vision, and her sexual objectification in both.

Thus, these visions pose serious and ironic questions for feminists in the postmodern world. The continuing emergence of the technologies of the indi-vidual, and particularly of computer technologies, is a defining condition of the postmodern, which, in turn, denies the possibility and/or theoretical and practical utility of essentialized subjectivities. For feminism this means the end of the "woman as subject," of the "feminist standpoint," and of the Truth of radical fem-inism as central to political strategy. The challenge for feminists in the post-modern world is to dare to construct a politics of resistance which is not based in Truths of the personal as political and in acts of recrimination. In the words of Wendy Brown:

> What postmodernity disperses and postmodern feminist politics requires are cultivated political *spaces* for posing and arguing about feminist political norms, for discussing the nature of "the good" for women.[36]

In this context, when Haraway writes: "I would rather be a cyborg than a god-dess,"[37] I read her to mean that "'the good' for women" must be found in the shaping of the evolving here and now of these technological visions, and not in some ethereal and static goddess world inscribed by a discourse of essential womanhood.

The problem that arises is how to recognize and know "'the good' for women" in a multiplicity of visions which threaten her extinction through various strategies: rendering her extraneous, absorbing her into the machine, reducing her to a sexually objectified cyborg shell, or simply acting as if she did not exist. It seems to me that this knowledge must arise from a new postmodern subjectivity which functions as both information and simulation and thereby reproduces/repli-cates the reality/realities of women. Borrowing terms from Haraway's chart of transitions from hierarchical to informatic domination, we can begin to characterize this new subjectivity; in particular, we consider these pairs:[38]

Representation	Simulation
Organism	Biotic component
Bourgeois novel, modernism	Science fiction, postmodernism
Depth, integrity	Surface, boundary
Perfection	Optimization

The female subjectivity of modernism purports to *represent* all women in a single realistic narrative; that "story" plumbs the depths of women's historical experience to construct "woman" as an organic unity in an organic system. The integrity of the story (and of the subjective woman) is related to the coherence, completeness, and perfection of this subjectivity as description and prescription for all women. As postmodern feminists point out, this subjectivity fails in several regards.

In contrast to the static, photographic, representational subjectivity of modernist feminism, I propose that in order to know about "'the good' for women" in the postmodern, women must come to know *ourselves* in the postmodern, that is, to have a *subjectivity which is postmodern*. This subjectivity cannot represent but simulates woman/women/womyn as ever changing, always becoming, responding to and/or resisting the inputs, outputs, and programs of domination. Like television with its instant replays, and computer simulations with their relational data bases, this subjectivity removes experience from the constraints of the linearity of time, the three dimensionailty of space, and the materiality of "legitimate" experience. The simulation as subjectivity is never complete, and neither true nor false nor falsifiable, but invites continual revisits and revisions; its ever changing boundaries and surfaces suggest new ways of construing and constructing femininst theory. In particular, they provide opportunity for interpretation(s) of the changing functions and positions of the minds and bodies of women in response to the inputs of postmodern reproductive and artificial intelligence technologies. "Woman's intuition" becomes a legitimate way of knowing, for example, and Luce Irigaray's questions concerning the sex of science can be turned to new modes of meaning for the bodies of women.[39]

A simulation as subjectivity provides opportunity both for multiple and conflicting representations of women and for "*spaces* for posing and arguing about feminist political norms."[40] Each of the visions described above can be seen as a "run" of the simulation under various conditions of input. These runs make visible boundaries and oppositions between the patriarchal inputs and "'the good" for women." Thus, these runs can specify sites for resistance both to the patriarchal visions and to the technical rationality that converts these visions to experience. In the absence of a postmodern subjectivity it is difficult to see how these sites for resistance would emerge. Without some sort of subjectivity, postmodern feminism might have an effect, but this effect would be without cause, that is, without women.

The real question(s) for feminists in the postmodern is not whether there *can* be feminism, but whether there *will* be women.

NOTES

Substantial parts of this article are revised from "Robots as Patriarchy: The Discourse of Artificial Intelligence," a paper presented at the meeting of the National Women's Studies Association, June 1990, Akron, Ohio.

1. Don Ihde, *Technology and the Life World: From Garden to Earth* (Bloomington: Indiana University Press, 1990), p. 73.

2. Ibid., p. 75.

3. Donna Haraway, "A Manifesto for Cyborgs: Science, Technology, and Socialist Feminism in the 1980s," in *Feminism/Postmodernism*, ed. Linda J. Nicholson (New York: Routledge, 1990), pp. 190–233 (reprinted from Socialist Review IS, no. 80 [1985]: 65–107).

4. Jean-François Lyotard, *The Postmodern Condition: A Report on Knowledge* (Minneapolis: University of Minnesota Press, 1984); Mark Poster, *The Mode of Information* (Chicago: University of Chicago Press, 1990); Seymour Papert, *Mindstorms* (New York: Basic Books, 1980).

5. Alan Turing, "Computing Machinery and Intelligence," *Mind* 59 (October 1950): 433–60; see also Joan Rothschild, *Teaching Technology from a Feminist Perspective* (New York: Pergamon, 1988), especially chap. 6.

6. Justin Lieber, *Can Animals and Machines Be Persons?* (Indianapolis: Hackett, 1985); Hubert L. Dreyfus, *What Computers Can't Do: The Limits of Artificial Intelligence* (New York: Basic Books, 1979); Hubert L. Dreyfus and Stuart E. Dreyfus, *Mind over Machine* (New York: Free Press, 1986); Pamela McCorduck, *Machines Who Think* (San Francisco: Freeman, 1979).

7. Allen Newell and Herbert S. Simon, *Human Problem Solving* (Englewood Cliffs, N.J.: Prentice Hall, 1972).

8. John Haugeland, ed., *Mind Design* (Cambridge: MIT Press, 1981); Jerry A. Fodor, *Modularity of Mind* (Cambridge: MIT Press, 1983).

9. Sherry Turkle, *The Second Self: Computers and the Human Spirit* (New York: Simon and Schuster, 1984). This theme is articulated more fully in Turkle, "Artificial Intelligence and Psychoanalysis: A New Alliance," in *The Artificial Intelligence Debate: False Starts, Real Foundations*, ed. Stephen Graubard (Cambridge: MIT Press, 1988), pp. 251–68.

10. Karel Čapek. *R. U. R. [Rossum's Universal Robots]* (London: Oxford University Press, 1928). Translated from the Czech by P. Selver and adapted for the English stage by Nigel Playfair, play in three acts and an epilogue.

11. Isaac Asimov, *I, Robot* (1941; reprint, New York: Fawcett, 1970)

12. Joseph Weizenbaum, "Limits in the Use of Computer Technology: Need for a Man-centered Science," in *Toward the Recovery of Wholeness: Knowledge, Education, and Human Values*, ed. Douglas Sloan (New York: Teachers College Press, 1984), pp. 149–58.

13. J. Aaron Hoko, "SIS: A Futuristic Look at How Computers and Classrooms Can Enhance Rather than Diminish Teachers' Pedagogical Power," *Computers in the Schools* 6, nos. 1 , 2 (1989): 135–43.

14. Suzanne K. Damarin, "Computers, Education, and Some Issues of Gender," *Journal of Thought* 25, nos. 1, 2 (1990): 81–98.

15. Cynthia Cockburn, *Machinery of Dominance: Women, Men, and Technical Know-How* (Boston: Northeastern University Press, 1988); Barbara Garson, *The Electronic Sweatshop: How Computers Are Transforming the Office of the Future into the Factory of the Past* (New York: Simon and Schuster, 1988); Robert Howard, *Brave New Workplace* (New York: Viking, 1985); Shoshanna Zuboff, *In the Age of the Smart Machine: The Future of Work and Power* (New York: Basic Books, 1988).

16. Karen J. Hossfield, "'Their Logic against Them': Contradictions in Sex, Race, and Class in the Silicon Valley," in *Women Workers and Global Restructuring*, ed. Kathryn Ward (Ithaca, N.Y.: ILR Press, 1990), pp. 149–78; Aihwa Ong, "Disassembling Gender in the Electronics Age," *Feminist Studies* 13, no. 3 (fall 1987): 609–26.

17. Ong, "Disassembling Gender," pp. 622–23.

18. See Garson, *Electronic Sweatshop*, p. 19.

19. Gerhard Fischer and Andreas C. Lemke, "Construction Kits and Design Environments: Steps Toward Human Problem-domain Communication," *HumanComputer Interaction* 3, no. 3 (1988): 179–222.

20. Sally Hacker, *Pleasure, Power, and Technology* (Boston: Unwin Hyman, 1989), p. 52; the term *pornotechnics* is attributed to Barton Hacker.

21. Gitte Moldrup Nielsen and Kristine Stougaard Thomsen, "Systems Development in a Women's Perspective," in *Women, Work, and Computerization: Opportunities and Disadvantages*, eds., Agneta Olerup, Leslie Schneider, and Elsbeth Monod (Amsterdam: North-Holland, 1985), pp. 187–94.

22. Haraway, "Manifesto for Cyborgs."

23. Andrew Ross, "Hacking Away at the Counterculture," in *Technoculture*, eds. Constance Penley and Andrew Ross (Minneapolis: University of Minnesota Press, 1991), pp. 107–34.

24. Haraway, "A Manifesto for Cyborgs," p. 192.

25. Advertisement in the *New Yorker*, September 11, 1989, 5.

26. Sally Shulman, "New Imaging—By Computer," *The Tab* (Wellesley, Mass.), May 30, 1989, 38.

27. Haraway, "A Manifesto for Cyborgs," p. 204.

28. Jan Raymond, *The Transsexual Empire* (London: Women's Press, 1983).

29. See ibid.; also, Marjorie Garber, "Spare Parts: The Surgical Construction of Gender," *Differences* 1, no. 3 (fall 1989): 137–59; and Carole-Anne Tyler, "The Supreme Sacrifice? TV, 'TV,' and the Renee Richards Story," *Differences* 1, no. 3 (fall 1989): 160–86.

30. See, for example, Rudy Rucker, *Software* (New York: Avon Books, 1982), and *Wetware* (New York: Avon Books, 1988).

31. Hans Morawec, *Mind Children: The Future of Robot and Human Intelligence* (Cambridge: Harvard University Press, 1988), p. 4.

32. Tom Stonier, "Machine Intelligence and the Long-term Future of the Human Species," *AI and Society* 2, no. 2 (1988): 133.

33. Ibid., p. 138.

34. Mary O'Brien, "Feminist Theory and Dialectical Logic," in *Feminist Theory: A Critique Ideology*, eds. Nannerl O. Keohane, Michelle Z. Rosaldo, and Barbara C. Gelpi (Chicago: University of Chicago Press, 1982), p. 100. See also O'Brien, *The Politics of Reproduction* (New York: Routledge, 1981).

35. Hacker, *Pleasure, Power, and Technology* (emphasis in original).

36. Wendy Brown, "Feminist Hesitations, Postmodern Exposures," *Differences* 3, no. 1 (spring 1991): 79.

37. Haraway, "A Manifesto for Cyborgs," p. 223.

38. Ibid., pp. 203–204.

39. Luce Irigaray, "Is the Subject of Science Sexed?" trans. Carol Mastrangelo Boye, *Hypatia* 2, no. 3 (1987): 65–87. In this paper, Irigaray considers the relation of science to desire and briefly points to the nonneutrality of various branches of science by describing aspects of sexuality represented or unrepresented in their stakes and projects.

40. Brown, "Feminist Hesitations."

Information and Reality

Albert Borgmann

THE LIGHTNESS OF BEING

The ecology of things used to enforce an economy of signs. The technology of information, however, has loosed a profusion of signs, and there is by now a rising sense of alarm about the flood of information that, instead of irrigating the culture, threatens to ravage it.[1] Technological information in the strict sense constitutes the distinctive and most energetic current of this development. But the more traditional stream of cultural signs has become roily and swollen as well. Information is about to overflow and suffocate reality.

Things will get worse before they get better. There is still a period of information expansion and integration ahead of us. What has been impounding information to some extent are the partitions between different carriers of information, of print vs. radio vs. television vs. the Internet vs. the telephone vs. video cassettes vs. compact discs vs. etc. One of the divides is structural—the division between analog and digital information. Elsewhere information flows are separated by their channels and the devices that present the information to us. Digitization will expand until it has breached the structural dams. Integration of hardware will advance as it already has in the "convergence technology" that fuses television sets with telephones and personal computers. But just as a contemporary home has innumerable clocks and timers rather than one central chronometer, so the household of the future will be saturated with all kinds of information devices. What matters is that clocks are synchronized and information appliances synchromeshed, that is, made to shake hands and converse with one another. Incompatibility will be the most enduring barrier.

Inevitably there will be some flood control too. The Internet in its presently free flowing luxuriance cannot survive. The selfless enthusiasm of hackers and the high-minded support of public institutions, so crucial to the first flowering of the Internet, will both decline. Hackers are getting tired, institutions will get stingy. Commerce will step into the breach, drain the swamps, channel the currents, erect dikes, build reservoirs, and install locks. All users will become customers and will have to pay either with money or with attention paid to commercials. Along with the commercial firming up of the information infrastructure, there will also be some beneficiating of information content. Individuals will come to realize that it is beyond their capacity to spot and evaluate information. Mechanical devices such as infobots or intelligent agents are of limited use. They can at most assure that the information with the right content or frequent readership is brought up. But they cannot guarantee that the stuff is accurate and worth viewing or reading. Thus media organizations that have earned our trust in collecting, editing, and warranting information will survive or reemerge. They will also satisfy our desire to know what everyone knows and so to have some sense of national awareness and belonging.

After a period of experimentation and dissolution, order will reassert itself in social organization as well. William Mitchell's idea that our tangible ways and places will become secondary to the orders and structures of cyberspace will be tried and tested in the developments of telecommuting to work and education.[2] No doubt there will be more flexibility to the times and places we learn and labor. But few people have the motivation or discipline to work well in solitary and seductive environments. Most of us fall victim to the combination of boredom and comfort that distinguishes today's domesticity. Once more, the requirements of human nature and the rigors of commerce will lead us to keep the tangible settings of home and work segregated so that we can continue to consume appropriately and produce efficiently.[3] Similarly, if the promoters of the virtual university are true to their promise of "competency verification," they will discover that many kinds of competence cannot be verified mechanically at all and that of those that can, few will be acquired by virtual students.[4]

Does this account of the expansion, integration, and organization of the new information culture do justice to the enormous innovations in hardware and the evolutions of software that are still to come? No doubt technologically literate observers will have occasion to be impressed, surprised, and astonished. At the same time they will have to recognize, if they are not convinced already, that universal and accurate speech recognition, complete automatic translation, deep syntactic and semantic analysis, and, a fortiori, artificial intelligence are unreachable goals for computers and will forever limit our ability to control information.

Most important, however, no matter how technically accomplished and admirable the breakthroughs in hardware and the advancements of software, the public will finally remain unimpressed. We are witnessing the ironical spectacle where on the public side of the stage futurists exhaust their powers of imagination and description to paint a captivating and amazing picture of the future, and where on the private side the ordinary Joe and Josephine sit on their couch having anticipated and discounted the greatest marvels of information technology. Regardless of how large, fine-grained, three-dimensional, and photorealistic the displays, no matter how universally accessible and smoothly integrated every imaginable piece of information, the rule of simple desire, having conceived of flying carpets, genies in bottles, and magic wands, has always and already preempted the most sophisticated feats of technology.

Thus with all its impending expansion, integration, organization, and innovation, the information revolution, if it stays on its present trajectory, will devolve into an institution as helpful and necessary as the telephone and as distracting and dispensable as television with an unhappily slippery slope between its cultural top and bottom. The characteristic mood of the information age begins to surface when the prophets of the new era try to get beyond grand generalities and reach for specificity. In the future, they tell us, we can get instant compliance to commands such as "List all the stores that carry two or more kinds of dog food and will deliver a case within sixty minutes to my home address."[5] And we will have the pleasure, before viewing a film on command, of being prompted "How about trying a delicious pizza from Marcello's?" and we will have the luxury of being able to "Click yes, and then select plain cheese, mushroom, sausage, or other toppings."[6] The crashing banality of such scenarios is matched by the generally dreary atmosphere that pervades gospels of cyberspace, be they science fiction or sober prognostication.[7]

The real effects of technological information will be subtle and light, but epochal all the same. For better or worse reality will become lighter, both more transparent and less heavy. We are still learning to see the world in the light of technological information, and as we do, expanses formerly too broad, structures once too fuzzy, matters at one time too dense are all becoming clear and bright. For humans who "by nature desire to know," this is a wonderful gain.[8] Where the luminous quality of technological information becomes specular, however, the remaining contingency of reality seems all the more dark and dense. The preternaturally bright and controllable quality of cyberspace makes real things look poor and recalcitrant in comparison. To be sure, reality at bottom remains inescapable and unfathomable. It is the ground on which the ambiguities of technological information can be resolved and its fragilities repaired. Yet on its surface reality appears lighter and often lite.

Information technology has begun to invade the plains and valleys of Montana's ranches. About half of the ranchers (often the women) use computers regularly for financial and cattle production records. They are significantly more satisfied with their performance of these chores than their paper-and-pencil neighbors.[9] There is still skepticism about the spreading of cyberspace under the big sky. "The first four ranchers that I know of that started using computers," said one rancher, "all went broke within five years."[10] But when tax accountants, breeders' associations, suppliers, and county agents all computerize, the ranches cannot afford to remain islands of natural and cultural information. Some ranchers look forward to cyberherding. "When scanners can read top," said one, "and use a scale under a working chute then we will gather the info. Data gathering needs to be automated."[11] To the agricultural information industry this is a meek request. Information technology is eager to deliver "precision agriculture" where everything under the sun will be measured, monitored, and controlled.[12] As in business generally, whether productivity will justify the investment in computers is an open question though not one individual ranchers are at liberty to answer according to their lights. Information technology is sweeping everything before it. What in any event is likely to get lost is the symmetry of natural information and human competence that is reflected in this observation: "My husband knows his cattle personally by working with them and has a memory for traits, problems and style. His father had that trait and so he does and our son seems to have it also."[13]

A similar loss is taking place in the wilderness of Montana. Smoke Elser, Missoula's revered outfitter, knows the Bob Marshall Wilderness as well as anyone and can tell his clients any time just where they are on their trip. But he has been shown up, at least as far as accuracy is concerned, by a know-nothing dude who carried a global positioning system (GPS) receiver that told him within fifty or so feet where he was. Such a device can also track his progress, tell him how far he has traveled from the trailhead and how long it will take to the camp. And if he liked the trip, he can store all this information and retrace his trip exactly a few years hence, stopping at all of Smoke's favorite campsites and fishing spots.[14] Soon it will take deliberate recklessness to lose cattle without a trace or to get lost on a wilderness trip. The last dark and dangerous recesses of the world and the remaining burdens of how to become intimate with the land will appear to have become things of the past.

There are grizzlies and wolves in Yellowstone Park, but you may not want to encounter the former and are at any rate unlikely to see either species when you travel through the park. What an embarrassment, however, to return to New York having to confess to your friends that you saw not hide nor hair of the two most charismatic members of the megafauna. To prevent such calamity, the Grizzly Discovery Center has established itself at the west entrance to the park and exhibits grizzlies and wolves, contented and playful to all appearances, and yet, much like their human spectators, cut off from the environment that once engaged their skills and warranted their ferocious power. The IMAX theater next door will hourly show you *Yellowstone*, the movie, on a screen five stories high and half a block wide. Enveloped by symphonic music pouring forth from the fourteen speakers of a six-channel stereo surround system, you glide over the sunny expanses of the park, move through centuries of human history, penetrate the geology of the geysers, come face-to-face with eagles and bears. The real park must appear dreary and boring in comparison.

But why drive all the way from the East Coast to begin with when the experience of the wild west is so much closer at Disney World's Wilderness Lodge?

Inside it, Silver Springs Creek emerges from the floor, flows past wildflowers, tumbles down falls, empties into pools, continues on the far side of the grounds, and ends in Fire Rock Geyser, shooting up 180 feet every hour on the hour—more faithfully than Old Faithful ever does.[15] Thus Yellowstone Park has become negligible, and so have traditional city libraries, concert halls, and street musicians now that we can have a splendid cultural and urban experience in our home theater.

Technological information is the consummation of a development that began a century ago. Describing the three decades that followed the Civil War, Alan Trachtenberg noted that "[i]n technologies of communication, vicarious experiences began to erode direct physical experience of the world."[16] And still more to the point, he observed that "the more knowable the world came to seem as *information*, the more remote and opaque it came to be seen as *experience*."[17] In Bertrand Russell's terms, our knowledge by description has displaced our knowledge by acquaintance, and with the displacement of the latter the difference between the focal area of the nearness of things and the peripheral area of information about things has dissolved.[18] It has done so in cognition as well as in emotion and in labor as well as in leisure. In our serious dealings with the world, we generate and possess more information than ever. But reality itself gets ever more deeply buried under all the information we have about it. In consumption we are getting so adjusted to the light fare of more or less virtual experiences and emotions that the reality of persons and things seems offensively heavy and crude.

One might find this account of our situation too optimistic. Based on information technology, our omniscience and omnipotence have achieved such transparency and control of information that there are no things any more to be discovered beyond the signs. Nothing any longer is buried beneath information. Behind the virtual self-representations there are no real persons left to be acknowledged. The philosopher Martin Heidegger, who throughout his life had struggled to articulate a sense of being beyond human machination, has been scolded for his nostalgic "Yearning for Hardness and Heaviness."[19] And as if taking this lesson to heart, Bill McKibben has movingly lamented the passing of nature's unsurpassable sovereignty.[20]

ADJUSTING THE BALANCE

There is no danger that technological information might entirely displace natural or cultural information. Inevitably we will spend much of our time navigating tangible environments, paying attention to the natural signs that tell us where we are going. Writing—the principal kind of cultural information—will remain indispensable as well. There is a real possibility, however, that natural and cultural information will decline to mere utilities, tools we need but fail to sustain as signs of irreplaceable kinds of excellence.

The succeeding kinds of information accomplish and in some ways surpass what their predecessors provide, and so it may seem as though technological information could give us the best of three worlds, the most powerful and refined information about, for, and as reality. But the successors do not fulfill their ancestors' task in quite the same way. Something does get lost when a later kind of information displaces an earlier. The loss is insignificant where natural or cultural information has irreversibly declined to a mere means or has reached the limits of its disclosive or transformative power. Most people, to be assured of a square meal, have no need of reading the tracks of animals or recognizing the edible fruits and plants in the wild. And no astronomer would want to be reduced to gathering information through optical telescopes and recording it by hand. It is otherwise when natural and cultural signs are at their highest.

For natural signs this is the case when they point us to a landmark and, having brought us face-to-face with it, mark off the presence and nearness of the focal area against the horizons of distance and the past. Nothing so engages the fullness of human capabilities as a coherent and focused world of natural information. No amount or sophistication of cultural or technological information can compensate for the loss of well-being we would suffer if we let the realm of natural information decay to one of resources, storage, and transportation. Analogously, nothing so concentrates human creativity and discipline as the austerity of cultural information, provided the latter again is of the highest order, consisting of the great literature of fiction, poetry, and music. Our power of realizing information and our competence in enriching the life of the mind and spirit would atrophy if we surrendered the task of realization to information technology. Perhaps what holds for the realization of information goes for its production too. The simplicity of the pen and the blankness of paper may be able to challenge, if they do not terrify, the resources of the writer and drafter in a way the obliging servility of the computer cannot match.[21]

Adjusting the kinds of information to one another and balancing information and reality cannot possibly be a return to earlier conditions. Modern technology and information technology particularly are tokens of a profound and irreversible change in the nature of reality. In the premodern world, the material force of reality issued in moral instruction and favored, though it did not guarantee, moral practices. Respectful attention to wild animals was a condition of survival for Native American hunters.[22] Marital fidelity, loyalty to one's band, and a tradition of courage allowed the Blackfeet to cope with harsh weather, scarce game, and injury or untimely death.[23] Within the postmodern lightness of being, to the contrary, the moral instruction of reality can restore its material force and with it our fundamental welfare—the full engagement of human capacities. We are essentially bodily creatures that have evolved over many hundreds of thousands of years to be mindful of the world not just through our intellect or our senses but through our very muscles and bones. We are stunting and ignoring this ancestral attunement to reality at our peril.[24]

It is impossible, however, to recover the full symmetry of humanity and reality through a virtual representation of reality. Information theory can be employed to show that, if virtual reality were to be made informationally exactly as rich as actual reality, the virtual would have to become a verbatim duplicate of the actual—a reduction to absurdity. But finally it is not the requirement of informational plenitude that leads us back to actuality but the moral eloquence of reality itself. As long as we remain in a cocoon of virtual reality or behold and control actual reality chiefly through information technology, the world out there seems light and immaterial. But once we take up the challenge of a natural area or the invitation of a truly urban space, material reality reappears in its commanding presence and engages our bodily exertion and spiritual pleasure to the limits of our capacities. But to repeat, the material vigor and splendor of reality will rest on its moral authority and our ability to respond to it. We can at any moment escape from the rigors of nature and the burdens of urbanity and surrender both city and country to neglect and abuse. The penalty of such evasiveness is no longer starvation or bodily injury but moral atrophy.

Righting the balance of information and reality is the crucial task. It amounts to the restoration of eminent natural information. A well-ordered realm of natural information in turn is both hospitable to practices of realizing cultural information and enlivened by such practices. As for technological information, there is no sense in trying to channel its development through narrow proscriptions or prescriptions. Nor does it make sense, of course, to let it run wild and overrun nature and culture. It is best allowed to develop freely within a world whose natural and cultural ecologies are guarded and engaged in their own right.

There are indications that people in this country are beginning to heed the voices of reality and not only are seeking engagement with things of nature and culture but are determined to renew what has been disfigured through neglect or brutality.

The Rattlesnake Valley in western Montana was settled by squatters and homesteaders only a hundred years ago. They found the valley in much the same condition it had been in for hundreds and perhaps a few thousand years. The settlers built log cabins and dug root cellars, cleared the valley bottom of trees and rocks, strung barbed wire fences, dug irrigation ditches, and did subsistence farming and ranching on thin and stingy soil. To make ends meet they logged the hillsides, pulling and shoving the logs down timber draws. The logs were used to make ties for the approaching Northern Pacific railroad and to supply the citizens of Missoula with firewood.[25]

Rattlesnake Creek was an early source of drinking water for Missoula. During the Great Depression the water company's desire to protect the watershed converged with the grinding poverty of the Rattlesnake farmers, and the company bought out the settlers in the upper region of the valley. The houses and barns were leveled, the timbers and implements hauled away; barbed wire and garbage is being removed to this day. Gradually, however, the water company shifted the source of its supply to the safer and more convenient wells in the Missoula Valley.

Development of the easily accessible and spectacular building sites in the upper
Rattlesnake would have been the normal course of events. But environmentalist
Cass Chinske and Congressman Pat Williams rallied the sense of the citizenry that
the area should be preserved for reasons the congressional act that secured the federal lands and provided for the acquisition of the private holdings put this way:

> This national forest area has long been used as a wilderness by Montanans and
> by people throughout the Nation who value it as a source of solitude, wildlife,
> clean, free-flowing waters stored and used for municipal purposes for over a
> century, and primitive recreation, to include such activities as hiking, camping,
> backpacking, hunting, fishing, horse riding, and bicycling; and certain other
> lands on the Lolo National Forest, while not predominantly of wilderness
> quality, have high value for municipal watershed, recreation, wildlife habitat,
> and ecological and educational purposes.[26]

The preservation of a natural area can no longer be a matter of simple human
withdrawal, nor can it seek guidance from a pretechnological norm of wilderness.
It must take the form of a conversation with nature that seeks answers to questions like these: Should the noxious Eurasian weeds that are invading the valley
be stopped by all available means? How far can wildfires be allowed to burn?
Should eroding trails be reconstructed? Does it or did it make sense to reintro-
duce mountain goats, bighorn sheep, moose, fishers, wolves, or grizzly bears?
Refusing to face these questions is merely one way of answering them.

Still, a natural area, however much it is informed by human decisions, leads
a life of its own, and its ecology of things and economy of signs most of the time
have a calmness and autonomy that set nature apart from culture. To enter a natural area is to be greeted and astounded by life in its own right. At the same time,
the life of nature engages you most deeply if you understand it in the context of
cultural information, that is, within the space of intelligibility that is circumscribed by information about the history of the place.

Thus walking north from the trailhead along Rattlesnake Creek, you come to
the abutments of a long-gone bridge that Sebastian Effinger, the first settler in
this area, threw across the river to get to the western part of his homestead, the
meadows at the confluence of Spring Creek and the Rattlesnake. The road used
to angle up from the bridge, and where it once reached the bench of the meadow
you walk across flat stones, nearly invisible now, that used to support the school-
house. Trees are now invading the meadows that Effinger and his neighbors
cleared. They felled the big ponderosas that had stood their ground when the peri-
odic fires cleared out the understory. And they loaded the scattered stones on low
sleds they called rock boats and dragged them off to the side where they now rise
from the grass in low, grey piles. They denuded the hillsides as well, crisscrossed
the valley bottom with dirt roads, and dumped their garbage and cans wherever
they were out of the way. Early in this century, the valley must have looked
ragged and disheveled.

Meanwhile, second growth ponderosa, Douglas fir, and western larch have come up among the few gigantic ponderosas that loggers have spared. Now the north facing slopes look lovely, dark, and deep. Those facing south are still open and sunny, but they too are "dougging in" as the foresters say. Walking on up along the Rattlesnake you find the remnants of a flume that used to carry water from a ditch across the creek to irrigate the meadows on the eastern side of the river. The ditches were dug by hand and where possible by mule-drawn scrapers, huge two-handled shovels that look like a wheelless wheelbarrow. The farmers used the same devices to raise the natural dams of the mountain lakes so that the water that accumulated in the spring could be released in the summer to replenish the Rattlesnake and the ditches it used to feed.

You undertake such backbreaking work in the hope of laying a foundation of prosperity for your children and grandchildren. The dams were helpful to Missoula's water supply for a couple of generations or so. But the ditches, roads, houses, root cellars, and barns turned out to be labor poorly spent. They have collapsed into scattered traces of desperately brave work, overgrown by the quiet of grass and shrubs, and shaded by apple trees gone wild on the abandoned homesteads. But mingled with admiration for the courage and with pity for the failures of one's forebears, there is gratitude to one's contemporaries for allowing nature to reclaim the once tortured land.

Moving farther north, the old road rises up the Hogback, the terminal moraine of the Rattlesnake glacier that sits astride the valley and forces the river to the east slope where the Rattlesnake has cut an opening through the glacial wash. And so on up the valley through a series of portals that open on meadows until after some twenty miles you reach the high country of the wilderness area. When on a summer's morning you run up the valley, the sun rises over the east slopes like a blast of trumpets, the canyon walls open like the doors of a cathedral. Humans have acceded to nature, nature graces humans.

The rising sensitivity to the claims of nature is matched by a growing recognition of the splendor of the city and of the injuries cities have suffered.[27] The automobile has been the bane of traditional urbanity. Automobility, of course, is only a symptom of our commitment to technological liberty and is now so deeply embedded in the culture that there is no rolling it back. But we can hope to counter the distended monotony of the postwar suburbs and the forbidding brutality of city centers with urban spaces diverse enough to be engaging and articulate enough to be intelligible. It has been the concern of the New Urbanism to re-create the compactness and comprehensibility of traditional towns in new developments outside the city. The results surely have some of the coherence and rhythm of the natural ecology of things and signs—a scale that can be bodily appropriated, an ordering of the parts that indicates the sense of the whole, a placement of landmarks that provides orientation.[28]

The crucial challenge, however, is the one that the New Urbanism has turned to just now, the restoration of the central city itself.[29] There are enough remnants of earlier order and splendor to indicate what an urban space on a grand scale should be, a place anchored by commanding landmarks such as theaters, set in streets or squares that are convenient for pedestrians and inviting for people who want to sit, eat, read, or watch. Such spaces too disclose their meaning readily and variously, exhibiting signs that inconspicuously become things so that the balance of presence and reference remains intact. But more is needed than arranging for signs and things to emulate the sane and measured world of natural information. Such arrangements all by themselves can become thin and lifeless. Worrying about this kind of problem, Herbert Muschamp has warned that "[l]ike the modernists, the new urbanists rely too much on aesthetic solutions to the social problems created by urban sprawl."[30] Most of these problems are distressingly urgent and obvious—crime, poverty, unemployment, decay, and dirt. Pragmatism, however, is a better route toward their solution than grand theory. Though morally urgent and hard to reach, solutions to these problems would still leave us with the specter of urban spaces without depth and vitality, "architecture for the Prozac age," as Muschamp calls it, "Potemkin villages for dysfunctional families."[31]

Nor is it the case that the morally urgent and the culturally subtle problems can only be, or should be, solved sequentially—first safe streets then vital urbanism. In fact as matters stand, the latter promotes the former. There is a large measure of truth in Jane Jacobs's thesis that engaging and healthy city life is the ground state and that crime and decay move in only after true urbanity has moved out.[32] Moreover urban vitality not only enlivens aesthetically pleasing spaces; it can also mark out a place in otherwise inconspicuous space. Daily urban life at its best is regular in the large and improvised in the small. Its daily rhythm provides a framework for the small variations of shopping, strolling, conversing, and watching. The fabric of daily city life in turn provides the backdrop for those festive events that lend cultural life significant structure. On such an occasion space comes alive, time is focused, and people are inspired. The convergence and florescence of these elements we call celebration. In our culture, public and communal celebrations are for the most part realizations of cultural information. Thus the enlivening moments in contemporary life are fusions of natural and cultural information, the skilled performance of a score or play in a well-ordered and intelligible setting.

One such monumental moment took place on June 23, 1993, when Luciano Pavarotti sang on the Great Lawn of Central Park in New York City before a crowd of half a million. Naturally the park lends itself, and does so every day, to less focused and massive celebrations. But the presence of so many riveted on just one person draws the city together and quickens its life like few other events. This particular part of Manhattan's emerald jewel is more dusty than green. Yet bordered by trees and lawns and surrounded by a familiar skyline, it leaves you in no doubt where you are. Pavarotti's singing that evening made it plain why you would want to be there and why there should be a place like the Great Lawn. It was filled with the joys, sorrows, and the consoling pleasures of the Italian tenor repertoire. Though the scores were familiar, the contingency of the occasion was poignant. Pavarotti's career had been in disarray and his voice in decline. Who would prevail in the struggle between tenor and time? There might have been a shadow of doubt in the opening Verdi aria "Quando le sere al placido," but soon Pavarotti rose to the soaring brilliance of his younger years. When at length he concluded with Puccini's "Nessun Dorma," jubilation swept across the Great Lawn from the stage to the last ranks of listeners.

The occasion was ringed by the ambiguities of technological information as much as by groves and high rises. To most of the audience Pavarotti's voice came through loudspeakers the size of outhouses, hoisted up on fork lifts, and his presence was reduced to something the size of one's thumb. Ironically a big screen behind him displayed his face in hyperreal size and color. One could have had a far superior view and sound in front of one's television set. Or, if sound alone and supremely mattered, one might have waited for the hyperreal studio renditions of London Records, "drawn from its rich catalog of Pavarotti recordings."[33] Yet many thousands were drawn to the occasion where music became real then and there.

Three days later another concert took place in Manhattan at the inauspicious corner of Fifty-first and Park. It was noon, and Randy and the Rainbows, the one-hit wonders of "Denise" fame, were doing golden oldies. Slowly people gathered, and soon there was a crowd of three hundred, overcome by the sweet sounds they once had poured their longings and passions into, "Only You," "Midnight Confession," "Don't Take Away the Music." Again they could have had better sound more readily from CDs and a stereo. But standing there, all more or less settled and accomplished, sadder than they once were and perhaps wiser, they could see this place and their lives within the space of their hopes and sorrows, and it was Randy's and the Rainbows' music that had summoned and enthralled them.

That evening, Lynn Redgrave once again presented *Shakespeare for My Father* in the Helen Hayes Theater on Broadway. Given the hypercharged entertainment we are used to getting from screens large and small, this seemed like an impossibly cool and austere affair, one actress telling stories and acting out scenes from Shakespeare. There was a particular grace to these hours, when a highly schooled and dedicated actress once again risked her stamina and reputation to gather for her audience a powerful strand of the theater's history. In her acting, pieces of literature, personages of the stage, and events of recent history were knitted together into a context that made sense of troubled circumstances, however implicitly and suggestively.

Much of Redgrave's material is available through information technology, and so are terabytes of closely connected information. Anyone familiar with the Web and the Net could have assembled an electronic portfolio that would be more varied and colorful than *Shakespeare for My Father*. But all this technological information is floating inconsequentially in cyberspace. It takes an actual time and place to gather an audience, it takes an audience to create a sense of expectation, it takes a real person to warrant such a concentration of reality and humanity, and finally it is such a warrant that sustains a place like the Helen Hayes Theater and made the architect Hugh Hardy restore the Victory Theater to its original splendor. Hardy is quite clear about the significance of his work and believes, as the *New York Times* had it,

> that it is the vitality of public spaces that keeps a city healthy enough to counter the mounting popularity of simulated experience—theme parks as well as enclosed shopping malls over the real thing.
>
> "People used to understand that gathering in public was good; that's what democracy meant," said Mr. Hardy.[34]

There are indications, then, that we are beginning again to recognize reality, to realize information, and to right the balance of signs and things. The symmetry of information and reality reaches its highest point in celebration when the ambiguities of signs and things complement and resolve each other. In celebration the austerity of a score or a text is redeemed through realization and the diffusion of reality is focused through a script. Information comes alive, reality becomes eloquent. When this happens, a celebration rises to the stature of a landmark in time, something that gives our lives coherence and significance. It marks a meaningful moment in the ravages of time. Like a landmark, such a moment of high contingency is both a sign and a thing. As a thing it has the presence of self-warranting clarity, as a sign it refers us to the darkness of contingency that constitutes its periphery.

A momentous celebration is suffused with enthusiasm and harmony. Half a million often surly New Yorkers gathered for pleasant and even artful picnics, made room for one another, exchanged amicable remarks, and were united in appreciation and affection when Pavarotti sang. But all this good will and pleasure were challenged, contradicted, and perhaps belied by the misery of the homeless nearby, by the muggings occurring in the run-down sections of Manhattan, by the strife, the diseases, and the starvation around the globe that every citizen of this country is somehow implicated in. The deplorable conditions that surround and question every celebration are no more distressing than the moral debilities that shadow individual joy. Surrounded by the peace and grandeur of the upper Rattlesnake Valley, I may feel the strength to forgive my enemies and to take on the labors of order and charity I have been avoiding. There is in fact no denying that resourcefulness and forbearance flow from natural and communal celebration. But at length I do fall prey again to anger and anxiety and unsay and undo the grace of nature or community. The warm and luminous moments of festivity are forever surrounded and threatened by darkness.

The darkness of contingency can open up in different ways. One is the violent ruination, the other the slow evaporation of meaning. The former is a holocaust and leaves us with ashes, the latter is oblivion and leaves us with nothing. The holocaust is the catastrophe of real ambiguity when reality more than disperses its meaning and energetically ruins it, most terrifyingly through those vital concentrations of reality we call persons. Oblivion is the failure of symbolic ambiguity, the human inability to meet the rush of time with the stability of memorable signs.

The common response to devastation and oblivion is remembrance. We cannot redeem the holocaust, but we must remember it, and we try to do so by erecting enduring things and seeing to truthful signs. Remembrance is the response to the evanescence of meaning as well. To be forgotten or to see events disappear through forgetfulness is deeply, if much less painfully, disturbing to human beings. To escape the raging darkness of oblivion Alexander the Great took a flock of historians on his campaign but must have been uncertain of their skill, for when he came upon the grave of Achilles, he exclaimed, "Fortunate young man, to have found in Homer the herald of your valor."[35] Similarly, Ben Sira urged his community to embrace and remember its forebears:

> Let us now sing the praises of famous men,
> our ancestors in their generations.[36]

And he makes clear the consequences of the failure of remembrance:

> Some of them leave behind a name,
> so that others declare their praise.
> But of others there is no memory;
> they have perished as though they had never existed;
> they have become as though they had never been born,
> they and their children after them.[37]

Writing, of course, is the great invention that produces enduring records where memory would fail. The power of this instrument particularly impresses itself on cultural awareness when it is first introduced on a wide scale as it was in Norman England. Accordingly Henry III, concerned about the memory of the celebration of Edward the Confessor in 1247, addressed a chronicler from the throne and ordered him "to write a plain and full account of all these proceedings, and insert them in indelible characters in a book, that the recollection of them may not in any way be lost to posterity at any future ages."[38]

Information technology has deeply influenced the ways we cope today with the threat of the devastation and loss of meaning. The challenge to the festive resolution of the ambiguity that rises from the surrounding injustice and misery we are inclined to meet with a version of virtual ambiguity, a loosening of the ties that should connect our celebrations with their real and entire context. While virtuality is our reply to the devastation of common meanings, hyperinformation is our response to the oblivion of individuals. Common hyperinformation is the huge amount of colorful information we accumulate through pictures and videos especially. But all the other records we keep and that are kept about us are part of hyperinformation. Utopian hyperinformation is the brainchild of scientists who, in the tradition of artificial intelligence, believe that the core of an individual is the information contained in the brain, and purport that software can and will be extracted from the wetware of neurons and transferred without loss to the hardware of a computer or some other medium forever and again in this way and that so that the core of individuals, their personal identity, will achieve immortality.[39]

All of these are desperate attempts. To the extent that we shield celebration through virtual ambiguity from the reproaches of a suffering world, we empty celebration of meaning. At the limit, when celebration is fully protected, it is no longer worth saving. The endeavor to be remembered through common hyperinformation is indistinguishable from a headlong rush into oblivion. Some of the information will be overtaken by its physical and social fragility. If we find a way to stabilize it, however, its sheer disorganized and imposing mass will excuse our offspring from taking it to heart.

The reach for utopian hyperinformation is perhaps the most telling and melancholy indication that no one wants to "pass into oblivion," as a medieval chronicler has it, "as hail and snow melt in the waters of a swift river swept away by the current never to return."[40] In ordinary people, atheists or not, this fear takes the form of the desire to be remembered and to be remembered well. Alexander wanted more than to be recorded by historians, he wanted to be transfigured by the poet. Ben Sira asks us not just to recall but to praise our ancestors. People seem to conceive of themselves as deeply ambiguous signs that call for resolution.

What is true of the microcosm of the person is true of the macrocosm of the universe. It too is a sign as much as a thing, something that refers to its beginning and end. In our culture the normative response to the unresolved references of the cosmos is astrophysics. Though it is not inevitably committed to a beginning and an end of all things, it does concern itself with the lawful structure of the world's past and future and aims at a final theory of all there is.[41]

As regards the middle region of the cosmos that is so artfully balanced between the structure of atoms and galaxies, the terrestrial realm of nature and culture, its welfare requires more calmness and lucidity of recollection.[42] As it is, contemporary culture may lapse into a condition where a surfeit of information is as injurious as the lack of information. Where in the latter case one is confined by the darkness of ignorance and forgetfulness, today we are blinded by the glare of excessive and confused information. To regain our sight for the coherence of the public world we must be able to count on our chroniclers—the journalists, essayists, and historians—and we must allow their work to come to rest and attention for a day at least, or a month, or some years. Newspapers, journals, and books have been the places of considered judgment, and these or some such focal points are needed if information technology, beyond its instrumental functions in science and industry, is to become a constructive strand in the texture of our lives.

To recover a sense of continuity and depth in our personal world, we have to become again readers of texts and tellers of stories. Books have a permanence that inspires conversation and recollection. When you read or recount a passage from a book to your loved one, the matter at issue envelops both of you and fills the place you occupy.[43] Stories are the spaces wherein pictures and mementos come to life and coalesce into a coherent picture of the past and a hopeful vision of the future. Records in turn keep stories straight and lend them detail. Thus the culture of the word can card, spin, and knit the mass of technological information into a tapestry that is commensurate with reality.

As for cosmic closure, I quite agree with Steven Weinberg that a final theory will be a noble and intellectually satisfying accomplishment, the crowning achievement in our search for structure and lawfulness.[44] But the world has a history as well as structure. History in the large and strict sense is the meaningful sequence of unpredictable events. It is contingency. Hence we face the question whether there can be cosmic closure of a historical as well as a structural sort. It may well turn out that at the beginning of the cosmos history and structure are indistinguishable, that the unfolding of the cosmos is simultaneously an unfolding of lawfulness. But in time structure and contingency must diverge in one way or another. There is no prospect of deducing the lightning that causes a devastating fire or the encounter that leads to a happy marriage from the laws of the universe alone.

History, then, requires its own kind of reading, one that must consist with the laws of nature but also attend to the givenness of things and events. The decline of meaning and the rise of information have kept contemporary readings of history weak and inconclusive. The recent burst of information technology has further, and fortunately, silenced the voices of overt misery, of disease, poverty, and violence, both here and around the globe. There is still unspeakable suffering in many parts of the world. But information technology is both the channel and the energy that is carrying the free-market economy and its blessings to every corner on earth.

As overt misery is waning, so is the inference that used to be drawn from it, namely, that suffering would not be what it is if it did not intimate salvation in the end. And similarly, as our celebrations are losing their context and contrast of poverty and violence, they also lose their reference, weak already, to the need for final salvation. But while information technology is alleviating overt misery, it is aggravating a hidden sort of suffering that follows from the slow obliteration of human substance. It is the misery of persons who lose their well-being not to violence or oblivion, but to the dilation and attenuation they suffer when the moral gravity and material density of things is overlaid by the lightness of information. People are losing their character and definition in the levity of cyberspace.

The engagement of reality is the proximate remedy for this condition, and yet many of us find it hard to face up and to be faithful to persons and things. Though we feel blessed by celebrations once we have been drawn into them, all too often we lack the strength or loyalty to enter them regularly. The moral paralysis people inflict on themselves through the abuse of technological information is miserable any way you look at it. The constructive responses are manifold, however, and not a matter of contestation but attestation. Christians, for example, owe what fidelity to persons and festive things they possess to a strong reading of cosmic contingency—the history of salvation. Whatever definition they attain as persons through their engagement with reality they see as precarious and in need of final resolution. The world as a sign makes them look forward to the event when

> Liber scriptus proferetur,
> In quo totum continetur,
> Unde mundus judicetur.[45]

> A written book will be brought forward
> Wherein everything is gathered
> Whence the world may be adjudged.

All of us will be remembered and more; our souls will be rocked in the bosom of Abraham.

NOTES

1. See, for example, Charles Leroux, "Drowned in Data?" *Chicago Tribune*, 15 October 1996; Richard Zoglin, "The News Wars," *Newsweek*, 21 October 1996, 58–64; Richard Harwood, "40 Percent of Our Lives," *Washington Post*, 30 November 1996; Louis Uchitelle, "What Has the Computer Done for Us Lately?" *New York Times*, 8 December 1996; and the 16 December 1996 cover of the *New Yorker* contrasting a Victorian and a contemporary Christmas.

2. William J. Mitchell, *City of Bits: Space, Place, and the Infobahn* (Cambridge: MIT Press, 1995).

3. Carol Levin, "Don't Pollute, Telecommute," *PC Magazine*, 22 February 1994, 32; Jim Ludwick, "Home at the Office," *Missoulian*, 17 November 1996; Hope Lewis, "Exploring the Dark Side of Telecommuting," *Computer World*, 12 May 1997, 37; Susan J. Wells, "For Stay-Home Workers, Speed Bumps on the Telecommute," *New York Times*, 17 August 1997.

4. "A Matter of Degrees: Colorado Governor Roy Romer on the Western Governors University," *Educom Review* (January/February 1997): 18–19, 23.

5. Bill Gates, *The Road Ahead* (New York: Viking, 1995), p. 80.

6. Ken Auletta, "The Magic Box," *New Yorker*, 11 April 1994, 42.

7. William Gibson, *Neuromancer* (1968; reprint, New York: Ace, 1984); Philip K. Dick, *Do Androids Dream of Electric Sheep?* (New York: Ballantine, 1996); Gates, *The Road Ahead*; Nicholas Negroponte, *Being Digital* (New York: Vintage, 1995).

8. Aristotle, *Metaphysics*, the first sentence (980a).

9. Lee Tangedahl and Jackie Manley, "Computer Cowboys," *Montana Business Quarterly* 34 (autumn 1996): 11–12.

10. Ibid., p. 14.

11. Ibid.

12. See, for example, http://pasture.ecn.purdue.edu/~mmorgan/PFI/graphic.html [26 June 1998].

13. Tangedahl and Manley, "Computer Cowboys," p. 13.

14. See, for example, the Adventure GPS Products site, http://www.gps4fun.com/gps_fun.html [10 June 1998].

15. Jennifer Cypher and Eric Higgs, "Colonizing the Imagination: Disney's Wilderness Lodge" (Vancouver: Centre for Applied Ethics) [online], http://www.ethics.ubc.ca/papers/invited/cypher-higgs.html [26 June 1998].

16. Alan Trachtenberg, *The Incorporation of America: Culture and Society in the Gilded Age* (New York: Hill and Wang, 1982), p. 122.

17. Ibid., p. 125.

18. Bertrand Russell, "Knowledge by Acquaintance and Knowledge by Description," *Proceedings of the Aristotelian Society*, New Series 11 (1910–11): 108–28.

19. Winfried Franzen, "Die Sehnsucht nach Härte und Schwere," in *Heidegger und die praktische Philosophie*, eds. Annemarie Gethmann-Siefert and Otto Pöggeler (Frankfurt: Suhrkamp, 1988), pp. 78–92.

20. Bill McKibben, *The End of Nature* (New York: Anchor, 1989).

21. Michael Heim, *Electric Language: A Philosophical Study of Word Processing* (New Haven, Conn.: Yale University Press, 1987).

22. Gordon G. Brittan Jr., "The Secrets of the Antelope," presented at the Claremont Graduate School, 25 November 1996. Brittan's paper will rekindle *our* respect for animals as well.

23. James Welch, *Fools Crow* (New York: Viking, 1986).

24. Robert Wright, "The Evolution of Despair," *Time*, 28 August 1995, 50–57; and *The Moral Animal: Evolutionary Psychology and Everyday Life* (New York: Pantheon, 1994). Evolutionary theory underdetermines culture and morality. The further back it reaches in evolution, the less pertinent are its findings to the human condition. Nonetheless, evolutionary psychology can uncover significant boundary conditions of human well-being as Wright amply demonstrates.

25. Forrest Poe and Flossie Galland Poe, *Life in the Rattlesnake*, ed. Mark Ratledge (Missoula, Mont.: Art Text Publication Service, 1992).

26. Public Law 96-476, 96th Congress, 19 October 1980.

27. Lawrence Haworth, *The Good City* (Bloomington: Indiana University Press, 1963); Daniel Kemmis, *The Good City and the Good Life* (Boston: Houghton Mifflin, 1995).

28. Vincent Scully, "The Architecture of Community," in *The New Urbanism*, ed. Peter Katz (New York: McGraw Hill, 1992), pp. 221–30.

29. Herbert Muschamp, "Can New Urbanism Find Room for the Old?" *New York Times*, 2 June 1996.

30. Ibid., p. 27.

31. Ibid.

32. Jane Jacobs, *The Death and Life of Great American Cities* (New York: Vintage, 1961), pp. 241–90.

33. *Pavarotti: Great Studio Recordings of His Central Park Program*, London Records 443 220–22.

34. Julie V. Iovine, "Tenacity in the Service of Public Culture," *New York Times*, 12 December 1995.

35. *Select Orations and Letters of Cicero*, ed. Francis W. Kelsey, 2d ed. (Boston: Allyn and Bacon, 1894), p. 155 (my translation).

36. Ben Sira (also called Sirach or Ecclesiasticus) 44:1 NRSV.

37. Ben Sira 44:8–9 NRSV.

38. *Matthew Paris's English History from the Year 1235 to 1273*, trans. J. A. Giles, 3 vols. (London: Bohn, 1852–54), 2: 243.

39. Hans Moravec, *Mind Children: The Future of Robot and Human Intelligence* (Cambridge: Harvard University Press, 1988); Frank J. Tipler, *The Physics of Immortality: Modern Cosmology, God, and the Resurrection of the Dead* (New York: Anchor, 1995).

40. *The Ecclesiastical History of Orderic Vitalis*, ed. and trans. Marjorie Chibnall, 6 vols. (Oxford: Clarendon, 1968–80), 3: 284.

41. Steven Weinberg, *Dreams of a Final Theory* (New York: Pantheon, 1992), pp. 211–40.

42. For pictures of the balance, see Kees Boeke, *Cosmic View: The Universe in Forty Jumps* (New York: John Day, 1957); and Philip Morrison and Phylis Morrison, *Powers of Ten: A Book about the Relative Size of Things in the Universe and the Effect of Adding Another Zero* (New York: Scientific American Library, 1982).

43. Wayne C. Booth, "The Company We Keep," *Daedalus* 111 (fall 1982): 33–59.

44. Weinberg, *Dreams*, pp. ix, 3–18.

45. From the sequence of the requiem, attributed to Thomas of Celano (1190–1260). The translation that follows is mine.